SIGNALING IN TELECOMMUNICATION NETWORKS

WILEY SERIES IN TELECOMMUNICATIONS AND SIGNAL PROCESSING

John G. Proakis, Editor
Northeastern University

Worldwide Telecommunications Guide for the Business Manager
Walter L. Vignault
Expert System Applications to Telecommunications
Jay Liebowitz
Business Earth Stations for Telecommunications
Walter L. Morgan and Denis Rouffet
Introduction to Communications Engineering, 2nd Edition
Robert M. Gagliardi
Satellite Communications: The First Quarter Century of Service
David W. E. Rees
Synchronization in Digital Communications, Volume I
Heinrich Meyr and Gerd Ascheid
Telecommunication System Engineering, 2nd Edition
Roger L. Freeman
Digital Telephony, 2nd Edition
John Bellamy
Elements of Information Theory
Thomas M. Cover and Joy A. Thomas
Telecommunication Transmission Handbook, 3rd Edition
Roger L. Freeman
Computational Methods of Signal Recovery and Recognition
Richard J. Mammone, Editor
Telecommunication Circuit Design
Patrick D. van der Puije
Meteor Burst Communications: Theory and Practice
Donald L. Schilling, Editor
Mobile Communications Design Fundamentals, 2nd Edition
William C. Y. Lee
Fundamentals of Telecommunication Networks
Tarek N. Saadawi and Mostafa H. Ammar, with Ahmed El Hakeem
Optical Communications, 2nd Edition
Robert M. Gagliardi and Sherman Karp
Wireless Information Networks
Kaveh Pahlavan and Allen H. Levesque
Practical Data Communications
Roger L. Freeman
Active Noise Control Systems: Algorithms and DSP Implementations
Sen M. Kuo and Dennis R. Morgan
Telecommunication System Engineering, 3rd Edition
Roger L. Freeman
Signal Processing Systems: Theory and Design
N. Kalouptsidis
Signaling in Telecommunication Networks
John G. van Bosse

SIGNALING IN TELECOMMUNICATION NETWORKS

John G. van Bosse

A Wiley-Interscience Publication

JOHN WILEY & SONS, INC.

New York / Chichester / Weinheim / Brisbane / Singapore / Toronto

This text is printed on acid-free paper.

Library of Congress Cataloging in Publication Data:

Van Bosse, John G.
Signaling in telecommunication networks / John G. van Bosse.
p. cm.
Includes index.
ISBN 0–471–57377–9 (cloth : alk. paper)
1. Telecommunication—Switching systems. 2. Signaling system 7.
3. Signal theory (Telecommunication) I. Title.
TK5103.8.V36 1997
621.382—dc20 96–43761

Printed in the United States of America

10 9 8 7 6 5 4 3 2 1

For
Mieps, Jacqueline, and Harold

CONTENTS

PREFACE

Telecommunications, which started as "telephony" before the turn of this century, has experienced a dramatic evolution in the past few decades. By pushing a few buttons on our telephones, we can make calls to almost any place in the world. Moreover, while the original purpose of these calls was to speak with a person at a distant location, we now also make calls to transmit written documents (facsimile), and other data. The introduction of computer-controlled exchanges has led to a host of new telecommunication services beyond "plain old telephone service," for example, "call waiting," "call forwarding," and intelligent network services such as "800 number calling." In addition, telecommunication networks are being converted from analog to digital technology, and Integrated Services Digital Networks (ISDN) have made their appearance in several countries. Finally, cellular-mobile communications have become widely accepted during the past few years.

This book is about "signaling" in telecommunication networks. It is intended as successor of the book with the same title, written by S. Welch in 1979. Broadly speaking, signals are instructions sent by a subscriber to the network—and between machines in the network—to set up and release connections. Signaling started as a part of Almond Strowger's invention of automatic telephony, which allowed the subscribers to call each other without the need of operator assistance. The subsequent developments in telecommunications have created the need for increasingly powerful and complex signaling systems and procedures.

In the evolution of signaling, we can distinguish three phases: dial-pulse signaling, multi-frequency signaling, and common-channel signaling. The signals of dial-pulse and multi-frequency signaling are carried on the lines and trunks that also transport speech and other subscriber communications, and are known as channel-associated signaling systems.

In the third-phase signaling systems, which became possible after the introduction of computer-controlled exchanges, lines and trunks do not carry signaling information. Instead, signaling consists of messages that are transferred by signaling data links. A data link is a common transfer channel for the signaling messages of a number of trunks, hence the name: common-channel signaling.

This book is intended both for students of telecommunications, and for engineers and technical managers in the telecommunication industry. Its main objective is to provide, within the confines of one book, a solid understanding of the main characteristics of several important signaling systems. Writing a technical book is somewhat similar to viewing a forest. In a ground-level view of a forest, we observe a few nearby trees in great detail, but tend to lose sight of the forest as a whole. From an airplane, the forest looks like a green glob without much detail. A view from a helicopter includes both the forest and its most important features. This book aims to present a helicopter view of signaling. It provides sufficient information to become familiar with the subject as a whole, and can also serve as an introduction to the study of the detailed specifications (ground-level views) of individual signaling systems.

Chapter 1 presents a broad-brush picture of telecommunication networks, including some hardware-oriented information on the principal entities in telecommunication networks. Chapter 2 introduces the most important signaling concepts. These chapters are included mainly for those readers who have little or no previous experience in telecommunications, and can be perused quickly by others. Chapters 3 and 4 outline channel-associated signaling.

Chapter 5 introduces common-channel signaling. It explores the reasons for going to this new type of signaling, and describes the networks of data links that carry the signaling messages. Chapter 6 covers the first common-channel systems: CCITT No.6 and Common-channel Interoffice Signaling (CCIS).

Chapters 7 through 9 and 11 through 17 constitute the largest part of the book, and describe Signaling System No.7 (SS7). This system is structured as a collection of parts, each of which is responsible for a particular set of signaling functions. Chapter 10 is an intermezzo that describes the Digital Subscriber Signaling System No.1 (DSS1), which is used by ISDN subscribers. Chapter 11 returns to SS7 and covers the ISDN user part of SS7. Chapters 12 through 17 discuss the signaling for transactions, and describe some applications of transactions in intelligent and mobile networks.

For brevity, the descriptions of signaling systems use many acronyms, and the resulting "alphabet soup" may initially be somewhat hard to digest. As a partial remedy, a list of acronyms appearing in the text and figures is included at the end of each chapter.

The material in this book is based largely on the specifications of various signaling systems. I gratefully acknowledge the International Telecommunication Union (ITU), Bellcore Inc., the Alliance for Telecommunications Industry Solutions (ATIS), the Electronics Industries Association (EIA), the Telecommunications Industry Association (TIA), and the European Telecommunications Standards Institute, for allowing me to reproduce a number of figures and tables. Some of these reproductions have been simplified, by omitting details that are beyond the scope of this book. The original versions can be found in the referenced source documents.

This book could not have been written without the help of many former colleagues at Bell Laboratories, especially Dave Barclay, Mary Brown, Fabrizio Devetak, Juli Federico, Chuck Ishman (now at Motorola), Joel Marks, Eugenia Mindlin, Jay Mitchell, Bob Multra, Gerald Novak, John Rosenberg, and Don Truax. Special thanks are due to Adrian de Vries, who read the entire manuscript, and suggested numerous clarifications and improvements.

Virginia Matthews, the librarian of Lucent Technologies' Bell Laboratories at Indian Hills, helped enormously by locating the various specification documents, which are scattered across numerous laboratory sites.

Last but not least, I would like to thank my wife Maria, for her encouragement at the times that I wondered whether this book would ever be completed, and her patience with the delays in "home" projects, which I put on the back burner while this book was being written.

JOHN G. VAN BOSSE

SIGNALING IN TELECOMMUNICATION NETWORKS

1

INTRODUCTION TO TELECOMMUNICATIONS

There are two types of communication networks: *circuit-switched* networks and *packed-switched* networks. In circuit-switched networks, a dedicated physical circuit between the calling and called party is set up at the start of a call, and released when the call has ended. Telephone networks are circuit-switched networks. Today, these networks are used for speech and other purposes, such as facsimile, and are usually referred to as *telecommunication* networks.

Initially, all communication networks were circuit-switched networks. *Data communication* networks made their appearance around 1970. In these networks, a call consists of short data bursts (packets) followed by relatively long silent intervals, and does not require a dedicated physical circuit. Internet is an example of a data communication network.

Today, the terms "telecommunication network" and "data communication network" usually imply circuit-mode and packet-mode networks, respectively.

This book is about signaling in telecommunication networks. To understand signaling, it is necessary to be familiar with some basic telecommunication concepts and terms. This chapter presents an overview of telecommunication networks. It is intended as an introduction, and sets the stage for the later chapters.

1.1 TELECOMMUNICATION NETWORKS

1.1.1 Introduction

Figure 1.1-1 shows a small part of a telecommunication network. It consists of *exchanges*, *trunks*, and *subscriber lines*. Trunks are circuits between exchanges,

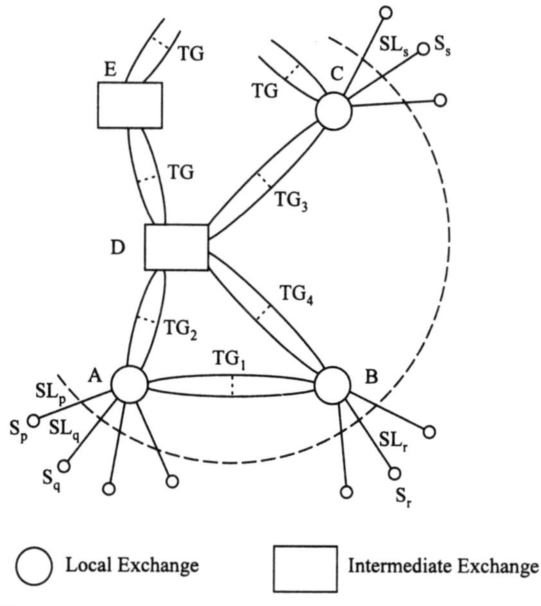

Figure 1.1-1 Partial view of a telecommunication network.

and the group of trunks between a pair of exchanges is known as a *trunk group* (TG). Subscriber lines (SL) are circuits between a subscriber S and his *local* exchange (A, B, C). Exchanges D and E do not have subscriber lines, and are known as *intermediate, tandem*, *toll,* or *transit* exchanges.

Calls. A call requires a communication circuit (connection) between two subscribers. Figure 1.1-2 shows a number of connections in the network of Fig. 1.1-1 that involve subscriber S_p. In Fig. 1.1-2(a), S_p is on a call with S_q who is attached to the same exchange. Calls of this type are known as *intraexchange* calls. The circuit for the call consists of the subscriber lines SL_p and SL_q, and a temporary path in exchange A. Cases (b) and (c) are calls between S_p and

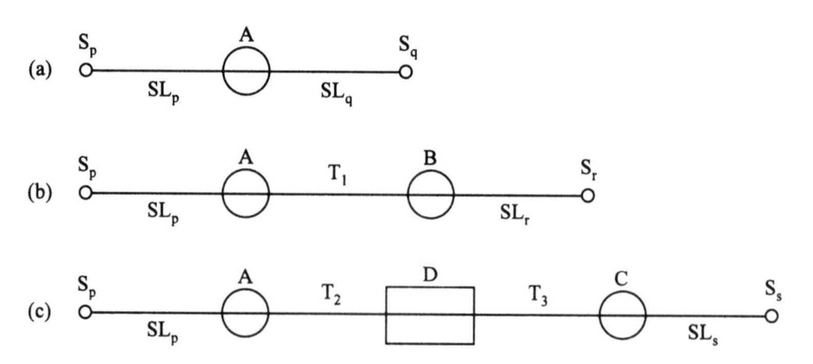

Figure 1.1-2 Connections involving subscriber S_p.

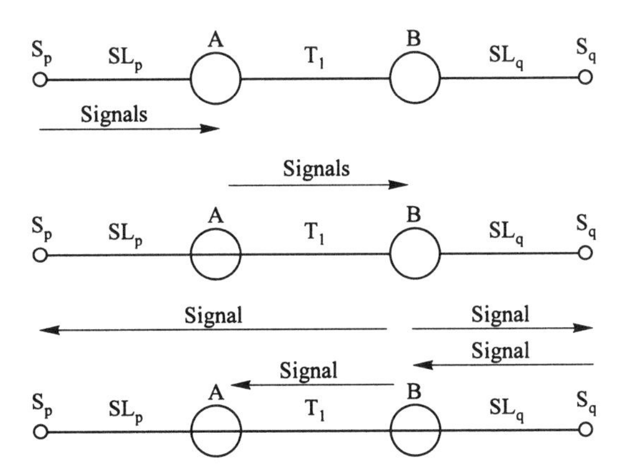

Figure 1.1-3 Set-up of a connection.

subscribers attached to other local exchanges (*interexchange* calls). The circuit in case (b) consists of SL_p, a temporary path across exchange A, trunk T_1, a temporary path across exchange B, and SL_r. The connections of Fig. 1.1-2 are set up (switched "on") at the start of a call, and released (switched "off") when the call ends.

Set-up and Release. The set-up and release of connections in telecommunication networks is triggered by *signals*. Starting and ending a call involves signaling between the subscribers and their local exchanges and, for interexchange calls, signaling between the exchanges along the connection.

Figure 1.1-3 shows the signaling for the set-up of the connection of Fig. 1.1-2(b). Subscriber S_p sends a request-for-service signal to exchange A (by lifting the handset of his telephone) and then signals the digits of the telephone number of S_q (with the dial or keyset of the telephone).

From the received number, exchange A determines that S_q is served by exchange B, and that the call is to be routed out on a trunk in group TG_1 (Fig. 1.1-1). It then searches for an idle trunk in this group, and finds trunk T_1. Exchange A now seizes the trunk, and sends a seizure signal, followed by signals that represent digits of the called number, to exchange B. It then sets up a path between SL_p and T_1.

When exchange B receives the seizure signal and the called number, it checks whether S_q is idle. If this is the case, it sends a ringing signal on SL_q, and a ringing-tone signal T_1, to inform S_p. When S_q lifts the handset of his telephone, an answer signal is sent to exchange B, which then stops the ringing signal and ringing tone, sets up a path between T_1 and SL_q, and signals to exchange A that the call has been answered.

The connection is now complete, and allows speech or other communications between the subscribers. At the end of the call, another signaling sequence takes place to release the connection.

One-way and Bothway Trunk Groups. In Fig. 1.1-1, there is at most one trunk group between two exchanges. Let us consider the group TG_1. The network should allow calls originating at A with destination B, and calls originating at B with destination A. Therefore, both exchanges are allowed to seize trunks in TG_1. A trunk group whose trunks can be seized by the exchanges at both ends is known as a *bothway* trunk group [1,2].

A pair of exchanges can also be interconnected by two *one-way* trunk groups. The trunks in one-way groups can be seized by one exchange only. For example, exchanges A and B could be interconnected by two one-way trunk groups TG_{1A} and TG_{1B}, whose trunks can be seized by A and B, respectively.

Both arrangements are used in actual networks. Two-way groups have an economic advantage because, for a given traffic intensity, the number of trunks of a bothway trunk group can be smaller than the total number of trunks in the one-way groups.

In bothway groups, it can happen that the exchanges at both ends of a trunk group seize the same trunk at the same time (double seizure). There are several alternatives to deal with a double seizure. For example, it can be arranged that one exchange continues the set-up, and the other exchange backs off (tries to seize another trunk for its call). The signaling on bothway trunks includes provisions to alert the exchanges when a double seizure occurs.

1.1.2 Networks

In everyday life, we think of "the" telecommunication network that allows us to speak, or send faxes and other data, to just about anybody in the world. In fact,

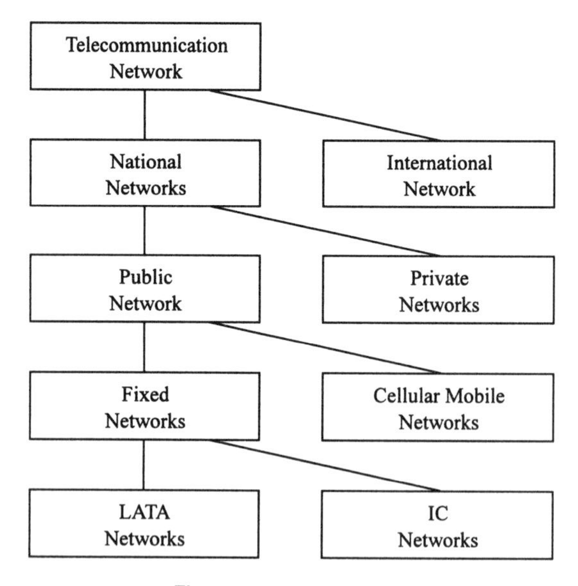

Figure 1.1-4 Networks.

the telecommunication network is an aggregation of interconnected networks of several types.

Networks can be classified as shown in Fig. 1.1-4. In the first place, the (global) telecommunication network consists of national networks and the international network. In turn, a national network is a combination of public and private networks. Public networks are for general use; private networks can be used only by employees of the organization (an airline company, the U.S. government, etc.) that owns the network. A public network consists of a "fixed" network and a number of "cellular mobile" networks. In the U.S., the fixed public network—known as the *public switched telecommunication network* (PSTN)—consists of about 150 LATA (*local access and transport area*) networks (the network of Fig. 1.1-1 is a LATA network), interconnected by networks that are known as IC (*interexchange carrier*), or long-distance, networks.

We now examine the interconnections of these networks. LATA and IC networks are interconnected by inter-network trunk groups—see Fig. 1.1-5. Some local exchanges (A) have a direct trunk group to an exchange of an IC, other exchanges (B, C, D, E) have access to the IC network via an intermediate (tandem) exchange in their respective LATAs.

A cellular network has one or more *mobile switching centers* (MSC)—see Fig. 1.1-6. Each MSC is connected by an inter-network trunk group to a nearby

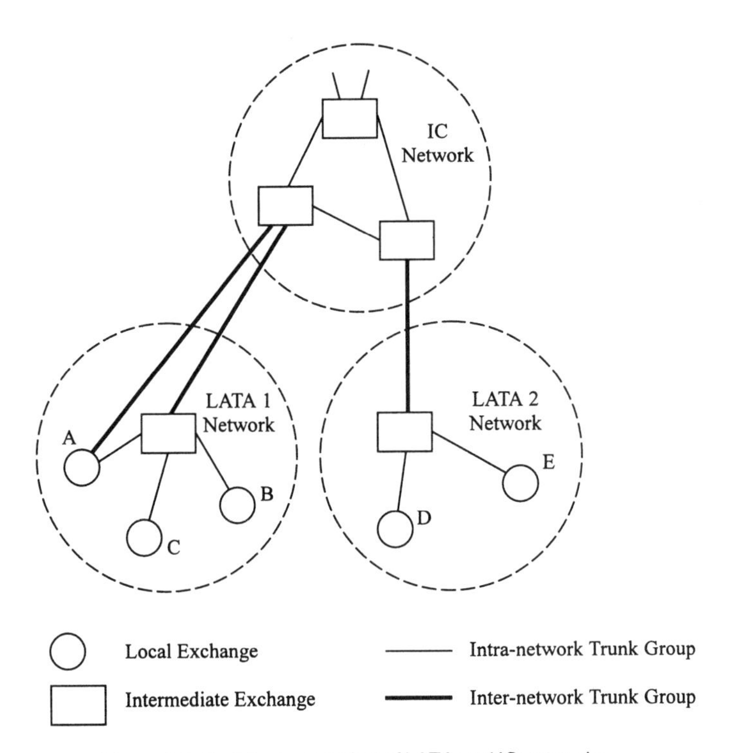

Figure 1.1-5 Interconnection of LATA and IC networks.

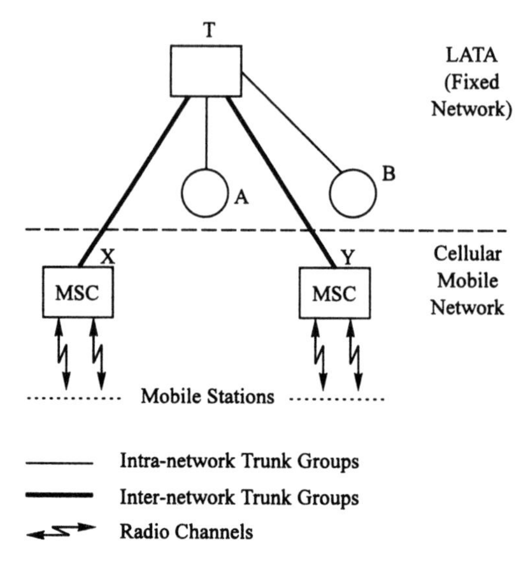

Figure 1.1-6 Interconnection of fixed and mobile networks. MSC: mobile switching center.

tandem exchange T of a fixed (LATA) network. When a mobile station is making a call, it uses a radio channel of a nearby MSC.

Private Branch Exchanges (PBX) are exchanges owned by government agencies, businesses, etc., and located in buildings that belong to these organizations. A PBX enables the employees in a building to call each other, and to make and receive calls from subscribers served by the public network. A PBX is connected by an *access line group* (ALG) to a nearby local exchange (Fig. 1.1-7).

An organization with PBXs in several cities can establish a *private network*

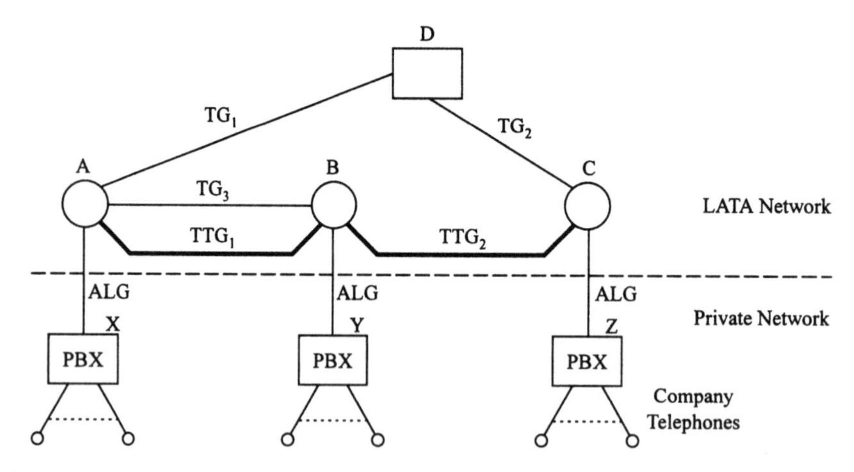

Figure 1.1-7 Interconnection of private network and a LATA network. ALG: access line group. TG: public trunk group. TTG: tie trunk group (private). PBX: private branch exchange.

that consists of the PBXs, and a number of *tie trunk groups* (TTG) between the public local exchanges to which their PBXs are attached. A TTG is a "private" group that is leased by the LATA operator, and dedicated to private-network calls. In Fig. 1.1-7, the connection for a call between public branch exchanges X and Y uses a trunk of TTG_1, and is switched in the public local exchanges A and B.

These days there are also *virtual private networks* (VPN). They appear to a business as a private network, but use the trunks of the public networks.

International Calls. Figure 1.1-8 shows the interconnection of long-distance (IC) networks in different countries. For a call from country A to country C, an IC network in country A routes the connection to an *international switching center* (ISC). An ISC has national trunk groups to exchanges of its IC, and international trunk groups to ISCs in foreign countries.

The term "international network" refers to the combination of the ISCs and their interconnecting trunk groups.

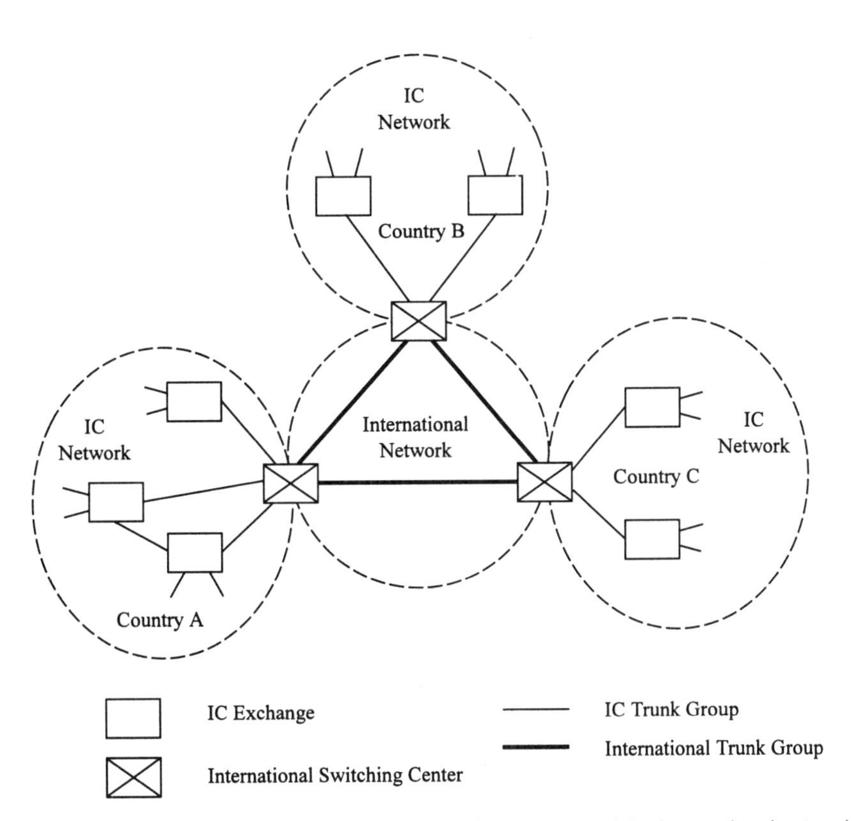

Figure 1.1-8 Interconnections between national IC networks and the international network.

1.1.3 Telecoms

We shall use the term "telecom" to denote a company that owns and operates a public telecommunication network. Until recently, the telecoms in most countries were government-owned monopolies that operated an entire national network. In recent years, a number of countries have started to privatize their telecoms, and to allow competition by newly formed telecoms.

The networks in the U.S. are operated by investor-owned telecoms. Until 1984, the Bell System was the largest telecom, operating practically the entire long-distance network, and many—but by no means all—local networks. *Independent* (non-Bell) telecoms, such as GTE, United Telecoms, and a host of smaller companies, operated (and continue to operate) the other local networks.

The division of the U.S. public network into LATAs and IC networks took place in 1984, as a result of government actions that broke up the Bell System [1]. LATAs are owned by *local exchange carriers* (LEC). Usually, a LATA network consists of a number of adjacent pre-divestiture local networks, and is a *regional* (rather than *local*) network. Long-distance networks and ISCs are owned by interexchange carriers. An international trunk group between the IC networks of two nations is usually owned jointly by the ICs.

1.1.4 Synonyms

The terms introduced in this section have several synonyms. This is because telecommunication terms originated rather independently, in different countries of the world. Today's telecommunications documents in English still use different terms for the same concept, depending on whether they have been written by workers in the U.S., the United Kingdom, or are translations of non-English documents. Some frequently used synonyms are listed below:

Subscriber, customer, user.

Subscriber line, line, loop.

Local exchange, local office, central office, end office.

Intermediate exchange, tandem exchange, toll exchange, transit exchange.

International switching center, gateway, international exchange.

Trunk, junction, circuit.

Telecom, administration, carrier, operating company, telephone company, telco.

1.2 NUMBERING PLANS

This section explores the formats of the numbers (sometimes called addresses) that identify the subscribers of telecommunication networks.

Subscriber Numbers (Directory Numbers). The geographical area of a nation is divided into several *numbering areas*, and subscriber numbers (SN) identify subscriber lines within a particular numbering area. A SN consists of an *exchange code* (EC) that identifies an exchange within a numbering area, followed by a *line number* (LN):

$$\text{SN} = \text{EC–LN}.$$

National Numbers. Within a country, a subscriber is identified by a *national number* (NN), consisting of an *area code* (AC), which identifies the numbering area, followed by a subscriber number:

$$\text{NN} = \text{AC–SN} = \text{AC–EC–LN}.$$

International Numbers. Worldwide, a subscriber is known by an *international number* (IN) that consists of a country code (CC), followed by a national number:

$$\text{IN} = \text{CC–NN} = \text{CC–AC–SN} = \text{CC–AC–EC–LN}.$$

When subscriber S_1 calls a subscriber located in the same numbering area, he dials a SN. If the called subscriber lives in the same country but in a different area, S_1 has to dial a NN and, if the called party lives in another country, S_1 needs to dial an IN.

National numbering plans define the formats of subscriber and national numbers. Most countries have individual numbering plans. However, the U.S., Canada, and a number of Caribbean countries are covered by a common plan that was introduced in the mid-1940s.

1.2.1 North American Numbering Plan [1]

The North American territory is divided into about 160 numbering areas known as *numbering plan areas* (NPA), which are identified by *three-digit area codes*, AC(3). Each area covers a state, or part of a state, but never crosses a state boundary. Lightly populated states (Nebraska, Arkansas, etc.) have one NPA, while more densely populated states (Illinois, New York, etc.) are divided into several NPAs. The territory of an NPA is not identical to the service area of a LATA network (LATA boundaries were established much later, after the break-up of the Bell System).

A subscriber number has seven digits: a three-digit exchange code, followed by a four-digit line number:

$$\text{SN(7)} = \text{EC(3)–LN(4)}.$$

The format of EC(3) is: *NXX*, where *N* ranges from 2 through 9, and *X* ranges from 0 through 9.

EC identifies a local exchange within a NPA. Each EC can cover maximally 10,000 line numbers. Since local exchanges can serve up to some 100,000 subscribers, more than one exchange code may have to be assigned to a particular exchange. For example, in a particular NPA, the subscriber numbers 357–*XXXX*, 420–*XXXX* and 654–*XXXX* might be served by the same local exchange.

National numbers consist of ten digits: a three-digit area code AC(3)—which also has the *NXX* format—followed by a seven-digit subscriber number:

$$\text{NN}(10) = \text{AC}(3)\text{–SN}(7) = \text{AC}(3)\text{–EC}(3)\text{–LN}(4).$$

These numbers are often shown as: NPA–*NXX–XXXX*. The U.S. numbering plan is an example of a closed (*uniform*) numbering plan. In these plans, the lengths of all subscriber numbers, and of all national numbers, are constant.

1.2.2 Other National Numbering Plans

Some countries have *open* numbering plans, in which subscriber numbers and area codes (sometimes called *trunk* or *city* codes) are not of fixed length. In these plans, the numbering areas usually have comparable geographical sizes. Heavily populated areas need subscriber numbers with six or seven digits, while four or five digits are sufficient in lightly populated areas. To limit the differences in length of national numbers, the area codes for areas with long subscriber numbers are usually shorter than those for areas with short subscriber numbers. An example of national numbers in an open numbering plan is shown below:

$$\text{NN}(8) = \text{AC}(2)\text{–SN}(6)$$
$$\text{NN}(9) = \text{AC}(2)\text{–SN}(7)$$
$$\text{NN}(8) = \text{AC}(4)\text{–SN}(4)$$
$$\text{NN}(9) = \text{AC}(4)\text{–SN}(5)$$

In order to allow exchanges to interpret national numbers in this plan, the two initial digits of a four-digit area code cannot be the same as those of a two-digit area code. For example, the two-digit area code 70 precludes the use of four-digit area codes 70*XX*.

1.2.3 Country Codes

The country codes have been established by CCITT [2], and consist of one, two, or three digits. The first digit indicates the world zone in which the called party is located:

World Zone

1: North America

2: Africa

3: Europe
4: Europe
5: Latin America
6: Australia and Southern Pacific Region
7: Former Soviet Union
8: China and Northern Pacific Region
9: Middle East

Country codes starting with 1 and 7 are one-digit codes, and represent respectively North America and the former U.S.S.R. Country codes starting with 2 through 9 can have two- or three-digit codes, and the combinations of the first and second digit determine which is the case. For instance, in world zone 3, all combinations except 35 are two-digit codes, as shown by the following examples:

31	The Netherlands
38	Yugoslavia
354	Iceland
359	Bulgaria

These rules enable exchanges to separate the country code from the national number in a received international number.

Country code 1 represents the U.S., Canada and a number of Caribbean countries. Most area codes represent areas in the U.S. Other codes represent areas in Canada and individual Caribbean nations.

1.2.4 Digit Deletion

In Fig. 1.2-1, calling party S_1 and called party S_2 are located in different NPAs of the U.S. S_1 therefore dials the national number (NN) of S_2. As a general rule, exchanges send subscriber numbers to exchanges that are in the NPA of the destination exchange, and national numbers to exchanges outside the destination NPA. In Fig. 1.2-1, exchange C is in the NPA of D, and exchanges

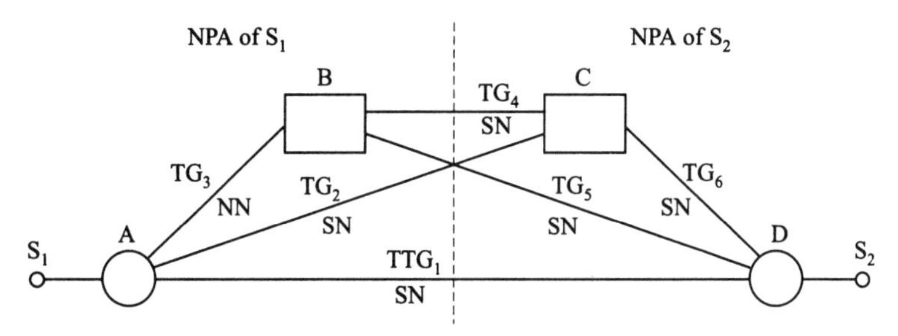

Figure 1.2-1 Called numbers. SN: subscriber number. NN: national number.

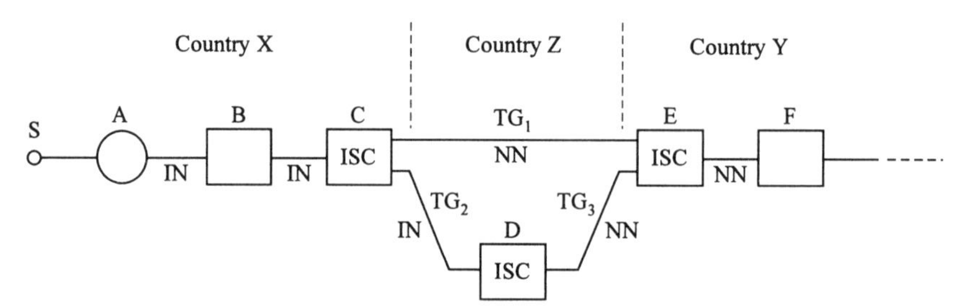

Figure 1.2-2 Called number formats on international calls. NN: national number. IN: international number.

A and B are not. Exchange A has received the NN of S_2. If A routes the call on a trunk of TG_1 or TG_2, it deletes the area code from the received number, and sends the SN of S_2. However, if A routes the call on TG_3, it has to send the NN of S_2. Exchanges B and C always send the SN.

A similar digit deletion occurs on international calls. Figure 1.2-2 shows a call from subscriber S in country X to a subscriber in country Y. An ISC sends a national or international number, depending on whether it routes the call on a direct trunk on an ISC in the destination country, or on a trunk to an ISC in an intermediate country.

1.3 DIGIT ANALYSIS AND ROUTING

1.3.1 Destinations and Digit Analysis

Connections for interexchange calls are set up along paths that have been predetermined by the network operator. A *route* is a path to a particular *destination*. An exchange determines the call destination by analyzing the called number, and then selects an outgoing trunk in a route to the destination.

We need to distinguish two destination types. The *final destination* (FDEST) of a call is the local exchange that serves the called party. An *intermediate destination* (IDEST) is an exchange where the call path enters another network, on its way to the final destination. For a connection, the destination at the exchanges of the the local network serving the called party is an FDEST. The destinations at exchanges in the other networks are IDESTS.

As an example, take a call from a calling party in local network $LATA_1$ to a called party in $LATA_2$. The interexchange carrier designated by the calling party is IC_D. In the exchanges of $LATA_1$, the call has an IDEST, namely an exchange in the network of IC_D, predetermined by the telecoms of the $LATA_1$ and IC_D networks. In the IC_D exchanges, the call also has an IDEST: an exchange in $LATA_2$ predetermined by the telecoms of IC_D an $LATA_2$. In the exchanges of $LATA_2$, the call has final destination.

Digit analysis is the process that produces an FDEST or IDEST from the called subscriber number (EC-LN), national number (AC-EC-LN), or international number (CC-AC-EC-LN).

In LATA exchanges, calls with subscriber and national numbers can have IDEST or FDEST destinations. Calls with international called numbers always have an IDEST. In IC exchanges, all calls have IDEST destinations. In calls with national called numbers, the IDEST is an exchange in the LATA network determined by the combination AC-EC.

For calls with international called numbers, the IDEST depends on whether the IC exchange is an ISC. If the exchange is not an ISC, the call destination is an ISC in the IC network, determined by the country code (CC) in the number. At an ISC, the destination is an ISC in the country identified by CC.

1.3.2 Routing of Intra-LATA Calls

Intra-LATA calls are handled completely by one telecom. In these calls, the FDEST is the local exchange of the called party. We examine a few routing examples for calls from a caller on local exchange A to a called party on local exchange D.

In Fig. 1.3-1(a), the telecom of the LATA has specified one indirect route, consisting of trunk groups TG_1 and TG_2:

Route A-TG_1-X-TG2-D

The fact that a TG belongs to a route does not mean that the TG is dedicated to the route. For example, TG_1 can also belong to routes from A to other destinations.

In Fig. 1.3-1(b), the telecom has specified a set of four routes:

A-TG_3-D
A-TG_4-Y-TG_6-D
A-TG_1-X-TG_2-D
A-TG_1-X-TG_5-Y-TG_6-D

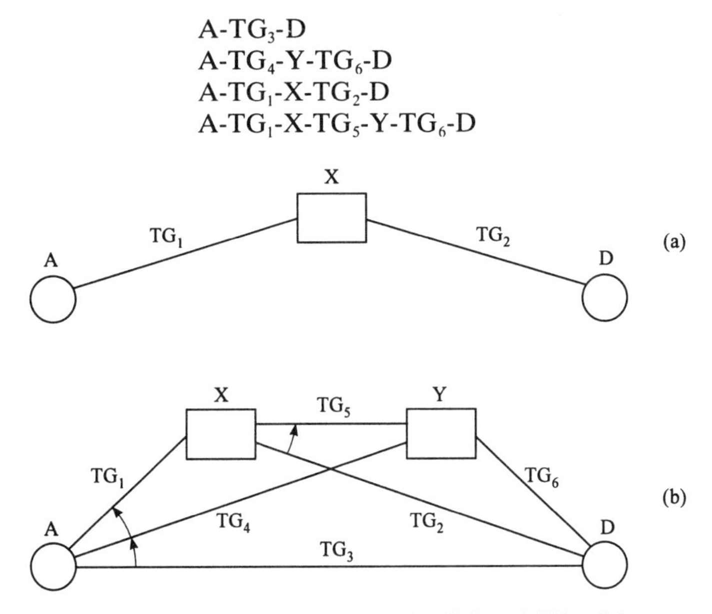

Figure 1.3-1 Routes for calls from A to D (intra-LATA calls).

As perceived by an exchange, a route to a destination is an outgoing trunk group (TG). In Fig. 1.3-1(b), the route set at exchange A for destination D consists of trunk groups TG_1, TG_3, and TG_4.

Each exchange has a list of routes that can be used for a destination. The lists for destination D at exchanges A, X, and Y are:

Exchange	Routes for Destination D
A	TG_3, TG_4, TG_1
X	TG_2, TG_5
Y	TG_6

Alternate routing is the procedure by which an exchange selects an outgoing trunk for a call when there are several routes to a destination. In this procedure, the order in which the routes are listed specifies the sequence in which an exchange checks the outgoing trunk groups for available trunks. In this example, exchange A first tries to find an available trunk in its first-choice route TG_3. If a trunk is available, A seizes the trunk. If not, it attempts to find an available trunk in its second-choice route TG_4, and so on. If none of these routes has an available trunk, exchange A aborts the set-up of the call.

In alternate routing, the TGs to a destination are ordered such that the first-choice route is the most direct one (passing through the smallest number of intermediate exchanges), the second-choice route is the most direct one among the remaining routes, etc. In Fig. 1.3-1(b), the arrows indicate the selection sequences at exchanges A and X.

1.3.3 Routing of Inter-LATA Calls

Figure 1.3-2 shows a routing example for a call originated by a subscriber attached to local exchange A of $LATA_1$, to a called party attached to local exchange Z in $LATA_2$.

In $LATA_1$, the IDEST of the call is exchange P or Q, in the IC network designated by the calling subscriber—Fig. 1.3-2(a). The routes for this destination at exchanges A, B, and C are:

Exchange	Routes for IDEST
A	TG_1, TG_2
B	TG_3, TG_4
C	TG_5, TG_6

In the IC network, the IDEST for the call is exchange K or L, in $LATA_2$—Fig. 1.3-2(b), and the routes for this destination at exchanges P, Q, R, S, and T are:

Exchange	Routes for IDEST
P	TG_7, TG_8
Q	TG_9, TG_{10}
R	TG_{11}, TG_{12}
S	TG_{13}, TG_{14}
T	TG_{15}

Finally, in $LATA_2$, the FDEST of the call is local exchange Z—Fig. 1.3-2(c), and the routes at exchanges K, L, and M are:

Exchange	Routes for IDEST
K	TG_{16}, TG_{17}
L	TG_{18}
M	TG_{19}

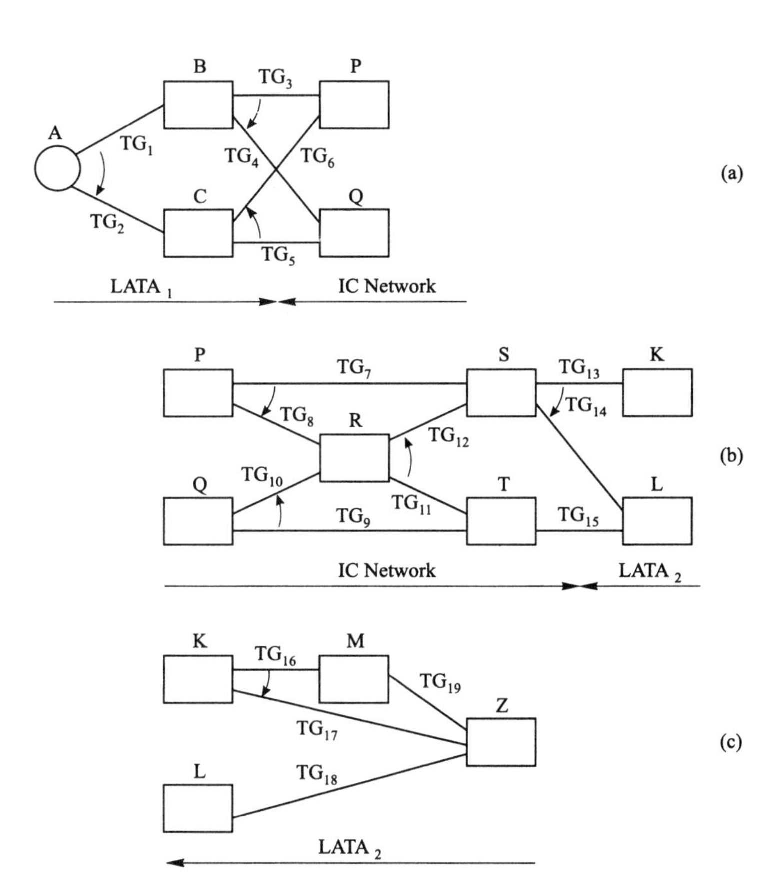

Figure 1.3-2 Routing of inter-LATA call. (a): routing in $LATA_1$. (b): routing in IC network. (c): routing in $LATA_2$.

1.3.4 Automatic Rerouting

Automatic rerouting (also called *crankback*) is a refinement of *alternate routing*. It is used in AT&T's long-distance network [3]. The procedure is illustrated with the example of Fig. 1.3-1(b).

Suppose that TG_3 is congested, and that A has seized trunk T in its second-choice route TG_4. The call set-up arrives at exchange Y, which has only one route (TG_6) to destination D. Under alternative routing, if no trunk is available in TG_6, exchange Y abandons the set-up, and informs the calling party with a tone or recorded announcement.

Under automatic rerouting, exchange Y signals to exchange A that it is unable to extend the set-up. A then releases trunk T, and tries to route the call on its final route (TG_1). If trunks are available in TG_1 and TG_2, and/or TG_5 and TG_6, the connection can be set up.

Automatic rerouting depends on the ability of an exchange (Y) to signal the preceding exchange (A) that it is not able to extend the set-up. We shall encounter signaling systems that have signals for this purpose.

1.4 ANALOG TRANSMISSION

Until 1960, analog transmission was the only form of transmission in telecommunication networks. Today, the telecommunication network is mostly digital, except for the subscriber lines.

This section outlines some basic aspects of the transmission of analog signals in telecommunications. In this section, the term *signal* refers to information (speech or voiceband data) exchanged between subscribers during a call (as opposed to *signaling*, which is the subject of this book).

1.4.1 Analog Circuits [4,5]

An analog signal is a continuous function of time. Telecommunication started out as telephony, in which a microphone (or transmitter, mouthpiece) produces an electrical *analog* signal, whose variations in time approximate the variations in air pressure produced by the talker's speech. A receiver (or earpiece) reconverts the electrical speech signal into air-pressure variations that are heard by the listener.

The pressure variations of acoustic speech are complex, and not easily described. "Average" acoustic speech contains frequencies from 35 Hz to 10,000 Hz. Most of the speech power is concentrated between 100 Hz and 4000 Hz. For good-quality telephony, only the frequencies between 300 and 3400 Hz need to be transmitted. Analog communication channels in the network, which historically have been designed primarily for speech transmission, therefore accommodate this range of *voiceband* frequencies (Fig. 1.4-1).

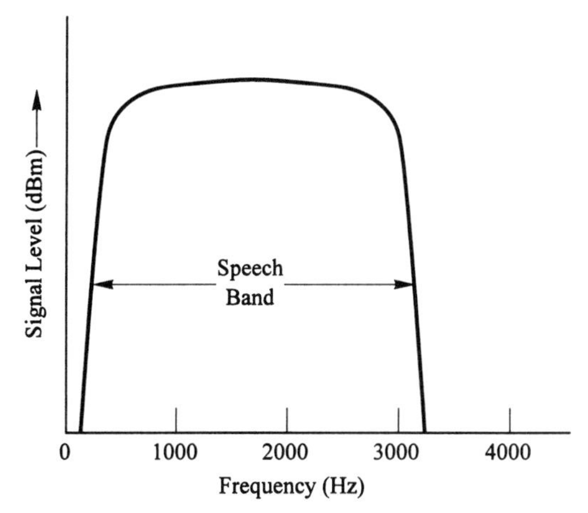

Figure 1.4-1 Frequency response of analog transmission circuits. (From R. L. Freeman, *Telecommunication System Engineering*, 2nd Ed., Wiley, 1992.)

Two-wire and Four-wire Circuits. Analog circuits can be two-wire or four-wire. Subscriber lines are two-wire circuits, consisting of a pair of insulated copper wires that transfer signals in both directions. Most analog trunks are four-wire circuits, consisting of two unidirectional two-wire circuits, one for each direction of transmission.

In the diagrams of this section, circuits are shown as in Fig. 1.4-2. The arrows indicate the directions of transmission. In general, bidirectional circuits (two- or four-wire) are shown as in (a), and a note indicates whether the circuit is two-wire of four-wire. When discussing the two unidirectional circuits of a four-wire circuit, representation (b) is used.

Data Transmission on Analog Circuits. In the 1960s, subscribers also began to use the telephone network for transmission of digital data, and the network has become a *telecommunication* network. Digital data are converted by *modems* into a form that fits within the 300–3400 Hz band of analog circuits. There are several modem types. *Frequency-shift keying* (FSK) modems convert the zeros and ones of the digital bit stream into two voiceband frequencies, for example, 1300 and 1700 Hz. With FSK, data can be sent at speeds of 600 or

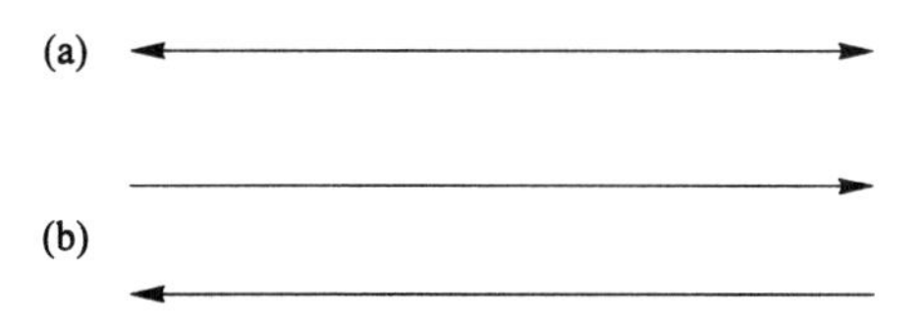

Figure 1.4-2 Circuit representations. (a): bidirectional two-wire or four-wire circuit. (b): two unidirectional two-wire circuits forming a bidirectional four-wire circuit.

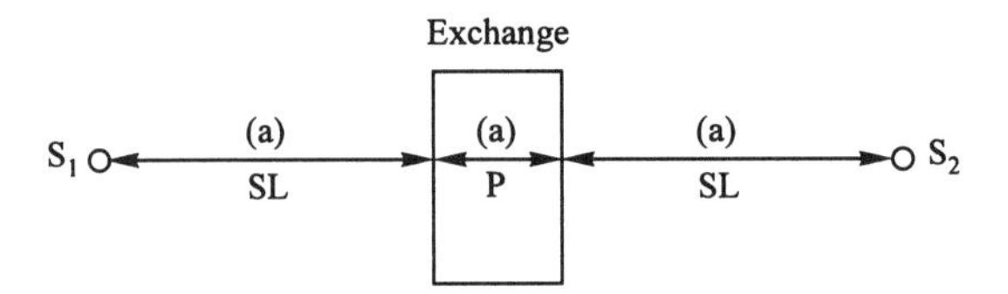

Figure 1.4-3 Circuit for intraoffice call. (a): two-wire analog circuits.

1200 bits/second. The signal produced by a *differential phase-shift keying* (DPSK) modem is a single frequency with phase shifts. In the widely used V.26 modem [6], the frequency is 1800 Hz, and the phase shifts occur at a rate of 1200 shifts/second. The phase shifts can have four magnitudes, each of which represents the values of two consecutive bits in the digital signal. The modem thus transfers 2400 bits/second. More recently developed modems have transfer rates of 4800, 9600, 14,400, and 28,800 bits/second.

1.4.2 Analog Subscriber Lines [4,5]

Figure 1.4-3 shows a connection between two subscribers served by an analog local exchange. The subscriber lines (SL), and the path (P) across the exchange, are two-wire circuits.

The power of an electrical signal decreases as it propagates along the circuit. This attenuation becomes more severe with increasing circuit length.

The characteristics of the microphones and receivers in the Western Electric type 500 telephones (which are still regarded as the "standard" for telephones in the U.S.) are such that a listener receives a sufficiently strong acoustical signal when at least 1% of the electrical signal power produced by the talker's microphone reaches the listener's receiver. This corresponds to the attenuation in a circuit of about 15 miles. Most subscriber lines are less than 4 miles long, and there are no signal-strength problems in intraexchange calls.

1.4.3 Two-wire Analog Trunks

Two-wire trunks are similar to subscriber lines, and have similar attenuation characteristics. This limits the trunk length to about 10 miles.

1.4.4 Four-wire Analog Trunks

Long-distance trunks require amplification to compensate the signal attenuation. Amplifiers are unidirectional devices, and this is why long-distance trunks are four-wire trunks. A four-wire circuit consists of two amplified unidirectional two-wire circuits. In Fig. 1.4-4, *hybrid* circuits (H) at both ends of the trunk convert a two-wire circuit into a four-wire circuit, and vice versa. Amplifiers (A) are located at regular intervals along the two unidirectional circuits. The unidirectional circuits at an exchange that transfer signals to and

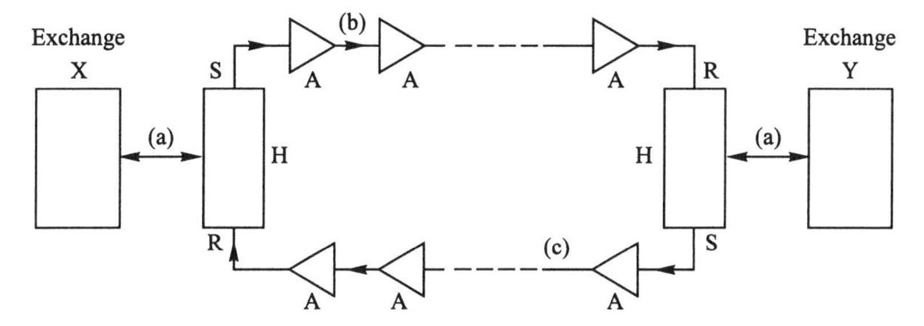

Figure 1.4-4 Amplified four-wire analog trunk. (a): two-wire bidirectional analog circuits. (b,c): pair of unidirectional analog circuits forming a bidirectional circuit.

from the distant exchange nearby exchange are known as the send circuit (S) and the receive circuit (R).

1.4.5 Frequency-division Multiplexing [4,5]

Frequency-division multiplexing (FDM) is a technique to carry the signals of a group of n analog four-wire trunks on a common four-wire analog transmission system. In each direction of transmission, the bandwidth of the transmission system is divided into n 300–3400 Hz channels, spaced at intervals of 4 kHz, for a total bandwidth of $n \times 4$ kHz (Fig. 1.4-5).

Figure 1.4-6 shows a group of n FDM trunks between exchanges X and Y. In this example, the trunks appear at the exchanges as two-wire circuits (a) that can transfer analog signals in the 300–3400 Hz range.

Frequency-division multiplexing equipment (FDM) is located at both exchanges. Hybrids (H) convert the two-wire circuits into four-wire circuits. For transmission in direction X → Y, multiplexer FDM-X shifts the 300–3400 Hz send channels (S) of the trunks "up" in frequency (by different amounts), and then combines the channels on an unidirectional multiplex transmission circuit (b). FDM-Y demultiplexes the signal received on (b), by shifting the channels downward in frequency. The demultiplexed channels are the 300–3400 Hz receive channels (R) of the trunks at exchange Y.

In the other direction, the send channels at exchange Y are multiplexed by FDM-Y, combined on multiplex transmission circuit (c), and demultiplexed by FDM-X. Amplifiers A along the multiplexed circuits (b) and (c) compensate the signal attenuation.

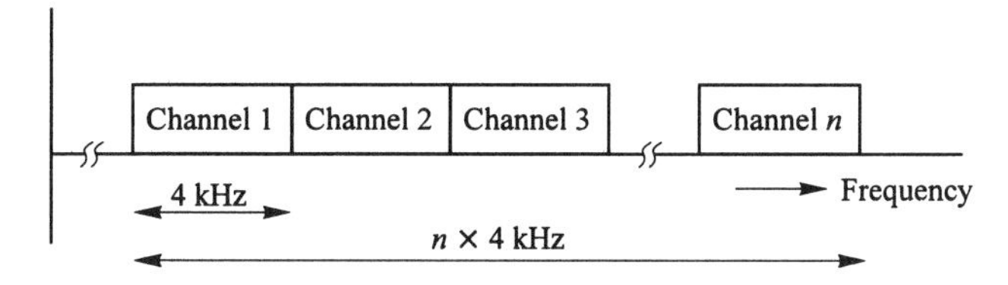

Figure 1.4-5 Frequency allocation on FDM transmission system.

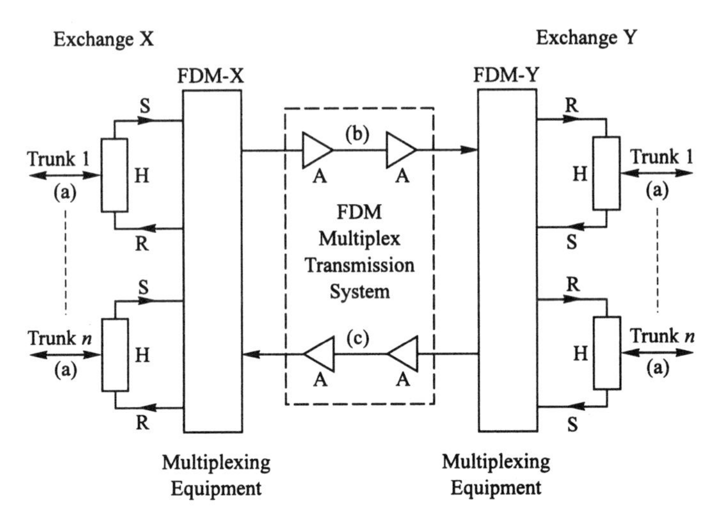

Figure 1.4-6 Multiplexed analog trunks. (a): two-wire bidirectional circuits. (b,c): multiplexed unidirectional circuits (*n* channels), forming a four-wire FDM transmission system. A: amplifiers. S,R: two two-wire unidirectional circuits, forming a bidirectional circuit.

Larger FDM multiplex systems can be formed by repeated multiplexing. This creates a FDM multiplex hierarchy, which has been standardized internationally [5]. In Fig. 1.4-7, the 60 two-wire trunks T are converted to four-wire

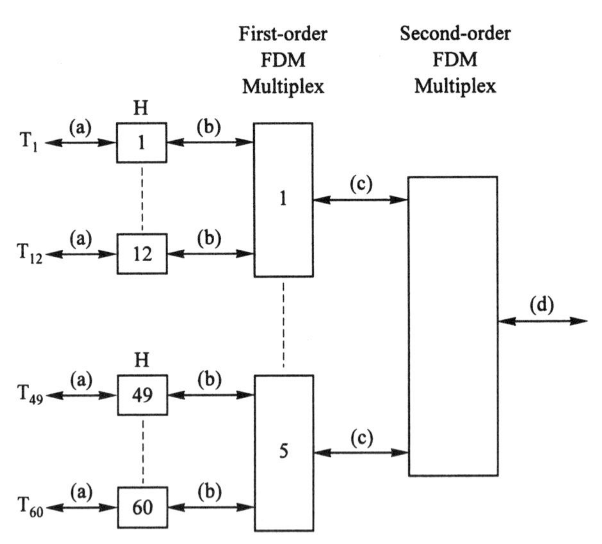

Figure 1.4-7 First- and second-order FDM multiplexes. (a): two-wire bidirectional circuits. (b): four-wire circuits. (c): multiplexed four-wire circuits (12 channels). (d): multiplexed four-wire circuits (60 channels).

circuits (b) by hybrids (H). The first-order FDM multiplexes convert 12 four-wire circuits into a four-wire *group* circuit (c) with 12 channels. The bandwidth of group circuits is $12 \times 4 = 48$ kHz in each direction (60–108 kHz).

The second-order FDM multiplex combines five group circuits (c) into one four-wire *supergroup* circuit (d) with 60 channels. The bandwidth of a supergroup circuit is 5×48 kHz $= 240$ kHz in each direction (312–552 kHz).

Continuing in this way, ten supergroups can be combined into a 600 channel *mastergroup*, and six mastergroups form a 3600 channel *jumbogroup* [5].

1.4.6 FDM Transmission Systems

Group circuits can be carried on two amplified wire-pairs, with amplifiers at regular spacings. Signals of higher order multiplexes are carried on transmission systems of several types.

Cable transmission systems consist of two coaxial cables, with amplifiers at regular spacings. Cable systems are used on overland and underwater (transatlantic, transpacific) routes.

Microwave radio systems consist of a pair of unidirectional radio-frequency (RF) transmission links in the microwave region (2, 4, 6, 11, or 18 GHz). The output signal of a FDM multiplexer modulates the frequency of the microwave carrier. The RF signal travels in a narrow beam from the transmitting antenna to the receiving antenna.

In terrestrial systems, the microwave links are divided into several sections (hops), of say 20–40 miles. The transmission in each section is on a line-of-sight path. Repeater stations at the section boundaries amplify the RF signal received from one section, and retransmit it to the next section. The repeater stations are land based, and terrestrial microwave systems can therefore be used on overland routes only.

Satellite microwave systems use communication satellites. These satellites are in *geosynchronous* orbits, some 22,300 miles above the equator. At this altitude, a satellite circles the earth at an angular velocity equal to the angular velocity of the earth's rotation. When observed from a point on earth, the satellite therefore appears in a fixed position, and only small and infrequent adjustments are needed to keep an antenna of a ground station aimed at the satellite. Each satellite link consists of an "up" link from a ground station to the satellite, and a "down" link from the satellite to another ground station. The satellite amplifies the signal received on the "up" link, and retransmits it on the "down" link.

Satellite microwave systems were introduced in the 1960s, and have been used extensively on transoceanic routes. A disadvantage of satellite transmission is the long signal propagation time (in excess of 250 ms from ground station to ground station), which can cause problems in telephone conversations.

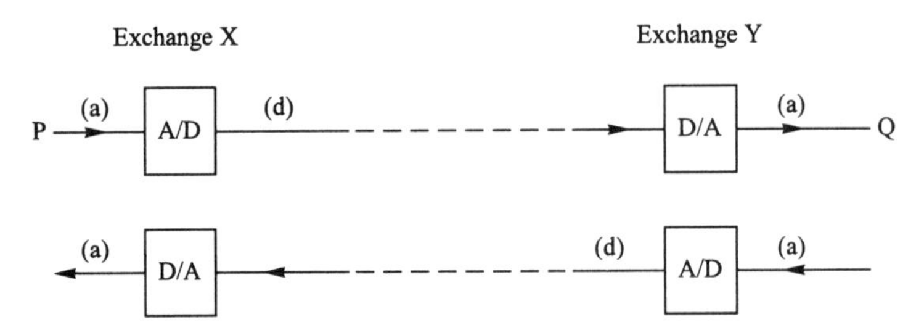

Figure 1.5-1 Digital transmission system. (a): analog signals. (d): digital bit stream. A/D: analog to digital converter. D/A: digital to analog converter.

1.5 DIGITAL TRANSMISSION

Analog signals (a) can be converted to digital bit streams (d), and transmitted on digital transmission systems (Fig.1.5-1).

Digital transmission is more robust and of better quality than analog transmission. While digital signals require more bandwidth, digital transmission systems are often less expensive than their analog counterparts, and digital transmission is rapidly replacing analog transmission in telecommunications.

1.5.1 Pulse Code Modulation

Pulse code modulation (PCM) is the earliest method for *analog-to-digital* (A/D) and *digital-to-analog* (D/A) conversions used in telecommunication networks [4,5]. The basic idea dates back to 1938 [7]. However, PCM requires a fair amount of digital signal processing, and only became technically feasible after the invention of semiconductor devices. The first PCM systems were installed in the U.S., by the Bell System, in 1962 [1,4].

PCM is designed such that its frequency response, measured from analog input signal (P) to analog output signal (Q) of Fig. 1.5-1 is the same (300–3400 Hz) as for FDM analog transmission—see Fig. 1.4-1.

In PCM, the analog input signal (a) of an A/D converter is sampled at t_s = 125 microsecond intervals (8000 samples/second)—see Fig. 1.5-2. The magnitude of each sample is then converted into an *octet* (an eight-bit binary number). The digital signal (d) of Fig. 1.5-1 is a sequence of octets, transmitted at $8 \times 8 =$ 64 kb/s (kilobits/second).

PCM Standards. In order to preserve the quality of low-level speech, the A/D and D/A conversions are non-linear. Two coding rules are in existence: mu-law coding is the standard in the U.S., Canada, and Taiwan, and A-law coding is the standard in the rest of the world [7, 8].

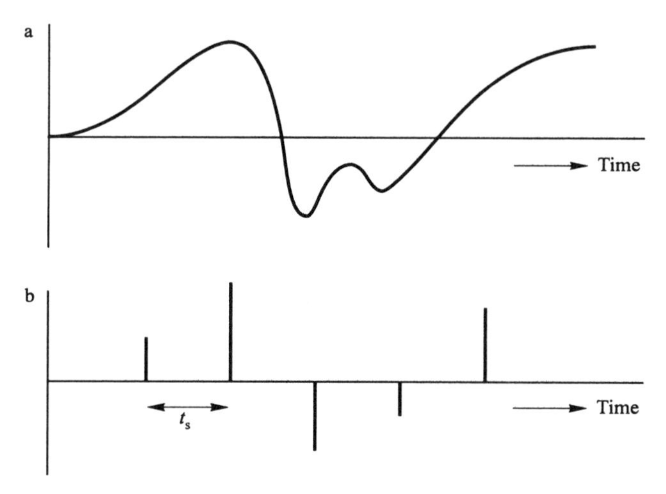

Figure 1.5-2 Sampler input (a) and output (b).

1.5.2 Time-division Multiplexing

PCM trunks are carried on four-wire *time-division multiplexed* (TDM) transmission systems [8]. Figure 1.5-3 shows a first-order pulse-code modulation multiplex PCM-X for m trunks. In the example, the trunks are attached to the exchange as two-wire analog trunks (a), and are converted to four-wire analog circuits by hybrids (H).

Multiplexer PCM-X does the A/D conversions of the signals in the analog *send channels* (S), and then multiplexes the resulting octets into the outgoing

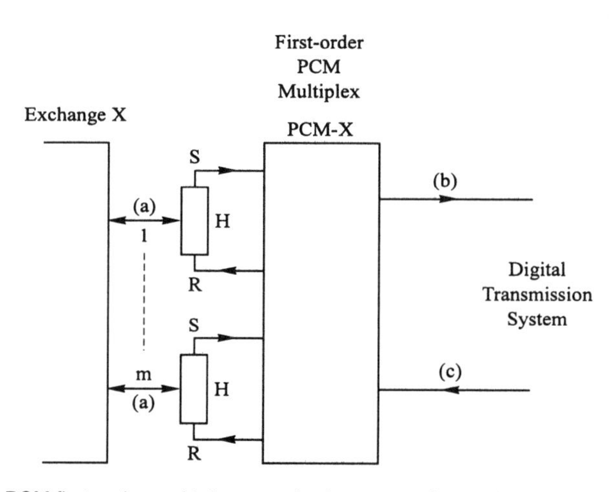

Figure 1.5-3 PCM first-order multiplex, attached to a two-wire analog exchange. (a): analog two-wire bidirectional circuits. S,R: send and receive circuits of analog four-wire circuits. (b,c): outgoing and incoming circuits of a four-wire multiplexed digital circuit (m channels). H: hybrids.

bit stream (b). PCM-X also segregates the octets of the individual channels in the incoming bit stream (c), and does the D/A conversions that produce the voiceband analog signals in the *receive channels* (R).

T1 and E1 Digital Transmission Systems. There are two standards for first-order transmission systems. The T1 system, a development of Bell Laboratories, is used mainly in the U.S. [1,8]. The E1 system has been defined by the European Conference of Postal and Telecommunications Administrations (CEPT), and is used in most other countries [8].

In both systems, the bit stream is divided in *frames* that are transmitted at a rate of 8000 frames/second. Each frame is divided into a number of eight-bit time-division channels, known as *time slots* (TS), that contain the PCM samples of the individual trunks.

The frame format of the T1 system (known as the *DS1 format*) is shown in Fig. 1.5-4(a) [4,5,8]. It consists of $m = 24$ time slots (TS_1 through TS_{24}), and a framing bit (F). The F bits in successive frames form a fixed *synchronization* sequence that repeats every 12 frames. This enables a multiplexer to identify the start points of the individual frames in the incoming bit stream. The frame length is $1 + 8 \times 24 = 193$ bits, and the transmission rate—at points (b) and (c) of Fig. 1.5-3—is 193×8000 bits/second = 1544 kilobits/second (kb/s).

The E1 frame contains 32 eight-bit time slots, numbered TS_0 through TS_{31}—see Fig. 1.5-4(b)—and serves $m = 30$ trunks. TS_1 through TS_{15}, and TS_{17} through TS_{31} hold the octets for the samples of the trunks. Time slot TS_0 has a

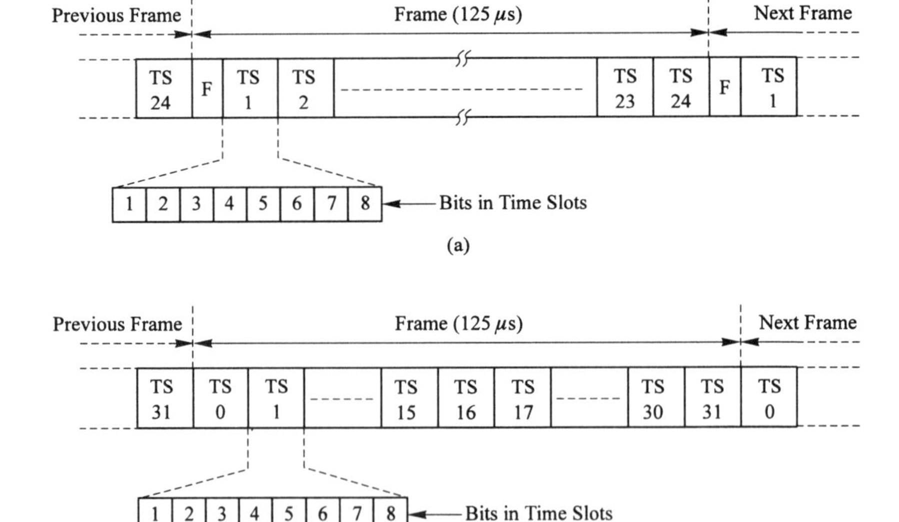

Figure 1.5-4 Formats of first-order PCM frames. (a): DS1 format ($m = 24$ channels). (b): E1 format ($m = 30$ channels). TS: time slot.

fixed eight-bit synchronization pattern, and time slot TS_{16} contains signaling information [4,8].

A E1 frame thus has $32 \times 8 = 256$ bits, and the bit rate on the transmission system is $8000 \times 256 = 2{,}048$ kb/s.

Higher Order Multiplexes. As in FDM, higher order PCM multiplexes are formed by repeated multiplexing. The North American and the CEPT system have separate hierarchies [8].

1.5.3 Digital Transmission Systems [4,5]

First-order PCM multiplexes are carried on two copper wire-pairs in a conventional cable. The pulses attenuate and stretch as they traverse the cable, and have to be regenerated by repeaters (R), located at intervals of 1 mile (Fig. 1.5-5). Since it is difficult to maintain transmission systems with large numbers of repeaters, the maximum length of these systems is about 200 miles.

The signals of higher order multiplexes can be carried on transmission systems of several types.

The North American T4M system carries a fourth-order PCM multiplex (4032 channels) on a pair of coaxial cables, with 1-mile repeater spacings. The T4M system is also limited to rather short trunks. Longer trunks are carried on microwave radio, or fiber-optic transmission systems.

In PCM microwave radio systems, the digital bit-stream modulates a microwave carrier signal. *Phase-shift keying* (similar to DPSK modems—see 1.4.1) is a frequently used modulation form. Both terrestrial and satellite microwave transmission systems are in use. Like their analog counterparts, digital terrestrial radio links are divided into sections, with repeater stations at the section boundaries.

In optical fiber systems, the digital signal is in the form of lightwave pulses that propagate through an extremely thin glass fiber (diameters in the order of 0.01 mm). At the sending end of the fiber, a laser diode converts the electrical pulses into "light" pulses. A photodiode at the receiving end converts the optical pulses back into electrical pulses.

These transmission systems offer high-speed transmission with low

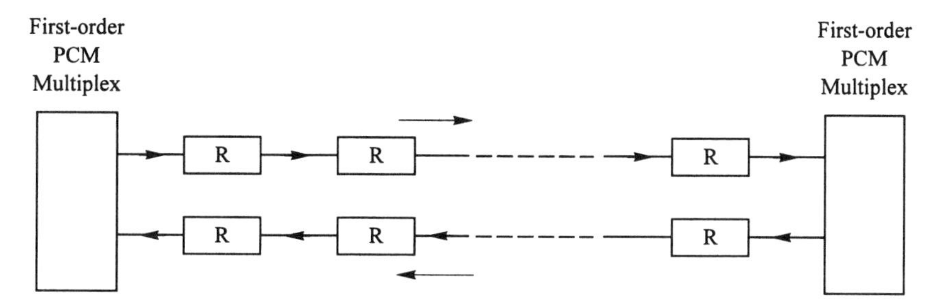

Figure 1.5-5 Transmission of first-order PCM multiplex on wire pairs. R: repeaters.

attenuation, and permit large repeater spacings. In 1988 it was already possible to carry 280 Mb/s (four thousand 64 kb/s channels) on a fiber, with repeaters spaced at 30 miles. The fiber technology continues to evolve, and experimental systems that transmit at speeds of 10,000 Mb/s have been demonstrated. Fiber-optic systems are rapidly being deployed in both medium- and long-distance (transcontinental and transoceanic) applications.

1.5.4 Adaptive Differential Pulse Code Modulation

Since the introduction of PCM, several other coding techniques have been developed. The main objective of these developments is to lower the bit rate while maintaining the transmission quality. One of these methods, *adaptive differential pulse code modulation* (ADPCM), is being introduced in some telecommunication networks.

In ADPCM, four-bit digital samples, transmitted at 8000 samples/second, represent the *differences* in magnitude of two consecutive samples of the analog input signal. The coding is "adaptive": the relation between the magnitude of the analog quantity and the associated digital code is not fixed, but adapts automatically to the characteristics of the signal being encoded.

An ADPCM channel for a trunk requires $8000 \times 4 = 32$ kb/s, and gives a speech quality almost equal to PCM. ADPCM doubles the channel (trunk) capacity of digital transmission systems. For example, by dividing the eight-bit octets of the E1 frame of Fig. 1.5-4(b) into two four-bit groups, a first-order E1 transmission system can serve 60 trunks.

1.6 SPECIAL TRANSMISSION EQUIPMENT

This section describes two types of special transmission equipment: *echo control* equipment and *circuit multiplication* equipment. Both of these exist in analog and digital versions.

1.6.1 Echoes

Figure 1.6-1 shows the transmission circuit (omitting the exchanges) for a typical long-distance connection with analog four-wire trunks. It consists of three parts. Parts 1 and 3 are two-wire circuits, containing the subscriber loops, and possibly two-wire trunks. Part 3 is a four-wire trunk. Hybrids H_1 and H_2 convert two-wire transmission to four-wire transmission.

When subscriber S_1 speaks, the speech signal leaves H_1 at port P, reaches port Q of H_2, and then travels on the two-wire circuit to listener S_2. However, a small part of the signal received at Q "leaks" to R, and thus returns to S_1, who hears an *echo* of his speech. The leakage occurs because hybrid circuits are "balancing" circuits. For leak-free operation, the impedance presented by the two-wire circuit (at port T of hybrid H_2) would have to match the design

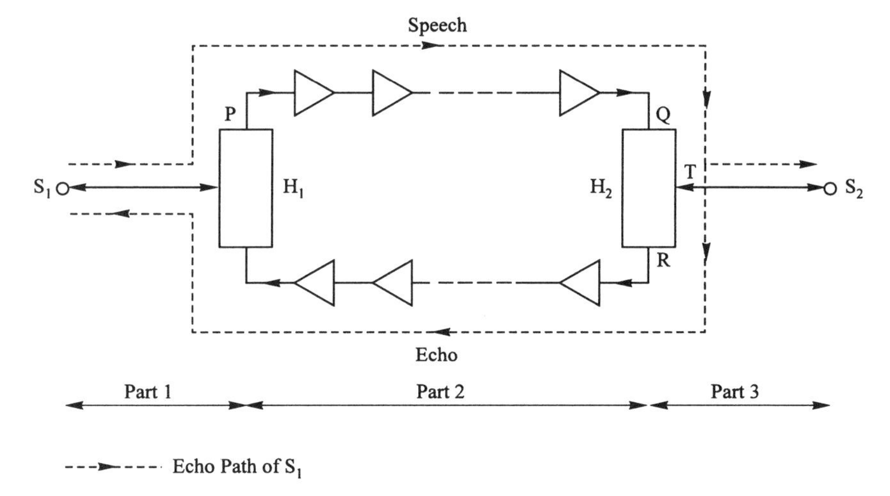

Figure 1.6-1 Long-distance connection.

impedance of the hybrid for all voiceband frequencies. In each call, H_2 is connected to a different two-wire circuit, and impedances of these circuits (which vary with circuit length and physical cable characteristics) cannot be precisely controlled. Complete balance therefore never occurs in practice, and some echo is always present. When S_2 speaks, a similar echo is caused by reflections at H_1.

The effect of echoes on a talker varies with the *echo delay* (the signal propagation time from talker S_1 to H_2 and back to S_1). Echoes delayed by less than 20 ms are barely noticeable, but echoes with larger delays rapidly become very disturbing to a talker. Echo delays increase with increasing length of the four-wire circuit, and trunks longer than 2000 miles (transcontinental, transoceanic and satellite trunks) are equipped with echo control devices.

1.6.2 Echo Suppressors [1,9]

Long four-wire analog trunks are equipped with echo suppressors. Figure 1.6-2 shows echo suppressor units ES-A and ES-B, located at both ends of a trunk between exchanges A and B (the other exchanges in the connection are not shown).

Each suppressor unit has detector (D) on its receive pair (R). When D detects the presence of a signal, the suppressor interrupts its send pair (S). For example, when S_1 speaks, ES-B interrupts its S path, and the leakage from hybrid H-B cannot return to S_1. ES-A protects S_2 from echoes in the same way.

When a suppressor has opened its send path, it "hangs over" (leaves the path open) for some 30–40 ms after the signal on the receive path has disappeared. This bridges the silent intervals between successive words of the distant speaker. Suppose now that S_1 has been speaking. The send path of ES-B is therefore open. When near-end subscriber S_2 starts to talk during a

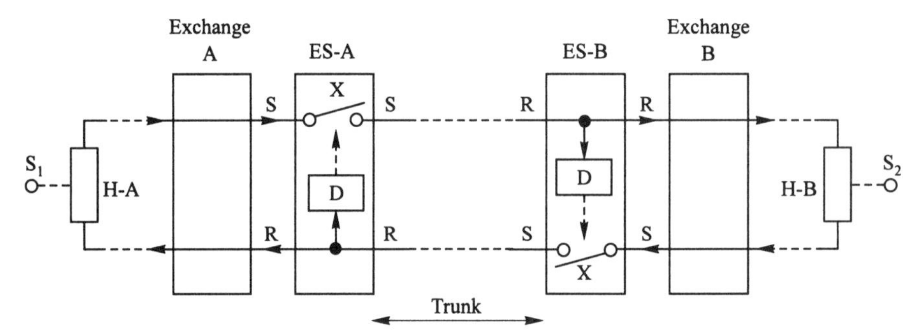

Figure 1.6-2 Echo suppressors on a long-distance analog trunk. D: voiceband signal detector. X: switch. H: hybrid circuit. ES: echo suppressor.

hangover, the initial part of his speech is clipped off. The effects of clipping become very noticeable on connections involving several echo-suppressed trunks in tandem. It is therefore desirable to avoid such connections. We shall see that some trunk signaling systems include indicators to make this possible.

Echo suppressor units are always installed in pairs—one at each end of a trunk. Some documents refer to the units as *half echo suppressors*.

1.6.3 Echo Cancelers [5,10]

Echo cancelers, which are used on long PCM trunks, also provide protection against echoes, but operate on a different principle. Figure 1.6-3 shows a pair of echo cancelers at the ends of a long-distance PCM trunk. The circuit between an echo canceler and its near-end subscriber is known as the "tail," and includes the subscriber line, hybrid circuit (H), an A/D and D/A converter, and can include other digital and analog trunks.

We assume that S_1 is speaking, and explore the operation of canceler EC-B. The signals r, e, c, and s are PCM signals (sequences of octets). When S_1 speaks,

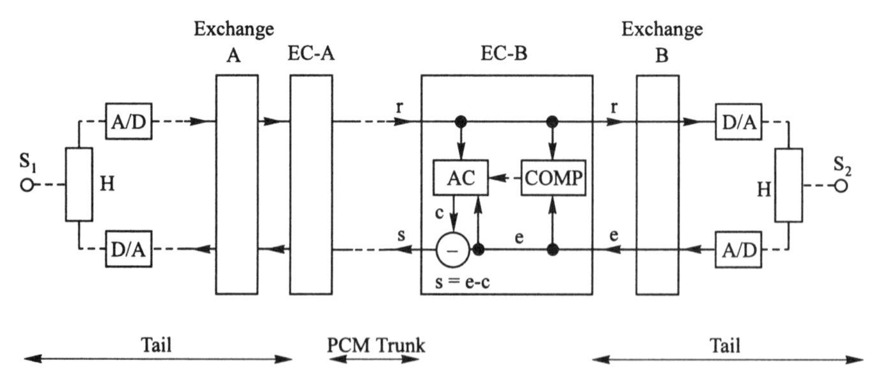

Figure 1.6-3 Echo cancelers on a PCM trunk. EC: echo canceler. AC: adaptive circuit. COMP: comparator. A/D: analog to digital converter. D/A: digital to analog converter. H: hybrid circuit.

a signal (r) is received on the receive pair of EC-B. During incoming speech, EC-B regards its tail as a two-port circuit that receives a signal (r) and returns an echo signal (e). Signal r also feeds the *adaptive circuit* (AC) in the canceler, which produces an output signal (c). Signal c is subtracted from signal e, and the result s = e – c is sent out on the send pair of EC-B.

The tail of an EC is different for each call. At the beginning of a call, the canceler automatically adjusts its adaptive circuit, AC, making the impulse transfer function of the circuit equal to that of the tail. This is done by comparing signals e and c. When the adaptation is complete, signals e and c are equal, and s = 0, which means that the echoes returning from the tail are being canceled. The adjustment takes place during the initial seconds of received far-end speech.

Comparator (COMP) compares the strengths of signals r and e. When r > e, the canceler is receiving speech from S_1, and can adjust circuit AC. However, when e > r, near-end subscriber S_2 is speaking, and no adjustments are made.

Echo cancelers have one significant advantage over echo suppressors: they do not clip the initial part of the near-end subscriber's speech. Therefore, more than one canceler-equipped trunk is allowed in a connection.

Digital Multiplexed Echo Cancelers. In the description above, the EC pair cancels echoes for one connection. Actual echo cancelers are digital devices that serve groups of *m* multiplexed PCM trunks. The canceler is connected to the output of a first-order PCM multiplex (Fig. 1.6-4) and processes the eight-bit PCM samples (octets) of the trunks, during their time slots in each PCM frame. The canceler has an adaptive circuit for each trunk, but its control circuitry is time-shared.

1.6.4 Circuit Multiplication

In a typical telephone conversation, each subscriber speaks during about 30% of the time, and both subscribers are silent during the remaining 40%. During the call, each channel of a four-wire trunk thus carries speech during only 30% of time. *Circuit multiplication equipment* allows a group of four-wire trunks to be carried on a four-wire transmission system with a smaller number of *bearer channels*, by making use of the "silent" intervals [5,9,11,12]. Circuit

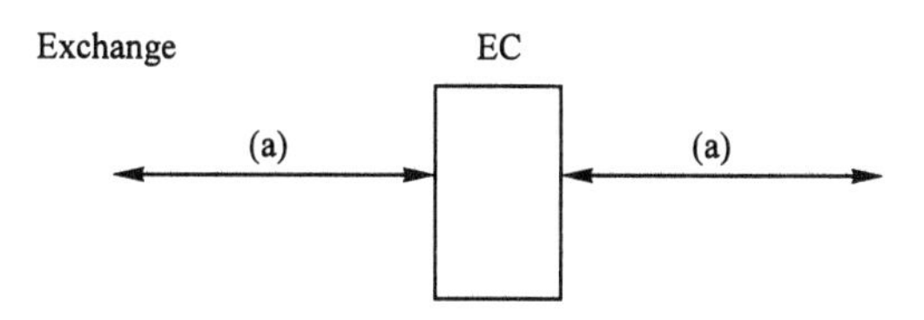

Figure 1.6-4 Echo cancelers for first-order digital time-division multiplexes (*m* channels). (a): four-wire multiplexed circuits.

multiplication is economically attractive in situations where transmission costs are high, for example, in transoceanic transmission systems.

Analog Circuit Multiplication Equipment. Circuit multiplication systems for analog four-wire trunks are generally known as *time assignment speech interpolation* (TASI) systems. Figure 1.6-5 shows two TASI units, at exchanges A and B, that concentrate m trunks to n (four-wire) bearer channels ($n < m$). In each TASI unit, the ($n \times m$) send switch allows up to n simultaneous unidirectional connections between the send circuits (S) of trunks and bearer channels. Likewise, the ($n \times m$) receive switch can connect up to n receive circuits (R) of bearer channels and trunks. Speech detectors are bridged across the S-pairs of the trunks. The controls in the TASI units communicate with each other on a pair of unidirectional control channels.

The assignments of bearer channels to trunks take place independently for each direction of transmission. The TASI operation for transmission from A to B is outlined below.

When a detector in TASI-A detects speech on the S-pair of a trunk, control-A seizes an available S-bearer channel, and connects the trunk and the channel. It also informs control-B, identifying the trunk and the channel, and control-B then sets up a path between R-pair of the trunk and the R-bearer channel. This establishes a unidirectional path between the S-pair of the trunk at exchange A, and the R-pair of the trunk at at exchange B. At the end of a speech burst, an *overhang* timer, with an expiration time of about 400 ms, is started. If new speech energy is detected on the bearer channel while the timer is running, the timer is stopped. Expiration of the timer indicates a 400 ms silent interval on the channel. Control-A then releases the channel, and informs control-B.

The set-up of connections is very fast (typically, within 20 ms), and only a

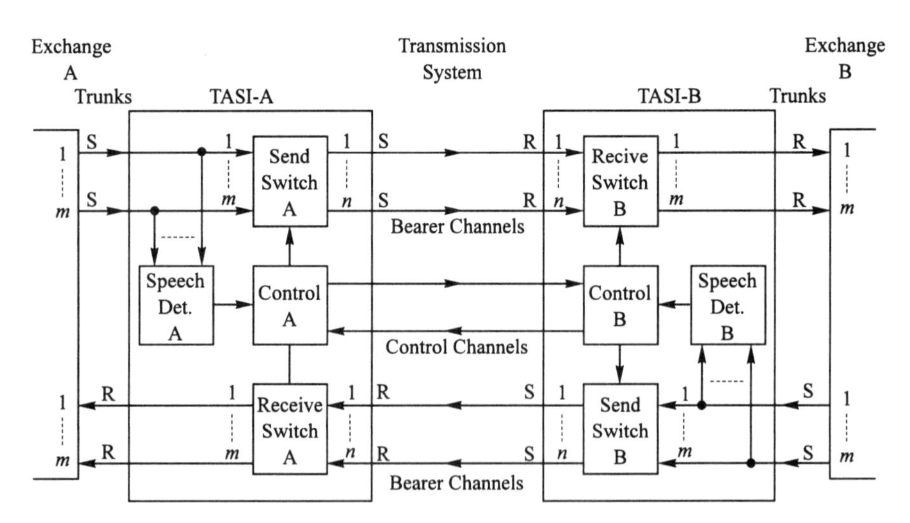

Figure 1.6-5 TASI equipment. R: analog receive circuits. S: analog send circuits.

small part of the initial syllable of the first word in a sentence is "lost". This is barely noticeable to the listener.

Freeze-out. It can happen that all S-bearer chanels are occupied at the time that speech is detected on a previously silent trunk. In that case, the system has to wait until a bearer channel becomes available, and the initial part of a speech burst is "frozen out" (lost).

The ratio of bearer circuits to trunks is designed such that the probability of freeze-outs lasting more than 60 ms is below 1%. In TASI systems, typical concentration ratios m:n are in the order of 2:1.

TASI systems were designed during the years when speech was the only form of subscriber communication. Today, a small but increasing fraction of calls is used for facsimile and data transmission. During these calls, modem signals can be present continuously in one or both directions, thus occupying one or two bearer circuits on a full-time basis. To keep freeze-outs on the other trunks to an acceptably low level, it has become necessary to reduce the number of trunks that can be served by a TASI system.

Digital Circuit Multiplication Equipment. Digital circuit multiplication equipment operates on the same principle, but has first-order multiplex interfaces with the exchange and the transmission system [12]. Moreover, most digital TASI systems convert the eight-bit PCM octets on the trunks to four-bit ADPCM groupings on the bearer channels. This doubles the channel capacity of the transmission system, and reduces the per-channel transmission cost by another 50%. Figure 1.6-6 shows a typical configuration, where 150 PCM trunks (five first-order E1 multiplexes) are concentrated to 60 ADPCM bearer channels, and carried on one E1 multiplex line.

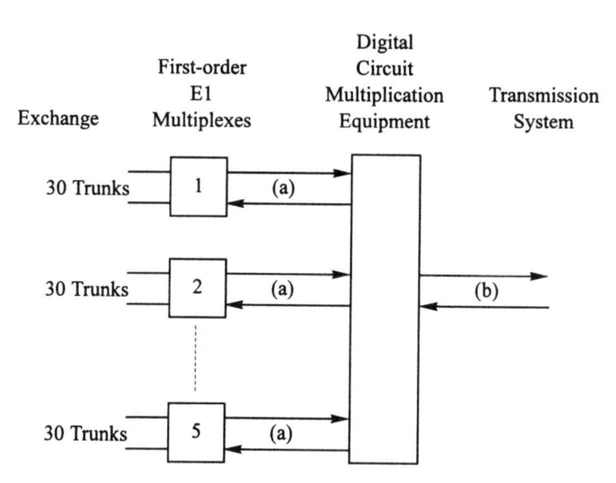

Figure 1.6-6 Digital circuit multiplication equipment. (a): 2048 kb/s E1 line (30 trunks). (b): 2048 kb/s E1 line (60 bearer channels).

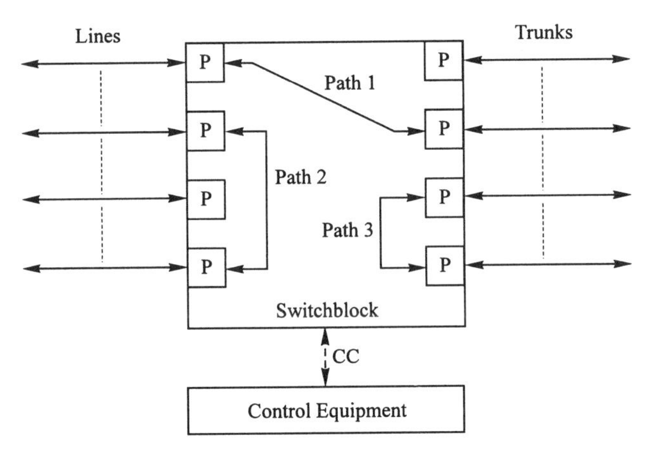

Figure 1.7-1 Exchange equipment. P: port. CC: control channel.

1.7 EXCHANGES

1.7.1 Exchange Equipment

The primary function of an exchange is to establish and release temporary paths between two subscriber lines, between a subscriber line and a trunk, or between two trunks [4].

The two major equipment units in an exchange are shown in Fig. 1.7-1. The *switchblock* (or *switching network*) has a large number (up to some 100,000) of *ports* (P) to which subscriber lines and trunks are attached. The ports are interconnected by arrays of switches. By closing or opening appropriate sets of switches, paths across the switchblock can be set up or released. At any point in time, a multitude of these paths can be in existence. In Fig. 1.7-1, path 1 connects a line and a trunk, path 2 connects two lines, and path 3 connects two trunks.

The switchblock paths are set up and released on command from the *control equipment* (CE) of the exchange. CE sends commands to, and receives responses from the switchblock, via a *control channel* (CC) (shown as a dashed line).

1.7.2 Switchblocks

Analog Switchblocks. Until about 1975, all exchanges had analog switchblocks, implemented with electromechanical switching devices. In local exchanges, the switchblocks were two-wire (Fig. 1.7-2). The temporary paths in the switchblock, and the circuits attached to ports P, were two-wire bidirectional analog voiceband (300–3400 Hz) circuits (a).

Subscriber lines and two-wire analog trunks (a) were attached to the ports of the switchblock via, respectively, two-wire *line circuits* (LC) and two-wire trunk circuits (TC2). Multiplexed analog and PCM trunks (b) were converted, by

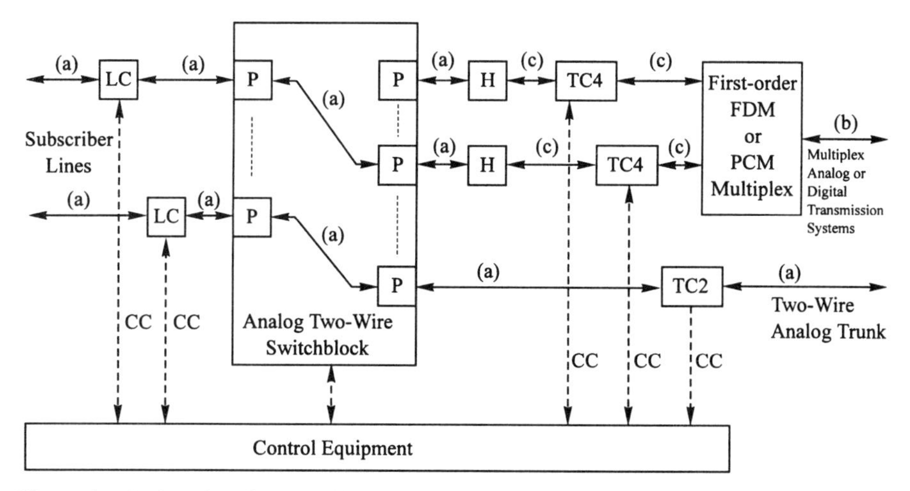

Figure 1.7-2 Local exchange with two-wire analog switchblock. P: port. LC: line circuit. H: hybrid circuit. TC2: analog two-wire trunk circuit. TC4: analog four-wire trunk circuit. CC: control channel.

first-order FDM and PCM multiplexes, to individual four-wire analog circuits (c). These circuits then passed through four-wire analog trunk circuits (TC4), and were converted by hybrids H into bidirectional analog two-wire circuits (a).

The switchblock, and the line and trunk circuits, were controlled by the control equipment of the exchange, via control channels (CC). The line and trunk circuits play a role in the transmission and reception of signaling information (see Chapters 3 and 4).

The analog switchblocks in intermediate exchanges (tandem exchanges, toll exchanges and international switching centers) provided four-wire analog paths between four-wire ports. This eliminated the need for hybrid circuits on multiplexed FDM and PCM trunks. Hybrid circuits are required on two-wire analog trunks, but the trunks attached to intermediate exchanges were predominantly four-wire trunks.

Digital Switchblocks. The introduction of digital time-division multiplex transmission systems has led to the development of exchanges with digital switchblocks. Most exchanges installed after 1980 have digital switchblocks which are implemented with integrated semiconductor circuits. These switchblocks are more compact, and more reliable, than their analog predecessors.

As shown in Fig. 1.7-3, the ports on digital switchblocks are usually four-wire digital first-order time-division multiplex ports (DMP). The frame formats at points (b) are as shown in Fig. 1.5-4. The DMPs of exchanges in the U.S. have the DS1 format (for $m = 24$ circuits). In most other countries, the DMPs have the E1 frame format (for $m = 30$ circuits).

The paths in the switchblock are 64 kb/s, four-wire circuits (a). The m multiplexed circuits entering a digital multiplex port DMP are segregated into

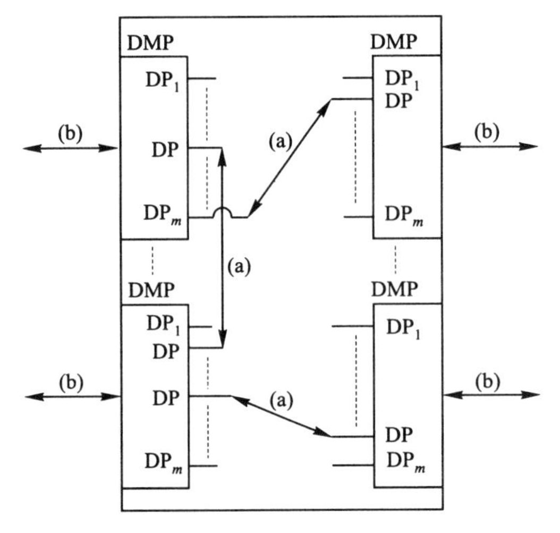

Figure 1.7-3 Paths in digital switchblock. (a): 64 kb/s four-wire paths. (b): first-order digital multiplex circuits (*m* channels). DMP: digital multiplex port.

m individual 64 kb/s digital ports (DP), and the switchblock paths connect pairs of these ports.

Attachment of Lines and Trunks (Fig. 1.7-4). T1 or E1 transmission systems for digital trunks (b) can be attached directly to the DMPs. Each DMP has a control channel (CC), on which the control equipment of the exchange sends and receives signaling information for the *m* multiplexed channels.

Subscriber lines arrive at the exchange as two-wire analog circuits (c). They pass through two-wire line circuits (LC) and hybrid circuits (H), and enter first-order PCM multiplexers (1.5.2). The outputs of the multiplexers are four-wire digital multiplexed circuits (b) that serve *m* lines, and are connected to DMPs.

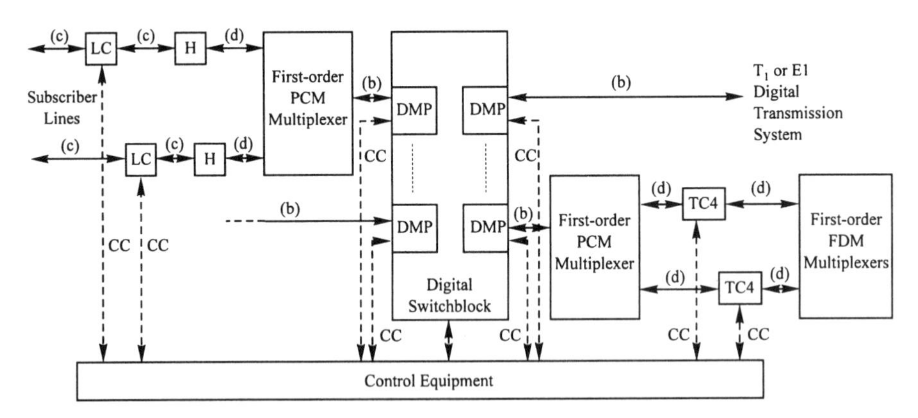

Figure 1.7-4 Exchange with digital switchblock. LC: line circuit. H: hybrid circuit. DMP: digital multiplex port. TC4: analog four-wire trunk circuit. CC: control channel.

Each line circuit also has a control channel (CC) to the control equipment of the exchange.

Analog trunks arriving on FDM transmission systems are first demultiplexed by FDM multiplexes (1.4.5) into individual four-wire analog circuits (d). These circuits then pass four-wire analog trunk circuits (TC4), and enter PCM multiplexes where they are converted to a first-order digital multiplex circuit (b).

1.7.3 Control Equipment [1,4]

The control equipment of early exchanges was implemented with electromechanical devices. Around 1955, it became clear that digital computers could be used to control exchanges. The speed and sophistication of computers at that time already greatly exceeded the capabilities of electromechanical exchange control equipment. The first *stored program controlled* (SPC) exchange was placed in commercial operation in 1965 [1].

The main elements of a stored-program exchange control system are shown in Fig. 1.7-5. A processor (or a group of processors) performs all exchange actions, controlling the switchblock, the line and trunk circuits, and other exchange equipment, by executing instructions in its program. The control channels for these equipment units are connected to a bus on the processor.

Processor memory is divided into semi-permanent memory and temporary memory. The semi-permanent memory stores the programs for call processing, charging, and exchange maintenance procedures, and tables with data for digit analysis, call routing and charging, etc.

Temporary memory stores data that are changed frequently. For instance, information on the status of the subscriber lines and trunks, records of currently existing calls, etc.

The increased logical power of SPC exchanges has allowed the inclusion of features that were not available in earlier exchanges. These fall in two main categories:

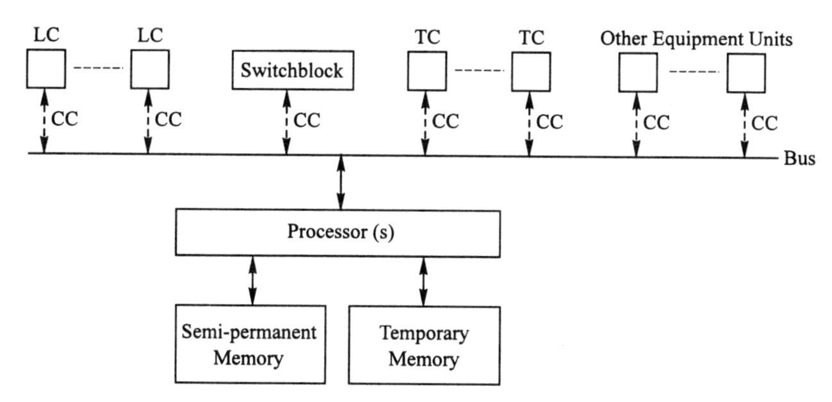

Figure 1.7-5 Stored-program exchange control. LC: line circuits. TC: trunk circuits. CC: control channels.

- New services for subscribers (call forwarding, call waiting, three-way calling, etc.).
- Procedures to facilitate the administration, operation, and maintenance of the exchange by the telecom. These include routine checks on the exchange equipment, periodic testing of attached lines and trunks, the generation of data on the traffic handled by the exchange and its trunk groups, the production of equipment trouble reports, and so on.

1.8 ACRONYMS

AC	Area code
AC	Adaptive circuit
A/D	Analog to digital converter
ADPCM	Adaptive differential pulse code modulation
CC	Control channel
CC	Country code
CCITT	International Telegraph and Telephone Consultative Committee
CEPT	European Conference of Postal and Telecommunications Administrations
D/A	Digital to analog converter
DMP	Digital multiplex port
DPSK	Differential phase-shift keying
DS1	North American first-order pulse-code modulation multiplex
EC	Echo canceler
EC	Exchange code
ES	Echo suppressor
FDM	Frequency-division multiplex/ing
FSK	Frequency-shift keying
H	Hybrid circuit
IC	Interexchange carrier
IN	International number
ISC	International switching center
ISDN	Integrated services digital network
LATA	Local access and transport area
LC	Line circuit
LE	Line equipment
LEC	Local exchange carrier
LN	Line number
MP	Multiplex port
MSC	Mobile switching center
NN	National number
NPA	Numbering plan area
P	Single port
PBX	Private branch exchange

PCM	Pulse-code modulation
PSTN	Public switched telecommunication network
RF	Radio frequency
R	Receive circuit of four-wire circuit
S	Subscriber
S	Send circuit of four-wire circuit
SL	Subscriber line
SN	Subscriber number
SPC	Stored program controlled
T	Trunk
TC	Analog trunk circuit
TC2	Analog two-wire trunk circuit
TC4	Analog four-wire trunk circuit
TASI	Time assignment speech interpolation
TDM	Time-division multiplex
TG	Trunk group

1.9 REFERENCES

1. *Engineering and Operations in the Bell System*, Second Edn, AT&T Bell Laboratories, 1984.
2. *International Telephone Service Operation*, CCITT Red Book, **II.2**, Rec. E.163-E.164, ITU, Geneva, 1985.
3. G.R. Ash, "Design and Control of Networks with Dynamic Nonhierarchical Routing," *IEEE Comm.*, **28**, Oct. 1990.
4. R.L. Freeman, *Telecommunication System Engineering*, Second Edn, Wiley-Interscience, New York, 1989.
5. *Transmission Systems for Communications*, Fifth Edn, Bell Telephone Laboratories, 1982.
6. *Data Communication over the Telephone Network*, CCITT Red Book, **VIII.1**, Rec. V.10-V.32, ITU, Geneva, 1985. Modems.
7. G. H. Bennett, *PCM and Digital Transmission*, Marconi Instruments Ltd, U.K., 1978.
8. *Digital Networks, Transmission Systems and Multiplex Equipment*, CCITT Red Book, **III.3**, ITU, Geneva, 1985.
9. J.M. Fraser, D.B. Bullock, N.G. Long, "Overall Characteristics of a TASI System," *Bell Syst. Tech. J.*, **51**, July 1962.
10. *General Characteristics of International Telephone Connections and Circuits*, CCITT Red Book, **III.1**, Rec. G.101-G.181, ITU, Geneva, 1985.
11. D.A. Bardouleau, "Time Assignment Speech Interpolation Systems," *British Telecom. Eng.*, **4**, Oct. 1985.
12. M. Onufry, "The New Challenges of DCME," *Telecomm. J.*, **55**, Dec. 1988.

2

INTRODUCTION TO SIGNALING

2.1 OVERVIEW

This section presents a brief historical outline of signaling. The earliest telephone exchanges were "manual" switchboards, in which all calls were set up and taken down by operators. Signaling between subscribers and operators was limited to *ringing*. To make a call, the subscriber would send a ringing signal. This alerted an operator, who would connect her telephone to the calling line, and ask for the called number. The operator then would connect her telephone to the called line, and ring the line. After answer by the called party, the operator would establish the connection.

Signaling as we know it today started around 1890, with the invention, by Almon B. Strowger (a Kansas City undertaker), of an automatic switchboard that could receive the called number dialed by the calling subscriber, and would then automatically set up the connection. During the past 100 years, signaling applications and technology have evolved in parallel with the developments in telecommunications.

2.1.1 Early Signaling

Signaling in the period from 1890 to 1976 had three main characteristics. In the first place, its application was limited to *plain ordinary telephone service* (POTS): the set-up and release of connections between two subscribers. In the second place, the signals were carried by the same circuit (subscriber line, trunk) that carried the speech during the call. This type of signaling is known as *channel-associated signaling* (CAS) [1,2]. Finally, signaling took place only between a subscriber and his local exchange (*subscriber signaling*), and between the exchanges at the two ends of a trunk (*interexchange signaling*).

Initially, automatic telephony was possible only for calls between subscribers served by the same exchange, which required subscriber signaling only. Later on, it became possible to dial calls between subscribers served by nearby exchanges. These calls also required interexchange signaling. National long-distance calls needed operator assistance until the 1950s, when *direct distance dialing* (DDD) was introduced. *International direct distance dialing* (IDDD), which requires signaling on international trunks, became possible in the 1960s [3].

Channel-associated call-control signaling is still widely used today. However, beginning in 1976, other forms and applications of signaling have made their appearance in telecommunication networks. They are briefly outlined below.

2.1.2 Common-channel Signaling

Common-channel signaling (CCS), introduced in 1976, was developed as an alternative form of call-control signaling for trunks [2,4,5]. In CCS, signaling information is not carried by the individual trunks. Instead, a *signaling network*, consisting of *signaling data links* (SDL) and *signal transfer points* (STP), transfers digital *signaling messages* between the exchanges. In Fig. 2.1-1, trunk groups TG_1, TG_2, and TG_3 have common-channel signaling, and the signaling network consists of one signal transfer point (STP), and the signaling data links SDL_A, SDL_B, and SDL_C. A call-control message from exchange A to exchange B for a trunk in TG_1 traverses SDL_A, the signal transfer point, and SDL_B. In this example, each signaling link carries messages for all CCS trunks that are attached to an exchange. For example, SDL_A carries messages for the trunks in

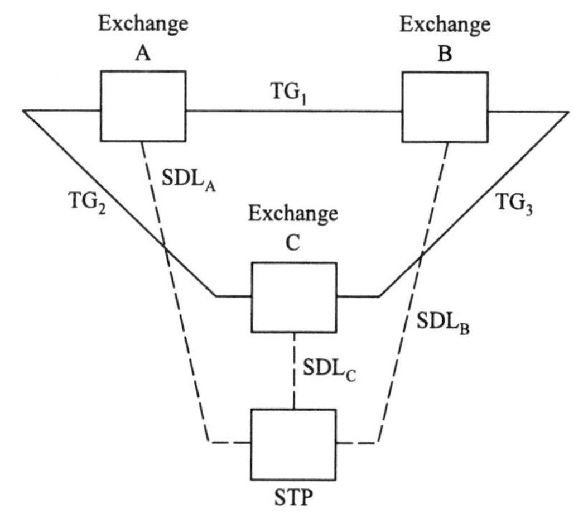

Figure 2.1-1 Common-channel call-control signaling. $TG_{1,2,3}$: trunk groups. $SDL_{A,B,C}$: signaling data links. STP: signal transfer point.

groups TG_1 and TG_2. We say that SDL_A is the "common" signaling channel for the trunks in these groups.

Common-channel signaling became possible after the introduction of stored-program controlled exchanges. We shall see that CCS signaling is more powerful, flexible, and also faster than CAS signaling. CCS is being introduced rapidly in national telecommunication networks, and in the international network. However, the replacement of CAS is not yet complete, and CAS and CCS signaling coexist in many networks.

2.1.3 Other Applications of CCS Signaling

Since 1980, CCS is also being used for other applications. In Fig. 2.1-2, a *service control point* (SCP) and an *operations, administration, and maintenance center* (OAM) also have signaling data links, and exchanges can send messages to, and receive messages from, these network entities. Procedures that involve signaling between an exchange and a SCP, or between an exchange and an OAM, are known as *transactions*.

The SCPs in a network support *intelligent network* (IN) services [5–7]. These services require information that cannot be stored conveniently in exchanges. A well-known IN service is 800 calling. The 800-numbers are not in the conventional format of national numbers, and cannot be used by the exchanges to route calls to their destinations. Therefore, when an exchange receives a call with an 800 number, it starts a transaction with the SCP, requesting the

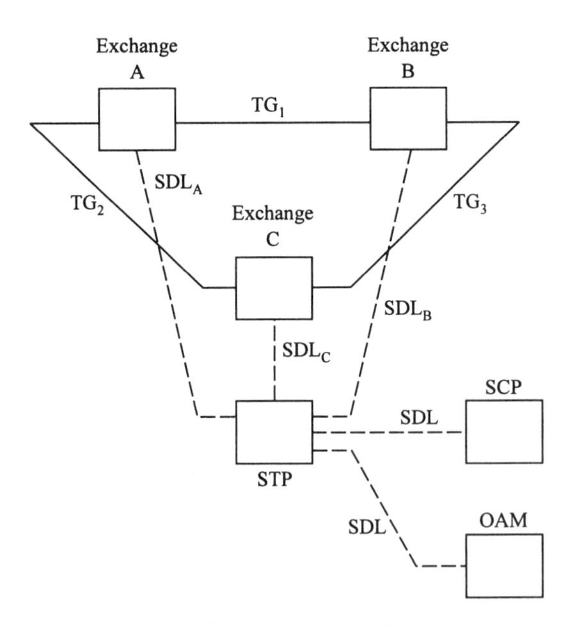

Figure 2.1-2 Common-channel signaling for transactions involving an exchange and a service control point (SCP), or an operations, administration, and maintenance center (OAM).

translation of the received 800 number. The SCP then translates the number into the national number of the called party, and returns that number to the exchange, which then proceeds to set up the connection.

The OAM centers allow centralized operation, maintenance and administration of the network. For example, there are transactions in which the OAM requests an exchange to test a particular trunk, and to report the test results. Other transactions enable the OAM to verify and change the subscriber and routing data that are stored in the exchanges.

2.1.4 Signaling in Cellular Mobile Networks

Cellular mobile telecommunications were introduced in the 1980s, and are widely used today [8]. Figure 2.1-3 shows a cellular mobile network. Exchanges in mobile networks are known as *mobile switching centers* (MSC). The MSC has one or more trunk groups (TG) to exchanges in the *public switched telecommunication network*. The service area of a MSC is divided into a number of cells. Each cell has a *base station* (BS) that contains microwave radio equipment, and has trunks and data links to the MSC.

A *mobile station* (MS) in a cell communicates with the base station of that cell on a radio-frequency channel, of which there are two kinds. A *voice channel* (VC) of a base station is permanently connected to a trunk between the BS and the MSC. A *control channel* (CC) of a base station is permanently connected to a *data link* between the BS and the MSC.

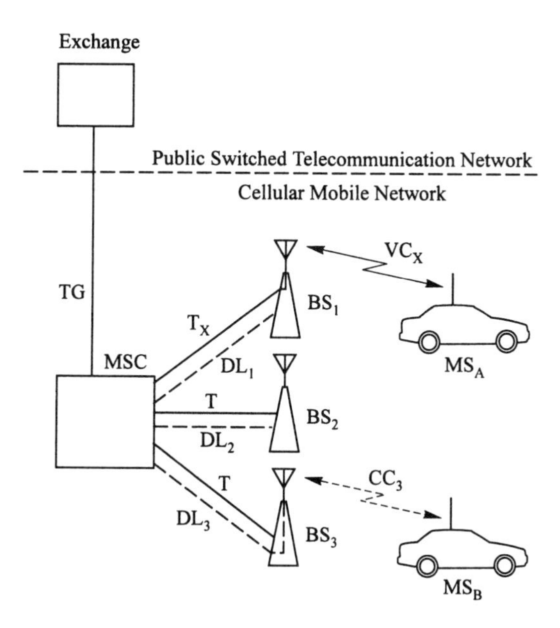

Figure 2.1-3 Signaling in cellular mobile networks. MSC: mobile switching center. MS: mobile station. VC: voice channel. CC: control channel. T: trunk. DL: data link. BS: base station.

In Fig. 2.1-3, mobile MS_A is engaged in a call. The MSC has allocated trunk T_x, and the associated voice channel VC_x, to the call. T_x and VC_x transfer the speech, and the signaling, between the mobile and the MSC.

The signaling for all mobiles in a cell that are not involved in a call is carried on a control channel of the cell, and its associated data link. For example, mobile MS_B is not involved in a call, and the signaling between the mobile and the MSC takes place on DL_3 and CC_3. The signaling between the mobiles and the MSC is thus a combination of channel-associated and common-channel signaling.

In addition, signaling in cellular mobile networks includes transactions between mobile switching centers and mobile-network databases. One group of these transactions allows a mobile to obtain service while it is *roaming* outside its "home" cellular system.

2.1.5 Digital Subscriber Signaling

The public switched telecommunication networks in some countries are being converted into *integrated services digital networks* (ISDN) [2,9,10]. An ISDN serves conventional (analog) subscribers and *ISDN users*. ISDN users can communicate with each other in two modes. In circuit-mode communications, the network sets up a dedicated connection for the call, which can be used for voice and data communications. In packet-mode communications (not covered in this book), the users communicate with short bursts of data, known as *packets*.

An ISDN user has 64 kb/s digital terminals TE of several types, for example digital (PCM) telephones, high-speed facsimile terminals, and high-speed computer modems, etc.—see Fig. 2.1-4. A *digital subscriber line* (DSL) connects the user's TEs to his local exchange. DSLs are two-wire or four-wire circuits that allow simultaneous information transmission at 144 kb/s in both directions. A DSL also carries a number of *overhead* bits, and the total transmission rate on the line is therefore somewhat higher.

The 144 kb/s information bit streams are divided into two 64 kb/s *B-channels*, and one 16 kb/s *D-channel*. The B-channels are used for circuit-mode communications, allowing two simultaneous calls. The D-channel is the

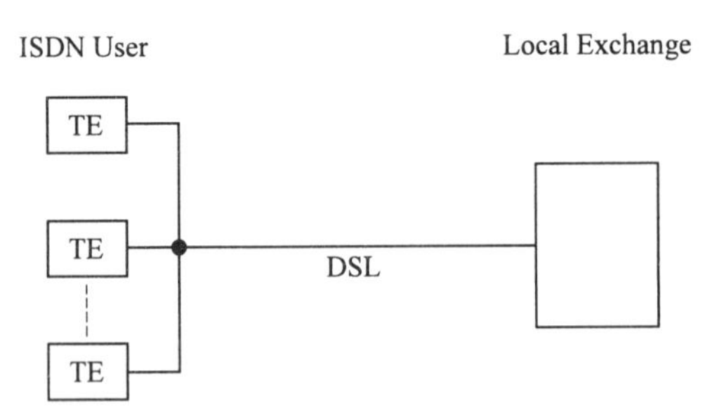

Figure 2.1-4 digntal subscriber line (DSL).

"common" channel for signaling between the user and the local exchange. The signaling is message-oriented, and known as *digital subscriber signaling system No. 1* (DSS1).

2.1.6 Multiple Interexchange Signaling Systems

For economic reasons, exchanges have to remain in service for about 20 years. This means that national networks, and the international network, contain exchanges of several vintages. The older exchanges are equipped to handle older interexchange signaling systems only—and it is difficult or impossible to upgrade them to accommodate newer forms of signaling.

Now, consider a telecom whose network currently uses two interexchange signaling systems (which we denote here as #1 and #2), and wants to introduce a new signaling system (#3). It can purchase a number of exchanges that are equipped to handle system #3, and interconnect them with "system #3" trunk groups. However, the new exchanges should also handle signaling systems #1 and #2, for use on trunk groups between an old and a new exchange. As a result, several signaling systems coexist in the network.

A connection routed via one or more intermediate exchanges may therefore involve trunks with different interexchange signaling systems. For example, at an intermediate exchange, the connection could involve a trunk with signaling system #2, and a trunk with signaling system #3. The call-control at the exchange has to include procedures for *signaling interworking* between the signaling systems. One aspect of interworking is the conversion of the formats of individual signals and/or messages. More difficult problems arise when a signal, or an information item in a CCS signaling message, exists in the new system #3, but not in the old system #2. Signaling interworking functions can be quite complex, and have to be designed with care.

2.2 STANDARDS FOR SIGNALING SYSTEMS

The equipment in a telecommunication network is usually purchased from several manufacturers. In order to ensure that equipment from different suppliers can be interconnected without problems, the telecoms have developed (and continue to develop) standards that are documented in *specifications*. Purchase orders for equipment are accompanied by a set of specifications which have to be met by the manufacturer's equipment. Specifications cover signaling, and many other aspects of telecommunications. The information in this book is based largely on signaling standards published by the organizations mentioned below.

2.2.1 Bell System

Prior to its divestiture in 1984, the Bell System was both the dominant telecom and equipment manufacturer, and set the *de facto* standards for the North American telecommunication network [11].

2.2.2 Exchange Carriers Standards Association

After the divestiture, the *Exchange Carriers Standards Association* (ECSA) was created to ensure that the North American network would continue to operate as an integrated entity [12]. A number of T1 committees within ECSA are responsible for generating drafts of telecommunication standards which are then submitted to the American National Standards Institute (ANSI), and published as *American National Standards for Telecommunications*.

2.2.3 Bellcore

Bellcore, the research and development arm of the former Bell System operating companies, has defined requirements for these companies, documented in its *Technical Advisories* (TA series), *Technical References* (TR series), and *Generic Requirements* (GR series).

2.2.4 TIA/EIA

Standards for cellular mobile systems in the U.S. are established by subcommittees of the TR.45 Committee. They are published jointly, by the Telecommunications Industry Association (TIA) and the Electronic Industries Association (EIA), as *TIA/EIA Interim Standards*.

2.2.5 CCITT/ITU-T

Historically, the telecoms that operate individual national networks established standards independently of each other. With the advent of international telecommunications, the need for international standards arose, and this has led to the establishment of international standards organizations.

The CCITT (International Telegraph and Telephone Consultative Committee) was established in 1956, as a part of the International Telecommunication Union (ITU), which has its headquarters in Geneva, Switzerland [13]. As a result of a major reorganization of ITU, the CCITT ceased to exist in 1993, and became the ITU-T, the telecommunications standardization sector of ITU.

During its life, CCITT activities took place in four-year *study periods*. Individual study groups met several times per year, and presented their results to the plenary assembly, in a meeting held at the end of the study period. Approved results became *recommendations*, which were published in a set of books that were updated and reissued at four-year intervals. The last set of books are the *CCITT Blue Books*, issued in 1989.

During the 1980s it became clear that the four-year study periods were too slow for the rapid developments that were taking place in telecomunications.

Therefore, ITU-T draft standards are now reviewed in meetings of the individual study groups, and recommendations are published individually. Signaling standards are the responsibility of Study Group 11.

While the CCITT/ITU-T recommendations are intended primarily for the international network, many national telecoms have adopted them—sometimes with country-specific modifications—for use in their respective networks.

2.2.6 CEPT and ETSI

European telecoms established the Conference of European Postal and Telecommunications Administrations (CEPT) in 1959 [14,15]. Its Coordination and Harmonization Committee (CHH) has issued a number of recommendations. The standardizing activities were transferred in 1988 to a new organization, the European Telecommunication Standards Institute (ETSI). The work is carried out by a number of technical committees, one of which deals with signaling protocols and switching (SDS). ETSI standards are published as *European Telecommunications Standards.*

2.3 ACRONYMS

ANSI	American National Standards Institute
CAS	Channel-associated signaling
CCIS	Common-channel interoffice signaling
CCITT	International Telegraph and Telephone Consultative Committee
CCS	Common-channel signaling
CEPT	European Conference of Postal and Telephone Administrations
DDD	Direct distance dialing
ECSA	Exchange Carrier Standards Organization
EIA	Electronic Industries Association
ETSI	European Telecommunication Standards Institute
IDDD	International direct distance dialing
ISDN	Integrated Services Digital Network
ITU	International Telecommunication Union
OAM	Operations, administration and maintenance center
POTS	Plain old telephone service
PSTN	Public switched telecommunication network
SCP	Service control point
SDL	Signaling data link
SSP	Service switching point
STP	Signal transfer point
TG	Trunk group
TIA	Telecommunications Industry Association

2.4 REFERENCES

1. S. Welch, *Signalling in Telecommunications Networks*, Peter Peregrinus Ltd, Stevenage, U.K., 1981.
2. R.L. Freeman, *Telecommunication System Engineering*, Second Edn, John Wiley & Sons, New York, 1989.
3. *A History of Engineering & Science in the Bell System*, Bell Telephone Laboratories, Inc, 1982.
4. Special Issue on CCIS, *Bell Syst. Tech. J.*, **57**, No.2, 1978.
5. R. Modaressi, R.A. Skoog, "Signaling System No.7: A Tutorial," *IEEE Comm. Mag.*, **28**, No.7, July 1990.
6. S.M. Boyles, R.L. Corn, L.M. Mosely, "Common Channel Signaling: The Nexus of an Advanced Communications Network," *IEEE Comm. Mag.*, **28**, No.7, July 1990.
7. R.B. Robrock, "The Intelligent Network Evolution in the United States," *Conference Proceedings, Singapore ICCS'90*, Elsevier, Amsterdam, 1990.
8. W.C.Y. Lee, *Mobile Cellular Telecommunications*, McGraw-Hill, Inc., New York, 1995.
9. P.K. Verma ed., *ISDN Systems*, Prentice Hall, Englewood Cliffs, 1990.
10. T.B. Bell, "Telecommunications," *IEEE Spectrum*, **29**, No. 2, 1992.
11. *Notes on the Network*, AT&T, New York, 1980.
12. A.K. Reilly, "A U.S. Perspective on Standards Development," *IEEE Comm. Mag.*, **32**, No. 4, Jan. 1994.
13. T. Irmer, "Shaping Future Telecommunications: The Challenge of Global Standardization", *IEEE Comm. Mag.*, **32**, No. 4, Jan. 1994.
14. J.J. Jacquier, ETSI-European Telecommunications Standards Institute, *Technische Mitteilungen PTT, Switzerland*, July 1990.
15. G. Robin, "The European Perspective for Telecommunication Standards", *IEEE Comm. Mag.*, **32**, No. 4, Jan. 1994.

3

SUBSCRIBER SIGNALING

The vast majority of the customers of telecommunication networks are subscribers who are attached to their local exchanges by analog subscriber lines. The signaling between subscriber and local exchange is known as *subscriber signaling* [1–4]. The original, and still predominant, application of subscriber signaling is *plain old telephony service* (POTS) calling. However, subscriber signaling today also supports supplementary services such as *call waiting*, *call forwarding*, *caller identification*, etc.

3.1 BASIC SUBSCRIBER SIGNALING

3.1.1 Signaling for an Intraexchange Call

Figure 3.1-1 shows the signaling for an intraexchange call between subscribers S_1 and S_2. The directory number of called subscriber S_2 is 347-9654.

Calling subscriber S_1 starts by going *off-hook* (lifting the handset of the telephone from its cradle). The off-hook is interpreted by the exchange as a *request-for-service* (a call origination, or the activation/deactivation of a subscriber service). In response, the exchange returns *dial-tone*, indicating that it is ready to receive digits. Subscriber S_1 then sends the digits of the called number, using the dial or the keypad of the telephone. After receipt of 3-4-7, the exchange recognizes one of its exchange codes, and thus knows that it is the destination exchange for the call.

The exchange can identify the called subscriber S_2 after receipt of the complete called number, and checks whether S_2 is free. Assuming that this is

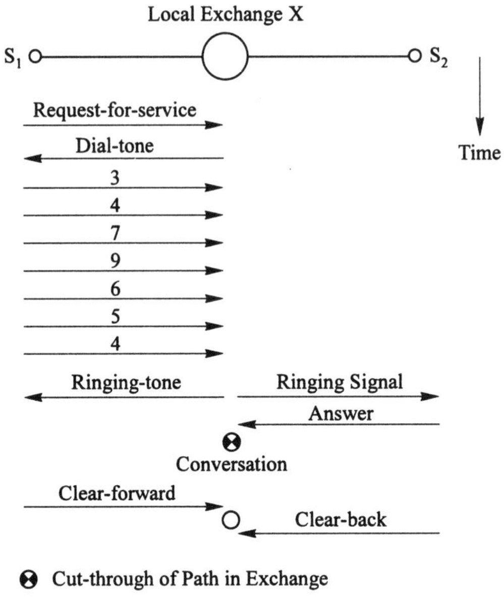

Figure 3.1-1 Signaling for an intraexchange call.

the case, it sends a *ringing signal* to alert S_2, and informs S_1 about the call progress with a *ringing-tone*.

When S_2 goes off-hook, an *answer* signal is generated. The exchange then *cuts through* (sets up a path in its switchblock between the subscriber lines). The conversation starts, and the exchange begins to charge S_1 for the call. At the end of the call, the subscribers put the handsets back in the cradles of the telephones. The signals generated by these actions from the calling and called subscribers are known as respectively the *clear-forward* and the *clear-back* signal.

Calling Party Control. The release of connections is usually under control of the calling party. In Fig. 3.1-1, the calling party clears first, and the exchange immediately releases the connection.

If the called party clears first, the exchange starts a timer of say 30–60 seconds, and releases the call when it receives a clear-forward, or on expiration of the timer, whatever occurs first. This is for the convenience of the called party, who may have picked up a phone in one room, but wants to have the conversation in another room. The called party can then hang up the first phone, move to the other room and pick up the other phone, while the connection stays up.

Forward and Backward Signals. Call-control signals are categorized as of *forward* and *backward* signals. Forward signals are sent in the direction in which the call is set up (from S_1 to S_2), and backward signals are sent in the opposite direction. The request-for-service signal, and the digits of the called number, are examples of forward signals. Dial-tone, ringing- tone, and answer are backward signals.

3.1.2 Groups of Subscriber Signals

The signals in the example can be divided into the following four groups.

Supervision Signals (or *line signals*) are signals sent by subscribers to local exchanges. The forward supervision signals (request-for-service, disconnect by calling party) request the start or end of a connection. The backward supervision signals (answer, disconnect by called party) change the state of a call.

Address Signals (also known as *digits* and *selection* signals) are forward signals that are sent by the calling subscriber when dialing the called party number.

Ringing. A forward signal sent by the exchange to the called subscriber, to indicate the arrival of a call.

Tones and Announcements. Audible backward signals (dial-tone, ringing-tone, busy-tone, etc.) sent by an exchange to the calling subscriber, and indicating the progress of a call.

3.2 SIGNALING COMPONENTS IN TELEPHONES

This section presents an overview of the circuitry in a telephone, focusing on the components for subscriber signaling.

3.2.1 Telephone

The major components in a telephone are shown in Fig. 3.2-1. The telephone is connected to a line circuit LC in the local exchange by a subscriber line that transfers the subscriber's speech and the subscriber signaling.

Transmitter (microphone) (TR) and receiver (RCV) convert acoustic speech signals to electrical analog signals, and vice versa. Transformer (T) and resistor (R) are part of the speech circuit.

The signaling functions of a telephone are: the generation (controlled by the subscriber) of supervision signals and digits; the conversion of received electrical tone, and announcement signals into acoustic signals, and the conversion of the electrical ringing signal into a high-level acoustic signal that can be heard at some distance from the telephone.

3.2.2 Supervision Signals

A telephone can be in two supervision states. When the telephone is not in use, its handset rests in a cradle, and depresses cradle switch (CS)—see Fig. 3.2-1.

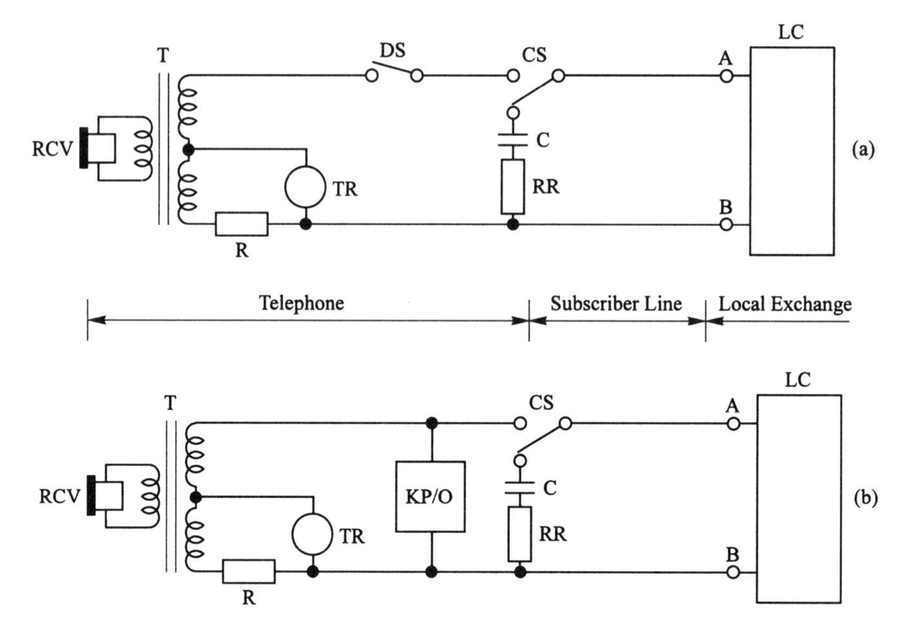

Figure 3.2-1 Components in a telephone, shown in the on-hook state. (a): dial telephone. (b): keypad telephone. C: capacitor. KP/O: keypad and oscillator. CS: cradle switch. DS: dial switch. LC: line circuit. RCV: receiver. TR: transmitter. R: resistor. T: transformer. RR: ringer.

In this state, the telephone is *on-hook* (this term has been carried over from the time when the receivers of telephones were resting on a hook). When the telephone is on-hook, switch (CS) connects ringer (RR), in series with capacitor (C), to the subscriber line.

When a subscriber starts to use the telephone, he lifts the handset out of its cradle. In this state, which is known as *off-hook*, switch (CS) connects transmitter (TR) and receiver (RCV) to the subscriber line.

When the telephone is off-hook, direct current can flow in the subscriber line. When the telephone is on-hook, capacitor (C) blocks direct current. At the exchange, line circuit (LC) determines the supervision state of the telephone from the presence or absence of direct current in the line.

3.2.3 Address Signals

Address signaling takes place while the telephone is off-hook (CS closed). There are two types of address signals (digits). Today's local exchanges handle both signal types.

Dial-pulse (DP) Address Signals. In early telephones, the address signals were generated by dials [1,2]. The dial switch (DS) in Fig. 3.2-1(a) is linked mechanically to the dial. When the dial is at rest, DS is closed, and the telephone

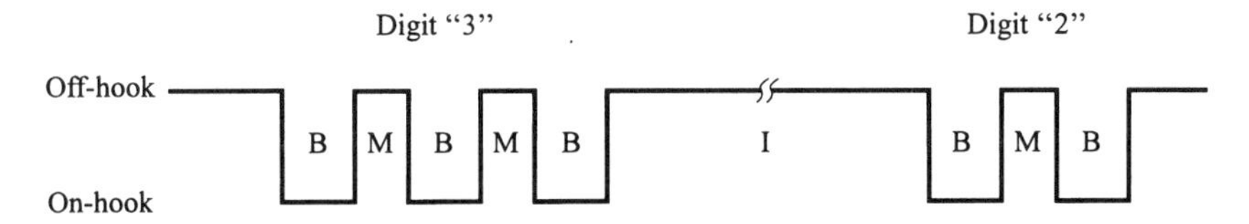

Figure 3.2-2 Dial-pulse address signals. B: break (60 ms). M: make (40 ms). I: interdigital interval (> 300 ms).

presents a path for direct current between points A and B. When the dial—after having been rotated by the subscriber—spins back to its rest position, DS opens and closes a number of times, producing a string of *breaks* in the d.c. path. The number of breaks in a string represents the value of the digit: one break for value 1, two breaks for value 2,..., ten breaks for value 0. The nominal length of a break is 60 ms (Fig. 3.2-2). The breaks in a string are separated by *make intervals* of nominally 40 ms. Consecutive digits are separated by an *interdigital interval* of at least some 300 ms.

Dual-tone Multi-frequency (DTMF) Address Signals. Around 1960, it became practical to place transistor oscillators in telephone sets, and this led to the development of DTMF address signaling [3–5]. Figure 3.2-1(b) shows a DTMF telephone, which includes a keypad (KP) that controls a dual-tone oscillator (O).

When a subscriber depresses one of the keys on the keypad (KP), oscillator (O) produces two simultaneous tones. A digit is represented by a particular combination of two frequencies; one selected from a *low* group (697, 770, 852, 941 Hz) and the other selected from a *high* group (1209, 1336, 1477, 1633 Hz). This allows 16 digit values, but only 12 of these are implemented on the keypads: digit values 1, 2, ..., 0, and the special values * and #.

The DTMF frequency combinations have been standardized by CCITT [6]:

Digit Value	Frequencies (Hz)
1	697 and 1209
2	697 and 1336
3	697 and 1477
4	770 and 1209
5	770 and 1336
6	770 and 1447
7	852 and 1209
8	852 and 1336
9	852 and 1477
*	941 and 1209
0	941 and 1336
#	941 and 1477

3.2.4 Ringing Signal

When the telephone is on-hook (Fig. 3.2-1) and the exchange sends an electrical ringing signal (an alternating current), ringer (RR) produces an audible signal that can be heard in the vicinity of the telephone.

In early telephones, the ringers were electromechanical devices. Modern telephones have electronic ringers.

3.2.5 Tones and Recorded Announcements

These signals have the same electrical characteristics as the speech received during a call. Like speech, they are converted into acoustic signals by receiver (RCV).

3.3 SIGNALING EQUIPMENT AT THE LOCAL EXCHANGE

This section gives an example of the equipment for subscriber signaling at local exchanges. We consider a local SPC (*stored program control*) exchange with a digital switchblock (see Figs. 1.7-4 and 1.7-5).

3.3.1 Overview

Figure 3.3-1 shows the local exchange and a number of subscriber lines. The lines are two-wire bidirectional analog circuits (c). They pass through their line

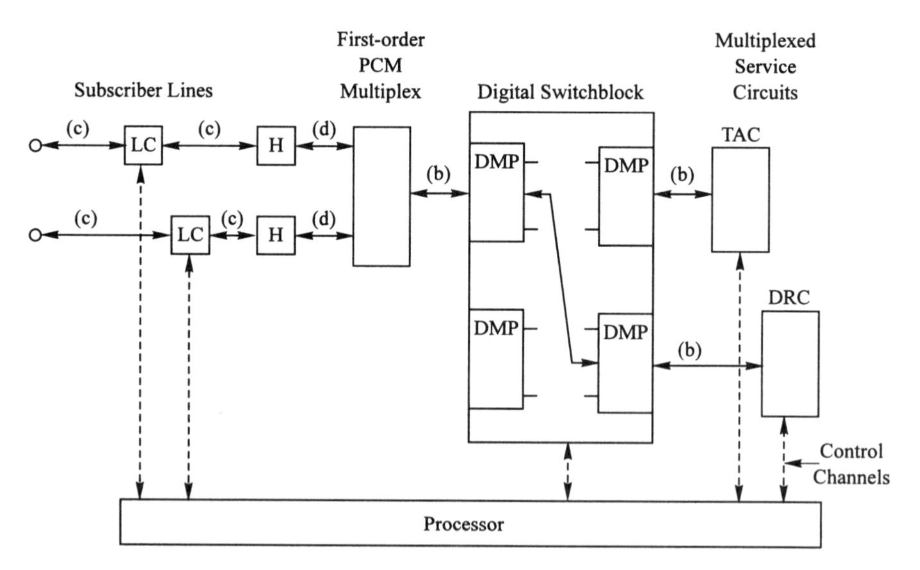

Figure 3.3-1 Equipment for subscriber signaling at a local exchange with a digital switchblock. LC: line circuits. H: hybrid circuits. DMP: digital multiplex port. TAC: tone and announcement circuits. DRC: DTMF receiver circuits.

circuits (LC), and are converted into four-wire analog circuits (d) by hybrids (H). First-order PCM multiplexes convert m of these circuits ($m = 24$ or 30) into a four-wire digital multiplex circuit (b) that carries PCM-coded speech (1.5.1). These circuits are attached to digital multiplex ports (DMP) of the switchblock. The bit format on (b) is as shown in Fig. 1.5-4. Also attached to DMP ports are a number of multiplexed digital *service circuits.*

The switchblock provides temporary 64 kb/s digital paths between a service circuit and a subscriber line—more exactly, a PCM multiplex channel (b) associated with the subscriber line. These paths transfer PCM-coded tones, DTMF frequencies, and announcements.

The switchblock, and the line and service circuits, have control channels (CC) to the exchange processor. This enables the processor to send commands to, and receive information from, these entities.

The implementation of subscriber-signaling functions is manufacturer-specific. In this example, we assume that the line circuits receive the supervision signals and dial-pulse digits from their lines, and send ringing signals to the lines.

We also assume that there are two types of service circuits. *Tone and announcement circuits* TAC have memories that store PCM sequences for all tones and announcements that can be sent to a subscriber. When for example a busy-tone has to be sent to a subscriber line, a switchblock path is set up between an available TAC circuit and the PCM channel associated with the line. The processor then orders the circuit to send busy-tone. The second type of service circuits are DTMF-digit receiver circuits (DRC). These circuits can provide dial-tone and receive DTMF digits.

3.3.2 Reception of Supervision Signals

Figure 3.3-2 shows a line circuit (LC) in some detail. The circuit can be in two states, which are changed on command from the processor. In the figure, LC is in its "normal" state. Switch (S) connects transformer (T) to the subscriber line. When the telephone is involved on a call, the transformer transfers the (analog) speech between the subscribers. The LC is set to the ringing state when the telephone has to receive a ringing signal.

Hook Status. In both LC states, the "hook" state of the telephone is monitored by current detector (CD), and reported to the exchange processor. When the telephone is on-hook (idle), there is no path for direct current between points A and B, and no current flows through CD, which then indicates to the processor that the telephone is on-hook. When the telephone is off-hook, there is a path for direct current between points A and B, and a current flows from ground—through transformer (T), the external path (A–B), and current detector (CD)—to the common battery (BAT). The CD then indicates that the telephone is off-hook.

Call States. The exchange processor keeps track of the "call" state of the

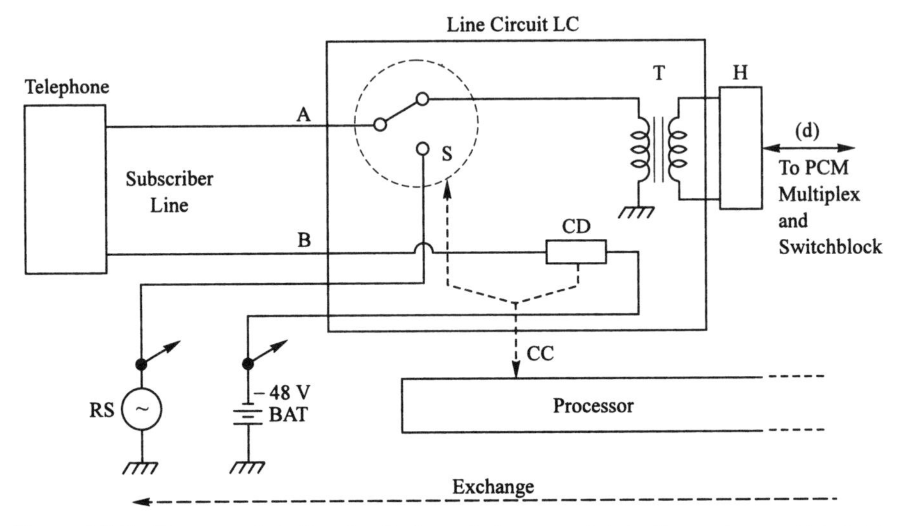

Figure 3.3-2 Components in line circuit, shown in the normal state. T: transformer. H: hybrid circuit. CD: current detector. CC: control channel. RS: ringing source. BAT: exchange battery. S: switch.

telephones, and stores these states in its temporary memory. We distinguish the following major call states:

Idle: not involved in a call.

Dialing: before and during the sending of sending address signals.

Calling: involved in a call as calling party, after having sent the address signals.

Ringing: receiving the ringing signal from the exchange (which indicates an incoming call).

Called: involved in a call as a called party, and having answered.

Determination of Signal Type. A change in the hook status of a telephone is a supervision signal. The processor determines the signal type, based on the present call state, and the type of hook-status change:

Present Call State	Change in Hook Status	Supervision Signal
Idle	To off-hook	Request for service
Dialing		(see 3.3.4)
Calling	To on-hook	Clear-forward
Ringing	To off-hook	Answer
Called	To on-hook	Clear-back

Recognition Time. Electrical disturbances on a subscriber line can result in brief off-hook (on-hook) pulses on lines that are on-hook (off-hook). The processor therefore takes no action until the new hook state has persisted for predetermined recognition time (in the order of 20–40 ms).

Hookswitch Flash (or Flash). This is a fifth supervision signal, sent by a subscriber who in is the calling or called state, to request an action from the local exchange.

The subscriber generates the flash by momentarily depressing the button of cradle switch (CS) (Fig. 3.2-1). This results in a temporary on-hook condition of the telephone. The length of a flash varies widely. Exchanges usually interpret on-hooks of 0.1–1.0 seconds as flashes, and consider longer on-hooks as clear-forward or clear-back signals. The uses of flash signals are discussed later in this section.

3.3.4 Reception of Address Signals

On receipt of a request-for-service from a subscriber line, the processor marks the line as "dialing," selects an idle digit receiver (DRC), orders the switchblock to set up a path between DRC and the line, and commands the DRC to send dial-tone.

If the calling subscriber is using a telephone with dial-pulse address signaling, she rotates the dial, and this generates the digits as strings of "break" and "make" pulses that are detected by current detector (CD) in the line circuit (Fig. 3.3-2), and reported to the processor. On receipt of the first break, the path between the line and DRC is released.

If the calling subscriber is using a telephone with DTMF signaling, she depresses the keys on the keypad. This generates DTMF digits that are received by DRC, and reported to the processor. On receipt of the first digit, the dial-tone is turned off. The path between the subscriber line and the DRC is released when the complete called number has been received.

Digit receivers have frequency-selective circuits that are tuned to the individual DTMF frequencies, and detect the presence of these frequencies on the subscriber line. The receivers accept a digit only if one frequency of the low group, and one frequency of the high group, are present simultaneously for at least 70 ms.

Digit Imitation. When a key on a keypad telephone is depressed, the transmitter is disabled. However, there are intervals (between digits) during DTMF signaling when no key is depressed. During these intervals, the transmitter is enabled, and may pick up speech, music, or noise in the vicinity of the calling subscriber. These sounds should not imitate DTMF digits. The DTMF frequencies have been chosen to minimize digit imitation, by making DTMF tone pairs distinguishable from naturally occurring sounds [5].

Naturally occurring sounds contain tone pairs whose frequency ratios are "simple" fractions (such as 1:2, 3:5, 2:3, 3:4, 4:5, etc.).

Suppose now that a sound has an 1336 Hz component. This is detected in the digit receiver by the frequency-selective circuit tuned to this frequency. The most likely companion frequencies in this sound are listed below:

Frequency (Hz)	Adjacent DTMF Low-Group Frequencies (Hz)
(1:2) × 1336 = 668	—697
(3:5) × 1336 = 801	770, 852
(2:3) × 1336 = 891	852, 941
(3:4) × 1336 = 1002	941, —
(4:5) × 1336 = 1069	941, —

Note that these companion frequencies are about mid-way between two adjacent low-group DTMF frequencies. Therefore, none of the selective circuits tuned to the low-group DTMF frequencies detects a signal, and the natural sound is not accepted as a digit. The same happens with natural sounds that contain one of the other DTMF frequencies.

3.3.5 Ringing Signals

The ringing signal is a high-level 20–25 Hz signal analog, of typically 100 V rms, and designed to drive the electromechanical ringers in the early telephones. This ringing signal cannot be provided by a service circuit, because there are no PCM codes to represent these high level voltages. The ringing signal is therefore injected into the line circuit (Fig. 3.3-2).

When a subscriber line has to be rung, the processor sets the line circuit to the ringing state. In this state, switch (S) connects a common ringing-voltage source (RS) to the A-wire of the subscriber line. The alternating ringing current passes through the external circuit [subscriber line, capacitor (C) and ringer (RR)—see Fig. 3.2-1], and detector (CD). However, this current is not large enough to cause CD to indicate "off-hook."

When the called telephone goes off-hook, the ringer current and the direct current from battery (BAT) flow through CD. An off-hook is reported to the processor, which then changes the LC state to normal.

3.4 TONES, ANNOUNCEMENTS, AND RINGING

A calling subscriber receives tones or announcements that inform him about the progress of the set-up of his call. These tones originate at the local exchange of the caller, at intermediate exchanges along the connection, or at the local exchange of the called party (Fig. 3.4-1).

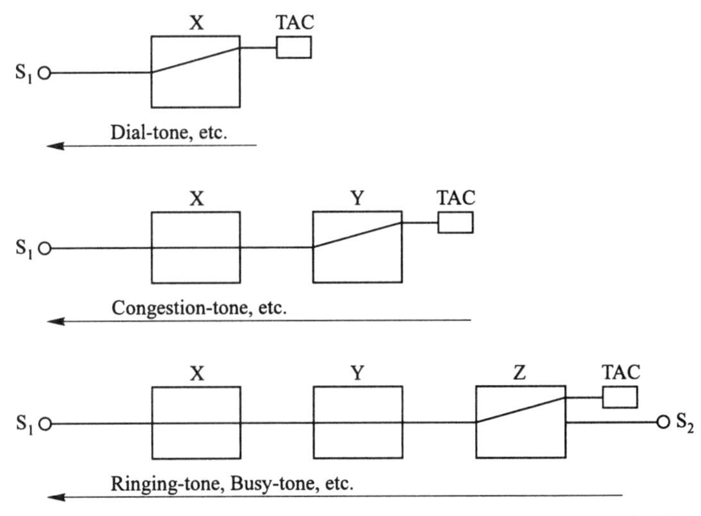

Figure 3.4-1 Call-progress tones and announcements. X: local exchange of calling party. Y: intermediate exchange. Z: local exchange of called party. TAC: tone and announcement circuit.

3.4.1 Tone Formats

In the early years of telecommunications, all call progress signals were tones, at frequencies of 400–600 Hz. These tones are either continuous, or repeating "on–off" cycles with a certain cadence. The frequencies and cadences were established by the telecoms of individual countries, well before the beginnings of international standards. The meanings of tones, and their frequencies and cadences, are therefore somewhat country-specific.

3.4.2 Basic Tones

The following functional tone signals are used in all countries.

Dial-tone. This is sent by the local exchange serving the calling subscriber, to indicate that the exchange is ready to receive the called number. In most countries, dial tone is continuous.

Ringing-tone (Audible Ring Tone). This is sent by the local exchange of the called subscriber to indicate that the subscriber is being alerted. In most countries, ringing-tone has a cadence, for example:

On (seconds)	Off (seconds)	
2	4	U.S., Canada,...
1	4	Brazil, Mexico,...

The U.K., Australia, and other countries that follow British telecommunication practices have the following ringing-tone cadence:

On (seconds)	Off (seconds)	On (seconds)	Off (seconds)
0.4	0.2	0.4	2

The ringing signals have the same cadences as the ringing-tones.

Busy-tone. This is sent by the local exchange that serves the called subscriber when that subscriber is busy. Examples:

On (seconds)	Off (seconds)	
0.5	0.5	U.S., Canada,....
0.25	0.25	Brazil, Mexico,...

3.4.3 Tones and Announcements in Other Failed Set-ups

Tones and announcements to indicate that a set-up has failed for a reason other than "called party busy" fall into the following broad categories:

Congestion (All Trunks Busy, ATB). The call set-up fails because it has reached an exchange at the time that the exchange has no available trunks in its route set to the called exchange.

Invalid or Non-working Called Number. An exchange along the route determines that the received called number contains invalid information (say an invalid area or exchange code), or a subscriber number that is presently not allocated.

In the early days of telephony, busy-tone was the universal signal for all set-up failures. However, it is important to give the calling party more detailed information, so that she can decide whether and when to repeat the call. For example, if the called subscriber is busy, it makes sense to repeat the call attempt after a few minutes. However, if the called number is invalid, all succeeding attempts are doomed to failure. Again, the tones and announcements for these failure cases are country-dependent. A few examples are outlined below.

Number Unobtainable Tone. This is used primarily in the U.K., and in countries that follow British telecommunication practices. It indicates that the called number is not a working number. In the U.K., the tone is continuous. Other countries use cadences, for example:

On (seconds)	Off (seconds)	
2.5	0.5	Australia, South Africa,..

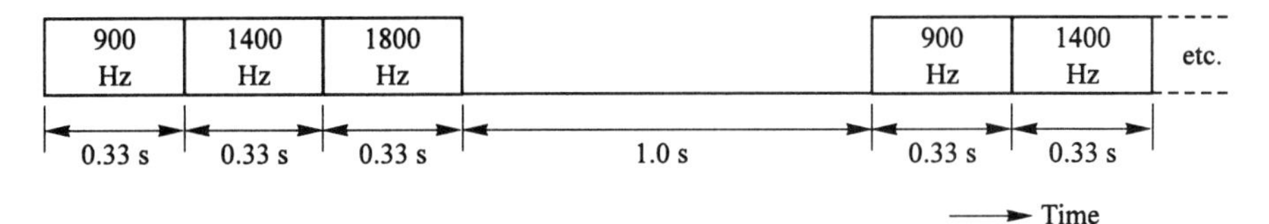

Figure 3.4-2 Special information tone.

Congestion-tone. This is sent by exchanges when the call cannot be set up because of congestion in the network (no outgoing trunk available). In the U.S. this tone is known as "reorder" tone. Some typical cadences are:

On (seconds)	Off (seconds)	
0.25	0.25	U.S., Canada,...
0.5	0.5	Belgium, Netherlands,...

In a number of countries (Austria, Brazil, Italy, Russia, etc.), congestion is still indicated by busy-tone.

Special Information Tone (SIT). A cadenced sequence of three frequencies, as shown in Fig. 3.4-2. It is used in many European countries to indicate that the received called number is not a working number.

SIT and Recorded Announcement. In the U.S., one cycle of SIT is often sent before an announcement that gives more details about the problem, for example: "Your call cannot be completed as dialed...." In case that the called number of a subscriber has been changed, the new number is usually included in the announcement.

An extensive survey of tones and recorded announcements used in various countries can be found in [7].

3.5 SUBSCRIBER SIGNALING FOR SUPPLEMENTARY SERVICES

Up to this point, we have discussed subscriber signaling for POTS (Plain Ordinary Telephone Service) call control. Today, customers in many countries can subscribe to *supplementary services*, for which they are charged a monthly fee. Every local exchange has a database with entries for each subscriber, listing the supplementary services available to that subscriber.

The supplementary services offered in the U.S. can be divided into two groups. *Custom calling* services became available in the late 1960s, with the introduction of *stored program controlled* local exchanges in the network. *Custom local area signaling services* (CLASS) are currently being introduced in the U.S.

This section examines some of these services, and outlines the additions to subscriber signaling to support them.

3.5.1 Custom Calling Services

Custom calling services do not require special signaling hardware in telephone sets. Subscribers have to invoke some of these services by dialing special digit sequences called *feature access codes.* These codes are distinguishable from called numbers because their initial digits include a * (asterisk) or a # (pound sign) when dialed from DTMF telephones, or special digit sequences (for example, 11) when dialed from dial-pulse telephones. Some services require the subscriber to send hookswitch flashes during the call.

The local exchanges on which these services are offered have to be equipped with tone/announcement circuits (TAC) that can provide a number of additional audible signals and messages to the subscriber.

Some widely used calling services are outlined below [3].

Call Waiting Service. Suppose that subscriber S_1 is marked at his local exchange for *call waiting service.* When S_1 is engaged in a call with subscriber S_2 and is called by another subscriber S_3, he is alerted by his local exchange with a *call waiting* tone. By sending a hookswitch flash, S_1 can put subscriber S_2 on hold, and be connected to S_3. From this point on, S_1 can alternate between S_2 and S_3 with subsequent flashes. Call waiting tone in the U.S. is a 440 Hz tone with a cadence of 0.3 s on–10 s off.

Call Forwarding Service. When S_1 is marked at the local exchange for *call forwarding service*, he can activate this service by dialing a *feature access* code. The exchange responds with dial-tone, after which S_1 enters the number of subscriber S_2 to whom his incoming calls are to be forwarded. The exchange acknowledges receipt of the number with two short "beeps" and then sets up a call from S_1 to S_2. Subscriber S_1 then hears ringing-tone. When S_2 answers, the call forwarding for S_1 is activated. If S_2 is busy or does not answer, S_1 disconnects, and repeats the activation request, which is then accepted by the exchange without setting up a call to S_2. When call forwarding has been activated, S_1 can still make outgoing calls, and his telephone still rings for incoming calls. However, these calls are forwarded until he deactivates the service, by dialing.

Three-way Calling. When S_1 is registered for *three-way calling service*, he can call S_2 and then add a call to S_3. Having established the call to S_2, subscriber S_1 flashes. The local exchange returns dial tone, and S_1 enters S_3's number. The exchange then sets up the call to S_3 while maintaining the call to S_2. If S_3 is busy or does not answer, S_1 can end the new call (but remain in conversation with S_2), by flashing twice.

3.5.2 Custom Local Area Subscriber Services

CLASS services are new supplementary services that are being introduced in the U.S. [8]. They depend on the availability of common-channel signaling system No.7 (see Chapter 11), which is presently the only signaling system in the U.S. that can transfer the calling number from the calling to the called exchange. Some CLASS services also require additions to subscriber signaling at the local exchanges, such as recognition of *feature access codes*, TAC circuits that include the required tones and announcements, and distinctive ringing. A few examples are outlined below.

Distinctive Ringing. A subscriber registered for this service is alerted by a ringing signal with a distinctive cadence when called by certain specified lines. The subscriber can list up to ten such calling numbers. To make a change in the list, the subscriber dials a feature access code. This initiates a dialogue with the exchange during which the subscriber is prompted by recorded messages, and enters the changes by sending digit strings that represent the special calling numbers. For this service, the local exchanges have to be able to generate the distinctive ringing signal, and to send the necessary recorded messages.

Selective Call Rejection. A subscriber registered for this service can enter a list of up to ten numbers from which she does not want to receive calls. Again, the subscriber can change this list by dialing a feature access code, and sending the digits of the calling numbers when prompted by messages from the exchange.

Caller ID (Calling Number Delivery). On calls to subscribers registered for this service, the local exchange signals the calling number, and the name of the calling subscriber, to a display device attached to the subscriber's telephone. The number is sent by a specialized type of service circuit in the local exchange.

The digits and characters are sent in a digital format in which 0s and 1s are sent as two voiceband frequencies (frequency-shift keying). The signaling takes place during the silent interval after the first burst of ringing.

A calling subscriber can prevent the presentation of his number by prefixing the called number with a feature access code.

3.5.3 Supplementary Services in Other Countries

Supplementary services are country-specific. Some of the services described above are also offered outside the U.S., others are not. There are also services that are available in other countries but not in the U.S. For example, *wake-up* service allows a subscriber to enter a feature access code and then dial a number of digits that specify the day, and time of day, when he wants to receive a wake-up call. Also, a subscriber can dial feature access codes to activate or deactivate the *do not disturb* service. When the service is activated, all incoming calls to the subscriber are diverted by her local exchange to a recorded announcement that indicates that the subscriber does not want to receive calls.

The feature access codes and signaling procedures to activate/deactivate or modify supplementary services vary from country to country. Many countries offer supplementary services to subscribers with DTMF telephones only, and use feature access codes that always include a # and/or *[9].

3.6 OTHER APPLICATIONS OF DTMF SIGNALING

DTMF address signaling was designed as a convenience feature for subscribers, providing a faster and more convenient way to send called numbers to their local exchanges.

Another aspect of DTMF signaling, which was not considered at the time, is that the frequency pairs of DTMF digits are in the voiceband range (300–3400 Hz), and enable a calling subscriber to send DTMF signal to other exchanges along the connection. This is not possible with dial-pulse signaling because most of the energy of dial-pulse signals is concentrated at frequencies below 300 Hz, and these signals cannot be transferred reliably across the network.

Some applications of the transfer of DTMF digits between the parties in a connection are outlined below.

3.6.1 Caller Interaction

Private branch exchanges (PBX) can be equipped with *caller-interaction service circuits* (CISC), which automate the handling of incoming calls. A CISC stores a number of spoken messages that are played on command of the PBX processor, and has a DTMF digit receiver. This enables the PBX to "interact" with the caller. When a call arrives at the PBX, the incoming trunk is first connected to a CISC, and plays an announcement that prompts the caller to send one or more digits, for example:

> Thank you for calling XYZ airlines. If you are calling from a pushbutton telephone, please press 2 to purchase tickets; press 3 to change your reservation; press 4 for flight arrival and departure information,... If you are calling from a dial telephone, stay on line, and you will be connected when an agent becomes available.

If the caller is using a pushbutton phone, he then sends the appropriate DTMF digit, and the PBX processor sets up a connection to an agent who is qualified to handle the requested service. This arrangement has become quite popular with business organizations, because it reduces the need for PBX attendants. Calling subscribers are less enthusiastic, especially when having to respond to a series of these prompts.

Telecoms are also equipping their local exchanges for caller interactions. This opens the door to *Intelligent Network* services which are described in Chapter 16.

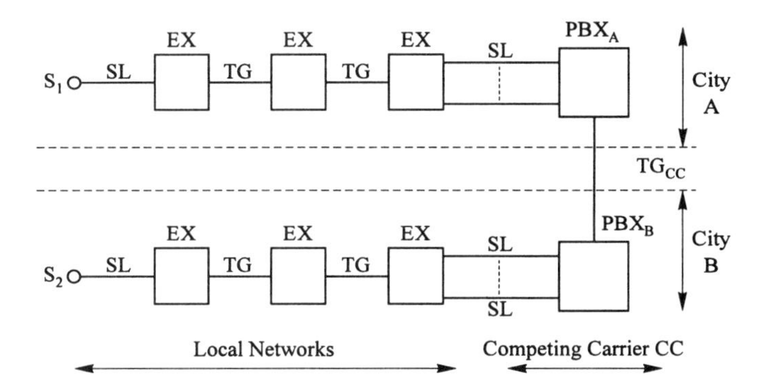

Figure 3.6-1 Original configuration for long-distance calls via competing carrier (CC). SL: subscriber lines. TG: trunk groups in local networks of cities A and B. TG_{CC}: group of tie trunks owned by CC. $PBX_{A,B}$: private branch exchanges owned by CC. EX: exchanges in the local networks of cities A and B.

3.6.2 Signaling to Competing Long-distance Carriers

An DTMF application that was not foreseen by the Bell System (the main developer and promoter of DTMF signaling) opened the door to competition in long-distance telecommunications in the U.S. An early form of this competitive service is shown in Fig. 3.6-1. A *competing carrier* (CC) maintained PBX exchanges in several cities, which were connected by subscriber lines to the respective local exchanges, and were interconnected by inter-city trunk groups TG_{CC}, owned by CC.

A subscriber (S_1) with a DTMF telephone, and located in city A, could call subscriber S_2 in city B by making a local call to CC's PBX_A in city A. When the connection was established, PBX_A would prompt S_1 to send—with DTMF digits—the called and calling numbers (the calling number was needed for charging). After this, PBX_A would extend the call to PBX_B in city B, on a trunk of TG_{CC}. The call would then be completed by PBX_B, as a local call in city B.

Dialing was cumbersome, and speech transmission was often substandard, because the connection involved four subscriber lines (SL) instead of two. However, the cost savings attracted a number of customers, and this allowed companies like CC to become established as "alternative" long-distance carriers.

The operators of the local telecommunication networks in the U.S. are now required to provide *equal access* to all interexchange carriers, and the connections shown in Fig. 3.6-1 are no longer needed.

3.7 DIALING PLANS

A dialing plan at a local exchange is a list of digit strings that can be dialed by its subscribers. A U.S. dialing consists of a *public office dialing plan* (PODP), and possibly a number of *customized dialing plans* (CDP).

Table 3.7-1 Public office dialing plan.

	Dialed Digits	Nature of Number
(a)	*NXX-XXXX*	Subscriber number
(b)	1+NPA-*NXX-XXXX*	National number
(c)	0+NPA-*NXX-XXXX*	National number, operator requested
(d)	1+700-*NXX-XXXX*	"700" number
(e)	0+700-*NXX-XXXX*	"700" number, operator requested
(f)	1+800-*NXX-XXXX*	"800" number
(g)	1+900-*NXX-XXXX*	"900" number
(h)	011+(7 to 12 digits)	International number
(j)	01+(7 to 12 digits)	International number, operator requested
(k)	0-(timeout)	No address present, LEC operator requested
(l)	00-(timeout)	No address present, IC operator requested
(m)	*N*11	Three-digit special numbers
(n)	10*XXX*+(a), (b),...or (j)	As above, and carrier specified by 10*XXX*
(p)	7*X* or 11*XX*	Feature access code, from dial telephone
(q)	7*X*# or **XX*	Feature access code, from DTMF telephone

X: digit with values 1,2,...9,0. *N*: digit with values 2,3,...8,9. *NXX*: exchange code. NPA: area code; also has *NXX* format.
Source: TR-NWT-001285. Reproduced with permission of Bellcore. Copyright © 1992.

3.7.1 Public Office Dialing Plan

PODP lists the digit strings that can be dialed by the general public. An example PODP is shown in Table 3.7-1. It covers called party numbers and feature access codes.

Called Party Numbers. A calling subscriber dials a subscriber number (a), a national number (b or c), or an international number (h or j), depending on whether the called party is in the NPA (*numbering plan area*) of the caller, outside the NPA but in the country of the caller, or in another country. On national and international calls, the subscriber can request operator assistance (c or j). A subscriber also can request operator assistance without dialing a called number. Code 0- (k) requests an operator of the *local exchange carrier* (LEC). Code 00- is a request for an operator of an *interexchange carrier* (IC).

If the called party is located outside the LATA (local area transport and access area) of the caller, the connection involves an IC. Each subscriber can designate a "regular" IC, whose identity is stored at his local exchange. On inter-LATA calls, the local exchange routes the call to an exchange of the designated IC, unless the caller has prefixed the called number with an 10*XXX* code (n), in which *XXX* specifies a different IC.

N11 codes (m) are dialed for calls to directory assistance (411), repair service (611), and police stations (911).

Prefixes. To expedite call-processing, an exchange must recognize the type of the dialed number as soon as possible. This is the reason for prefixes. In the

U.S., all prefixes start with a 0 or a 1. Since called numbers always start with *NXX* ($N > 1$), the exchange recognizes a prefix on receipt of the first digit.

Subscriber numbers (a) and N11 codes (m) are dialed without prefix. They become distinguishable on receipt of the third digit.

National numbers (b through g) are prefixed by 1 (no operator assistance) or 0 (operator assistance required). International numbers (h, j) have a 011 or 01 prefix. The exchange thus knows the called number type, and the need for operator assistance, after at most three digits.

Patterns (a) through (j) can be prefixed with a carrier-designation prefix 10XXX, which is recognized by the exchange on receipt of the second digit.

Timeouts. Some dialed patterns, for example: 0-(call to a local exchange operator), and 7*X*- (feature access codes from dial telephone), require a timeout for recognition. For example, after the exchange receives a 0, it starts a 5-second timer. If a next digit is received before the timer expires, an operator-assisted national number or an international number is being dialed. If the timer expires and no additional digits have been received, the caller is requesting a connection to an operator of his LEC.

Feature Access Codes are dialed to activate, or deactivate, a custom-calling feature. Their formats are 11*XX* or 7*X* (dialed on a dial telephone) and **XX* or 7*X*# (dialed on a pushbutton telephone). For example:

72 or 72#	Activates call-forwarding
73 or 73#	Cancels call-forwarding
1170 or *70	Deactivates call-waiting for one call
1167 or *67	Deactivates caller ID for one call

3.7.2 Customized Dialing Plans

At the request of a multi-location business, a telecom can provide a CDP for calls between employees of the organization. A typical customized plan allows the caller to dial an access code followed by a short number. The access code indicates whether the subsequent digits belong to the CDP, or to the public office dialing plan. Usually, access code 8 indicates a CDP, and access code 9 indicates the PODP.

3.8 ACRONYMS

ATB	All trunks busy
BAT	Exchange battery (48 V)
C	Capacitor
CC	Control channel
CD	Current detector

CDP	Customized dialing plan
CISC	Caller-interaction service circuit
CLASS	Custom local area subscriber services
CS	Cradle switch
DP	Dial pulse
DMP	Digital multiplex port
DRC	DTMF digit receiver
DS	Dial switch
DTMF	Dual-tone multi-frequency
H	Hybrid circuit
IC	Interexchange carrier
KP/O	Keypad and DTMF oscillator
LATA	Local area transport and access network
LC	Line circuit
LEC	Local exchange carrier
MCT	Malicious call tracing
PBX	Private branch exchange
PODP	Public office dialing plan
POTS	Plain old telephone service
R	Resistor
RAC	Recorded announcement circuit
RCV	Receiver (earpiece) of telephone
RR	Ringer
RS	Ringing source
S	Subscriber
S	Switch
SC	Service circuit
SIT	Special information tone
SL	Subscriber line
T	Transformer
TAC	Tone and announcement circuit
TG	Trunk group
TR	Transmitter (mouthpiece) of telephone

3.9 REFERENCES

1. S. Welch, *Signalling in Telecommunications Networks*, Peter Peregrinus Ltd, Stevenage, U.K., 1981.
2. R.L. Freeman, *Telecommunication System Engineering*, Second Edn, Wiley-Interscience, New York, 1989.
3. *Engineering and Operations in the Bell System*, Second Edn, AT&T Bell Laboratories, Murray Hill, 1983.
4. *A History of Engineering and Science in the Bell System*, Bell Telephone Laboratories, Inc., 1982.

5. L. Schenker, "Pushbutton Calling with a Two-Group Voice-Frequency Code," *Bell Syst. Tech. J.*, **38**, 1960.
6. *General Rec. on Telephone Switching and Signalling,* Recommendations Q.23-Q.24, CCITT Red Book, **VI.1**, ITU, Geneva, 1985.
7. *International Telephone Service Operation*, Recommendations E.180-E.183, and Suppl. Nos 2 and 4., CCITTT Red Book, **II.2**, ITU, Geneva, 1985.
8. S.M. Boyles, R.L. Corn, L.R. Moseley, "Common Channel Signaling: The Nexus of an Advanced Communications Network," *IEEE Comm. Mag.*, **28**, July 1990.
9. *International Telephone Service Operation*, Rec. E.131-E.132, CCITT Red Book, **II.2**, ITU, Geneva, 1985.

4

CHANNEL-ASSOCIATED INTEREXCHANGE SIGNALING

Channel-associated interexchange signaling (CAS)—also known as *per-trunk signaling*—has been in existence from the beginning of automatic telephony, and was the only form of interexchange signaling until 1976. It is still used in telecommunication networks, but is gradually being replaced by *common channel signaling.*

Early CAS systems were developed independently by individual equipment manufacturers, and exist in many varieties. Later CAS systems, notably those developed after the Second World War, show the increasing influence of national and international standards.

This section describes three important CAS systems, and their use on *frequency-division multiplexed* (FDM) analog trunks and *time-division multiplexed* (TDM) digital trunks (1.4.5 and 1.5.2).

The acronyms in the figures of this chapter are explained in Section 4.5.

4.1 INTRODUCTION

4.1.1 Interexchange Signaling Example

Figure 4.1-1 shows a typical interexchange signaling sequence for a call from subscriber S_1 to subscriber S_2. The subscriber signaling for the call is not shown.

After exchange A has received the called number from S_1, it decides to route the call via intermediate exchange B. Exchange A seizes an available trunk T_1, and sends a *seizure* signal on the trunk. Exchange B responds with a *proceed-to-send* (or *wink*) signal, indicating that it is ready to receive the digits of the

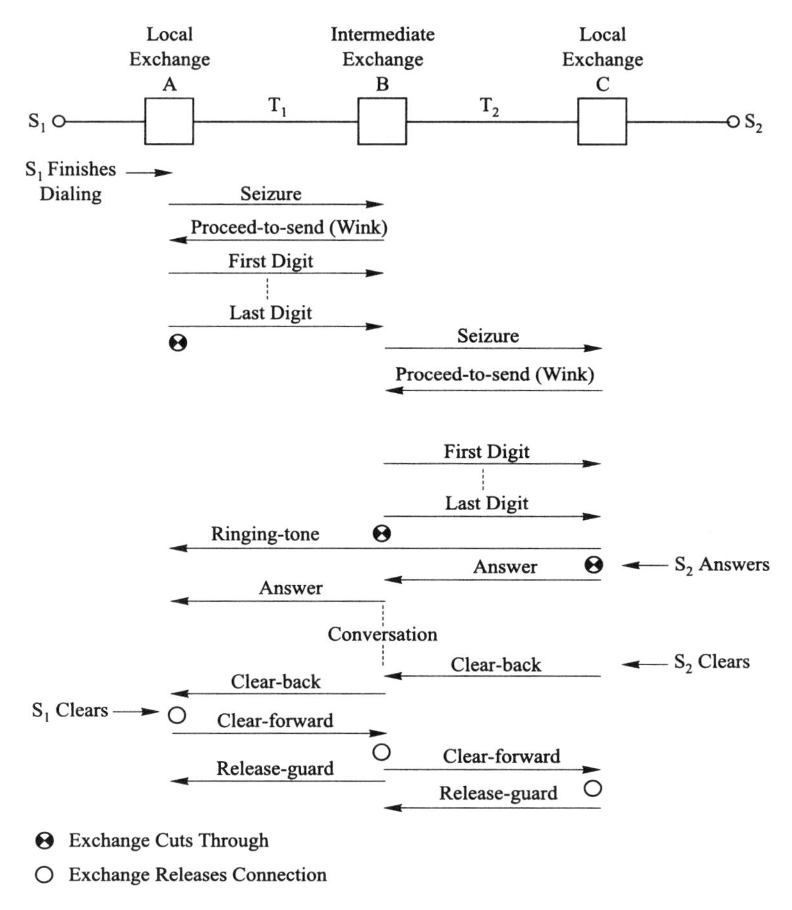

Figure 4.1-1 Interexchange signaling.

called number. Exchange A sends the digits, and then *cuts through* (sets up a path in its switchblock between the subscriber line of S_1 and T_1).

When exchange B has received the complete called number, it seizes an available trunk T_2 to destination exchange C, and sends a seizure signal on the trunk. Exchange C responds with a wink signal, after which exchange B sends the digits of the called number, and cuts through a path between trunks T_1 and T_2.

Exchange C then checks whether called subscriber S_2 is idle. If this is the case, C sends a ringing signal to S_2, and ringing-tone on trunk T_2. Because exchanges A and B have cut through, there is a connection between the calling subscriber S_1 and exchange C, and subscriber S_1 hears a ringing-tone.

When S_2 answers, exchange C cuts through a path between trunk T_2 and subscriber S_2. It also sends an *answer* signal on T_2, and exchange B repeats the signal on trunk T_1. Assuming that originating exchange A is responsible for charging the call, it establishes a billing record that includes the calling and called numbers, the date, and the time of answer.

The conversation now begins. In this example, called party S_2 hangs up first. Exchange C sends a *clear-back* signal to exchange B, which repeats the signal to exchange A. Like intraexchange calls, interexchange calls are usually controlled by the calling party (3.1.1). On receipt of the clear-back, exchange A stops charging, and enters the time when it received the clear-back in the billing record of the call. It also starts a 30–60-second timer. It then awaits a clear-forward from calling party S_1, or the expiration of the timer, and initiates the release of the connection when one of these events occurs.

The release takes place in the following way. Exchange A releases its path between S_1 and trunk T_1, and sends a clear-forward signal to exchange B, which releases its path between T_1 and T_2, and repeats the clear-forward to exchange C. This exchange then clears its path between T_2 and called subscriber S_2.

When exchanges B and C have completed the release of respectively T_1 and T_2, they send *release-guard* signals (to respectively exchanges A and B). When A and B receive the release-guard, they know that they can again seize respectively T_1 and T_2 for new calls.

This scenario is typical for CAS signaling in general, but we shall encounter variations that are specific to individual CAS signaling systems.

4.1.2 Groups of Interexchange Signals

Of the four groups of subscriber signals discussed in Section 3.1.2, three also exist in channel-associated interexchange signaling.

Supervision Signals (also known as *line signals*). The signals in this group represent events that occur on the trunk, such as, seizure, proceed-to-send, answer, clear-forward, etc. While the majority of supervision signals is used in all CAS systems, there are system-specific differences in the sets of supervision signals.

Address Signals (also known as *selection signals*, *digits*, or *register signals*). The digits are used primarily to indicate the called number, but can also have other meanings. Like dial telephones, early CAS systems used dial-pulse address signals. The systems described in this chapter have *multi-frequency* (MF) address signaling, similar to the DTMF (dual-tone multi-frequency) signaling in today's pushbutton telephones. The MF frequencies are system-specific, and different from those used in DTMF.

Tones and Announcements. (Ringing-tone, busy-tone, etc.). The tones and announcements in interexchange signaling are the same as in subscriber signaling (3.2.5).

4.1.3 CAS Signaling Equipment at the Exchanges

We consider a *stored program controlled* (SPC) intermediate exchange with a

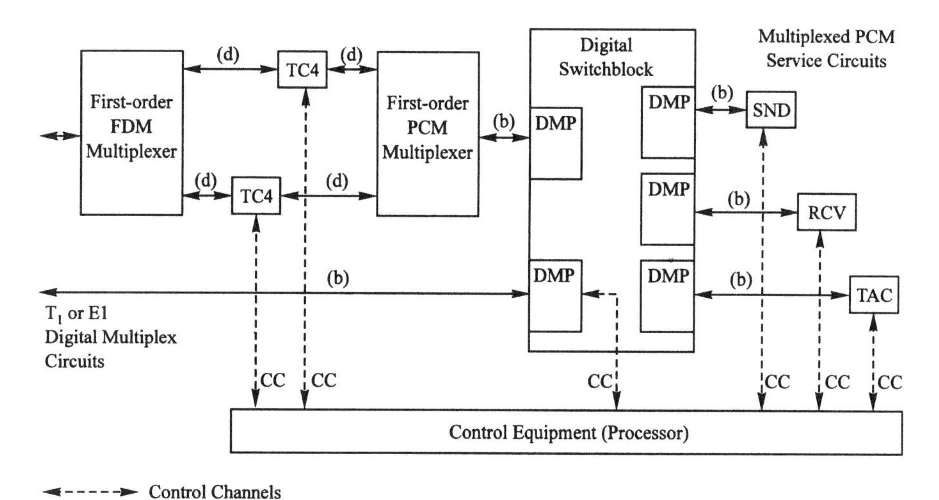

Figure 4.1-2 Exchange equipment for channel-associated interexchange signaling.

digital switchblock (Fig. 4.1-2). The switchblock provides temporary bi-directional 64 kb/s paths. The *digital multiplexed ports* (DMP) are first-order TDM circuits (b) with frame formats as described in Section 1.5.2.

All trunks have CAS signaling. The attachment of analog and digital (PCM) trunks to the switchblock is as described in Section 1.7.2. First-order TDM multiplexes carrying PCM trunks (b) are directly connected to DMPs. Analog trunks on first-order FDM transmission systems are first demultiplexed into individual four-wire analog circuits (d), which pass through trunk circuits (TC4) and then enter a first-order PCM multiplexer, where they are converted into TDM-multiplexed digital circuits (b).

Supervision Signaling. The TC4 circuits of analog four-wire trunks inject outgoing supervision signals into the trunk, and detect supervision signals received on the trunk. Each TC4 has a control channel (CC). It is used by the processor to order the TC4 to send signals, and by the TC4 to report received signals.

The supervision signaling information for the trunks in first-order digital trunk multiplexes is in certain bits of the bit streams (b). Control channel (CC) to the switchblock has two functions. In the first place, it transfers processor commands to set up and release switchblock paths. In the second place, it is used for communications between the processor and those DMPs that serve groups of multiplexed digital trunks. The processor can order a DMP to send out a supervision signal on a specified trunk in the multiplex, and a DMP reports the supervision signals received from the trunks in its multiplex to the processor.

Service Circuits. Address signaling, and the sending of tones/announcements, is done with service circuits. The exchange has pools of digital multiplexed service circuits which are attached to DMPs on the switchblock. Tone and announcement circuits (TAC) send tones and announcements. Digit-senders (SND) and digit-receivers (RCV) send and receive MF address signals to and from the trunks. These circuits are signaling-system specific. If an exchange has trunks with two CAS signaling systems, it needs separate groups of senders and receivers for each system. The circuits are controlled by the control equipment (processor) of the exchange, via control channels (CC).

To send address signals or tones/announcements on a trunk, or to receive address signals from a trunk, the processor seizes an available service circuit of the proper type, and commands the switchblock to set up a path between the circuit and the trunk. It then orders a SND or TAC circuit to send specific digits or tones, or orders a RCV circuit to report the received digits. When the sending or receiving has been completed, the path is released, and the circuit is returned to its pool.

The numbers of TAC, and RCV circuits in an exchange are small compared to the number of CAS trunks on the exchange, in the order of about one circuit of each type for every 10–20 trunks.

4.1.4 Interexchange Call-control Definitions

It is useful to introduce some terms that are frequently found in descriptions of call-control signaling, with the aid of Fig. 4.1-1.

Outgoing and Incoming. These terms can be applied to trunks, and to exchanges. An *outgoing* exchange seizes *outgoing trunks*, sends *forward* signals, and receives *backward* signals, on its outgoing trunks. An *incoming* exchange receives forward signals, and sends backward signals, on its *incoming* trunks. When discussing the signaling between exchanges A and B in Fig. 4.1-1, exchange A is the outgoing exchange, and B is the incoming exchange. When describing the signaling between exchanges B and C, exchange B is the outgoing exchange, and C is the incoming exchange.

Trunk T_1 (which is seized by exchange A), is regarded at A as an outgoing trunk, and at B as an incoming trunk. In the same way, trunk T_2 is an outgoing trunk at exchange B, and an incoming trunk at exchange C. If T_1 is a one-way trunk, it is always an outgoing trunk at A, and an incoming trunk at B. However, if T_1 is a bothway trunk, its role at exchanges A and B varies per call, depending on whether it has been seized by A or B. The same holds true for trunk T_2.

Originating and Terminating Exchanges. The *originating* exchange in a call is the local exchange serving the calling subscriber, and the *terminating* (or *destination*) exchange is the local exchange of the called subscriber. In the example of Fig. 4.1-1, exchange A is the originating exchange, and C is the terminating exchange.

Overlap and En-bloc Address Signaling. In the example of Fig. 4.1-1, exchange A seizes trunk T_1 after it has received the complete called number from subscriber S_1. Likewise, exchange B seizes trunk T_2 after receiving the complete number from exchange A. This means that, once the exchanges receive a proceed-to-send, they send out the complete called number in one uninterrupted stream. This mode of address signaling is called *en-bloc* register signaling.

However, exchanges can generally make route decisions after receipt of just the initial part of the called number. For example, if the called number in the example is a subscriber number, consisting of an exchange code followed by a line number (EC-LN), exchange A can seize trunk T_1 after receipt of EC from the calling subscriber, and can then send EC to exchange B. This exchange again can seize its outgoing trunk T_2 after receipt of the EC, and send the EC to exchange C. Exchanges A and B thus send out the initial digits of the called number while still receiving the later digits of the number, which are sent as soon as they have been received from the calling subscriber, or from the preceding exchange in the connection. This mode of register signaling is called *overlap* register signaling. The decision to use *en-bloc* or overlap address signaling is made by the individual telecoms. Overlap address signaling results in faster call set-ups.

Link-by-link and End-to-end Signaling. Signaling by two exchanges at the two ends of a trunk is called link-by-link signaling. In Fig. 4.1-1, supervision and address signaling are link-by-link. In general, supervision signaling is always link-by-link. Address signaling is link-by-link in most, but not all, CAS signaling systems.

In end-to-end address signaling, the digit sender in the originating exchange sends address signals successively to digit receivers in the second, and later exchanges in the connection.

4.1.5 CAS Signaling Systems

The CAS signaling systems discussed in this chapter are:

Bell System multi-frequency (MF) signaling,
CCITT No.5 signaling, and
R2 signaling.

4.2 BELL SYSTEM MULTI-FREQUENCY SIGNALING

This section describes the multi-frequency signaling system that was introduced by the Bell System after the Second World War [1–3]. It is still in use today, mostly in local U.S. networks. A nearly identical signaling system, known as the R1 signaling system [4] and defined by CCITT, is used on international trunk groups in the North American network (for example, groups between the U.S. and Canada).

Supervision and address signaling are link-by-link. The system can be used on one-way two-wire trunks, one-way and bothway FDM analog trunks, and one-way and bothway TDM digital trunks (see Sections 1.4.5 and 1.5.2).

The supervision signaling is described in Sections 4.2.1 through 4.2.3. The MF address signaling is discussed in 4.2.4.

4.2.1 Supervision Signaling

In Chapter 3 we have discussed the on-hook and off-hook states of a telephone. In Bell System MF signaling, we speak of the on-hook (idle) and off-hook (in use) states of a trunk. These states can be different at the exchanges connected by the trunk. Each exchange continuously sends the trunk state at its end to the other exchange. This is known as *continuous two-state* signaling. Changes in trunk state are supervision signals.

Supervision Signals. The repertoire of supervision signals, and the corresponding state changes are:

Forward Signals	State Change
Seizure	On-hook to off-hook
Clear-forward	Off-hook to on-hook

Backward Signals	State Change
Answer	On-hook to off-hook
Clear-back	Off-hook to on-hook
Proceed-to-send (wink)	Off-hook pulse, 120–290 ms

Consider a trunk between exchanges A and B. If the trunk is a one-way trunk that can be seized only by exchange A, this exchange is the outgoing exchange for all calls, and sends forward signals. Likewise, exchange B is always the incoming exchange, and sends backward signals. If the trunk is a bothway trunk, the exchange that seizes the trunk for a call is the outgoing exchange for that call.

Figure 4.2-1 shows the states, and the forward and backward supervision signals, for a typical call. Exchange A is the outgoing exchange.

Initially, the trunk is on-hook at both ends. Exchange A seizes trunk T, and sends a forward off-hook (seizure signal). Exchange then connects a digit receiver to the trunk, and sends a (backward) wink signal. After receipt of the wink, exchange A sends the digits of the called number. When the call is answered, exchange B sends an off-hook (answer signal). During the conversation, both exchanges are sending off-hook.

In this example, the called party clears first, and exchange B starts sending on-hook (clear-back signal). When the calling party clears, exchange A releases the trunk, and sends on-hook (clear-forward). In response, exchange B clears the trunk at its end.

The signaling system does not include a release-guard signal. Therefore, when outgoing exchange A releases the trunk, it starts a timer that expires after 0.75–1.25 s. The exchange does not seize the trunk for a new call until the timer has expired. This gives incoming exchange B the time to release the trunk at its end.

Double Seizures. On bothway trunks, double seizures of the trunk by both exchanges can occur. After exchange A seizes a trunk (Fig. 4.2-1), it expects to receive a backward change to off-hook that represents the leading edge of the wink signal. However, this change may also mean that exchange B (at the distant end of the trunk) is sending a seizure signal. After sending a seizure signal, the outgoing exchange thus has to time the duration of the received off-hook. The nominal length of the wink is 140–290 ms. Therefore, the exchanges are arranged to recognize a return on-hook within say 100–1000 ms as wink. If the off-hook duration exceeds 1 s, a double seizure has been detected.

There are several ways to deal with a double seizure. For example, both exchanges can be programmed to release the trunk, and make a second attempt to set up their calls, trying to seize a trunk in the same trunk group, or a trunk in a later-choice group.

4.2.2 Supervision Signaling on FDM Analog Trunks

FDM analog trunks can transfer frequencies between 300 and 3400 Hz. The exchanges indicate the states of the trunk with a 2600 Hz signaling tone. The tone is *in band* (audible), and should be off when the trunk is carrying a call. Therefore, *tone-on-idle* signaling is used:

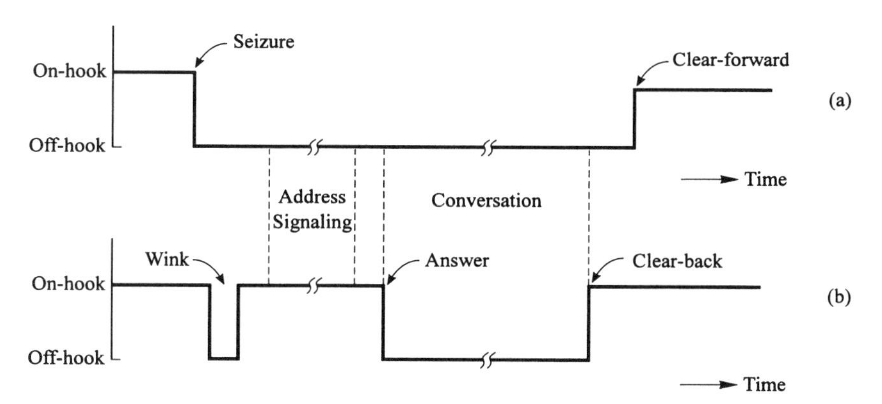

Figure 4.2.1 Supervision signals for a call. (a): sent by outgoing exchange A. (b): sent by incoming exchange B.

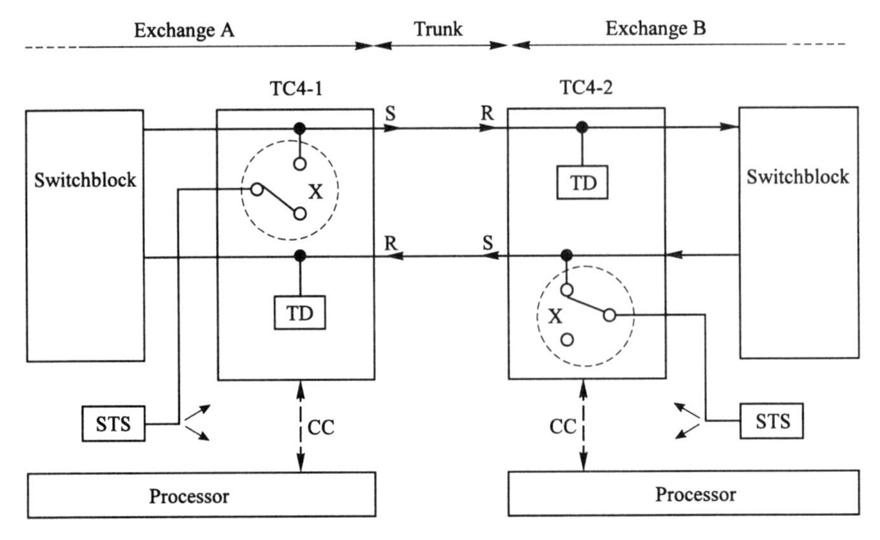

Figure 4.2-2 Four-wire analog trunk circuits (TC4).

Trunk State	Tone
On-hook (idle)	Tone on
Off-hook (in use)	Tone off

Sending and Receiving the Signaling Tone. Figure 4.2-2 is a simplified presentation of the signaling circuitry in the four-wire trunk circuits TC4 of a trunk. At each exchange, a *signaling tone source* (STS) supplies all trunk circuits with the 2600 Hz signaling tone. The sending of the tone is controlled by the exchange processor, which controls switch X in the TC4. In the figure, the processor at exchange A has sent an off-hook command to TC4-1, and no tone is sent on the send channel (S) of the trunk. At exchange B, the processor has sent an on-hook command to TC4-2, and switch X connects the tone on the S channel of the trunk. In the TC4s, a 2600 Hz tone detector (TD) is bridged across the receive channels (R). In this example, the TD in TC4-2 detects no tone, and reports to its processor that the trunk is off-hook at exchange A. The TD in TC4-1 detects the tone, and reports that the trunk is on-hook at exchange B.

There are several aspects of in-band supervision signaling that pose additional requirements on the circuits of the TC4s.

Blocking Received Signaling Tone. During the conversation, both exchanges indicate off-hook, and no signaling tone is present in either direction. However, in other call states, the signaling tone and other voiceband signals can be present simultaneously on a trunk. Consider a call from subscriber S_1, served by local exchange A, to a subscriber served by local exchange C. The connection passes through intermediate exchange B. Figure 4.2-3(a) shows the

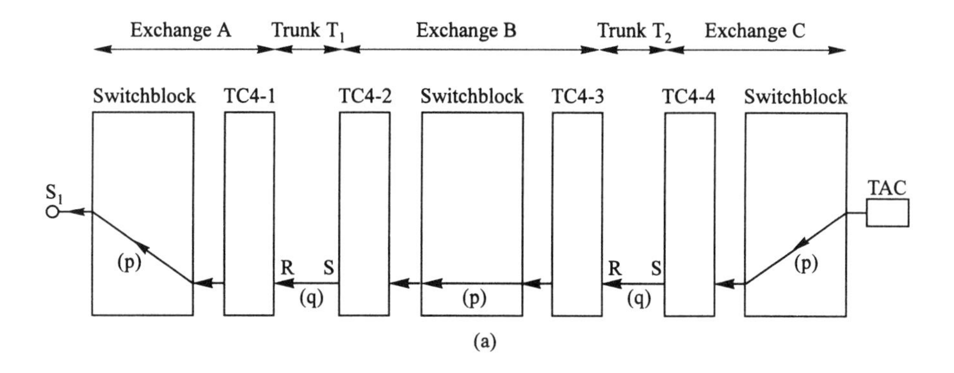

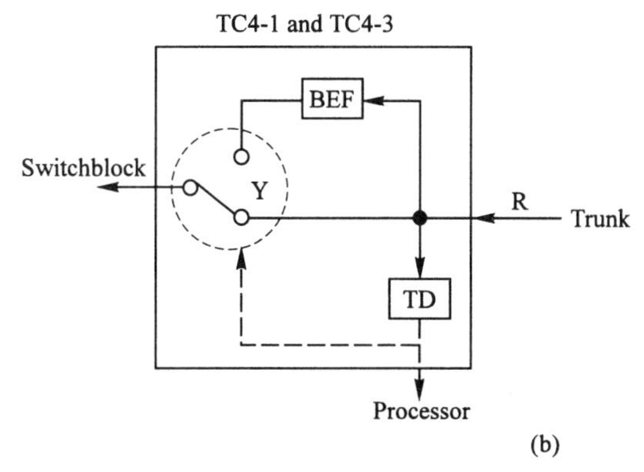

(b)

Figure 4.2-3 Blocking received signaling tone. (a): transmission path in the direction C → A. (b): band elimination filter (BEF).

transmission path in direction C → A only. The called party has not yet answered, and exchange C has attached tone/announcement circuit (TAC) to trunk T_2. The circuit is sending ringing-tone. At points (p), only ringing-tone is present. However, trunk T_2 is on-hook at exchange C, and trunk T_1 is on-hook at exchange B, and at points (q) the ringing-tone and signaling tone are both present.

There are two reasons why TC4-1 and TC4-3 should pass the ringing-tone (or other voiceband signals), but block the signaling tone. In the first place, S_1 should hear ringing-tone only. In the second place, supervision signaling is link-by-link. This means that the signaling tone from B to A on trunk T_1 should be controlled by exchange B. Therefore, TC4-3 has to block the received signaling tone, which otherwise would "leak" into T_1.

Figure 4.2-3(b) shows how received signaling tone is blocked. *Band-elimination filter* (BEF) blocks 2600 Hz, but passes other voiceband frequencies. The filter is inserted in receive channel R by switch Y, which is controlled by TD. When TD is not receiving signaling tone, switch Y is in the position shown, and BEF is not in the receive channel. However, when TD receives signaling tone, it sets Y in the other position, and BEF is inserted into the channel.

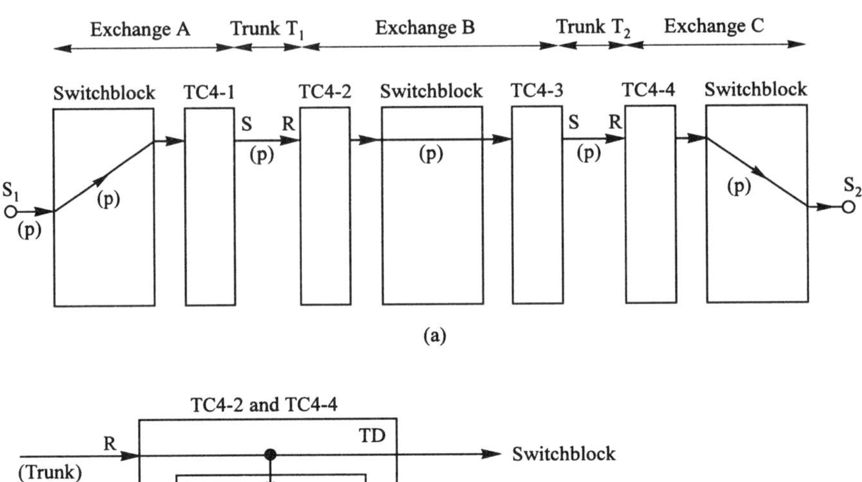

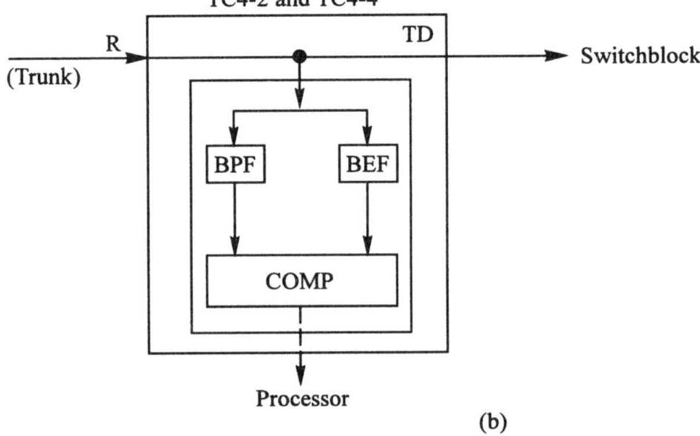

Figure 4.2-4 Protection against talkoff. (a): forward transmission path during conversation. (b): tone detector (TD).

Protection Against Talk-off. Figure 4.2-4(a) shows the transmission path in direction A → C for the connection mentioned above. The called party has answered and signaling tone is absent at points p. Speech—or other subscriber communications—can include components around 2600 Hz, which could simulate the signaling tone. In early in-band signaling, a speech fragment from calling subscriber S_1 would occasionally be interpreted by the tone detectors in TC4-2 or TC4-4 as a forward tone-on (on-hook), which is a clear-forward signal. Exchange B or C would then clear the connection. This was known as *talk-off*.

In-band supervision signaling is protected against signal simulation by a combination of two techniques.

In the first place, a tone detector (TD) does not simply detect the presence of the 2600 Hz tone. As shown in Fig. 4.2-4(b), the signal received from the trunk is fed into band-pass filter BPF that passes a narrow band of frequencies around 2600 Hz only, and to band-elimination filter (BEF) that passes all voiceband frequencies, except the frequencies passed by BPF. Comparator (COMP) compares the signal strengths at the outputs of the filters.

When signaling tone is being received, the output of BPF is stronger than the output of BEF, even when speech is present at the same time. TD thus indicates tone-on (on-hook) to the exchange processor. When only speech is received, the output of BEF is stronger than the output of BPF, because speech power is mainly concentrated in the 300–1000 Hz range, and only a small part of it lies around signaling frequency. In this condition, TD indicates off-hook.

As a second safety measure, exchange processors do not recognize received changes to on-hook or off-hook as a valid signals until the new state has persisted during a *recognition time* of a certain length. With the exception of winks, the recognition time for most signals is 30 ms. However, as an extra protection against talk-off, the recognition time for clear-forward signals is 300 ms.

Blue-box Fraud. In the mid-1960s, the Bell System became aware of "blue box" fraud on FDM analog trunks [3]. In principle, all signaling systems with in-band supervision signaling are vulnerable to this type of fraud, in which a subscriber generates in-band supervision and address signals with a device—called "blue box"—attached to the subscriber line. The box can generate the 2600 Hz supervision tone, and the MF register signals. With the box, a fraudulent subscriber can manipulate the network to avoid charges on long-distance calls.

An example is shown in Fig. 4.2-5. Suppose that subscriber S, served by local exchange A, has placed a call to an operator at exchange B, and that the connection has been set up on trunk analog trunk (T). Since S has dialed 0 for operator assistance, exchange A does not charge the call, expecting the charging to be done at exchange B.

Once the connection has been set up, exchange A cuts through and there is a voiceband path from S to B. This allows S to send blue-box signals to B. Local exchange A is unaware of this because, once the connection has been set up, it

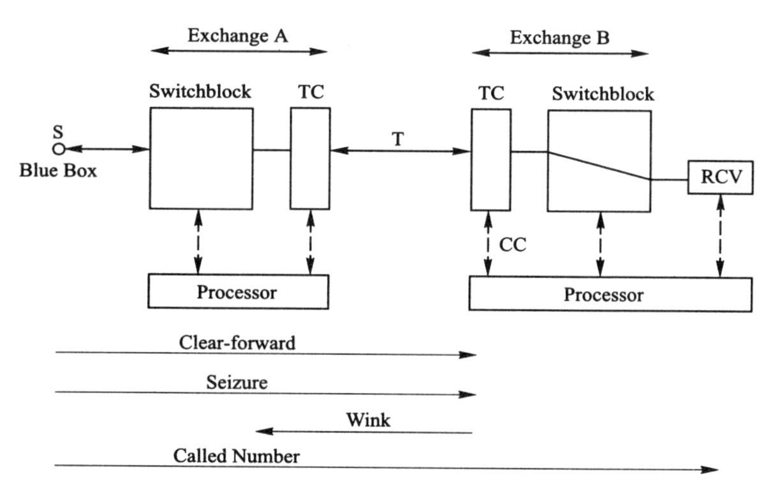

Figure 4.2-5 Blue-box signaling. Exchange B connects trunk T to digit receiver RCV after receiving the wink signal.

monitors only the on-/off hook condition of the subscriber line. However, the blue-box signals are accepted in good faith by exchange B.

Before the operator has answered, S sends a burst (say 1–2 s) of signaling tone. This is interpreted by exchange B as a clear-forward signal, and the exchange releases trunk T at its end. The end of the burst is interpreted at exchange B as a new seizure signal. Subscriber S waits about one second after the end of the signaling tone, and then sends the called party's number. The call is set up and, since exchange B has not received a dialing pattern that indicates an operator-assisted call, it assumes that exchange A will charge the call.

The Bell system has curbed blue-box fraud by vigorous legal prosecution of fraudulent subscribers, and by implementing protective procedures at the exchanges.

4.2.3 Supervision Signaling on Digital (PCM) Trunks

In the example of Section 4.1.3, the supervision signaling for trunks in first-order PCM multiplexes is sent and received by the digital multiplex ports (DMP) that attach the PCM multiplexes to the switchblock of the exchange. The DMPs have a control channel to the exchange processor, and report the received state (on-hook, off-hook) of the trunks in their multiplexes to the processor. The processor commands the state to be sent out on the trunk.

The DS1 frame format of the North American T1 first-order digital transmission systems (1.5.2) is shown in Fig. 4.2-6. It consists of consecutive frames. Each frame has 24 eight-bit time slots (TS), and one F bit. Each time slot is associated with a particular trunk.

Twelve frames form a superframe, in which the frames are numbered from 1 through 12. The DMPs maintain frame and superframe synchronization with the incoming bit stream by locking onto the F bits of the frames, which exhibit a repeating 12-bit pattern:

	Superframe											
Frame #	1	2	3	4	5	6	7	8	9	10	11	12
F bit	1	0	0	0	1	1	0	1	1	1	0	0

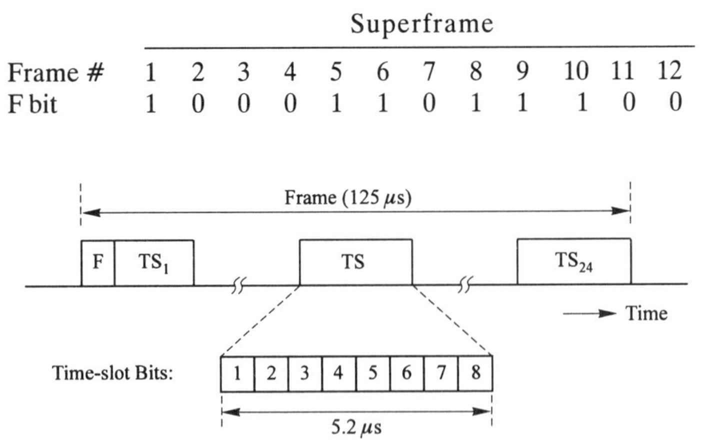

Figure 4.2.6 North American DS1 frame format. Signaling bits are in position 8 of each time slot (TS) during frames 6 and 12.

When locked on to this pattern, a DMP can determine the start of each frame, and of each superframe, in the bit stream. In frames 6 and 12, bits 8 (the least important bits of the 8-bit PCM codes) in the 24 channels are used for supervision signaling. This is known as *bit robbing*. The effect of bit robbing on the quality of PCM-coded speech is negligible.

The signaling bits in frames 6 and 12 are known as the S_a and S_b bits. The DMPs update their outgoing signaling bits every 1.5 ms (once per superframe), which is sufficiently fast for supervision signaling. The combinations of an S_a and S_b bit could indicate four trunk states. However, the S_b bit in each time slot is set equal to the previous S_a bit, resulting in two-state continuous supervision signaling. The bit values 0 and 1 represent respectively on-hook and off-hook.

The signaling bits cannot be heard by the subscribers, and the subscriber's speech, or a blue box, cannot corrupt the supervision signals. This avoids the problems associated with in-band signaling.

4.2.4 Address Signaling

The MF address signals are combinations of two voiceband frequencies—chosen from a set of six frequencies [1,2,5]. The frequency assignments are shown in Table 4.2-1. Only 12 of the 15 possible two-out-of-six codes are used in CCITT-R1 signaling. Bell MF signaling uses some of the remaining codes in calls that are set up with operator assistance [5].

Address signaling sequences start with a KP (start-of-pulsing) signal, and end with an ST (end-of-pulsing) signal. Signaling sequences received without KP or ST are considered to be mutilated, and discarded by the incoming exchange.

The KP signal has a duration of 90–110 ms. The duration of the other signals is 61–75 ms. The originating exchange sends the complete called number (*en-bloc* address signaling), with silent intervals between signals of 61–75 ms. The intermediate exchanges can use overlap address signaling.

Table 4.2-1 Bell system multi-frequency address signals.

Signal	Frequencies (Hz)
Digit 1	700 and 900
Digit 2	700 and 1100
Digit 3	900 and 1100
Digit 4	700 and 1300
Digit 5	900 and 1300
Digit 6	1100 and 1300
Digit 7	700 and 1500
Digit 8	900 and 1500
Digit 9	1100 and 1500
Digit 0	1300 and 1500
KP	1100 and 1700
ST	1500 and 1700

Address Signaling Sequences. In its simplest form, an address signaling sequence conveys the called number only. The called number can be a subscriber number, or a national number. Let AC(3), EC(3), and LN(4) represent a three-digit area code, a three-digit exchange code, and a four-digit line number (1.3.1). The address signaling sequences are then: KP-EC(3)-LN(4)-ST (subscriber number), or KP-AC(3)-EC(3)-LN(4)-ST (national number).

A calling subscriber dials subscriber numbers for calls inside his numbering plan area (NPA), and national numbers for calls outside his NPA. An originating exchange that receives a national called number from the calling subscriber, or an intermediate exchange that receives a national called number from the preceding exchange in the connection, sends out the called number as either a national number, or a subscriber number. This depends on whether the exchange has seized an outgoing trunk to an exchange outside, or inside the NPA of the called party .

Automatic Number Identification. In the early years of subscriber-dialed long-distance calling, most local exchanges were not equipped to produce billing records for these calls, and the billing records were generated by the first intermediate (toll) exchange in the connection. The local exchange would send both the called and calling numbers, and the toll exchange would handle the charging for the call. After the end of the call, the toll exchange would generate a billing record that included both numbers, the date of the call, and the times of answer and call clearing. The sending of calling numbers is known as *automatic number identification* (ANI).

After the break-up of the Bell system, calls between subscribers in different local (LATA) networks are billed by either the LATA of the calling party, or by the interexchange carrier (IC). This is a matter of mutual agreement between the LATA and IC carrier.

Figure 4.2-7 shows the supervision and address signals for an inter-LATA call originated by subscriber S. The call is to be billed by IC exchange B.

A subscriber can designate a "default" IC for his inter-LATA calls. This information is stored at the local exchange. If a subscriber just dials a called number, the exchange routes the call to an exchange of his default IC. If subscriber S desires a different carrier for the call, he dials a prefix 10*XXX*, where *XXX* identifies the IC.

After the subscriber has dialed:

1-AC(3)-EC(3)$_1$	(Called national number, default IC), or
10*XXX*-1-AC(3)-EC(3)	(Called national number, specified IC) or
EC(3)	(Called subscriber number, default IC), or
10*XXX*-EC(3)	(Called subscriber number, specified IC)

exchange A knows the desired IC, and the nearest exchange (B) of that IC. This example assumes that a direct trunk group connects exchanges A and B. Exchange A seizes a trunk T in this group, and sends a seizure signal. After

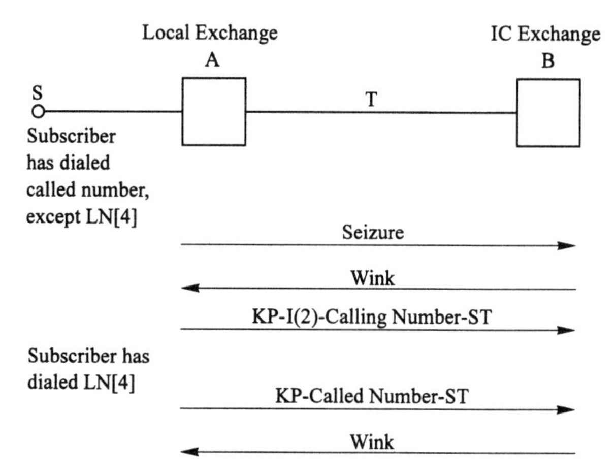

Figure 4.2-7 Transfer of calling and called numbers to exchange B of interexchange carrier.

receiving the wink signal, A first sends the national number of the calling subscriber:

$$\text{KP-I(2)-AC'(3)-EC'(3)-LN'(4)-ST.}$$

When the subscriber has finished dialing LN(4), the last four digits of the called number, exchange A sends a second digit sequence, which identifies the called national- or subscriber number: KP-AC(3)-EC(3)-LN(4)-ST, or KP-EC(3)-LN(4)-ST. Exchange B then acknowledges the receipts of both numbers with a wink signal.

Sending the calling number first minimizes the elapsed time from the end of address signaling by the calling subscriber to the end of address signaling by the local exchange.

The codes in the two information digits I(2) ahead of the calling number characterize the calling line:

I(2) = 00	Identified subscriber line
I(2) = 02	ANI failure (calling number not included)
I(2) = 06	Call from hotel without room identification
I(2) = 10	Test call

When I(2) = 02 or 06, an operator at exchange B verbally obtains the calling number.

4.2.5 Failed Set-ups

Bell MF signaling does not include backward signals to indicate that the set-up of a connection has failed. The exchange where the failure occurs sends a tone (busy-tone, reorder-tone) or announcement, and the calling party disconnects (sends a clear-forward to the originating exchange). The originating exchange then initiates the release of the connection.

4.3 CCITT NO.5 SIGNALING

This signaling system has been developed jointly by the U.K. Post Office and Bell Laboratories, and is similar to Bell System MF signaling. It was adopted in 1964 by CCITT for use in the international network, and is documented in CCITT Recommendations [6].

CCITT No.5 has been designed especially to operate on TASI-equipped analog trunks (1.6.4). It has been used extensively on long international trunks (underwater transoceanic trunks, satellite trunks). Despite its age, it is still in use on several international trunk groups.

Supervision and address signaling are both link-by-link and in-band.

4.3.1 Supervision Signaling

Supervision signals consist of one or two in-band signaling tones. TASI compatibility requires a special form of supervision signaling. The continuous two-state signaling described in 4.2.1 would defeat the purpose of TASI equipment, because signaling tones would be present in both directions when a trunk is idle, and each idle trunk would therefore occupy a pair of TASI bearer channels.

A number of CAS signaling systems use pulsed supervision signals, in which the signals are short bursts (typically 50–150 ms) of signaling tone [1]. However, this form of signaling is also incompatible with TASI trunks, because *freezeouts* (1.6.4) could shorten the pulses beyond recognition, or suppress them completely.

Therefore, *compelled* (sometimes called *continuous-compelled*) supervision signaling is used. An exchange that sends a supervision signal keeps the signaling tone(s) "on" until it receives an acknowledgment signal (also consisting of one or two signaling tones) from the exchange at the distant end of the trunk.

Figure 4.3-1 shows a supervision signal on a trunk T between international switching centers (ISC-A) and (ISC-B), as observed on the S (send) and R (receive) channels of the trunk at ISC-A. At t_1, ISC-A turns on the signaling tone(s). When ISC-B receives the signal it responds with an acknowledgment. At t_2, the signaling tone(s) of the acknowledgment arrive at ISC-A, and this exchange turns off its signaling tone(s) at t_3. ISC-B then notices that the signal from ISC-A has ended, and stops sending its acknowledgment signal. Finally, ISC-A notices the end of the acknowledgment at t_4. Compelled signaling thus successfully transfers a supervision signal, even when the signal or its acknowledgment is subjected to a TASI freeze-out.

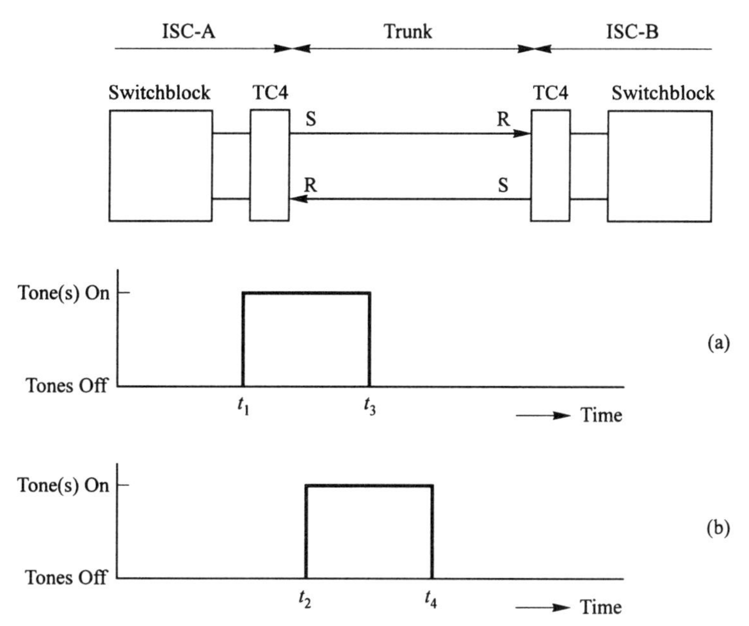

Figure 4.3-1 Supervision Signals of CCITT No. 5. (a): signal on send channel at ISC-A. (b): signal on receive channel at ISC-A.

Signaling Frequencies. Two in-band signaling frequencies are used: f_1 = 2400 Hz, and f_2 is 2600 Hz. There are three signals: f_1, f_2, and a composite signal that contains f_1 and f_2.

Signaling Circuitry in Trunk Circuits. The functions and circuitry for supervision signaling in the TC4 circuits of analog CCITT No.5 trunks are similar to those discussed in 4.2.2. The tone detectors (TD) have to detect the presence of the signaling frequencies f_1 and/ or f_2 on the receive (R) channel of the trunk. Since the signaling frequencies are in-band, a TD that detects their presence has to block them from entering the switchblock of the exchange. In addition, the tone detectors include circuitry that prevents the simulation of signaling tones by the speech (or other voiceband communications) of the subscribers.

Because of its in-band supervision signals, CCITT No.5 signaling is vulnerable to "blue box" fraud.

4.3.2 Supervision Signals and Call Handling Procedures

Table 4.3-1 lists the supervision signals, their acknowledgments, and the signaling frequencies. A particular physical signal can have several logical meanings, depending on its direction (forward or backward), and the state of the call. The supervision signals for a typical call on a trunk seized by ISC-A in Fig. 4.3-1 are outlined below.

Table 4.3-1 CCITT No.5 supervision signals and their acknowledgments.

Signal	Direction	Frequency
Seizure	fwd	f_1
Proceed-to-send	bkwd	f_2
Answer	bkwd	f_1
Acknowledgment	fwd	f_1
Clear-back	bkwd	f_2
Acknowledgment	fwd	f_1
Clear-forward	fwd	f_1 and f_2
Release-guard	bkwd	f_1 and f_2
Busy-flash	bkwd	f_2
Acknowledgment	fwd	f_1
Forward-transfer	fwd	f_2

f_1 = 2400 Hz. f_2 = 2600 Hz. fwd: forward. bkwd: backward.
Source: Rec. Q.141. Courtesy of ITU-T.

Seizure and Proceed-to-send. On receipt of the seizure signal, ISC-B attaches a MF digit receiver to the trunk, and then acknowledges with a proceed-to-send signal. When ISC-A receives the acknowledgment, it sends the called number in a sequence of address signals.

Answer and Clear-back. These backward signals have their conventional meanings, and are simply acknowledged at outgoing exchange ISC-A.

Clear-forward and Release-guard. At the end of the call, outgoing ISC-A releases trunk (T) at its end, and sends a clear-forward signal. ISC-B clears the trunk at its end, and then acknowledges with a release-guard signal. This indicates that the trunk can be seized again for a new call.

Busy-flash is a backward signal sent by ISC-B to indicate that it cannot extend the call set-up (for example, because no trunk to/towards the call destination is available). On receipt of the signal, ISC-A acknowledges. It then releases the trunk, and sends a clear-forward signal to ISC-B.

If the call arrived at ISC-A on a trunk that has a signaling system which includes signals or messages to indicate the failure of a call set-up, ISC-A informs the preceding exchange, which then releases its trunk to ISC-A.

Otherwise, ISC-A connects its incoming trunk to a tone or announcement circuit. This alerts the calling subscriber, who then sends a clear-forward to her local exchange, and that exchange initiates the release of the connection. The busy-flash reduces the amount of time that international trunks are held in failed set-ups.

Forward-transfer is used on calls that are set up with operator assistance (Section 4.3.5). The signal is a pulse of nominally 850 ms, and is not acknowledged.

Double Seizures. Suppose that ISC-A seizes a bothway trunk, and starts sending the seizure signal. The seizure may arrive at ISC-B with a delay of up to 600 ms (on satellite circuits). During that time ISC-B may also have seized the trunk.

A double seizure is detected when an exchange that has sent a seizure signal receives a f_1 (seizure signal) instead of the expected f_2 (proceed-to-send) acknowledgment. Both ISC-A and ISC-B detect the double seizure. In response, they release the trunk at their respective ends. Clear-forward signals are not sent. International exchanges are usually arranged to make a repeat attempt to set up a connection that has failed because of a double seizure.

Recognition Times. An exchange recognizes—and accepts—a received supervision signal only when it has persisted for a certain amount of time. The nominal recognition time for the seizure and proceed-to-send signals is 40 ms. The recognition times for all other signals are 125 ms.

4.3.3 Multi-frequency Address Signaling

CCITT No.5 address signals are combinations of two frequencies, selected from the same set of frequencies (700, 900, 1100, 1300, 1500, and 1700 Hz) that is used in Bell System MF signaling.

To avoid clipping of the address signals by TASI equipment, the address signals are sent *en bloc*. The silent interval between the end of the seizure signal and the start of the first address signal is at most 80 ms, and the silent intervals between address signals are 55 ms. These intervals are shorter than the 400 ms TASI "overhang" intervals (1.6.4), and the TASI bearer channel that was assigned to the trunk for the seizure signal is therefore not released until all address signals have been sent.

Address Signals. Table 4.3-2 lists the frequencies of the address signals. All address signaling sequences start with a KP signal. The KP1 signal indicates that the called number is in the form of a national number, and KP2 indicates that the called number is an international number.

Code 11 and Code 12 are one-digit addresses for international operators at an ISC.

All address signaling sequences include a Z-digit (or: Language digit, Discriminating digit). The value of the digit indicates the calling party category:

0: Subscriber
1: French-speaking operator
2: English-speaking operator
3: German-speaking operator
4: Russian-speaking operator
5: Spanish-speaking operator

TABLE 4.3-2 CCITT No.5 address (register) signals.

Signal	Frequencies (Hz)
Digit 1	700 and 900
Digit 2	700 and 1100
Digit 3	900 and 1100
Digit 4	700 and 1300
Digit 5	900 and 1300
Digit 6	1100 and 1300
Digit 7	700 and 1500
Digit 8	900 and 1500
Digit 9	1100 and 1500
digit 0	1300 and 1500
Codc 11	700 and 1700
Code 12	900 and 1700
KP1	1100 and 1700
KP2	1300 and 1700
ST	1500 and 1700

Source: Rec. Q. 152. Courtesy of ITU-T.

The principal function of the Z-digit is to inform the incoming exchange whether the international call is subscriber-dialed, or set up with operator assistance.

4.3.4 Address Signaling Sequences

Some basic address signaling sequences for a call that originates in country A, and has a destination in country C, are shown in Fig. 4.3-2 [7]. The called party is identified by the national number NN (AC-EC-SN) in country C, and the country code of country C will be denoted by CC. If ISC-A routes the call on a direct trunk to ISC-C, it sends the sequence KP1-Z-NN-ST. This informs ISC-C that its country is the destination country of the call, and that the called number is a national number.

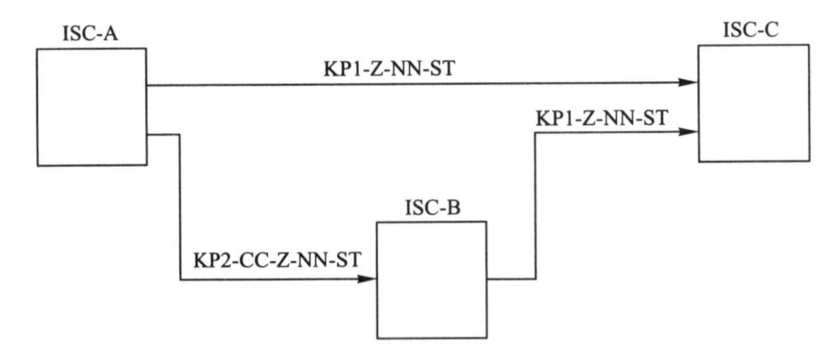

Figure 4.3.2 CCITT No. 5 address signaling.

If ISC-A routes the call on a trunk to transit country B, it sends KP2-CC-Z-NN-ST. This informs ISC-B that country B is not the destination of the call, and that the called number is an international number (CC-NN). If ISC-B routes the call on a direct trunk to ISC-C in the destination country, it deletes the country code, and sends KP1-Z-NN-ST.

Outgoing international exchange ISC-A, which has to send its address signals *en bloc*, may have received the call from its national network on a trunk whose signaling system allows overlap address signaling. ISC-A thus has to determine when it has received the complete international number. This is usually done with the use of stored data that indicate the maximum lengths of national numbers in the various foreign countries. After ISC-A has received the country code of the destination country, it thus knows the maximum number of expected address digits. It seizes an outgoing international trunk (and starts address signaling) when it has received the expected maximum number of digits, or when 5 s have elapsed after receipt of the most recent digit.

International operators have a "start" key which they depress after dialing an international number. When the ISC receives the "start" signal, it immediately seizes the outgoing trunk (5 s timing not necessary).

4.3.5 Signaling Features for International Operators [7]

An international switching center usually includes a number of consoles for several groups of operators. *Outgoing* operators assist in-country subscribers on outgoing international calls [7]. *Incoming* and *assistance* operators provide help to foreign outgoing operators on incoming international calls. ISCs provide several special call-handling procedures for operators. An ISC knows, from the value of the Z-digit, whether a received call has been dialed by a subscriber or an operator. A number of operator procedures and corresponding CCITT No.5 signaling sequences are outlined below.

Forward Transfer. Suppose that originating country A in Fig. 4.3-2 is an English-speaking country, and that an outgoing operator at ISC-A has dialed a call to a subscriber in country C, at the request of the calling subscriber. The address signaling sequence sent by ISC-A is one of those described above, with $Z = 2$. The call is handled by ISC-C as a call to an in-country subscriber. In addition, ISC-C tags the incoming trunk as carrying a connection that is being set up by an English-speaking operator. If the operator at ISC-A runs into a set-up problem, he depresses the "forward-transfer" key on his console, causing ISC-A to send a forward-transfer (line signal). On receipt of the signal, ISC-C bridges an English-speaking *assistance* operator on to the call.

Code 11 Calls. An outgoing operator in country A can call an incoming operator in country B by dialing the country code (CC) of B, and then depressing the "code 11" key on his keypad. After seizing a trunk, ISC-A then sends the sequence KP1-Z-"code 11"-ST (on a direct trunk to ISC-C), or KP2-CC-Z-"code 11"-ST (on a trunk to a transit country). ISC-C connects the trunk on which the call arrives to an available incoming operator speaking the language indicated by the value of *Z*.

Code 11 calls are typically made when the outgoing operator cannot dial the called party because the calling subscriber is unable to provide a correct called number.

Code 12 Calls. The code 11 call may result in a successful call set-up. However, there are cases where the call cannot be completed immediately. The operator at ISC-C may then offer to "try later". In this case, the operator at ISC-A identifies herself by a number, say "operator 27." When the operator at ISC-C succeeds in reaching the called party, she recalls the outgoing operator at ISC-A by dialing the country code of ISC-A, followed by code 12 and the operator number. ISC-C then sets up a connection to ISC-A, sending the sequence: KP1-Z-"code 12"-2-7-ST, or KP2-CC-Z-"code 12"-2-7-ST, and ISC-A connects the call to operator 27. The operator, who has kept a record of the calling party number, then completes the connection by calling that party.

4.4 R2 SIGNALING

R2 signaling was known originally as *multi-frequency code* (MFC) signaling. It was developed cooperatively by European telecommunication equipment manufacturers and the CEPT (European Conference of Postal and Telecommunications Administrations), and was introduced in the 1960s. It is still used in many national networks in Europe, Latin America, Australia, and Asia.

A few years after the introduction of MFC, CCITT defined a version for use in the international network. This international version is known as CCITT-R2 signaling. Today, the national MFC systems (which exist in several country-specific versions) are also referred to as R2 systems [1,4,8].

R2 signaling can be used on two-wire analog trunks, and on four-wire analog and digital trunks. It cannot be used on TASI-equipped trunks, or on trunks carried by satellite transmission systems. This limits the application of CCITT-R2 to relatively short international trunks.

Compared with Bell System MF signaling and CCITT No.5 signaling, the most important difference of R2 is its *register* (address) signaling.

4.4.1 Supervision Signaling on FDM Analog Trunks

This section describes the supervision signaling of CCITT-R2 and (national) R2, for four-wire FDM analog trunks [4].

The signaling is *out-of-band*: the bandwidth of the channels in FDM transmission systems is divided into a 300–3400 Hz band for the subscriber's speech (or other communications), and a narrow band centered at the signaling

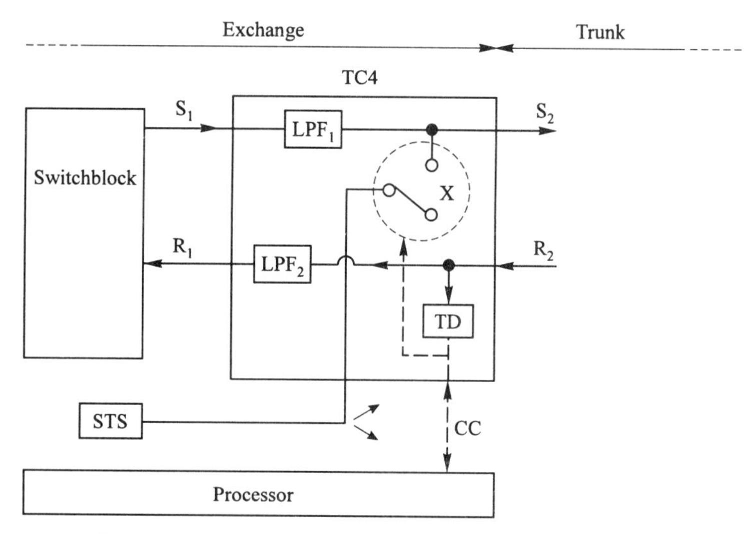

Figure 4.4-1 Four-wire analog trunk circuit (TC4) for out-of-band supervision signaling.

frequency f = 3825 Hz. This separates signaling tone and speech, and avoids the problems associated with in-band supervision signaling (4.2.2).

Trunk Circuit. Figure 4.4-1 shows a four-wire trunk circuit TC4 for out-of band supervision signaling. *Low-pass filters* (LPF) that block frequencies above 3400 Hz separate speech from signaling tone, and vice versa.

Switch X inserts the signaling tone into the send (S_2 channel on command from the exchange processor. and tone detector (TD) reports the presence or absence of the signaling tone on the receive channel R_2 to the processor.

CCITT-R2 Supervision Signals. The supervision signaling of CCITT-R2, intended for one-way analog trunks only, is two-state, tone-on-idle (as in Bell System MF signaling). For this reason, R2 signaling cannot be used on TASI-equipped trunks. Nominal signal recognition times are 40 ms. The signals and the corresponding changes in trunk state are:

Forward Signals	State Change
Answer	Tone-on to tone-off
Clear-back	Tone-off to tone-on

Backward Signals	State Change
Answer	Tone-on to tone-off
Clear-back	Tone-off to tone-on
Release-guard	450 ms Tone-off pulse, or tone-off to tone-on
Blocking	Tone-on to tone-off

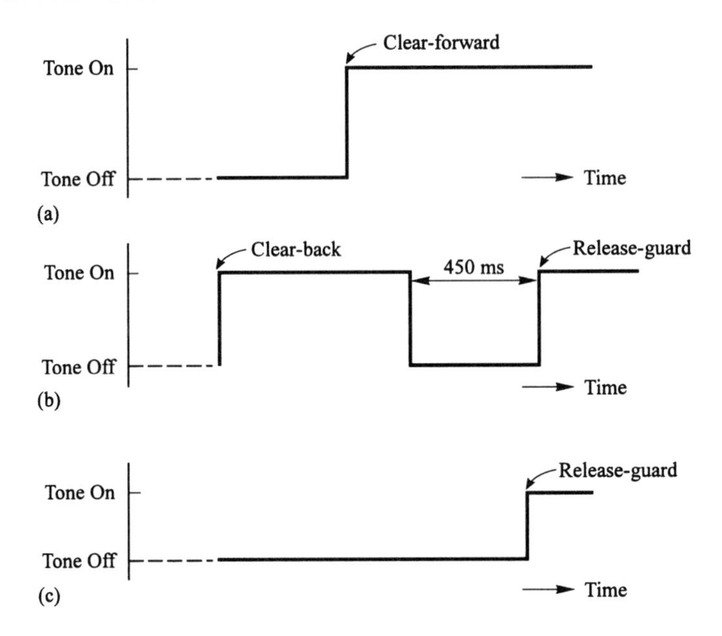

Figure 4.4-2 Release-guard signals. (a): clear-forward. (b): release-guard signal when clear-back signal has been sent. (c): release-guard signal when clear-back signal has not been sent.

The system does not include a proceed-to-send signal. The blocking signal is a backward transition to tone-off when the trunk has not been seized by the outgoing exchange. It is sent by the incoming exchange, and requests the outgoing exchange to suspend seizing the trunk for new calls, because the incoming exchange is performing maintenance on the trunk.

The release-guard signal is sent by the incoming exchange after it has received a clear-forward signal. It indicates that the incoming exchange has released the trunk at its end, and that the outgoing exchange can therefore seize the trunk for a new call. As shown in Fig. 4.4-2, if the incoming exchange receives a clear-forward after it has sent a clear-back, the release-guard signal is a 450 ms tone-off pulse (b). If the incoming exchange has not sent a clear-back when it receives the clear-forward, the release-guard signal is a change to on-hook, 450 ms after the receipt of the clear-forward (c).

National R2 Supervision Signals. Most national R2 systems use pulsed out-of-band supervision signals. There are several country-specific pulsed signaling systems, for example:

Forward Signals Pulse	Duration (ms)
Seizure	150
Clear-forward	600

Backward Signals	Pulse Duration (ms)
Answer	150
Clear-back	600
Release-guard	600
Blocking	Continuous

4.4.2 Supervision Signaling on Digital Trunks

Networks that use national R2 or CCITT-R2 signaling use E1 first-order transmission systems for digital trunks [4]. The bit streams on these multiplexes are organized in frames that are transmitted at a rate of 8000 frames/s. Each frame has 32 eight-bit time slots, numbered from TS_0 through TS_{31}—Fig. 1.5-4(b). Time slots TS_1 through TS_{15}, and TS_{17} through TS_{31} carry PCM-encoded speech, or 64 kb/s subscriber data, for 30 trunks. For frame alignment, bits 2 through 8 of TS_0 have the fixed pattern 0011011.

A superframe consists of 16 consecutive frames, numbered from 0 through 15. For superframe alignment, bits 1 through 4 in TS_{16} of frame 0 are coded 0000. TS_{16} in frames 1 through 15 carries four status bits (a,b,c,d) bits for the trunks:

	<Bits in Time Slot 16>							
	1	2	3	4	5	6	7	8
	a	b	c	d	a	b	c	d
Frame 1	<Trunk 1>				<Trunk 17>			
Frame 2	<Trunk 2>				<Trunk 18>			
Frame 3	<Trunk 3>				<Trunk 19>			
⋮	⋮				⋮			
Frame 15	<Trunk 15>				<Trunk 31>			

The supervision signaling for digital CCITT-R2 trunks is continuous, with two forward, and three backward trunk states that are represented by bits a_f, b_f and a_b, b_b respectively. Bits c and d are not used, and are set to 0 and 1.

The signaling can be applied to one-way and bothway trunks. On one-way trunks, only one exchange can seize a trunk, and sends forward bits a_f and b_f, and the other exchange sends backward bits a_b and b_b. On bothway trunks, the roles of the exchanges vary from call to call, depending on which exchange seizes the trunk.

The idle state at both ends of a trunk is represented by a,b = 1,0. The supervision signals are represented by changes in bit patterns:

Forward Signals	Change
Seizure	a_f, b_f: 1,0 → 0,0
Clear-forward	a_f, b_f: 0,0 → 1,0

Backward Signals	Change
Seizure acknowledgment	a_b, b_b: 1,0 → 1,1
Answer	a_b, b_b: 1,1 → 0,1
Clear-back	a_b, b_b: 0,1 → 1,1
Release guard	a_b, b_b: 1,1 → 1,0
or	a_b, b_b: 0,1 → 1,0

Blocking. An exchange can block an idle trunk by changing its status bits from a,b = 1,0 to a,b = 1,1. Exchanges do not seize trunks that are in this state at the distant end. To end blocking, the exchange returns the bits to a,b = 1, 0 (idle).

Double Seizure. After sending a seizure signal, the outgoing exchange expects to receive a change to a_b,b_b = 1,1 from the incoming exchange. The response a_b,b_b = 0,0 indicates a double seizure. Both exchanges then abort their call set-ups and—depending on the procedures of the telecom—either make a second attempt or abort the call set-ups, sending congestion indications to the calling subscribers.

4.4.3 Interregister Signaling

In R2 signaling, the equipment units at the exchanges that send and receive digits, and the signaling between these units, are usually referred to as *registers*, and *interregister signaling.*

R2 uses forward- and backward in-band MF (multi-frequency) signals. On

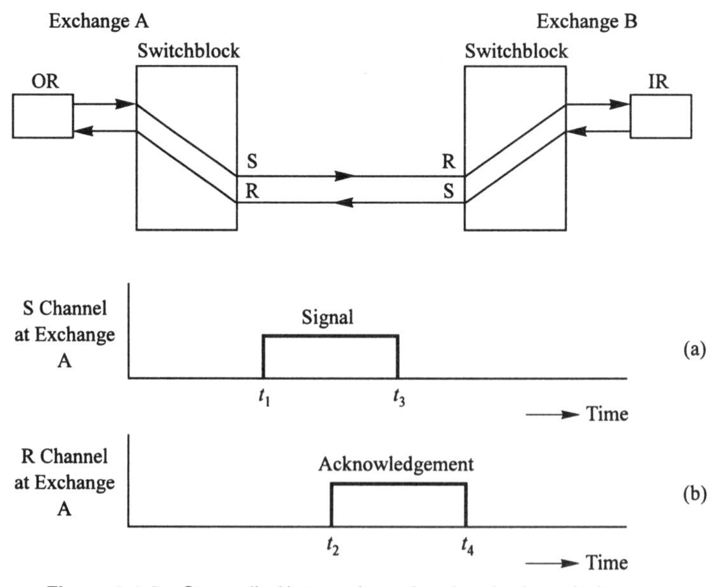

Figure 4.4-3 Compelled interregister signal and acknowledgment.

digital trunks, these signals are PCM-coded. Signaling is *compelled*: a forward MF signal, sent by an *outgoing* register (OR) at outgoing exchange A, is held "on" until the receipt of a backward MF acknowledgment from incoming register (IR) at incoming exchange B—see Fig. 4.4-3. R2 registers are transceivers: they both send and receive register signals. The compelled procedure is identical to the supervision signaling of CCITT no.5, but its purpose is entirely different.

Compelled interregister signaling is not practical for trunks carried on satellite transmission systems, because of their long propagation times (about 600 ms). The time required for the transmission of one digit and its acknowledgment would be in the order of 1.2 s, and this would result in extremely slow signaling.

Interregister Signals. The forward and backward signals consist of two voiceband frequencies, selected from a set of six—see Table 4.4-1. The forward and backward frequency sets are different. This is necessary to make the register signaling suitable for two-wire analog trunks.

Because of the two frequency sets, two register types are needed. In Fig. 4.4-3, outgoing register (OR) sends forward frequencies, and receives backward frequencies. The reverse holds for incoming register (IR).

Groups of Interregister Signals. A particular forward or backward signal can have several meanings. We speak of group I, group II, and, in some countries, group III *forward* signals, and of group A and group B *backward* signals. A signal is denoted by its "group" and its "value". For example, we denote a group II signal with value 7 by II-7.

Table 4.4-1 National R2 and CCITT-R2 interregister signal frequencies.

Signal Value	Forward (Hz)	Backward (Hz)
1	1380 and 1500	1140 and 1020
2	1380 and 1620	1140 and 900
3	1500 and 1620	1020 and 900
4	1380 and 1740	1140 and 780
5	1500 and 1740	1020 and 780
6	1620 and 1740	900 and 780
7	1380 and 1860	1140 and 660
8	1500 and 1860	1020 and 660
9	1620 and 1860	900 and 660
10	1740 and 1860	780 and 660
11	1380 and 1980	1140 and 540
12	1500 and 1980	1020 and 540
13	1620 and 1980	900 and 540
14	1740 and 1980	780 and 540
15	1860 and 1980	660 and 540

Source: Rec. Q.441. Courtesy of ITU-T.

The group I forward signals represent the digits of the called party number. Group II signals indicate the category of the calling party, and group III signals, which are used in some national networks only, represent the digits of the calling party's number.

In R2 and CCITT-R2 signaling, the *incoming exchange* controls the signaling sequence. A group A signal requests a particular next forward signal, or indicates that register signaling has ended. Group A signals can be sent by intermediate exchanges, and by the terminating local exchange.

The group B backward signals are sent by the terminating local exchange only. They acknowledge a forward signal, and convey call-charging instructions, and called-party status.

The incoming and outgoing exchanges must know the type of the signal that is being received. To accomplish this, a R2 register signaling sequence follows certain rules:

- The first signal received by an incoming exchange is a group I signal.
- The outgoing exchange interprets received backward signals as group A signals, until it receives a group A signal that indicates that the next backward signal will be a group B signal. The receipt of a group B signal always ends the signal sequence.

4.4.4 National R2 Interregister Signaling Sequences

We now explore a few register signaling sequences in a national network. In these examples, the meanings of the forward and backward register signals, which vary somewhat from country to country, are as listed in Tables 4.4-2 and 4.4-3.

Figure 4.4-4 shows the signaling on trunk T, for a call from subscriber S_1 to S_2, whose subscriber number is 34-5678. Exchange X has received the called number from S_1. It seizes trunk T, and sends a seizure signal. The exchange also connects an outgoing register (OR) to the trunk, and orders it to send I-3 (the first digit of the called number). When exchange Z receives the seizure signal, it connects an incoming register (IR) to the trunk. The register receives the I-3, and acknowledges with A-1 (send next digit). The acknowledgment of the first digit indicates to exchange X that an incoming register has been connected to the trunk (this is why R2 signaling does not include a proceed-to-send signal). Exchange X sends the subsequent digits, and exchange Z acknowledges the second through fifth digits with A-1. On receipt of the sixth digit (I-8), exchange Z knows that the called number is complete, and acknowledges with A-3, which requests the calling party category, and indicates that the next backward signal is a group B signal. After receiving calling category (II-2), exchange Z sends a group B signal that contains information on the called party status and on charging. The group B signal ends the interregister signaling.

Table 4.4-2 Example of national R2 forward interregister signals.

Group I: Digits in the called number

I-1 digit 1
I-2 digit 2
I-3 digit 3
I-4 digit 4
I-5 digit 5
I-6 digit 6
I-7 digit 7
I-8 digit 8
I-9 digit 9
I-10 digit 0
I-15 end of called number
I-11 through I-14 are not used

Group II: Calling party category

II-1 Operator with trunk-offering
II-2 Subscriber
II-3 Pay-phone
II-4 through II-15 are not used

Group III: Digits in the calling number

III-1 digit 1
III-2 digit 2
III-3 digit 3
III-4 digit 4
III-5 digit 5
III-6 digit 6
III-7 digit 7
III-8 digit 8
III-9 digit 9
III-10 digit 0
III-15 end of calling number
III-11 through III-14 are not used

End-to-end Interregister Signaling. Now consider the signaling for a call from S_1 to S_2 that is routed via intermediate exchange Y—see Fig. 4.4-5. The subscriber number of S_2 is again 34-5678. In some countries, R2 register signaling is link-by-link. This means that exchange Y receives the entire called number from exchange X, then seizes trunk T_2, and sends the number to exchange Z.

However, in most national networks, R2 register signaling is end-to-end. In this mode, outgoing register (OR) in originating exchange X communicates successively with incoming registers (IR) in exchanges Y and Z. The initial register signaling is between exchanges X and Y. Having received the called

Table 4.4-3 Example of R2 national backward interregister signals.

Group A	
A-1	Send next digit of called number.
A-2	Resend first digit of called number.
A-3	Send calling line category, and prepare to receive a group B signal
A-4	Congestion
A-5	Send calling line category
A-7	Send next digit of calling number
A-6	Send next digit of calling number
A-8	Digit *n* of called number received; resend digit (*n*-1)
A-9	Digit *n* of called number received; resend digit (*n*-2)
A-10	through A-15 are not used
Group B	
B-1	Called subscriber idle, charge
B-2	Called subscriber busy
B-5	Called subscriber idle, do not charge
B-6	Called party idle, call to be held under control of called subscriber
B-7	Vacant number received
B-3, B-4, and B-8 through B-15 are not used	

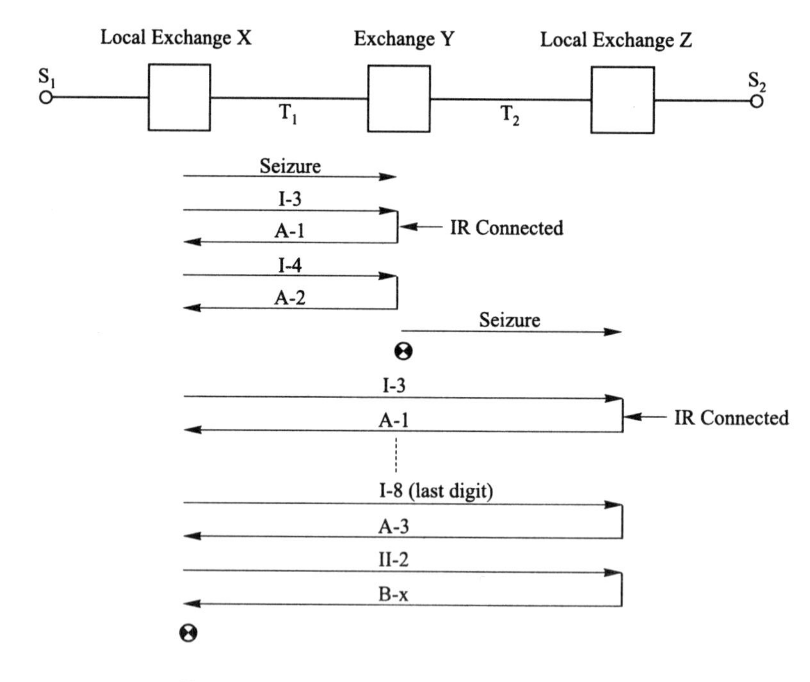

Figure 4.4-4 R2 interregister signaling example.

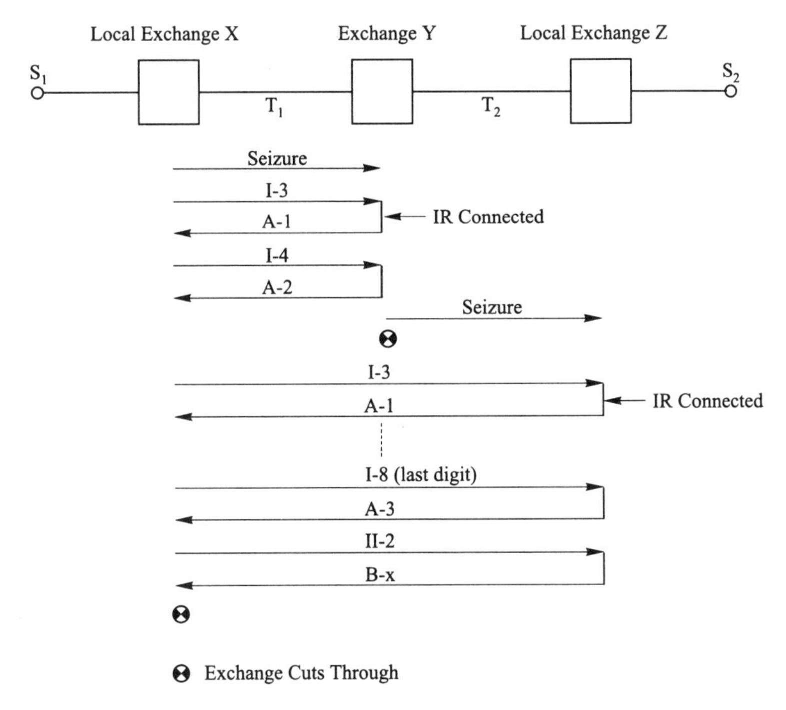

Figure 4.4-5 End-to-end interregister signaling.

number 34-5678 from S_1, exchange X seizes trunk T_1, and starts sending the called number. After receiving the exchange code EC = 34, exchange Y knows that Z is the terminating exchange for the call. It therefore acknowledges the I-4 with A-2, which is a request to restart sending the called number.

Exchange Y then disconnects its IR from T_1, seizes trunk T_2, sends a seizure signal on T_2, and cuts through a path between the trunks. From this point on, the register signaling is between exchanges X and Z. Exchange X responds to the A-2 by sending I-3 (the first digit of the called number). Exchange Z connects an IR to T_2 and, when the register receives the I-3, it requests the next digit by acknowledging with A-1. The second through fifth digits are acknowledged in the same way, and the sixth (final) digit is acknowledged with A-3. Exchange X then sends the calling party category (II-2), and exchange Z ends register signaling with a group B signal.

R2 end-to-end signaling in national networks is also possible on connections that pass through several intermediate exchanges.

4.4.5 R2 Supported Features in National Networks

With end-to-end R2 interregister signaling, information about the status and nature of the called party can be sent from the terminating to the originating local exchange. In addition, the calling party category (subscriber, operator, etc.) is sent from the originating- to the terminating exchange. The combination of these procedures supports a number of national network features that cannot be provided by Bell System MF signaling. Some of these features are outlined below, using the call of Fig. 4.4-5 as an example.

Free Calls. Certain lines on a local exchange may be marked as "free" destinations. During the register signaling on a call to a free destination at exchange Z, this exchange can indicate to originating exchange X that the call should not be charged (signal B-5).

Called Party Hold. Specific lines on a local exchange can be marked as having "called party holding." On calls to such lines, terminating exchange Z ends the signaling with B-6. This requests originating exchange X to hold the connection until it receives a clear-back (line signal). Lines of police and fire departments are often marked as lines with called party holding.

Malicious Call Tracing. Lines on a local exchange can be marked as "subjected to malicious calls". On calls to such lines, terminating exchange Z requests the digits of the calling party number from exchange X, with a series of A-7 signals, and stores the number. If the call is malicious, the called party alerts the local exchange, usually by a hookswitch flash—see 3.3.2. In response, exchange Z prints a record that includes the date and time-of-day of the call, and the numbers of the calling and called parties.

Trunk Offering. This allows an operator to break in on a busy line. Suppose that a subscriber, say S_1, calls a subscriber S_2 on exchange Z, and that this subscriber is busy. In an emergency situation, S_1 can call a "trunk offering" operator for assistance. The operator then places a call to S_2. Say that this call arrives at exchange Z on a trunk T. Exchange Z receives the calling party category signal II-1, and recognizes that the calling party is a "trunk-offering" operator. It then bridges trunk T onto the existing call of subscriber S_2. The operator then informs the subscriber about the emergency call.

Release of Connection When Set-up Fails. When intermediate exchange Y or terminating exchange Z cannot extend the call set-up, it sends a backward register signal that indicates the nature of the problem. For example, intermediate exchange Y can indicate that no trunk to Z is available (A-4), and exchange Z can indicate that the called line is busy (B-2), or that the called number is not in use (B-7), etc. In response, originating exchange X initiates the release of the connection, and connects the calling party to a suitable tone or announcement source. This expedites the release of trunks T_1 and T_2. The "busy-flash" in CCITT no.5 signaling has the same purpose, but does not include information on the failure cause (4.3.2).

4.4.6 CCITT-R2 International Signaling

CCITT-R2 signaling is the international version of R2 signaling, specified by CCITT [4] for use in the international network on one-way four-wire analog (FDM) trunks, and on one-way and bothway digital (PCM) trunks [4].

CCITT-R2 supervision signaling is as described in Sections 4.4.1 and 4.4.2. The supervision signaling on analog trunks is continuous "tone-on-idle." This makes the system unsuitable for TASI-equipped trunks (1.6.4) because, when a trunk is idle, the tone-on condition would cause the occupation of a pair of bearer channels.

CCITT-R2 is also used very rarely on satellite trunks, because each compelled MF interregister signal would involve two one-way propagation delays of about 600 ms each, which would result in very slow register signaling.

4.4.7 CCITT-R2 Interregister Signaling

The signaling is an adaptation of R2 national signaling to the requirements of the international network, and includes a number of specific international signals, similar to those of CCITT No.5.

The forward register signals are divided into three groups—see Table 4.4-4.

Group 0 Signals. The first signal received by an incoming exchange is a group 0 signal. One of the functions of these signals is similar to the KP1 and KP2 signals in CCITT No.5: they differentiate calls that terminate in-country from transit calls. The first signal in a *terminal* seizure is 0-1 through 0-5, or 0-10 (which represent Z-digit values), or 0-13 (indicating a call from test equipment at the outgoing ISC).

Signals 0-11, 0-12 and 0-14 indicate a *transit* seizure. These signals imply that the called number includes a country code, and are also known as *country code indicators*. In addition, the signals contain information for the control of echo suppressors.

Group I Signals. The forward signals I-1 through I-10 represent the digits of the called number. I-11 and I-12 are addresses for groups of international operators (as in CCITT-No.5), and I-13 is the first digit of addresses that specify a particular type of test equipment at the ISC.

Group II Signals. These forward signals indicate the calling party category, and are sent to exchanges in the destination country when end-to-end signaling is possible. In practice, only II-7 is used.

The backward signals are listed in Table 4.4-5.

Group A Signals. Signal A-1 requests the next address digit. Signals A-2, A-7, A-8, A-11, and A-12 are sent, just before cut-through, by a transit ISC that has selected a CCITT-R2 outgoing trunk. They indicate the next signal to be sent by the originating ISC.

Table 4.4-4 CCITT-R2 forward interregister signals.

Group 0	
0-1	French-speaking operator
0-2	English-speaking operator
0-3	German-speaking operator
0-4	Russian-speaking operator
0-5	Spanish-speaking operator
0-10	Subscriber
0-11	Country code indicator; outgoing half-echo suppressor required
0-12	Country code indicator; no echo suppressor required
0-13	Call by automatic test equipment
0-14	Country code indicator; outgoing half-echo suppressor included
0-6 through 0-9, and 0-15 are not used	
Group I: Digits in called number	
I-1	Digit 1
I-2	Digit 2
I-3	Digit 3
I-4	Digit 4
I-5	Digit 5
I-6	Digit 6
I-7	Digit 7
I-8	Digit 8
I-9	Digit 9
I-10	Digit 0
I-11	Code 11
I-12	Code 12
I-13	Address code for test equipment at incoming exchange
I-14	Incoming half-echo suppressor required
I-15	End of called number
Group II: Calling party category	
II-7	Subscriber, or operator without forward transfer
II-8	Data transmission
II-9	Subscriber with priority
II-1	through II-6, and II-10 through II-15 are not used

Source: Rec. Q. 441. Courtesy of ITU-T.

A-3 and A-6 indicate that the complete called number been received. A-6 ends the register signaling, and A-3 requests the calling party category, which will be acknowledged by a group B signal.

A-4 and A-15 indicate that the call cannot be set up, and are requests to the outgoing ISC to release the connection.

Group B Signals. These signals convey information on the nature and status of the called subscriber. These signals can be sent only by an ISC that receives this information from a terminating local exchange in its national network.

Table 4.4-5 CCITT-R2 backward interregister signals.

Group A	
A-1	Send next digit (n+1)
A-2	Resend digit (n-1)
A-3	Address complete, prepare to receive a group B signal
A-4	Congestion in national network
A-5	Send calling party category
A-6	Address complete, end of register signaling
A-7	Resend digit (n-2)
A-8	Resend digit (n-3)
A-11	Send Country Code Indicator (transit seizure)
A-12	Send Z-digit (terminal seizure)
A-13	Send nature of circuit
A-14	Send echo suppressor information
A-15	Congestion at International Switching Center, or all trunks busy
A-9	and A-10 are not used
Group B	
B-2	Send Special Information Tone
B-3	Subscriber line busy
B-4	Congestion
B-5	Unallocated number
B-6	Subscriber idle, charge
B-7	Subscriber line idle, do not charge
B-8	Subscriber line out of service
B-9	through B-15 are not used

Source: Rec. Q. 441. Courtesy of ITU-T.

4.4.8 CCITT-R2 Interregister Signaling Example

We consider the international call shown in Fig. 4.4-6. The call is routed via transit country B. Trunks T_1 and T_2 have CCITT-R2 signaling. Even if the national networks in countries A and C use R2 signaling, there are three signaling sections in the connection. This is necessary because local exchanges

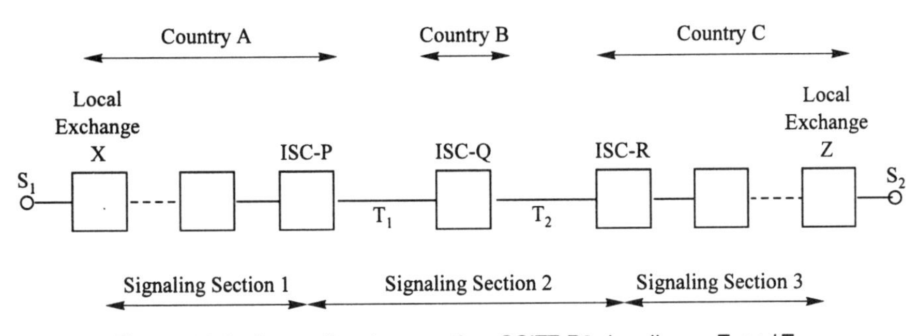

Figure 4.4-6 International connection. CCITT-R2 signaling on T_1 and T_2.

(for example, X) are not equipped to handle international signaling procedures, and because some national R2 register signals are not identical to the CCITT-R2 register signals. If the networks of countries A and C use R2 signaling, end-to-end signaling is possible in signaling section 1 (between local exchange X and ISC-P), in section 2 (between ISC-P and ISC-R), and in section 3 (between ISC-R and the terminating local exchange Z).

The signaling in the international network is shown in Fig. 4.4-7 (see Tables 4.4-4 and 4.4-5). The called international number is 34-67-412-1093, where 34 is the country code of C, and 67-412-1093 is the national number of S_2.

Outgoing ISC-P seizes trunk T_1 to ISC-Q, and starts by sending country code indicator O-12, indicating a "transit" seizure (the call destination is not in country B). ISC-Q acknowledges with A-1, requesting the first digit of the called international number. We assume that ISC-R is the only ISC in country C. Therefore, ISC-Q can make a route decision after receipt of the country code

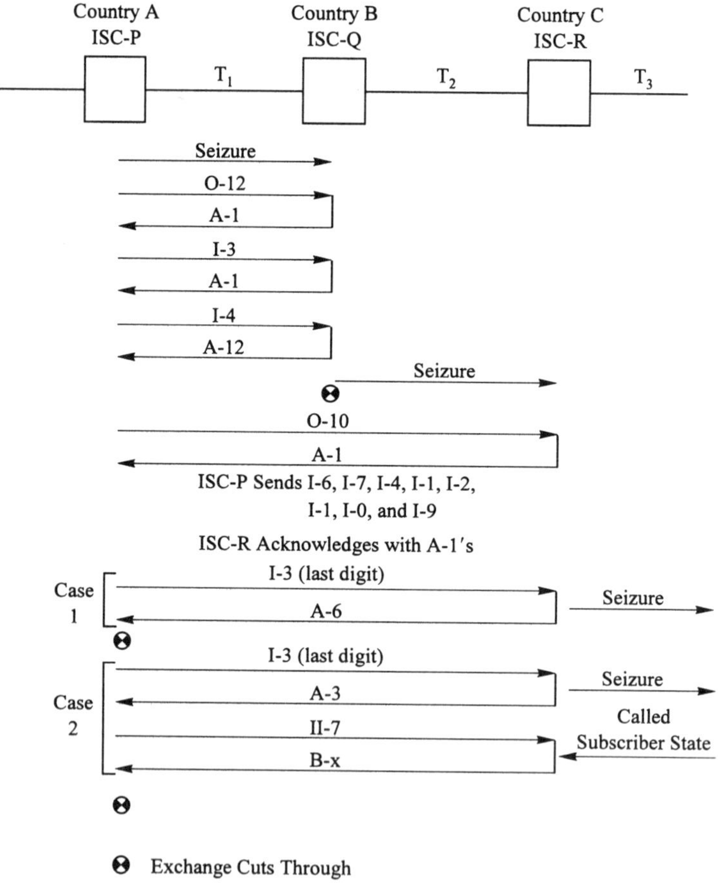

Figure 4.4-7 Signaling in the international section of the connection of Fig. 4.4-6.

CC (34). In this example, it seizes a direct trunk T_2 to ISC-R. Since this ISC is in the destination country, it should receive a terminal seizure signal, followed by the called national number. ISC-Q therefore acknowledges the final digit of the country code with A-12, and then cuts through. ISC-P starts its register signaling with ISC-R by sending O-10 (terminal seizure, call originated by a subscriber) and, prompted by successive A-1 signals, sends the national number. ISC-R knows the lengths of national numbers in its country, and acknowledges all received digits except the last one with A-1.

The acknowledgment of the final digit of the called number depends on whether ISC-R can obtain information about the called subscriber from its national network.

Case 1. The national network of the destination country cannot signal the status of the called party to its incoming ISC-R. This was the case in the U.S. prior to the introduction of common-channel signaling. After receiving the last digit (I-3) of the called number, ISC-R seizes national trunk T_3, and acknowledges the I-3 with A-6. This ends the register signaling in the international network. ISC-P cuts through, and the calling subscriber eventually receives ringing-tone, busy tone, or an announcement from an exchange in country C.

Case 2. ISC-R can obtain the condition of the called subscriber (for example, when country C uses national R2 with end-to-end register signaling). In this case, ISC-R seizes trunk T_3 after receipt of the last digit of the called number, and acknowledges with A-3. This requests the calling party category, and informs ISC-P that the next acknowledgment it will receive is a group B signal.

ISC-P then sends II-7 (call originated by a subscriber). When the connection in the destination country has been set up, and ISC-R has received information on the status of the called subscriber, it sends the appropriate group B signal. This ends the register signaling between ISC-P and ISC-R.

4.4.9 Pulsed Group A Signals

Up to this point, a backward group A signal always acknowledges a received forward signal. National R2 signaling and CCITT-R2 also use pulsed group A signals. The pulsed signals (pulse duration: 100–200 ms) do not acknowledge a received forward signal, and are sent when the incoming exchange has information for the outgoing exchange at a time that the outgoing exchange is not sending a forward signal.

The use of pulsed signals is illustrated with the international call of Fig. 4.4-6. We assume that originating country A has R2 signaling, and consider the signaling between originating exchange X and ISC-P. The forward and backward interregister signals of Tables 4.4-2 and 4.4-3 are used.

Originating local exchanges cannot determine whether an international number received from the calling party is complete. Moreover, early ISCs were

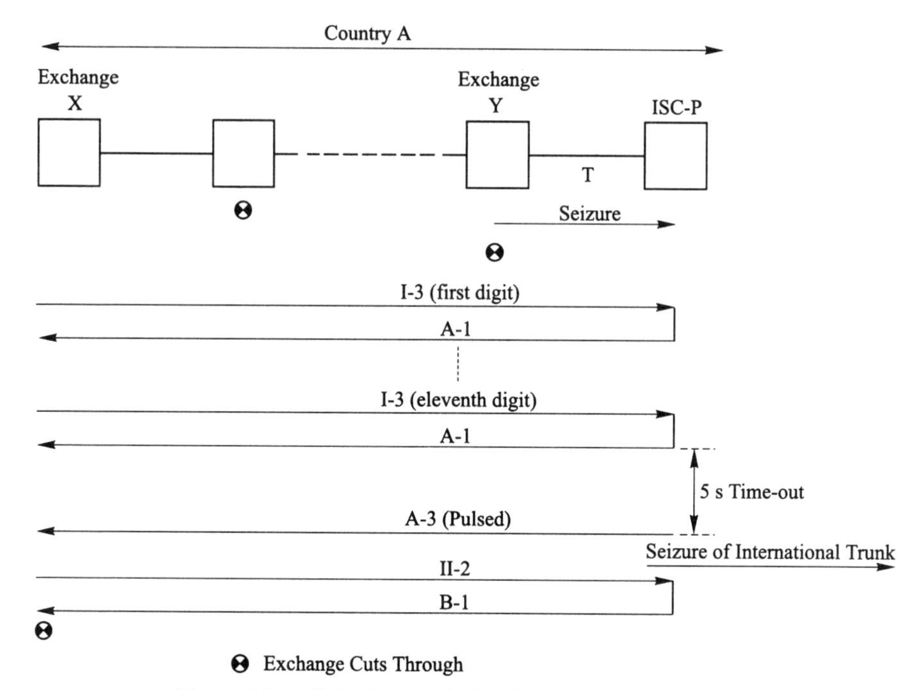

Figure 4.4-8 Pulsed group A signal in section 1 of Fig. 4.4-6.

not equipped with stored information on the lengths of international numbers. We assume that ISC-P is such an exchange. Figure 4.4-8 shows the end-to-end interregister signaling between X and ISC-P, from the time that exchange Y has seized trunk T. Exchange X then starts to send the digits of the international called number (34-67-412-1093) received from S_1. Since ISC-P cannot determine when the received number is complete, it acknowledges each group I signal with an A-1, and also starts (or restarts) a 5-s timer on the receipt of each signal. When X receives the A-1 acknowledgment of the eleventh digit (I-3), it falls silent, because it has sent all received digits. The timer at ISC-P then times out, and ISC-P assumes that the called number is complete.

It now sends a pulsed A-3 signal to exchange X, requesting the calling party category. It also seizes an outgoing trunk in the international network and, since the A-3 also indicates that the next acknowledgment will be a group B signal, acknowledges the II-2 with a B-1. This ends the register signaling between X and ISC-P, and originating exchange X cuts through.

The B-1 (called subscriber free, charge) is sent by convention: ISC-P does not know whether it will receive information about the call set-up, or the status of the called subscriber. If ISC-P ends its register signaling in the international network without obtaining this information, or receiving an indication that the called subscriber is free, it cuts through. The calling subscriber then receives an audible signal from an intermediate exchange, or from terminating local exchange Z. If ISC-P receives an indication during its international register signaling that the call cannot be set up (called subscriber busy, no trunks available, etc.), it connects its incoming trunk T to a tone or announcement source.

4.5 ACRONYMS

AC	Alternating current
AC	Area code
ANI	Automatic number identification
BEF	Band elimination filter
BPF	Band pass filter
CAS	Channel-associated signaling
CC	Control channel
CCITT	International Telephone and Telegraph Consulative Committee
CEPT	European Conference of Postal and Telephone Administrations
COMP	Comparator
DC	Direct current
DDD	Direct distance dialing
DMP	Digital multiplex port
DS1	Frame format of American first-order digital multiplex
EC	Exchange code
FDM	Frequency division multiplex
IC	Interexchange carrier
IR	Incoming register
ISC	International switching center
KP	First signal in MF register signaling sequence
LATA	Local access and transport area
LN	Line number
LPF	Low-pass filter
MF	Multi-frequency
MFC	Multi-frequency code signaling
NN	National number
OR	Outgoing register
PCM	Pulse code modulation
POTS	Plain ordinary telephone service
R	Receive channel
RCV	Digit receiver
S	Send channel
SPC	Stored program controlled
SN	Subscriber number
SND	Digit sender
ST	Final signal in MF register signaling sequence
STS	Signaling tone source

TAC	Tone and announcement circuit
TASI	Time assignment speech interpolation
TC4	Four-wire analog trunk circuit
TD	Signaling tone detector
TDM	Time-division multiplex
TS	Time slot
X	Switch
Y	Switch

4.6 REFERENCES

1. S. Welch, *Signalling in Telecommunications Networks*, Peter Peregrinus Ltd, Stevenage, U.K., 1981
2. C. Breen and C.A. Dahlbom, "Signalling Systems for Control of Telephone Switching," *Bell Syst. Tech. J.*, **39**, 1960.
3. *A History of Engineering and Science in the Bell System*, Bell Telephone Laboratories, Inc., 1982.
4. *Specifications of Signalling Systems R1 and R2*, Rec. Q.310–Q.331 (R1 signaling) and Q.400–Q.480 (R2 signaling), CCITT Red Book, **VI.4**, ITU, Geneva, 1985.
5. *Notes on the BOC Intra-Lata Networks*, AT&T, New York, 1983.
6. *Specifications of Signalling Systems No.4 and No.5*, Rec. Q.140–Q.164, CCITT Red Book, **VI.2**, ITU, Geneva, 1985.
7. *General Specifications on Telephone Switching and Signalling*, Rec. Q.101–Q.107, CCITT Red Book, **VI.1**, ITU, Geneva, 1985.
8. M. den Hertog, "Interregister Multifrequency Code Signalling for Telephone Switching in Europe," *Elec. Comm.*, **38**, 1963.

5

INTRODUCTION TO COMMON-CHANNEL SIGNALING

In *channel-associated signaling (CAS)* systems, the signaling information for a trunk is carried by the trunk itself. In *common-channel signaling* (CCS), a common *signaling link* (SL) carries *signaling messages* for a number of trunks. Just as multi-frequency (MF) signaling became feasible with the introduction of the second-generation (common-control) switching systems, CCS was developed for the third-generation (stored program controlled, SPC) exchanges that were introduced in the 1960s.

There are several reasons for the move from multi-frequency signaling to CCS [1]:

1. It is often less costly to interface the processing equipment of SPC exchanges with a relatively small number of signaling links than to provide pools of MF registers and line-signaling hardware for the individual trunks.
2. Common-channel signaling is much faster than multi-frequency signaling. Early CCS systems already reduced post-dialing delays on long-distance calls from 10–15 s to around 3 s.
3. New telecommunication technology and services require the transfer of additional signaling information for processing a call. In Chapter 4, we have already encountered some additional information items (echo suppressor information, calling line category, etc.). In CCITT No.5 and CCITT-R2 signaling, this information is in the form of digits with special meanings (see Section 4.3.3 and Tables 4.4-4 and 4.4-5). Common-channel signaling messages provide a more flexible way to transfer both

the classical supervision and address signals, and other types of call-control information.

4. Subscribers cannot access the CCS signaling links. This avoids the "blue-box" fraud problems that have plagued many *frequency-division multiplexed* (FDM) trunk groups that use channel-associated signaling with in-band signal frequencies (Section 4.4.6).
5. In channel-associated signaling, the signals on a trunk necessarily relate to that trunk, and are used for call control. In CCS the messages can—but do not have to be—related to individual trunks. Call control on trunks was the original application of CCS signaling, and still is the predominant one. However, CCS signaling links have now become a common transport facility for call control and other applications (see Sections 2.2.3–2.2.5) .

The first-generation common-channel signaling system, known as *signaling system No.6* (SS6), was introduced in the 1970s. It exists in two versions. *Common-channel interoffice signaling* (CCIS), defined by the Bell System, has been deployed in the U.S. network, but is now being replaced. An international version (CCITT No.6) is still used in the international network.

SS6 was followed, about 10 years later, by *signaling system No.7* (SS7). This system also exists in several versions. The version specified by CCITT is in service on a number of trunk groups in the international network and, with country-specific modifications, in a number of national networks. A version defined by the American National Standards Institute and Bellcore is in operation in the U.S.

This chapter introduces a number of basic CCS concepts, setting the stage for more detailed discussions on individual CCS systems in later chapters.

5.1 SIGNALING NETWORKS

Telecommunication networks that employ CCS signaling require, in addition to the network of trunks and exchanges, a *signaling network*. This network consists of *signaling points* (SP), interconnected by *signaling links* (SL). We start with a few definitions.

Signaling Point (SP). A signaling point is an entity in the network to which CCS signaling links are attached. For example, an exchange that serves CCS trunk groups has CCS signaling links, and is therefore a signaling point. Likewise, a network database that is accessed via CCS signaling links is a signaling point.

Signaling Link (SL). A signaling link is a bidirectional transport facility for CCS signaling messages between two signaling points.

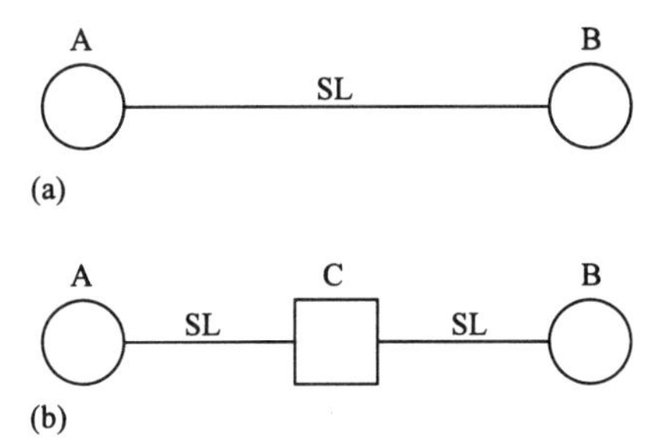

Figure 5.1-1 Signaling routes for relation (A, B). (a): associated signaling. (b): quasi-associated signaling.

Signaling Relation. A signaling relation exists between any pair of signaling points that need to communicate by CCS signaling. For example, when two exchanges, say A and B, are interconnected by a group of CCS trunks, there is a signaling relation between these signaling points. In what follows we denote a signaling relation between points A and B by (A,B).

Signaling Route. A signaling route is a predetermined path for the CCS messages of a particular relation. Usually, there is a *signaling route set*, consisting of several routes, for each signaling relation.

Associated and Quasi-associated Signaling. When messages for relation (A,B) are carried on a signaling route that consists of a direct signaling link (SL) between A and B, we speak of *associated* signaling—Fig. 5.1-1(a). When A and B have a signaling relation, but are not directly interconnected by a SL, a signaling route for relation (A,B) consists of two or more SLs in tandem—Fig. 5.1-1(b). This signaling mode is called *quasi-associated signaling*.

Types of Signaling Points. In Fig. 5.1-1(a) and 5.1-1(b), signaling points A and B, which originate and receive (and process) signaling messages for relation (A,B), are known as the *signaling end points* (SEP) for that relation. In Fig. 5.1-1(b), signaling point C transfers messages for relation (A,B), but does not originate or process messages for that relation. We say that signaling point C is a *signal transfer point* (STP) for signaling relation (A,B).

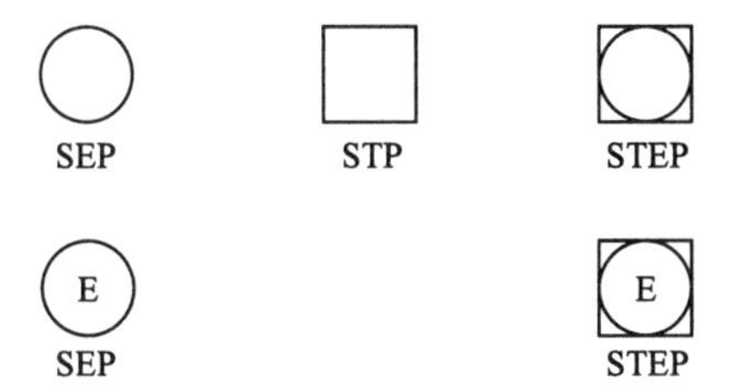

Figure 5.1-2 Signaling points. SEP: signaling end point. STP: signal transfer point. STEP: combined end/transfer point. E: signaling point and exchange.

Some signaling networks include signaling points that function as end points for some relations, and as transfer points for other relations. These dual purpose signaling points are called *signal transfer and end points* (STEP).

In documents on signaling networks, the various types of signaling points are usually shown as in Fig. 5.1-2. The letter "E" is used in this section to indicate signaling points (SEP or STEP) that are exchanges with CCS trunks.

5.1.1 Basic Signaling Networks

As a start, we explore some alternative signaling networks to handle CCS signaling between exchanges A, B, and C that are interconnected by CCS trunk groups TG_1, TG_2, and TG_3—see Fig. 5.1-3. The signaling network therefore requires routes for relations (A,B), (B,C), and (C,A).

Associated Network. Figure 5.1-3(a) shows a network in which each route consists of one signaling link that is *associated* with one signaling relation. For example, route SL_1 is associated with signaling relation (A,B), and transports the signaling messages for the trunks in trunk group TG_1 only. All exchanges in the figure are signaling end points.

Signaling links in associated operation are often poorly used, because a link

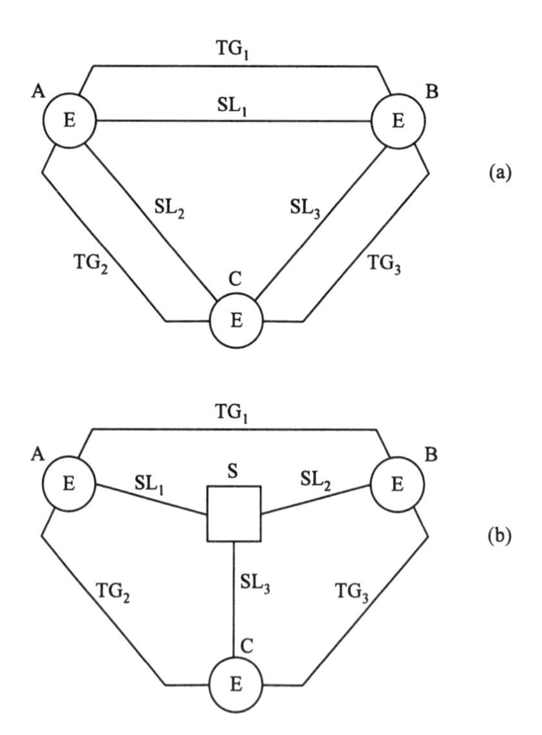

Figure 5.1-3 Signaling networks (a) for associated operation and (b) for quasi-associated operation.

can handle the signaling messages for several thousand trunks, while most trunk groups consist of fewer than 100 trunks.

Quasi-associated Network. Figure 5.1-3(b) shows a configuration for quasi-associated operation of the signaling links. None of the exchanges are directly connected by a signaling link. Instead, each exchange has a link to signal transfer point S. Each link carries messages for several relations. All signaling routes are indirect, traversing two links in tandem, and passing through signal transfer point S. All exchanges are again signaling end points.

The structures of associated and quasi-associated signaling networks resemble respectively the mesh- and star-configurations of trunk groups in telecommunications networks.

5.1.2 Signaling Reliability and Load Sharing

In the networks of Fig. 5.1-3 there is one signaling route for each relation. A failure of a signaling link disables the signaling route(s) for which it carries CCS messages, and this severely affects the service in a telecommunications network. For example, a failure of SL_1 in Fig. 5.1-3(a) stops all signaling for relation (A,B), and thus shuts down all trunks in trunk group TG_1. Also, on failure of SL_1 in Fig. 5.1-3(b), the trunks in groups TG_1 and TG_2 are disabled, and signaling point A becomes isolated.

Actual signaling networks are therefore designed with redundancy, such that signaling for all relations remains possible when a link failure occurs.

Redundancy can be obtained in several ways. For example, the signaling routes in the associated signaling network of Fig. 5.1-3(a) can be replaced by route sets containing two direct routes each (see Fig. 5.1-4). Then, if say SL_1 fails, signaling for relation (A,B) is still possible, using SL_4.

In the quasi-associated configuration of Fig. 5.1-3(b), redundancy can be obtained by a adding a second signal transfer point with links to each signaling end point. This creates a network in which the route sets for each signaling relation consist of two quasi-associated routes (Fig. 5.1-5). Under normal

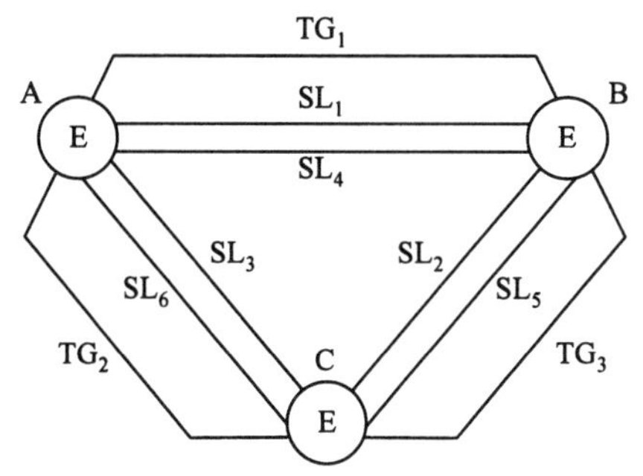

Figure 5.1-4 Signaling network with two routes for each signaling relation (associated signaling).

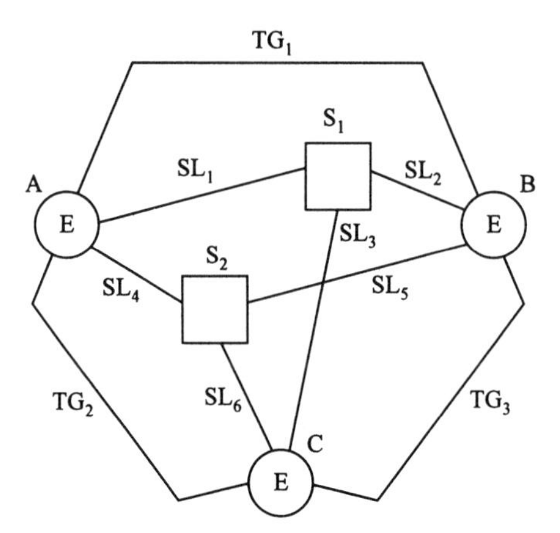

Figure 5.1-5 Signaling network with two routes for each signaling relation (quasi-associated signaling).

conditions, the signaling traffic for a trunk group is divided across the signaling routes in a route set. For example, signaling for the odd numbered trunks in TG_1 is on route SL_1–SL_2, and signaling for the even numbered trunks is on route SL_4–SL_5. When SL_1 fails, route SL_1–SL_2 is disabled, and all signaling traffic for TG_1 is carried by the other route.

A third alternative is shown in Fig. 5.1-6. This arrangement differs from Fig. 5.1-3(a) in that exchanges A, B and C are now *signaling transfer and end points* (STEP). Under normal conditions, all signaling is associated. However, when for example SL_1 fails, the messages related to trunks of group TG_1 are sent via C, which then acts as the STP for the signaling traffic of relation (A,B).

We conclude this section by exploring two well known signaling networks.

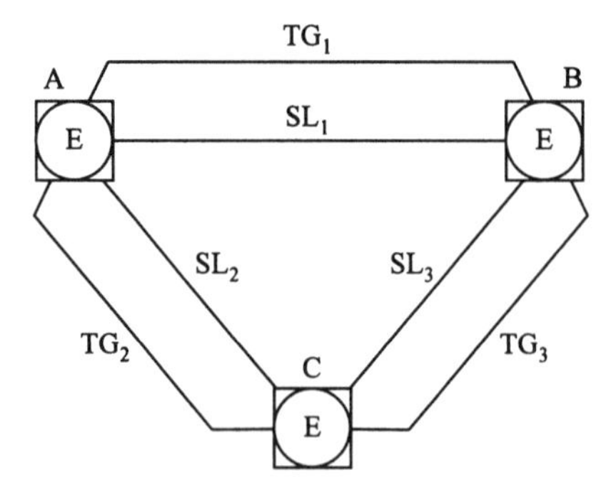

Figure 5.1-6 Signaling network with combined signal transfer and end points (STEP).

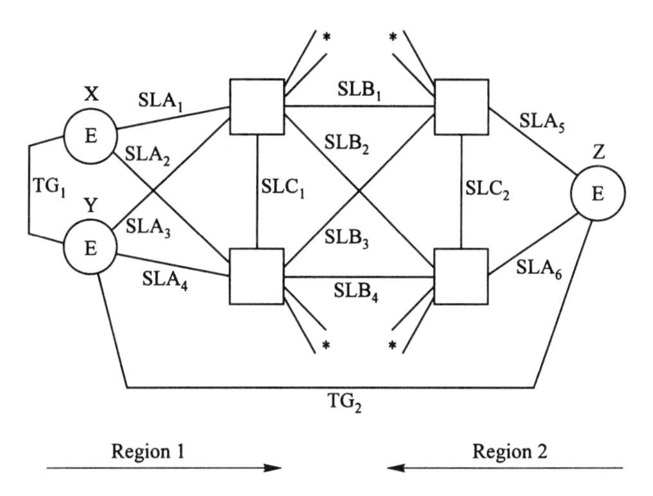

Figure 5.1-7 Bell System signaling network. * Signaling links to other regions. (From *IEEE Comm. Mag.* **28.7**. Copyright © 1990 IEEE.)

5.1.3 The Bell System Signaling Network

Figure 5.1-7 shows part of the quasi-associated signaling network deployed by the Bell System for CCIS signaling [2, 3]. The basic structure has been retained for SS7 signaling in AT&T's present long-distance network [4].

The territory of the U.S. is divided into a number of regions, and each region is equipped with a pair of STPs (only two regions are shown in Fig. 5.1-7). Each exchange with CCS trunks is a signaling end point, and has an *A-link* (SLA) to the two STPs in its region. The *B-links* (SLB) interconnect STPs of different regions, and the *C-links* (SLC) interconnect STP pairs of individual regions. The STPs in the network are thus interconnected by a complete mesh of signaling links. The C-links normally do not carry signaling traffic, and are used only under certain failure conditions.

In normal operation, the signaling network has two signaling routes R for relations between two SEPs located in the same region. For example, the route set for relation (X,Y) consists of:

$$R(X,Y)_1: \quad SLA_1\text{-}SLA_3$$
$$R(X,Y)_2: \quad SLA_2\text{-}SLA_4$$

Also, in normal operation, the signaling route set for a relation between SEPs located in different regions consists of four routes. For example, the route set for relation (Y,Z) consists of:

$$R(Y,Z)_1: \quad SLA_3\text{-}SLB_1\text{-}SLA_5$$
$$R(Y,Z)_2: \quad SLA_3\text{-}SLB_2\text{-}SLA_6$$
$$R(Y,Z)_3: \quad SLA_4\text{-}SLB_3\text{-}SLA_5$$
$$R(Y,Z)_4: \quad SLA_4\text{-}SLB_4\text{-}SLA_6$$

In this arrangement, the routes of a route set again share the message load for a signaling relation. For example, trunk group TG_2 can be divided (for signaling purposes only) into four subgroups. When all routes for relation (Y,Z) are operational, the message traffic for a particular subgroup is carried by one of the four routes. When a route is disabled, the signaling traffic for the affected subgroup is diverted to one of the remaining routes.

Simultaneous failures of signaling links in a route set can disable all normal signaling routes for a relation. In Fig. 5.1-7, simultaneous failures of SLA_1 and SLA_4 disable both signaling routes for (X,Y). In this case, signaling is maintained by using the route SLA_2-SLC_1-SLA_3, which does not belong to the normal route set for (X,Y).

5.1.4 Mixed Signaling Networks

Some networks use associated signaling for some relations, and quasi-associated signaling for the others relations. Figure 5.1-8 shows an example that is often used between international exchanges (ISC) in two countries. Exchanges X and Y in country A are connected to exchanges Z and U in country B by international CCS trunk groups TG_1,...,TG_4, and by two international signaling links (SL_1,SL_2). Each pair of in-country exchanges is also interconnected a by national signaling link (SL_3,SL_4). All exchanges are combined signal transfer and end points.

The normal (non-failure) signaling routes can be assigned as follows:

R(X,Z):	SL_1	(Associated signaling)
R(Y,U):	SL_2	(Associated signaling)
R(X,U):	SL_3-SL_2	(Non-associated signaling)
R(Y,Z):	SL_3-SL_1	(Non-associated signaling)

In this example, there is only one normal route for each signaling relation. However, signaling for all relations can be maintained on failure of a signaling

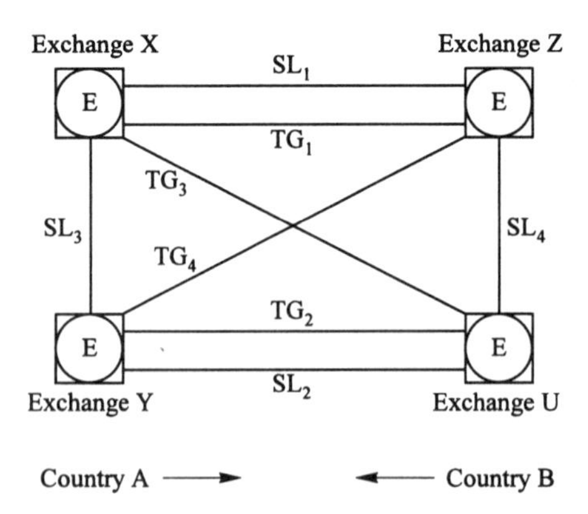

Figure 5.1-8 Mixed-mode signaling network.

link, by using routes that are not used for signaling relations under normal conditions. For example, when SL_1 fails, the message traffic for relation (X,Z) can be diverted to the route SL_3–SL_2–SL_4, and the traffic for relation (Y,Z) can be diverted to the route SL_2–SL_4.

5.2 SIGNALING LINKS AND SIGNAL UNITS

This section introduces some fundamental aspects of CCS signaling links and signal units. The signaling links are described using the hardware-oriented terms of the literature on SS6 [3,4,5]. The characteristics of the signaling links in SS7 are quite similar, but are described in a more abstract manner (see Chapter 8).

5.2.1 Signaling Link

Figure 5.2-1 shows a signaling link between signaling points A and B. It consists of two *signaling terminals* (ST), and a bidirectional *signaling data link* that transfers digital data. The primary function of the signaling link is to provide a reliable transfer of signaling messages between processors P_A and P_B.

In a signaling point, there are four interfaces between the processor and a signaling terminal. The processor enters its outgoing messages M_O into—and retrieves its incoming messages M_I from—the ST. In addition, the processor can send commands (COM) to the ST—for example, to activate or deactivate the link. A ST sends indications (IND) to alert the processor about certain conditions on the link (excessive errors in received messages, overload, etc.).

5.2.2 Signal Units

Information is transferred across the signaling data link in *signal units* (SU) (groups of consecutive bits). We distinguish two SU types—see Fig. 5.2-2. *Message* SUs (a) transfer processor messages (b). *Link* SUs (c) transfer information originated by the ST at one end of the signaling link, and intended for the ST at the other end.

A processor message is an ordered set of digitally coded *parameters*

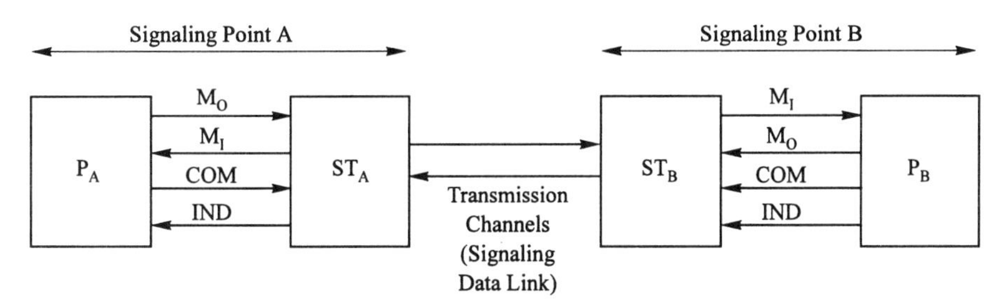

Figure 5.2-1 Signaling link.

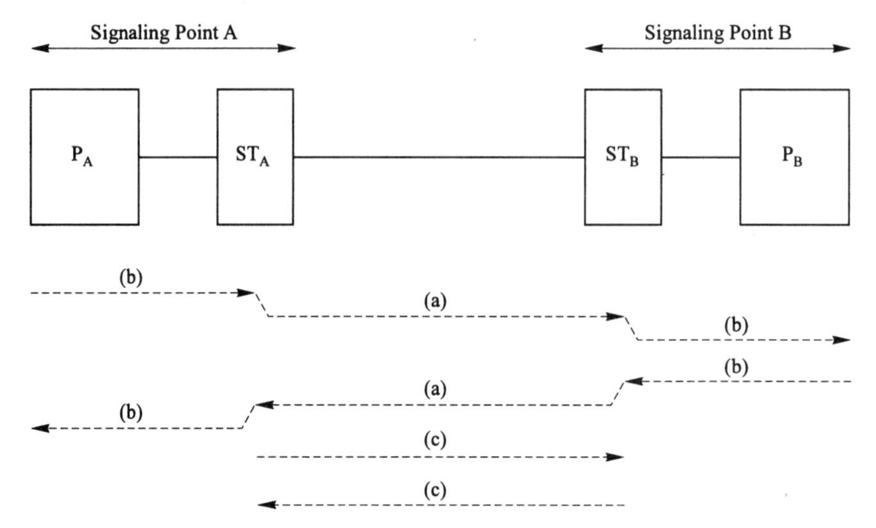

Figure 5.2-2 Signal units and processor messages. (a): message signal unit. (b): processor message. (c): link signal unit.

(information elements). The initial parameters identify the message type, and imply the meanings and locations of the later parameters in the message.

The length (number of bits) of SUs is fixed in SS6, and variable in SS7. The contents of SUs are shown in Fig. 5.2-3. All SUs have an information field (INF) and a check-bit field (CB). In SS6, the INF fields of message SUs holds message parameters only. Most messages fit in one message SU, but some messages require two or more SUs. In SS7, the INF field of a message SU holds all parameters of a message, and several link parameters.

In SS6 and SS7, the INF fields of link SUs contain link parameters only.

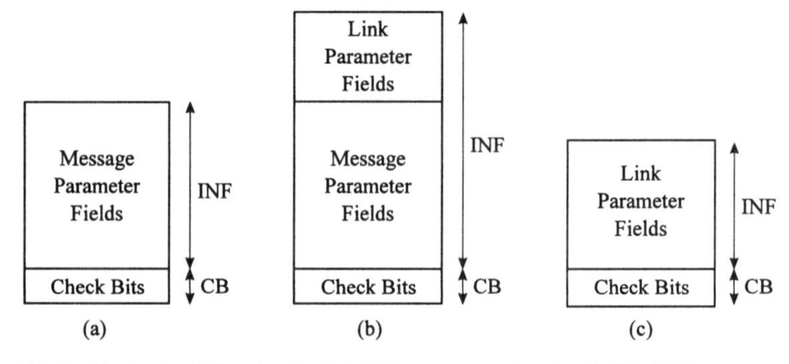

Figure 5.2-3 Contents of signal units. (a): SS6 message signal unit. (b): SS7 message signal unit. (c): SS6 and SS7 link signal unit.

5.2.3 Signaling Terminal Functions

The most important signaling terminal functions are synchronization, the transfer of SUs (including error control for message SUs), and monitoring of the signaling link.

Synchronization. A working signaling link conveys an uninterrupted bit stream of adjacent SUs in each direction. When no message SU needs to be transmitted, a ST sends certain types of link SUs (*filler* or *synchronization* SUs). The "receive" part of a ST has to be *synchronized* (*aligned*) with its incoming bit stream, so that it can determine the start points of the individual SUs. Synchronization needs to be acquired when a link is turned "on," and reacquired when a disturbance on the link causes a terminal to lose its alignment. In SS6, alignment and realignment procedures rely on a specific bit pattern in synchronization SUs. In SS7, adjacent SUs are separated by "flags" that have a bit pattern that does not occur in SUs.

Error Control. Electrical disturbances on a signaling link can introduce errors in signal units, changing one or more 1s into 0s, and vice versa. Errors in message SUs can cause more serious problems than errors in the signals of channel-associated systems. For example, errors in a particular call-control message for a particular trunk can change the message into a message for another trunk, or into a different message for the intended trunk, and thus cause the processor to take a wrong call-control action.

In addition, the signaling speed on data links is in the order of several thousand bits per second. A disturbance on a signaling link of just a few milliseconds can therefore severely mutilate a SU, while a similar disturbance on a trunk with channel-associated signaling (in which signal recognition times are in the order of 50–200 ms) is usually harmless.

Signaling terminals therefore execute error control procedures when sending and receiving SUs. The procedure for SUs sent by ST_A to ST_B is described below (the same holds for transmission in the other direction).

Error Detection makes use of the CB fields of SUs. The contents of CB are calculated by ST_A just before the SU is sent, and enable ST_B to determine whether a received SU is error-free.

Acceptance and Acknowledgment of Message SUs. ST_B determines whether a received SU is free of errors. Error-free SUs are accepted, and SUs with errors are discarded. Moreover, ST_B "positively" acknowledges error-free message SUs and "negatively" acknowledges message SUs received with errors. In SS6, ST_B sends acknowlegements in special link SUs (acknowledgment SUs). In SS7, ST_B sends acknowlegement information by including link parameters in message and link SUs it sends to ST_A. Received link SUs are not acknowledged.

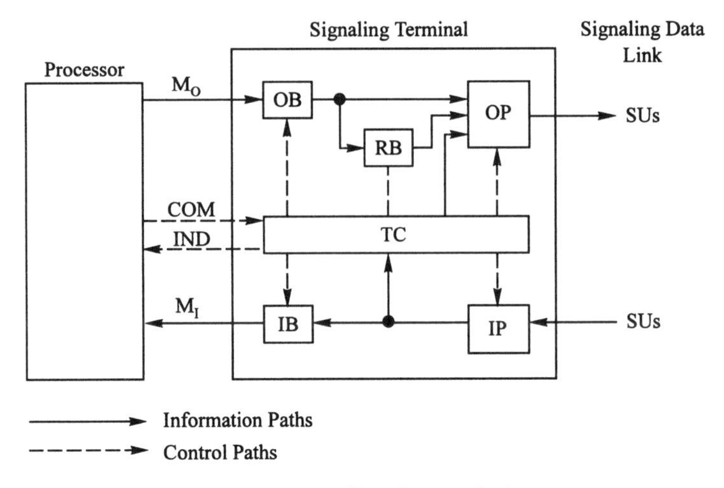

Figure 5.2-4 Signaling terminal.

Retransmission of Message SUs. When ST_A receives a negative acknowledgment of a sent message SU, it retransmits the SU.

Monitoring. The signaling terminals at both ends of a signaling link monitor the condition of the link by keeping track of alignment losses, SUs received with errors, etc. When a misalignment lasts too long, or when the fraction of SUs received with errors becomes too high, the ST alerts its processor with an indication (IND) (Fig. 5.2-1). The processor may then decide to take the signaling link out of service.

5.2.4 Signaling Terminal Elements and Operation

The major functional elements of an SS6 or SS7 signaling terminal are shown in Fig. 5.2-4. *Terminal control* (TC) controls the ST, originates the link parameters of outgoing SUs, and processes the link parameters of incoming SUs.

The processor at a signaling point places its outgoing messages M_O in *output buffer* (OB), and retrieves incoming messages M_I from *input buffer* (IB). *Retransmission buffer* (RB) stores messages that have been sent out, but have not yet been acknowledged positively.

Sending SUs. Information to be sent out can come from three sources: output buffer (OB), retransmission buffer (RB), or terminal control (TC) (link information).

After a SU has been sent, the selection of the next SU to be sent is made according to the following priority scheme:

Priority 1: A link SU, if TC needs to alert the distant TC about an event on the link.

Priority 2: Retransmission of a message in RB that is marked as "to be retransmitted".

Priority 3: Initial transmission of a message in OB.

Priority 4: A "filler" link SU (if nothing else is waiting to be sent).

Initial Transmission of a Message in OB. The message is removed from OB, and is entered into RB, and into *output processing* (OP). The SU is assembled in OP, which first forms the INF part of the SU (in SS7, this involves adding the link parameters to the message parameters). OP then calculates the contents of CB, and appends it to INF.

Retransmission of a Message in RB. A copy of the message is entered into OP, where the SU is formed as described above. The message itself remains in RB.

Receiving a Signal Unit. All received SUs are checked for errors by *input processing* (IP). SUs with errors are discarded. Error-free SUs are processed further. If the SU is a message SU, its message parameters, which constitute the processor message, are entered into buffer IB. In SS7, the link parameters of a message SU are passed to TC. In SS6 and SS7, the parameters in link SUs are passed to TC. Some link parameters acknowledge received message SUs. When TC receives a positive acknowledgment of a message in RB, the message is deleted. When TC receives a negative acknowledgment of a message in RB, the message is marked as "to be retransmitted."

5.2.5 Message Sequencing

We have seen that the retransmission of messages in buffer RB has priority over the initial transmission of messages in OB. We now examine the selection for transmission of a message in a buffer.

CCS messages belong to several signaling applications, such as call-control, management of trunks, etc. Each application has a priority class. Buffers OB and RB are operated "first-in, first out" for each application: the message selected for transmission is the "oldest" message with the highest priority class.

Let us focus two call-control messages, M_1 and M_2, in buffer OB of signaling terminal ST_A (Fig. 5.2-5). M_1 was entered in buffer OB by processor P_A before message M_2.

Since M_1 is "older," its initial transmission takes place before the initial transmission of M_2. Assuming that both messages are received error-free and accepted by ST_B, M_1 is entered in buffer IB of ST_B before M_2. Buffer IB is also operated "first-in, first out" for each application priority, and processor P_B retrieves and processes M_1 before M_2—Fig.5.2-5(a).

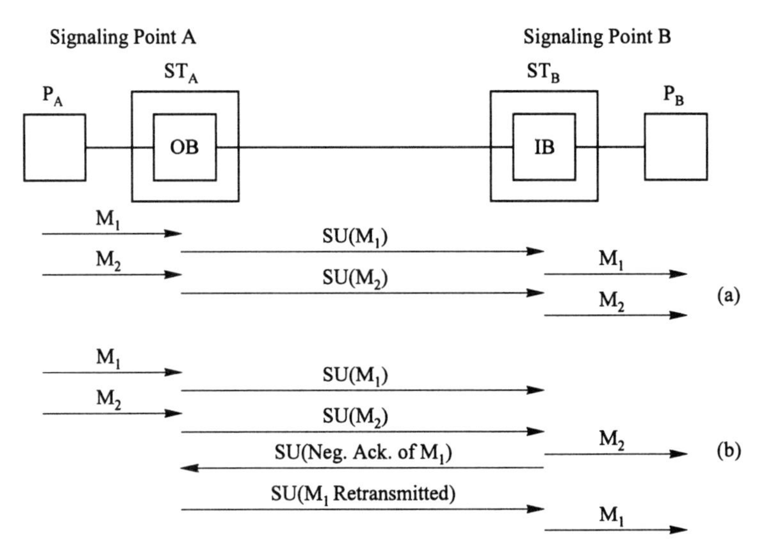

Figure 5.2-5 Transfer of call-control messages M_1 and M_2.

We speak here of "in-sequence" delivery of messages. In-sequence delivery is of importance when messages M_1 and M_2 pertain to the same call.

Now suppose that message M_1 is received with errors by ST_B, and therefore discarded—Fig.5.2-5(b). ST_B informs ST_A by sending a *negative* acknowledgment of M_1.

In SS6, ST_A retransmits message M_1 only, and processor P_B therefore receives M_1 *after* M_2—Fig. 5.2-5(a). When these messages pertain to same the call, this out-of-sequence (O-S) delivery causes a problem for P_B in the processing of the call.

In SS7, when signaling terminal ST_B receives message M_1 with errors, it discards the message—and all following messages, until it receives an error-free retransmission of M_1. ST_A, after receiving a negative acknowledgment of message M_1, retransmits all sent messages (starting with M_1) before sending out any new message. In this way, SS7 eliminates the major cause of O-S message deliveries.

We shall see later that certain failures of signaling links can also result in O-S deliveries. SS7 thus reduces the probability of such deliveries, but does not eliminate them. Therefore, the call-control procedures in both SS6 and SS7 have to take into account the possibility that messages are not delivered in their proper sequence.

5.2.6 Cyclic Redundancy Checking

Error detection of SUs in SS6 and SS7 is done by *cyclic redundancy checking* (CRC), a technique used in many data communication systems. The CRC procedure for SUs from ST_A to ST_B is described below (Fig. 5.2-6).

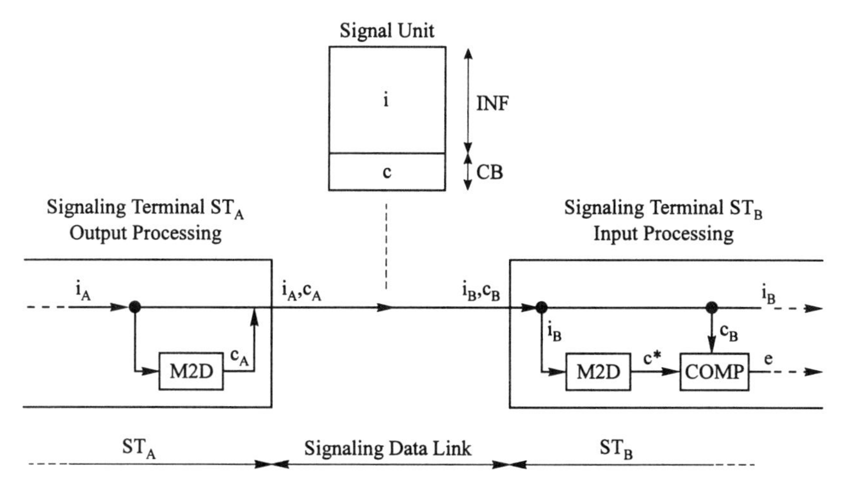

Figure 5.2-6 Cyclic redundancy checking.

In the final step of output processing at ST_A, the contents of INF is regarded as a binary number (i_A). This number is fed into a *modulo 2 divider* (M2D), where it is divided (mod 2) by a divisor d. The remainder of the division (c_A) is entered into CB of the SU.

Suppose now that the SU arrives at ST_B, and that the contents of INF and CB are i_B and c_B. In the initial step of input processing at ST_B, i_B is fed into M2D, where it is divided (mod 2) by the divisor d mentioned above. The division yields a remainder c^*.

Comparison circuit COMP compares c_B and c^*. If there are no errors in INF ($i_B = i_A$), then $c^* = c_A$. If there also are no errors in CB ($c_B = c_A$), the inputs to COMP are equal, and its output $e = 0$. This is taken as the indication that the SU is error-free. Otherwise, the SU is deemed to contain errors, and is discarded.

Cyclic redundancy checking does not catch all transmission errors but, with a properly selected divisor d, the probability of an undetected SU error can be made very low (in the order of 1 in 10^8).

Modulo 2 Division. Mod 2 arithmetic is binary arithmetic without "carries" or "borrows". For example, consider the mod 2 subtraction (or addition):

$$\begin{array}{l} 101101 \\ \underline{111001} - (\text{or } +) \\ 010100 \end{array}$$

The result is obtained by performing "exclusive or" operations on the bits in corresponding positions of both numbers.

Mod 2 division consists of mod 2 subtractions, and shifts. The example below shows the calculation of the remainder c that results from mod 2 division of i = 1101011 by divisor d = 1011. By long division we have:

```
1011 | 1101011
       1011
       ----     ← mod 2 subtraction
        1100
        1011
        ----    ← mod 2 subtraction
         1111
         1011
         ----   ← mod 2 subtraction
          1001
          1011
          ----  ← mod 2 subtraction
Remainder c: 010
```

For more information on CRC checking, including probabilities of undetected errors, the reader is referred to [6] and [7].

5.3 ACRONYMS

ANI	Automatic number identification
ANSI	American National Standards Institute
CAS	Channel-associated signaling
CB	Check-bit field of signal unit
CCIS	Common-channel interoffice signaling
CCITT	International Telephone and Telegraph Consultative Committee
CCS	Common-channel signaling
COM	Command
COMP	Comparison circuit
CRC	Cyclic redundancy check
FDM	Frequency-division Multiplex
IB	Input buffer
IND	Indication
INF	Information field of signal unit
IP	Input processing
ISC	International switching center
M	Message
M2D	Mod 2 division circuit
MF	Multi-frequency
OB	Output buffer
OP	Output processing
P	Processor
RB	Retransmission buffer
SEP	Signaling end point
SL	Signaling link
SP	Signaling point

SPC	Stored program controlled
SS6	Signaling System No.6
SS7	Signaling System No.7
ST	Signaling terminal
STEP	Signaling transfer and end point
STP	Signal transfer point
SU	Signal unit
TC	Terminal control
TG	Trunk group

5.4 REFERENCES

1. *A History of Engineering and Science in the Bell System*, Chapter 12, Bell Telephone Laboratories, Inc., 1982.
2. *Notes on the Network*, AT&T, New York, 1980.
3. C.A. Dahlbom, J.S. Ryan, "History and Description of a New Signaling System," *Bell Syst. Tech. J.*, **57**, 1978.
4. A.R. Modaressi, R.A. Skoog, "Signaling System No.7: A Tutorial," *IEEE Comm. Mag.*, July 1990.
5. *Specifications of Signalling System No.6*, CCITT Red Book, Rec. Q.251, **VI.4**, ITU, Geneva, 1985.
6. W. Stallings, *ISDN*, MacMillan, New York, 1989.
7. W. Peterson, D, Brown, "Cyclic Codes for Error Detection," *Proc. IRE*, January, 1961.

6

SIGNALING SYSTEM NO. 6

This chapter describes signaling system No.6, the first-generation common-channel signaling system. There are two versions of this system, both of which were first deployed in the mid-1970s. CCITT No.6 [1, 2] is still in use in the international network, on a number of transatlantic and transpacific trunk groups. *Common-channel interoffice signaling* (CCIS), defined by the Bell System, has been used in the U.S. toll network [3,4], but has now been replaced by the North American version of signaling system No.7.

There are many similarities between the CCITT No.6 and CCIS. In this chapter, we use the acronym SS6 when discussing matters that are common to both versions.

The definition of SS6 took place at the time that *telecommunications* was still synonymous with *telephony*, and long-distance trunks were carried on analog voiceband transmission channels of FDM multiplexed transmission systems. In SS6, a signaling data link usually consists of a pair of these analog voiceband channels. However, SS6 signaling data links can also be implemented as pairs of digital (PCM) transmission channels.

SS6 signaling can be used for FDM (analog) and PCM (digital) trunks, which may be equipped with *circuit-multiplication equipment* (CME—see Section 1.6.4). However, the signaling data links have to be carried on transmission channels without CME.

SS6 was designed originally for call control applications only. Around 1980, the Bell System expanded CCIS to include query–response transactions between exchanges and centralized network databases.

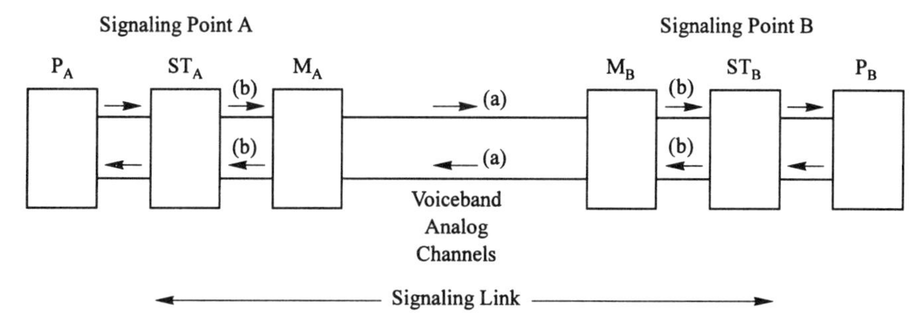

Figure 6.1-1 SS6 signaling link. (a): analog signal. (b): digital bit stream.

6.1 SIGNALING LINKS

6.1.1 Signaling Link Components

Figure 6.1-1 shows a SS6 signaling link in which the data link consists of a pair of analog voiceband (300–3400 Hz) transmission channels [5]. At signaling points A and B, modems M_A and M_B provide the interface between the signaling data link and the signaling terminals ST_A and ST_B. For transmission from A to B, modem M_A converts the digital bit stream (b) from ST_A into an analog signal (a) that is suitable for transport on the voiceband channel, and modem M_B converts signal (a) back into a bit stream (b). A similar conversion sequence takes place in the other direction.

SS6 originally used V.26-bis modems, defined by CCITT [6]. The relation between analog signal (a) and bit stream (b), at respectively the input and output of a modem, is shown in Fig. 6.1-2. Signal (a) is an 1800 Hz sine wave that changes its phase at intervals T_B of 1/1200 s. The power spectrum of this signal is essentially contained within a 600–3000 Hz frequency band. There are four phase changes (with respect to the previous interval T_B), each of which represents a specific combination of two consecutive data bits in the bit stream:

Phase Change (degrees)	Bits
+ 45	0 0
+ 135	0 1
+ 225	1 1
+ 315	1 0

The bit transfer rate of the V.26-bis modem is therefore 2400 bits/s. In the U.S., the V.26-bis modems were replaced later on by modems that operate at 4800 bits/s.

Signaling Link Capacity. Signaling links with 2400 bits/s modems can carry call-control messages for up to 3000 trunks. Under normal conditions, the signaling links carry messages for 1500 trunks. The 3000 trunk load occurs only in failure conditions, when the link carries its normal signaling traffic, and the traffic

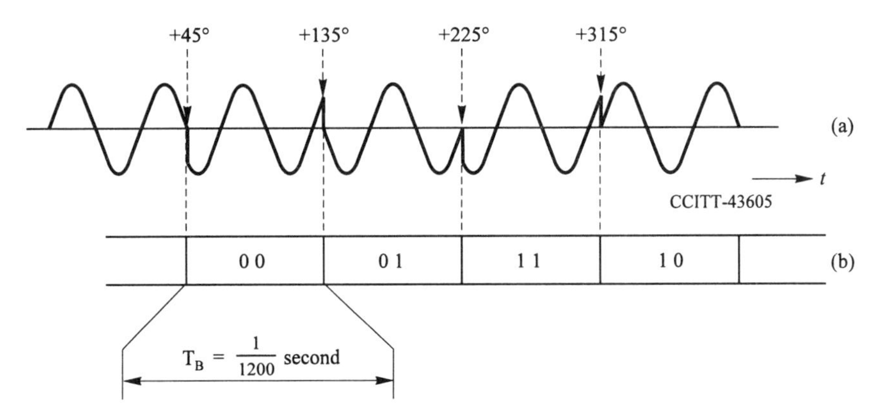

Figure 6.1-2 Relation between analog signal and digital bit stream. (a): analog signal (shown as modem input). (b): bit stream (shown as modem output). (From Rec. V.26.bis. Courtesy of ITU-T.)

of a failed companion link. Operation at this load increases the queuing delays (the time spent in output buffer OB) of the messages. On signaling links with 4800 bits/s modems, the capacities are 3000 trunks and 6000 trunks, respectively.

6.1.2 Signal Units and Blocks

Working signaling links transmit a continuous stream of adjacent signal units (SU) in each direction. Each SU has a 20-bit information (INF) field, followed by an eight-bit check bit (CB) field for error detection (Fig. 6.1-3). Most SUs start with a *heading field* (H) and a *signal information field* (SI). Usually, H identifies a group of SU types, and SI defines a SU type within this group. Some SU types are defined completely by their H field, and do not have a SI field. The lengths of H fields are different in CCITT No.6 and CCIS.

We distinguish "message" signal units (MSU) that carry information between the processors P_A and P_B of Fig. 6.1-1, and "link" signal units that convey information originated by signal terminal ST_A and intended for ST_B, and vice versa (5.2.2).

Twelve consecutive signal units form a *block*. The last SU in a block is an "acknowledgment" signal unit (ACU)—see Fig. 6.1-4. The ACUs are "link" SUs that contain acknowledgments of the signal units in a received block.

Signal units SU_1 through SU_{11} are either MSUs or *"synchronization" SUs* (SYU). A SYU is a "link" SU that is sent by a signaling terminal when no MSU

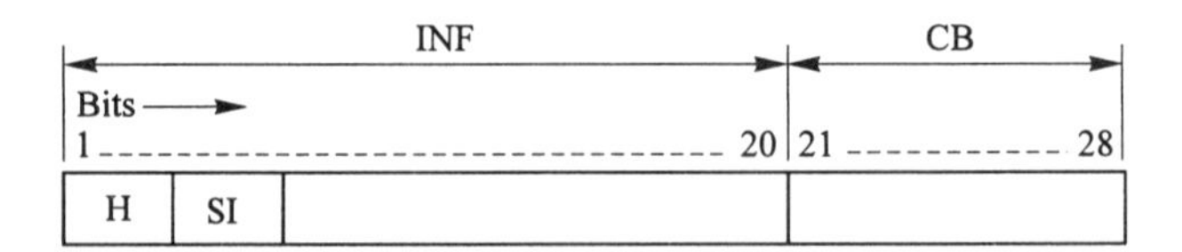

Figure 6.1-3 Signal unit. (From Rec. Q.257. Courtesy of ITU-T.)

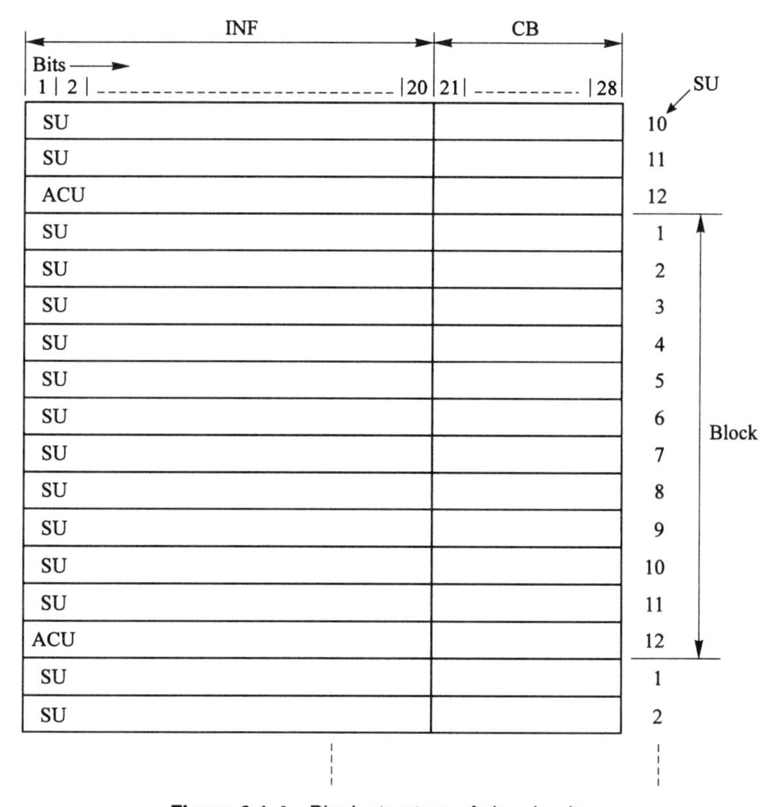

Figure 6.1-4 Block structure of signal units.

is waiting to be transmitted. SYUs are also sent when a signaling link has to be synchronized (6.1.4).

6.1.3 Error Control

SS6 error control consists of error detection, acknowledgments, and retransmission of MSUs received with errors, and takes place in the signaling terminals at both ends of the links [7].

Error Detection. Cyclical redundancy checking (see Section 5.2.3) is used. The check bits for each SU are calculated by mod 2 division of the number in INF by the divisor = 1 0001 0111. The eight bits representing the result (remainder), are inverted, and placed in the CB field of the SU. Signal units received error-free are accepted by the signaling terminal, and positively acknowledged. Signal units received with errors are discarded, and negatively acknowledged.

Acknowledgments. The 11 SUs in a block sent by terminal ST_A are acknowledged by the acknowledgment unit (ACU) in a block sent by ST_B, and vice versa.

Bits →

1 -----	4 ------------- 14	15–17	18–20	21 ----------- 28
H = 011		BA	BC	CB

Figure 6.1-5 Acknowledgment signal unit (ACU). (From Rec. Q.259. Courtesy of ITU-T.)

The layout of the ACU is shown in Fig. 6.1-5. The ACU heading is 011. Bits 4–14 indicate positive- or negative acknowledgments (0 or 1) of SU_1–SU_{11} in a previously received block.

A signaling terminal identifies each outgoing block by a *block completed number* BC (bits 18–20). This number is incremented cyclically, from 0 through 7, for consecutive transmitted blocks. When terminal ST_B has received a block with say BC = 3 from terminal ST_A, it acknowledges the SUs in this block in an ACU in which it sets the *block acknowledged* number (bits 15–17) to BA = 3.

Retransmission. Each ST retains all sent messages in its retransmission buffer until they have been positively acknowledged. Most SS6 messages are one-unit messages (fitting in one message SU). However, there are also multi-unit messages that are transferred in a number of consecutive message SUs. When a terminal receives a negative acknowledgment of a message SU that has been sent, it retransmits the entire message of which the negatively acknowledged SU is a part.

In SS6, a MSU transmission error results in out-of-sequence message delivery (5.2.5). Out-of-sequence delivery of call-control messages pertaining to the same call causes call-processing problems.

6.1.4 Link Synchronization

SS6 signaling terminals have to be *synchronized* (or *aligned*) with their incoming bit streams, so that they can determine the start points of SUs and blocks.

Synchronization units are used to acquire synchronization when the link is turned on, and re-synchronization after a disturbance on the link [8]. The format of a SYU in CCITT No.6 is shown in Fig. 6.1-6. The combination H = 11101, SI = 1101 identifies the signal unit as a SYU. Bits 6–16 are coded 1100011. Bits 17–20 contain a number (*N*) that indicates the position of the SYU within a block. SYUs in CCIS have a slightly different format.

When a signaling link is turned on, both terminals start by sending blocks with

Bits →

1 ---- 5	6 ---- 9	10 ------ 16	17 ------ 20	21 ----------- 28
H = 11101	SI = 1101	1100011	N	CB

Figure 6.1-6 CCITT No. 6 synchronization signal unit (SYU). (From Rec. Q..259. Courtesy of ITU-T.)

11 SYUs and one ACU. The "receive" part of a ST has a counter that steps up one unit with each bit in the incoming bit stream, and recycles when it has counted 12 × 28 = 336 bits (the number of bits in a block). When a terminal is aligned, the counter value should be 1 on receipt of the first bit of the blocks. In the initial part of the synchronization procedure, the *signaling terminal* (ST) looks for the SYU bit pattern 11101-1101-1100011. When it recognizes this pattern, it knows that it has received bits 1 through 16 of a SYU. Moreover, from the value of N, it knows the place of the SYU in the block, and thus can initialize the counter. In the second step of the procedure, both terminals inform each other about their progress by sending acknowledgments of the received SYUs. When the procedure completes successfully, the signaling link is put in service.

On a working signaling link, the STs keep verifying their alignment by looking for the fixed pattern 011 (the ACU heading) during steps 309–311 of the counter.

6.2 MESSAGES, LABELS, AND ROUTING

6.2.1 Message Structure

As shown in Fig. 6.2-1, there are two SS6 message sizes. A one-unit message occupies a single signal unit, called a *lone signal unit* (LSU). *Multi-unit messages* (MUM) require several consecutive SUs, and consist of an *initial signal unit* (ISU) followed by one or more *subsequent signal units* (SSU).

A call-control message pertains to a particular trunk, which is identified by

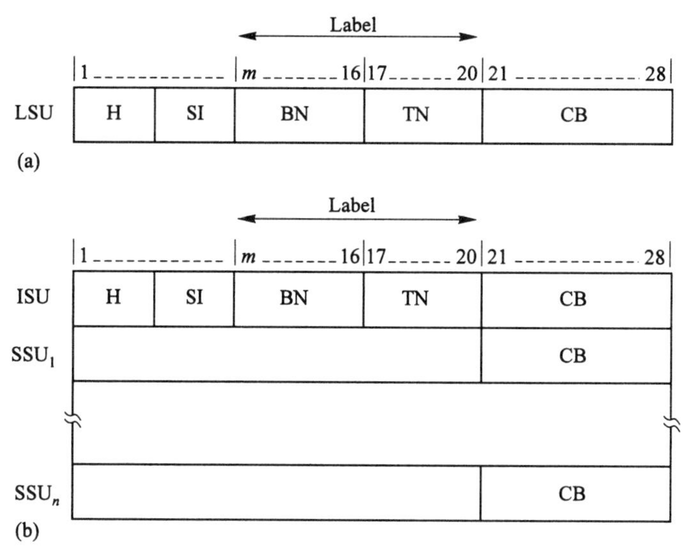

Figure 6.2-1 SS6 call-control messages. (a): one-unit message. (b): multi-unit message. Note: In CCITT No. 6, m = 10; in CCIS, m = 8. (From Rec. Q.257. Courtesy of ITU-T.)

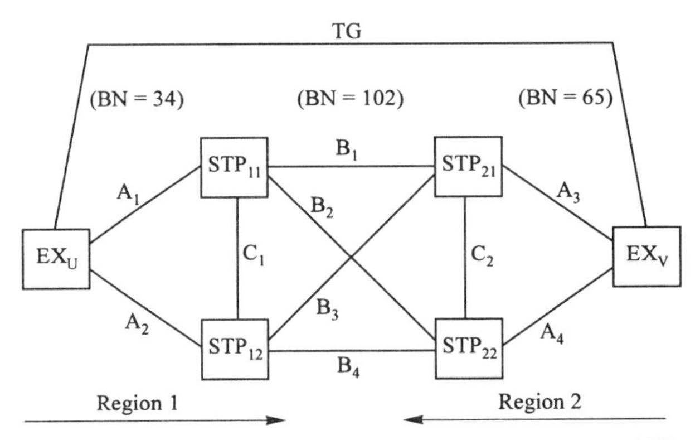

Figure 6.2-2 Band number assignments for trunk group (TG).

the *label* in the message. The label consists of two parts: a *band number* (BN) identifies a "band" of trunks, and a *trunk number* (TN) identifies a trunk within a band. The band and trunk numbers in CCITT No.6 occupy seven bits and four bits respectively. A CCITT No.6 label can thus identify up to 128 bands with up to 16 trunks in each band, for a theoretical label capacity of 2048 trunks. CCIS has nine-bit band numbers, and can identify up to 512 bands of up to 16 trunks (theoretical label capacity: 8192 trunks).

The BN in a message is also used by *signal transfer points* (STP) to route the message to its destination (the exchange at the distant end of the trunk). A BN thus cannot be shared by trunks in different trunk groups. As a consequence, a group of for example 20 trunks requires two band numbers, but uses only 20 of the 32 possible (BN,TN) combinations. This reduces the number of trunks that can be identified on a signaling link in actual networks.

Since the number of available BNs is limited, they have to be "reused" in large signaling networks, such as the CCIS network in the U.S. Therefore, a particular BN value usually identifies different bands of trunks in different parts of the signaling network, and a particular band of trunks is usually identified by different band numbers on the various links in its signaling route. When a call-control message arrives at a signaling point, the affected band of trunks is determined from two data items: the BN in the message, *and* the identity of the signaling link on which the message came in.

Figure 6.2-2 shows an example for a trunk group TG with at most 16 trunks (one band). On the "A" signaling between EX_U and the STPs of region 1, the group is identified by BN = 34. On the "B" signaling links between the STPs in regions 1 and 2, the BN = 102. On the "A" signaling links between the STPs of region 2 and EX_V the BN = 65. This is because these BNs were available at the time that the trunk group was installed.

6.2.2 Band Number Translation and Outgoing Link Selection

We now explore the BN translations and outgoing link selections at the signaling points for messages from Ex_U to Ex_V, which relate to a specific trunk T of group TG, in Fig. 6.2-2 [4]. The trunk has TN = 5 (this number does not change when the messages traverse a STP). We assume that all signaling links are operational, and that the messages for TG are to be load-shared evenly by the four normal signaling routes for the group:

R_1: A_1-B_1-A_3
R_2: A_1-B_2-A_4
R_3: A_2-B_3-A_3
R_4: A_2-B_4-A_4

All messages relating to a specific trunk have to use the same route. Otherwise, two consecutive messages for that trunk, say M_1 (sent first, on route R_1) and M_2 (sent later, on route R_2) could arrive out-of-sequence at EX_V because, at the time that the messages are sent, the queuing delays at the signaling links in R_1 are large, and small on R_2. Out-of-sequence delivery of messages relating to a trunk can cause problems in call processing.

The selection of outgoing signaling links that accomplishes load sharing of signaling links, and also associates a signaling route with a particular trunk, can be done in several ways, for example:

At exchanges, the primary outgoing A link for messages with odd or even TN go to respectively the odd- and even-numbered STPs in the region of the exchange. If the primary link fails, the messages are diverted to the other (alternative) A link. In this example, the messages for trunk T normally go out on A_1, and reach STP_{11}.

Every STP has a table for each attached signaling link, with entries for all incoming band numbers. Each entry contains the outgoing band number (BN_O), and the identity of a primary and alternative outgoing link.

For messages received by an STP on its A links, the value of BN_O indicates the destination region for the message. In addition, when BN_O is odd, the primary and alternative outgoing B links go to respectively the odd- and even-numbered STP of that region. If BN_O is even, the primary and alternative B links go to respectively the even- and odd-numbered STP. In this example, the entry at STP_{11} for messages received on link A_1 with BN = 34 thus indicates:

Outgoing band number: BN_O = 102,
Primary outgoing SL: B_2 (to STP_{22}),
Alternative outgoing SL: B_1 (to STP_{21}).

For messages received by an STP on its B links, the BN identifies an exchange in the region of the STP. The primary outgoing link is the A link to the exchange, and the alternative link is the C link to the other STP in the region.

At STP_{22}, the table entry for messages received on B_2 with BN = 102 thus indicates:

Outgoing band number: BN_O = 65,
Primary outgoing SL: A_4,
Alternative outgoing SL: C_2.

The normal route for messages from Ex_U to Ex_V for trunk T is thus A_1–B_2–A_4. At destination exchange Ex_V, the label (BN = 65, TN = 5) indicates that the message concerns trunk T.

Applying the same rules to messages relating to trunk T, and sent by exchange Ex_V, the signaling route is A_3–B_3–A_2. Messages relating to a trunk, and sent in opposite directions thus may traverse different signaling routes.

When a primary signaling link fails, messages normally sent out on that link are diverted to the alternative link. Consider again the messages sent by Ex_U, relating to trunk T. On failure of link A_1, Ex_U diverts the messages to link A_2. On failure of link B_2, STP_{11} diverts the messages to link B_1, and on failure of link A_4, STP_{22} diverts the messages to link C_2 (and STP_{21} sends the messages on A_3).

6.3 CCITT NO.6 CALL CONTROL

The CCITT No.6 signaling system is used for international transatlantic and transpacific trunk groups. The call-control features of CCITT No.6 are comparable to those of CCITT No.5 and R2 international signaling (Chapter 4). All signaling is link-by-link.

The most important call-control messages [9] are described in Sections 6.3.1 through 6.3.4. In CCITT No.6 documents, these messages are usually denoted by three-character acronyms. The coding of the heading and signal information fields of the messages are listed in Table 6.3-1.

Signaling procedures are outlined in Sections 6.3.5 and 6.3.6.

6.3.1 Initial Address Message

The *initial address message* (IAM) is the first forward message in a call. It indicates the seizure of a trunk, and contains the initial digits (in overlap address signaling), or all digits (in *en-bloc* address signaling), of the called number, and parameters that affect the routing and processing of the call.

The IAM layout is shown in Fig. 6.3-1. Label L in the initial signal unit (ISU) identifies the trunk for which the message is intended. The subsequent signal units (SSU) have a heading code H = 00, and a length indicator (LI) that represents the number of SSUs beyond SSU_2.

SSU_1 contains a number of international routing indicators, which we have already encountered in No.5 and international R2 signaling.

Country Code Indicator (C). This indicates whether the called number is an international number (C = 1; country code included), or a national number(C = 0; no country code included).

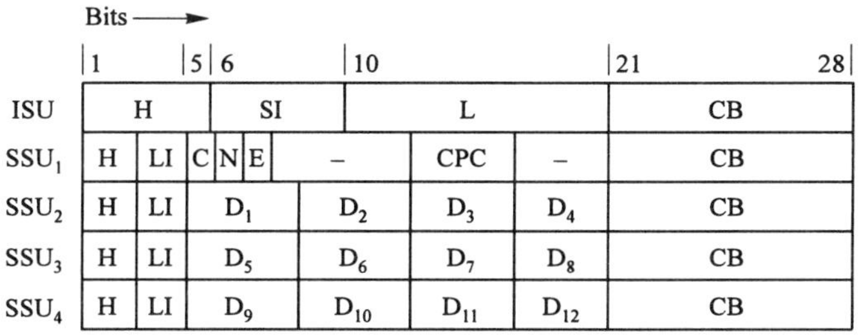

Bits →

	1–5		6–10		10–21				21–28
ISU	H		SI		L				CB
SSU_1	H	LI	C	N	E	–	CPC	–	CB
SSU_2	H	LI	D_1		D_2		D_3	D_4	CB
SSU_3	H	LI	D_5		D_6		D_7	D_8	CB
SSU_4	H	LI	D_9		D_{10}		D_{11}	D_{12}	CB

Figure 6.3-1 CCITT No. 6 Initial address message (IAM). Note: CCIT No. 6 one-unit messages have the format of ISU. (From Rec. Q.258. Courtesy of ITU-T.)

Table 6.3-1 Coding of heading and signal information fields in CCITT No.6 call-control messages and signals.

Acronym	Name	Heading	Signal Information
ADC	Address complete, charge	11011	1010
ADI	Address incomplete	11011	1101
AFC	Address complete, subscriber free, charge	11011	0001
AFN	Address complete, subscriber free, no charge	11011	0010
ANC	Answer, charge	11000	0010
ANN	Answer, no charge	11000	0011
BLA	Blocking acknowledgment	11010	1101
BLO	Blocking	11010	1011
CFL	Call failure	11001	1000
CGC	Circuit group congestion	11001	0100
CLB	Clear-back	11000	0100
CLF	Clear-forward	11010	0010
COF	Confusion	11001	1110
COT	Continuity	11010	0001
FOT	Forward-transfer	11010	0011
IAM	Initial address message	10000	0000
LOS	Line out of service	11011	0110
RAN	Reanswer (after CLB)	11000	0101
RLG	Release-guard	11000	0001
SAM	Subsequent address message	10001	0000
SEC	Switching equipment congestion	11001	0011
SSB	Subscriber busy	11011	0100
UBA	Unblocking acknowledgment	11010	1110
UBL	Unblocking	11010	1100
UNN	Unallocated national number	11011	0101

Source: Rec. Q.257. Courtesy of ITU-T.

Nature of Circuit Indicator (N). This indicates whether the connection built up so far includes, or does not include, a satellite trunk (N = 1, or N = 0). This indicator is used by incoming exchanges. When a call is received with N = 1, the exchange avoids routing the call on another satellite trunk.

Echo-suppressor Indicator (E). This indicates whether the connection built up so far includes, or does not include, an outgoing half-echo suppressor (E = 1, or E = 0). The incoming exchange uses this indicator to determine whether to insert an echo suppressor on its incoming or outgoing trunk.

Calling Party Category Indicator (CPC). This parameter is coded as follows:

0001	French-speaking operator
0010	English-speaking operator
0011	German-speaking operator
0100	Russian-speaking operator
0101	Spanish-speaking operator
1010	Subscriber
1101	Test Equipment

Address Digits (D). The second and later SSUs contain up to four address digits each. Digit values 1, 2 ,..., 0 are coded as 0001, 0010,..., 1010. International operator access codes "11" and "12" are coded 1011 and 1100.

An IAM can include up to 4, 8, 12, or 16 address digits (depending on the number of SSUs). The final SSU may contain one or more unused digit slots, which are coded as 0000 (filler code).

On international calls dialed by outgoing operators (who can indicate "end of dialing" from their consoles), the last digit of the called number is followed by an *end of address* (ST) digit (coded 1111). On subscriber-dialed international calls, the ST digit is not included because outgoing international exchanges (ISC) have no information about the lengths of foreign national numbers.

If overlap address sending is used, the outgoing ISC sends an IAM, which includes enough digits (usually four) to allow the incoming ISC to make a route decision. The later digits of the called number are sent in a *subsequent address message* (SAM).

Test Addresses. CCITT No.6 includes procedures for making test calls on trunks between international exchanges. The outgoing ISC sends an IAM with calling party category (CPC) =1101 (test equipment). In this case, the called address consists of one digit (D_1) that specifies a test termination at the incoming ISC. For example:

0001	ATME 2 (automatic transmission measurement equipment no. 2) signaling and transmission test line
0010	ATME 2 signaling test line
0011	Quiet termination test line
0100	Echo suppressor test line
0101	Loop around test line

6.3.2 Other Forward Messages

Subsequent Address Message (SAM). Used in overlap address sending. It consists of an ISU, and one or more SSUs, with the format of SSU_2 in Fig. 6.3-1.

All other forward call-control messages are one-unit messages, often called "signals." They are carried in LSUs.

Continuity Signal (COT). This is used to report a successful continuity test (see 6.3.5).

Clear-forward Signal (CLF). This requests the release of the trunk.

Forward Transfer Signal (FOT). This is used by outbound operators, to request the assistance of an inbound operator in a foreign country.

6.3.3 Backward Messages

All backward messages are one-unit messages (signals) and are carried in LSUs.

Three groups of backward signals can be distinguished. The first group (heading code: 11000) consists of the backward signals for successful calls, and are the equivalents of the backward line signals in channel-associated signaling systems.

Answer, Charge (ANC). This indicates that the call has been answered, and should be charged.

Answer, No charge (ANN). The call has been answered, and should not be charged.

Clear-back (CLB). The called party has disconnected (gone on-hook).

Reanswer (RAN). This is sent after a CLB, when the called party has gone off-hook again.

Release-guard (RLG). This is sent by an exchange that has received a CLF for a trunk. It indicates that the exchange has released the trunk at its end.

Signals of the second group (heading code: 11011) indicate that the call can not be set up.

Switching Equipment Congestion (SEC). This indicates that the ISC is congested.

Circuit Group Congestion (CGC). No outgoing trunk to the call destination is available.

Call Failure (CFL). The call has failed (reason not specified).

Confusion (COF). The response to a received message that is not "reasonable," given the state of the call.

On receipt of one of these backward signals, the ISC releases the outgoing trunk, and sends a CLF. Depending on the received signal, and on agreements between the involved countries, the ISC either aborts the call (repeating the received signal to the previous exchange), or makes a second attempt.

Second attempts can be made using a trunk that terminates at the same ISC, or on a trunk that terminates at a different ISC. The latter procedure, called *automatic rerouting* (see Section 1.3.6), is used, by bilateral agreement, when an ISC receives a SEC or CGC signal.

The backward signals in the third group (heading code: 11011) are sent by the last ISC in the CCITT No.6 section of an international connection, after it has seized an outgoing trunk with a different signaling system.

Address Complete, Subscriber Free, Charge (AFC). This indicates that the called party is free, and that the call should be charged.

Address Complete, Subscriber Free, No Charge (AFN). As above, but the call should not be charged.

Subscriber Busy (SSB).

Unallocated National Number (UNN). The called number is not a working number.

Line Out of Service (LOS).

Address Complete, Charge (ADC). The entire called number has been received; call should be charged.

Address Incomplete (ADI). The received digits do not constitute a called national number.

The signals AFC, AFN, SSB, UNN, and LOS can be sent only by an ISC in a destination country that uses a signaling system (for example, national R2 signaling) in which the status of the called subscriber can be made known to the ISC.

When the last ISC in a CCITT No.6 section of an international connection is not in the destination country, or is in a destination country where the signaling does not include backward information about the status of the called party, it sends an address complete (ADC) signal. The signal is repeated backwards to the first ISC in the international connection, and indicates to the ISCs along the connection that their activities for call set-up are complete. The ISCs then erase the called number, and other routing information, from memory.

The ADC signal is sent when one of the following events occurs:

1. Receipt of an ST digit in an IAM or SAM.

2. Receipt of the maximum number of digits used in the national numbering plan of the country.
3. Receipt of sufficient digits to route the call.
4. Receipt of an "address complete" signal from the national network.
5. 15–20 seconds have elapsed since the receipt of an IAM or a SAM.

6.3.4 Blocking and Unblocking Signals

The exchanges at both ends of a trunk T can send a *blocking signal* (BLO) for the trunk. This requests the other exchange to suspend seizing the trunk for new calls. The other exchange acknowledges the receipt of BLO with a *blocking acknowledgment signal* (BLA). When the trunk can be unblocked, the exchange that sent the BLO signal sends an *unblocking signal* (UBL). On receipt of this signal, the other exchange sends an *unblocking acknowledgment signal* (UBA), and resumes seizing the trunk for new calls.

An exchange sends a BLO signal when it wants to test the trunk, or when a continuity check of the trunk fails.

6.3.5 Continuity Check

In channel-associated signaling, the signals for a trunk are carried by the trunk itself. If the transmission of a seized outgoing trunk is not working properly, the signaling for the call set-up fails.

In common-channel signaling, successful signaling during the set-up of a call is no indication that the transmission of the seized trunk is in working order.

Therefore, when an exchange seizes an outgoing CCS trunk, it makes a *continuity check* [10] on the trunk—see Fig. 6.3-2. Exchange A connects the "send" and "receive" channels of the trunk to a 2010 Hz tone generator (TG) and tone detector (TD), and sends the IAM. On receipt of the IAM, exchange B connects the "send" and "receive" channels of the trunk (loop-back). If the

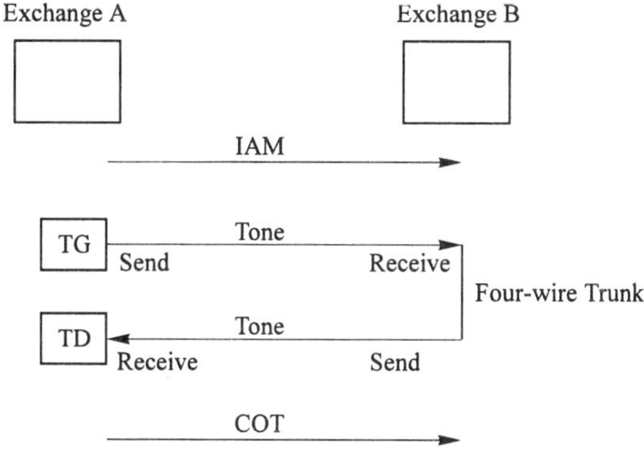

Figure 6.3-2 Continuity test.

transmission is working in both directions, TD receives the 2010 Hz tone, and reports this to the processor of exchange A. The exchange then sends a COT, and disconnects TG and TD from the trunk. On receipt of COT, exchange B terminates the loop-back, and proceeds with the set-up.

If, after about one second, no tone has been detected by TD, exchange A sends a BLO signal for the trunk, and exchange B responds with a BLA signal. Exchange A then releases the trunk, and sends a CLF signal. Exchange B, after releasing the trunk at its end, sends a RLG signal.

6.3.6 Signaling Sequence

Figure 6.3-3 shows the signaling sequence for an international call originating in country A, with a destination in country C, and routed via an ISC in country B [11]. Trunks T_1 and T_2 have CCITT No.6 signaling. *En-bloc* address signaling is assumed at ISC_A and ISC_B:

1. ISC_A has received an outgoing international call from its national network. After receiving the complete called address, it seizes trunk T_1, attaches a check-tone generator and detector for the continuity check, and sends an IAM to ISC_2. Since ISC_B not in the destination country of the call, the called number in the IAM is an international number.
2. ISC_B receives the IAM, and loops the receive and send channels of T_1.

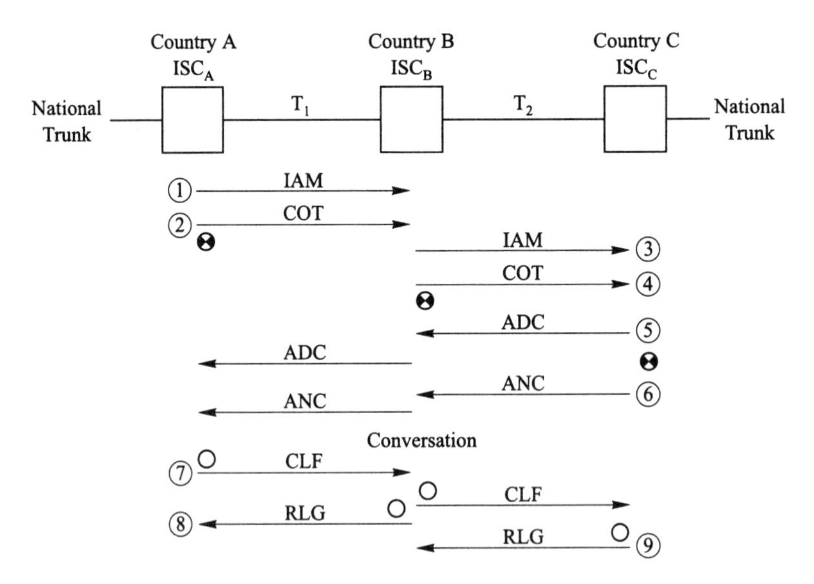

Figure 6.3-3 Signaling sequence for an international call.

ISC_A receives the test tone, and sends a COT signal. It also cuts through (sets up a path between the incoming national trunk and T_1).

3. On receipt of COT, ISC_B proceeds with the set-up of the call. It analyzes the IAM, seizes trunk T_2, attaches a check-tone generator and detector, and sends an IAM to ISC_C. Since this ISC is in the destination country, the called number in the IAM from ISC_B is a national number. On receipt of the IAM, ISC_C loops the send and receive channels of trunk T_2.
4. ISC_B detects the check tone. It disconnects the tone equipment, and cuts through (sets up a path between T_1 and T_2). A COT signal is sent to ISC_C.
5. On receipt of COT, ISC_C ends the loopback on T_2, analyzes the received IAM, and seizes a trunk in its national network. After completing the set-up signaling for this trunk, it cuts through, and sends an ADC signal to ISC_B, which repeats the signal to ISC_A.
6. When ISC_C receives the answer signal from its national network, it sends an ANC signal to ISC_B, which passes it to ISC_A, and this exchange then starts to charge the call. The conversation begins.
7. Assuming that the calling party clears first, ISC_A receives a clear-forward signal from its incoming trunk. The ISC then releases trunk T_1, and sends a CLF signal to ISC_B. This ISC then clears the connection, releases T_1 and T_2, and sends a CLF to ISC_C.
8. When ISC_B has released trunk T_1, it sends a RLG signal, to indicate that T_1 can now be selected for a new call.
9. When ISC_C has released T_2, it sends a RLG to ISC_B, indicating that T_2 is now available for a new call.

Set-up Failures. When an exchange determines that it cannot extend the set-up of a call that has arrived on a SS6 trunk, it signals to the preceding exchange, which then initiates the release of the trunk without waiting for a disconnect

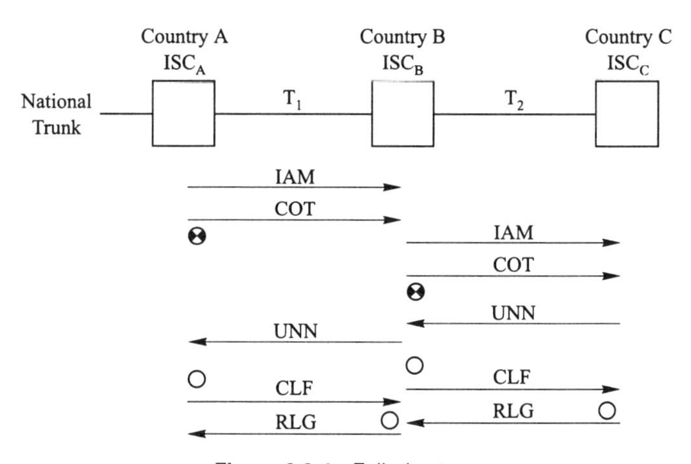

Figure 6.3-4 Failed set-up.

indication from the calling party. This procedure is the counterpart of a procedure in R2 signaling (Chapter 4).

As an example, in Fig. 6.3-4 ISC_C has determined that the area code of the called national number, received in the IAM, is not in use in its country. It then sends an unallocated number (UNN) signal to ISC_B, which then initiates the release of T_2, and returns a CLF signal. ISC_B also repeats the UNN signal to ISC_A, and this exchange starts the release of T_1.

ISC_A provides an appropriate audible tone or announcement to the calling party, by connecting the incoming national trunk to a tone or announcement source. In addition to expediting the release of the international trunks, the procedure has the advantage that the caller receives audible information that originates in his national network, and thus does not hear an unfamiliar tone, or an announcement in a foreign language.

Unreasonable and Superfluous Messages. Two call-control messages for a trunk can be received out-of-sequence (6.1.3). An incoming message can therefore appear to be unreasonable (inappropriate for the state of the call). Actions to be taken when an unreasonable message has been received have been specified by CCITT [12].

6.4 CCIS CALL CONTROL

Common-Channel Interoffice Signaling is the North American version of SS6. CCIS call-control messages are similar to those of CCITT No.6, and have the same acronyms. However, there are some differences in message formats and contents [3,4].

6.4.1 Formats of CCIS Signal Units

The main reason for the format changes was the need to address a larger number of trunks in the CCIS signaling network. The band number field (BN) of the label was therefore increased to nine bits, allowing 512 band numbers (Fig. 6.2-1). This increase was balanced by reducing the heading field (H) to three bits.

Figure 6.4-1 shows a CCIS signaling sequence. The LSUs of call-control signals have headings H = 000, 001, 010, 100, or 111, followed by the four-bit SI field and the 13-bit label.

6.4.2 Initial Address Message

The initial SU in CCIS multi-unit messages has heading H = 101. The heading is followed by a discrimination bit D. The combination H = 101, D = 0 identifies an IAM—see Fig. 6.4-1. CCIS address signaling is always en-bloc (IAM includes the complete called number). The called number in the figure is a ten-digit national number NPA-*NXX-XXXX*. The last digit slot contains a filler code.

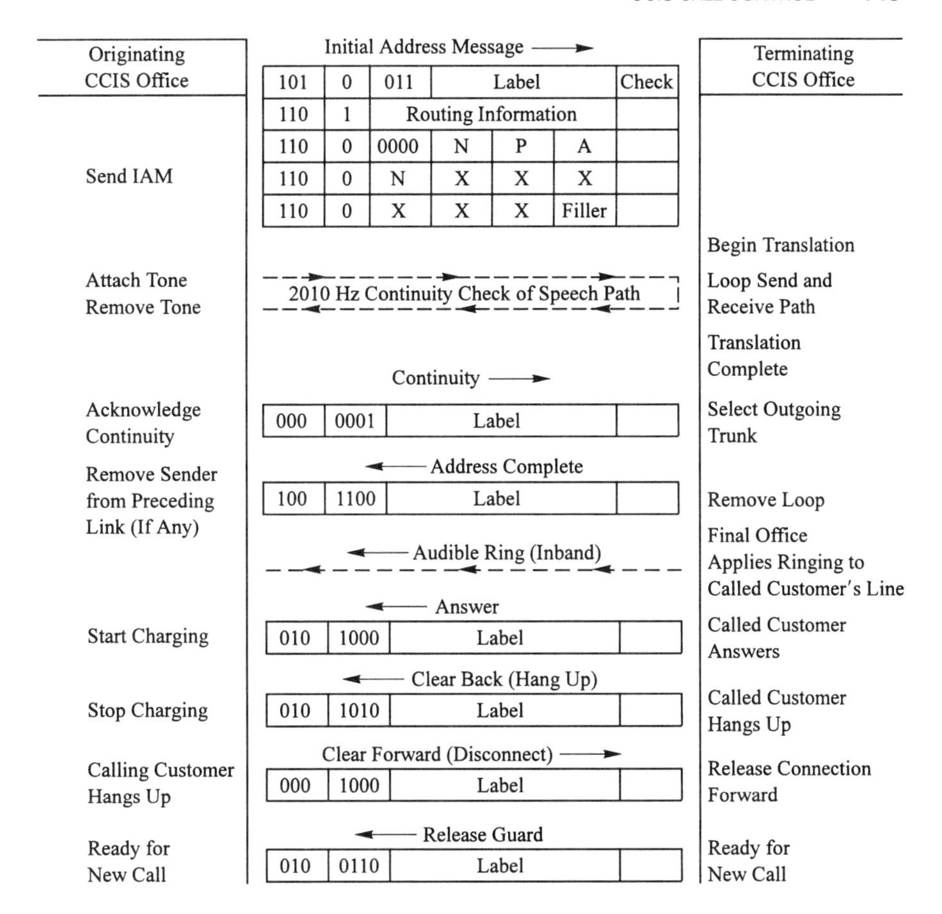

Figure 6.4-1 CCIS signaling sequence. (From *Bell Syst. Tech. J.* **57**, No. 2. Copyright © 1975 AT&T.)

The subsequent SUs in the message have a heading H = 110, followed by a discrimination bit D. In the SSUs that contain the called number, D = 0.

Routing Information. An IAM can contain *basic*, or *extended*, routing information (a set of routing indicators). In the figure, the first SSU has D = 1. This indicates that the IAM has extended routing information, in bits 5–20 of the SSU. In this case, bits 5–8 of the next SSU are set to zero. If the IAM includes basic routing information, it does not include the first SSU, the second SSU in the figure is now the first SSU, and bits 5–8 hold the basic routing information.

Some routing indicators, such as, nature of circuit (indicating whether there are satellite circuits in the connection), echo-suppressor indicator (inserted/not inserted), and calling-party category (subscriber/operator) are the same as in CCITT No.6. Other indicators are specific to CCIS. A few examples are outlined below.

Routing Category. This indicator categorizes the connection in terms of its origination and destination:

(a) Call originated in U.S.
(b) Inbound (to U.S.) international call.
(c) International outbound call, overflowing from an international switching center in the U.S.

The information is used by exchanges for routing and call processing decisions. For example, calls with routing categories (a) and (b) need different backward indications (CCIS messages or audible signals) when the call cannot be extended.

The use of routing category (c) is illustrated in Fig. 6.4-2, showing two international centers (ISC_1 and ISC_2) in the U.S. Suppose that TG_1 and TG_2 are the only trunk groups for calls from the U.S. to country C. Outbound international calls to C arrive from the U.S. national network with routing category (a). When ISC_1 receives an IAM for a call to C when all trunks in group TG_1 are busy, it routes the call on overflow trunk group TG. ISC_2 then attempts to route the call on TG_2. Likewise, a call to country C that arrives with routing category (a) at ISC_2 is routed to ISC_1 when all trunks in TG_2 are busy. In these cases, the first ISC sets the routing category indicator to (c). This informs the second ISC that the call has already overflowed from the other ISC, and should not overflow again. Otherwise, when TG_1 and TG_2 are busy at the same time, the set-up would shuttle back and forth between ISC_1 and ISC_2, tying up all trunks in the overflow trunk group.

Continuity Check Canceled Indicator. When an exchange processor is overloaded, it may decide to omit the continuity check on an outgoing trunk. The continuity check indicator informs the other exchange whether it should loop back the incoming trunk, and expect a COT signal.

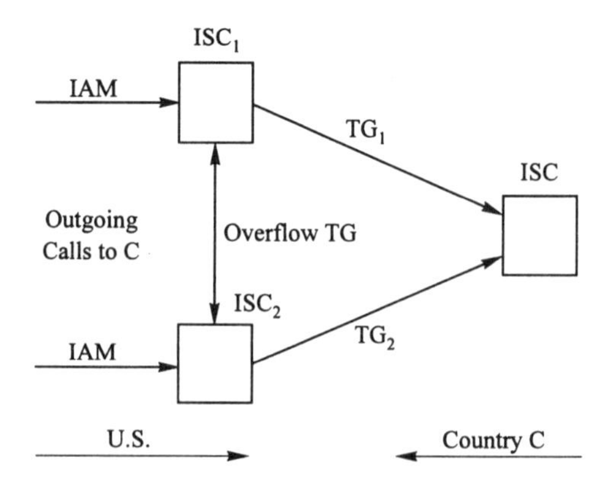

Figure 6.4-2 Overflowing outgoing international calls.

6.5 CCIS DIRECT SIGNALING

The term *direct signaling* covers the procedures and CCIS messages that were developed by the Bell System for signaling between exchanges and signaling points of other types, for example centralized databases and maintenance centers [4, 13, 14].

6.5.1 "800" Calls

This section explores direct signaling, using 800-number calling (an early application) as example. In Fig. 6.5-1, subscriber S has dialed a call to the number 800-234-5678. An 800-number identifies a called party (usually a business) who pays for calls dialed to the number. 800-numbers do not have the format of a regular national number: in the example, 800 is not an area code, and 234 is not an exchange code. To route 800-calls to their destinations, the 800-numbers have to be translated into *routing numbers*, which have the format of national numbers.

The databases with 800-number translations are located in *service control points* (SCP). The SCPs receive queries (requests for translations) from exchanges that are known as *service switching points* (SSP). The queries from the SSPs and responses from the SCPs, are transferred in CCIS direct-signaling messages.

Local exchange EX_X routes the 800-234-5678 call to a nearby SSP_Y. This SSP then queries SCP_Z, and receives the routing number, say 201-654-7789. This number is then used by SSP_Y, and other exchanges along the connection, to route the call to its destination.

6.5.2 Addresses and Routing of Direct Signaling Messages

In a SS6 call-control message, the band number (BN) is used by the STPs as the address for routing the message to its destination. Direct-signaling messages do not pertain to trunks, and therefore do not have BNs. The address in these

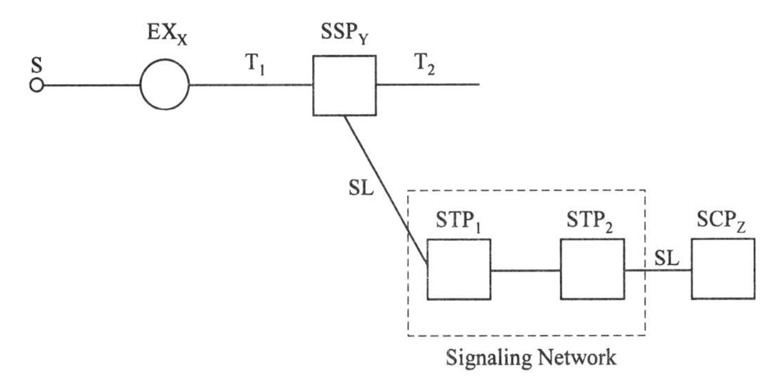

Figure 6.5-1 Handling of a call to an "800" number.

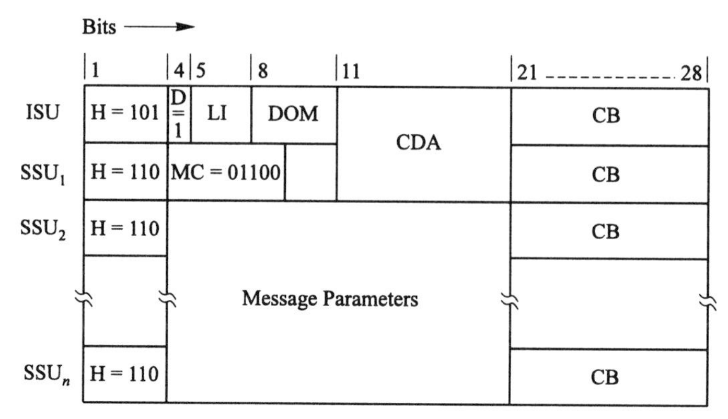

Figure 6.5-2 General format of CCIS direct signaling messages. (From *Bell Syst. Tech. J.* **61**, No. 7. Copyright © 1982 AT&T.)

messages can be a "function number address", or a "symbolic address". Figure 6.5-2 shows the general format of direct-signaling messages. The address is in fields DOM (domain) and CDA (called address). If DOM = 0, the CDA contains a function number. In symbolic addresses, DOM has a non-zero value.

Function Number Addresses. A function number (FN) identifies a particular function in a particular CCIS signaling point. The function in SSP_Y that handles 800-number queries has an FN, but the same function in a different SSP, and other functions in SSP_Y, have other FNs. The 800-number functions in SSPs and SCPs have FN addresses.

Symbolic Addresses. SCP functions also have "symbolic" addresses SA. The 800-number functions in an SCP have symbolic addresses in which DOM = 1, and CDA holds the initial six digits dialed by the subscriber. In our example, the address in the query message from SSP_Y has DOM = 1, CDA = 800-234. This identifies the 800-number translation function in an SCP that stores translations for the numbers 800-234-*XXXX*.

Routing of Query Messages. The signal transfer points STP store two sets of routing tables for direct-signaling messages. A "SA-to-FN" table translates a symbolic address into a primary and a backup FN. A "FN-to-SL" table gives the outgoing signaling links for a destination identified by FN.

In Fig. 6.5-1, signal transfer point STP_1 translates the symbolic address in the query from SSP_Y, and normally selects the primary FN. The backup FN is used when the function associated with the primary FN is out of service. STP_1 then replaces the symbolic address in the message by the FN address, determines an outgoing signaling link, and sends the message. STP_2 determines an outgoing signaling link for the message from the FN address. The message reaches SCP_Y, which delivers it to its 800-number translation function.

The use of symbolic addresses concentrates all routing information in the

STPs, instead of in the SSPs. This has a significant advantage when changes in this information have to be made. At the time of the development of direct signaling for 800-number calls, there were about 20 STPs in the U.S. signaling network, versus about 1000 anticipated SSPs. With symbolic addresses, changes in routing information require updating the STP data, but do not affect the SSPs.

6.5.3 Message Format

Returning to Fig. 6.5-2, the DOM and CDA fields have been described earlier. In the ISU, the combination H = 101, D = 1 indicates a multi-unit message other than an IAM (for which D = 0). *Message category* (MC) indicates the direct-signaling application (in this example "800-number service"). *Length indicator* (LI) indicates the message length (the number of SSUs).

SSU_2–SSU_n include the other message parameters. The originator of a query message includes its *return address* (a function number), for use by the responder in its return message. The originator also includes a reference number assigned to the query, which is copied by the responder in the return message. In this way, the originator can correlate a return message to a particular query, several of which may be in progress simultaneously.

Finally, the query message includes the two input parameters for the translation: the complete number to be translated (800-234-5678), and the area code of the calling party. Translations of 800-numbers can be arranged to give different routing numbers for different calling area codes. This is of importance for a business that has an 800-number, and has offices in several locations. The business owner can specify that calls from subscribers on the west coast should be routed to his office in say Los Angeles, calls from the east coast to his office in New York, etc.

The return messages sent by the responder have FN addresses, and contain the reference number, the routing number, and charging instructions for the SSP, which has to produce a billing record for the call. In case the received 800-number cannot be translated, the return message includes the reason for the failure, for example "number no longer in use."

6.6 ACRONYMS

ACU Acknowledgment signal unit
ADC Address complete, charge signal
ADI Address incomplete signal
ADN Address complete, no charge signal
AFC Address complete, subscriber free, charge signal
AFN Address complete, subscriber free, no charge signal
ANC Answer, charge signal
ANN Answer, no charge signal
ATME 2 Automatic test and measurement equipment

BA	Block acknowledged number
BC	Block completed number
BLA	Blocking acknowledgment signal
BLO	Blocking signal
BN	Band number
C	Country code indicator
CB	Check bit field
CCIS	Common-channel interoffice signaling
CCITT	International Telephone and Telegraph Consultative Committee
CDA	Called address (called number)
CFL	Call failure signal
CGC	Circuit group congestion signal
CLB	Clear-back signal
CLF	Clear-forward signal
CME	Circuit-multiplication equipment
COF	Confusion signal
COT	Continuity signal
CPC	Calling party category
D	Digit
D	Discriminator bit
DB	Database
DOM	Domain
E	Echo suppressor indicator
EX	Exchange
FN	Function number
FOT	Forward-transfer signal
H	Heading field
IAM	Initial address message
INF	Information field
ISC	International switching center
ISU	Initial signal unit
LI	Length indicator
LOS	Line out of service signal
LSU	Lone signal unit
M	Modem
MC	Message category
MFC	Multi-frequency code signaling
MI	Management information parameter
MUM	Multi-unit message
N	Nature of circuit indicator
P	Processor
PCM	Pulse code modulation
RAN	Reanswer signal
RLG	Release-guard signal
SAM	Subsequent address message

SCP	Service control point
SEC	Switching equipment congestion signal
SI	Signal information field
SL	Signaling link
SS6	CCIS and CCITT No.6 signaling
SSB	Subscriber busy signal
SSP	Service switching point
SSU	Subsequent signal unit
ST	End of address digit
ST	Signaling terminal
STP	Signal transfer point
SU	Signal unit
SYU	Synchronization signal unit
TD	Tone detector
TG	Tone generator
TG	Trunk group
TN	Trunk number
UBA	Unblocking acknowledgment signal
UBL	Unblocking signal
UNN	Unallocated number

6.7 REFERENCES

1. S. Welch, *Signalling in Telecommunications Networks*, Peter Peregrinus Ltd., Stevenage, U.K., 1981.
2. R. Manterfield, *Common Channel Signalling*, Peter Peregrinus Ltd., London, 1991.
3. C.A. Dahlbom, J.S. Ryan, "History and Description of a New Signaling System," *Bell Syst. Tech. J.*, **57**, No.2, Feb. 1978.
4. *Notes on the Network*, Section 6, AT&T, New York, 1980.
5. *Specifications of Signalling System No.6*, Rec. Q.272-Q.276, CCITT Red Book, **VI.4**, ITU, Geneva, 1985.
6. *Data Communication over the Telephone Network*, Rec. V.26, CCITT Red Book, **VIII.1**, ITU, Geneva, 1985.
7. *Specifications of Signalling System No.6*, Rec. Q.277, CCITT Red Book, **VI.4**, ITU, Geneva, 1985.
8. *Ibid.*, Rec. Q.278.
9. *Ibid.*, Rec. Q.257–Q.258.
10. *Ibid.*, Rec. Q.271.
11. *Ibid.*, Rec. Q.261–Q.264.
12. *Ibid.* Rec. Q.267.
13. D. Sheinbein, R.P. Weber, "800 Service Using SPC Network Capability," *Bell Syst. Tech. J.*, **61** No.7, Sept. 1982.
14. R.F. Frerkin, M.A. McGrew, "Routing of Direct Signaling Messages in the CCIS Network," *Bell Syst. Tech. J.*, **61** No.7, Sept. 1982.

7

INTRODUCTION TO SIGNALING SYSTEM NO. 7

CCITT began the specification of the second-generation common-channel signaling system, known as signaling system No.7 (SS7), in the mid-1970s. The system was developed initially for telephony call control, and the first recommendations were published in the CCITT *Yellow Books* of 1981. During the past decade, the applications of SS7 applications have expanded, and now also include call control in the *integrated services digital network* (ISDN), and operations in the network that aeennot related to individual trunks. Most of this work has been published in the CCITT *Blue Books* of 1989. More recent additions have been issued as individual International Telecommunications Union (ITU) Recommendations.

CCITT No.6 and CCIS were used only in the international network, and in the AT&T long-distance network. The objective for SS7 was to develop a signaling system that can be used worldwide. However, this objective has not been met completely. Several national versions have been defined, notably a version specified by American National Standards Institute (ANSI) and Bellcore (Bell Communications Research) for the U.S. networks, and British Telecom's *national user part* (NUP), for the network in the U.K.

In what follows, we speak of SS7 when discussing the general aspects of the system that essentially apply to all versions, and refer to the CCITT-defined system as *CCITT No.7*, and to the North American version as *ANSI No.7.*

SS7 Signaling Links. The signaling links are carried on the digital transmission channels that were developed in the 1960s for digital (PCM) trunks. In the U.S., these signaling links generally transmit at 56 kb/s. In most other countries the transmission is at 64 kb/s [1].

These bit rates greatly exceed the 2400–4800 bits/s rate of SS6 signaling links. SS7 call-control messages usually include more information than their counterparts in SS6, and support new telecommunications services. SS7 labels are larger than those in SS6, and contain information that has a network-wide scope. SS7 messages can therefore be routed to their destinations without the band-number translations that are necessary in SS6 (Section 6.2).

7.1 SS7 STRUCTURE

In Chapter 6, SS6 signaling has been described in terms of a *processor* at a signaling point that performs call-control and other functions, and *signaling links* that allow the processor to communicate with processors at other signaling points.

SS7 is defined in terms of messages and functions; hardware architecture issues are left to tee equipment manufacturers. The signaling system is divided into a number of *protocols* (or *parts*), each of which handles a group of related functions. The interfaces between these parts were defined in the early stages of the specification work. The various parts could then be specified simultaneously, and independently of each other. This has made the overall specification effort more manageable.

7.1.1 SS7 Hierarchy

The parts of SS7 are organized in a four-level hierarchy [1–4]. We say that a higher-level part is a *user* of services provided by a lower-level part. This arrangement is similar to the seven-layer structure of the *open systems interconnection* (OSI) protocols for data communications that have been specified by the International Standards Organization (ISO).

Several efforts have been made to align the SS7 levels and OSI layers. However, this has been only partially successful. OSI layers 1 and 2 correspond with SS7 levels 1 and 2, but things start to differ in the higher parts of the hierarchies. In what follows, the four-level SS7 hierarchy will be used. It works well for trunk-related applications, and less well for other applications.

7.1.2 SS7 Protocols

The protocols of SS7, and their levels, are shown in Fig. 7.1-1.

Message Transfer Part (MTP). This provides message transfer services for its *users*. It is divided into three parts—denoted as MTP1, MTP2, and MTP3—that occupy levels 1,2 and 3 of the SS7 hierarchy.

A MTP-user passes its outgoing messages to—and receives its incoming messages from—the MTP3 at its signaling point. A signaling point has one MTP3.

A combination of a MTP2 and a MTP1 represents a signaling link at a

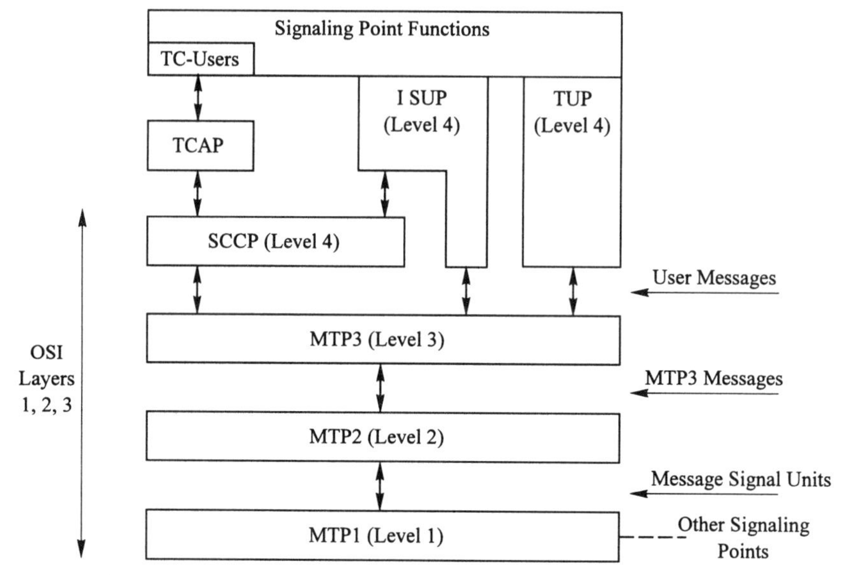

Figure 7.1-1 Structure of SS7. (From Rec. Q.700. Courtesy of ITU-T.)

signaling point (a signaling link between two signaling points consists of a MTP1/MTP2 combination each point). A signaling point that terminates n signaling links has n of these combinations.

Telephone User Part (TUP) (a MTP user) is a protocol for telephony call control, and for trunk maintenance. It is very similar to SS6, but includes a number of additional features.

Integrated Services User Part (ISUP) (another MTP user) is a protocol for call-control and trunk-maintenance procedures in both the telephone network and the ISDN.

TUP and ISUP messages are trunk-related: they contain information concerning a particular trunk with TUP or ISUP signaling, and are sent by the exchange at on end of the trunk to the exchange at the other end.

Signaling Connection Control Part (SCCP). This protocol (a MTP user) provides functions for the transfer of messages that are not trunk-related. Its users are ISUP and TCAP. SCCP does not fit neatly into the four-level hierarchy, because it is at the same level as ISUP.

The combination of MTP and SCCP corresponds to OSI layers 1, 2, and 3, and is known as the *network services* part of SS7.

Transaction Capabilities Application Part (TCAP). Transactions are operations that are not related to individual trunks, and involve two signaling points (2.2.4). The TCAP protocol provides a standard interface to *TC-users* (functions at a signaling point that are involved in transactions of various kinds). In turn, TCAP is a user of SCCP.

7.1.3 Messages and Message Transfer

In Fig. 7.1-1 we disinguish messages of several types in a signaling point. *User messages* are messages between a level-4 protocol and the MTP3, and are named after the level-4 protocol: we speak of TUP, ISUP and SCCP messages. *MTP3 messages* are messages between the MTP3 and a MTP2. *Message signal units* (MSU) are messages between the MTP2 and MTP1 of a signaling link at a signaling point, and between the MTP1s at both ends of a signaling link. A MSU contains a message originated by MTP3, or by a MTP3 user.

User messages are transferred between two "peer" level-4 protocols at two signaling points. Figure 7.1-2 illustrates the transfer of a TUP message from TUP-A to TUP-C (at signaling points A and C, respectively) that is routed via signal transfer point B.

At signaling point A, TUP-A passes the TUP message *downwards* to its MTP3, which expands it into a MTP3 message, and passes it to the MTP2 of the signaling link to B. MTP2 expands the MTP3 message into a message signal unit (MSU) and passes it to its MTP1.

The MSU traverses the signaling link, and arrives at the MTP1 of signaling point B, where MTP2 extracts the MTP3 message, and passes it to its MTP3. MTP3 transfers the MTP3 message to the MTP2 of the signaling link to C.

The second leg of the message transfer is similar: is passed and expanded downwards in signaling point B, traverses the signaling link between B and C (as a MSU), and is passed upwards in signaling point C. It finally arrives as a TUP message at TUP-C.

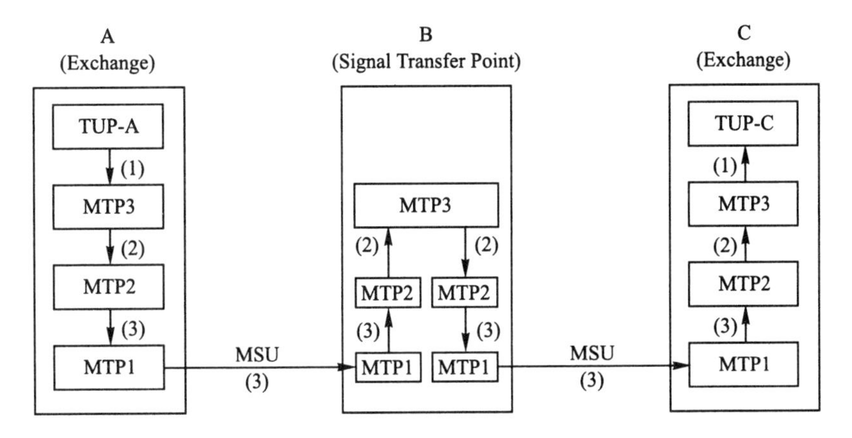

Figure 7.1-2 Transfer of a TUP message. (1): TUP message. (2): MTP3 message. (3): message signal unit.

7.2 IDENTIFICATION OF SIGNALING POINTS AND TRUNKS

Signaling system No.7 identifies signaling points (exchanges, service control points, etc.) and trunks with parameters that have a nationwide scope.

7.2.1 Point Codes

Each SS7 signaling point in a network is identified by a *point code* (PC). Most signaling points have only one (national) point code. However, an *international switching center* (ISC) is identified in the network of its country by a national PC, and in the international network by an international PC.

7.2.2 ANSI No.7 Point Codes

The format of ANSI No.7 point codes (used in the U.S. only) are shown in Fig. 7.2-1(a). The PC has three eight-bit fields that contain the parameters N, C, and M [1]. Parameter N identifies the network of a particular telecom in the U.S. (Ameritech, Sprint, AT&T, etc.). Parameter C represents a cluster of signaling points within a network. For example, a particular value of C may identify the exchanges with signaling links to a particular pair of STPs. Parameter M identifies a signaling point within a cluster. The ANSI No.7 point codes are assigned by Bellcore.

7.2.3 International Point Codes

Point codes for the international network are assigned by CCITT [5], and consist of parameters Z and V (3 bits each), and U (8 bits)—see Fig. 7.2-1(b).

Z identifies six major geographical *world zones*. In decimal representation:

2 Europe
3 North America
4 Mid-east and most of Asia
5 Australia and part of Asia (Indonesia, Malaysia, Thailand, Guam, etc)
6 Africa
7 Latin America

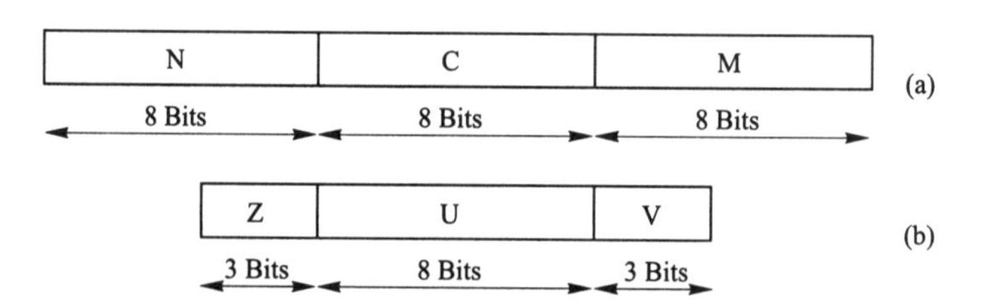

Figure 7.2-1 Point code formats. (a): ANSI No. 7. (b): CCITT No. 7. (From Rec. Q.708. Courtesy of ITU-T.)

These world zones are different from the world zones represented by the initial digits in country codes (1.2.3).

Parameter U identifies an area of a national telecommunication network within a world zone. Most countries have only one network that covers the entire country, and are therefore represented by one value of U. For example, the combination $Z = 2, U = 168$ represents the national network of Bulgaria.

V identifies a particular international switching center in the network—or network area—specified by Z and U. In countries with one national network, up to seven ISCs can be identified.

Countries with more than one national network have several U codes. For example, the U.K. national networks operated by British Telecom and Mercury Telecommunications are identified by $Z = 2, U = 068$, and $Z = 2, U = 072$.

In the U.S., values of U are assigned to particular areas of the long-distance networks operated by the vatious international carriers (AT&T, MCI, Sprint, etc.). For example, a value of U identifies the east-coast area of AT&T's long-distance network, and V identifies an ISC in that area of the network.

CCITT has allocated the range $U = 020\text{-}059$ in zone $Z = 3$ to the U.S.

7.2.4 National Point Codes in Other Countries

Countries other than the U.S. use 14-bit national point codes. The code assignments are made by the individual national telecoms.

7.2.5 Identification of Trunks.

A trunk group with TUP or ISUP signaling is identified uniquely in a national network (or in the international network) by the point codes of the exchanges that are interconnected by it.

The *circuit identification code* (CIC) identifies a trunk within a trunk group. The CIC field has a length of 12 bits, and thus can identify trunks in groups of up to 4095 trunks.

7.3 SS7 SIGNAL UNITS AND PRIMITIVES

SS7 signal units (SU) have different lengths, but always occupy an integral number of octets (groupings of eight bits). A SU consists of an ordered set of fields with parameters. In the documentation of the early SS7 parts (MTP and TUP), the SUs are shown as in Fig. 7.3-1(a). Par.1, Par.2, etc. denote the first, second, etc. parameter fields of the SU. In the more recently defined parts of SS7, a SU is shown as a stack of octets—see Fig. 7.3-1(b). Bits in the octets are numbered from right to left, and bit 1 of octet 1 is the first bit sent out. For uniformity, this representation is used throughout this book.

The lengths of SS7 parameter fields are not limited to integral multiples of octets. For example, Par.2 and Par.3 in Fig. 7.3-1(b) have a length of 12 bits.

A working signaling link carries a continuous SU stream in each direction.

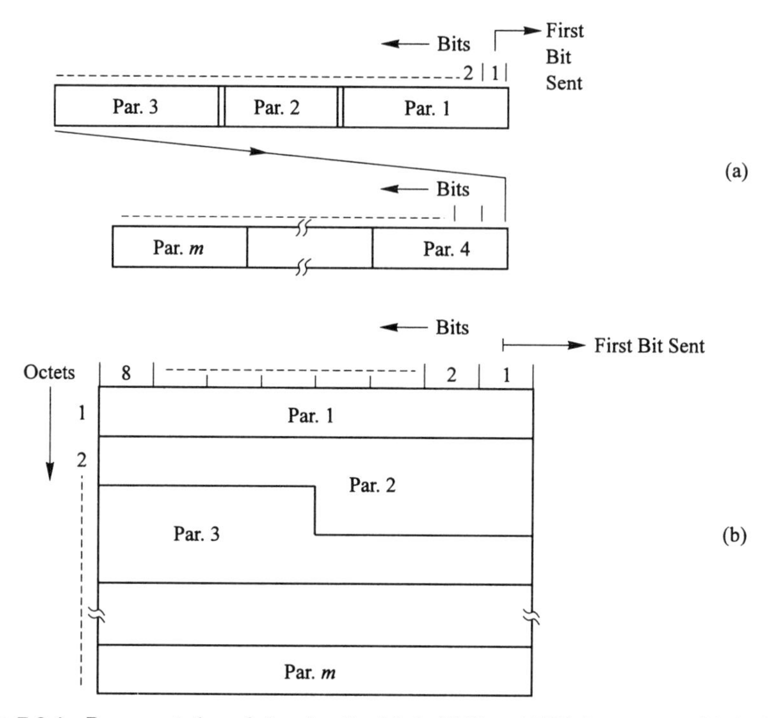

Figure 7.3-1 Representation of signal units. (a): in MTP and TUP documents. (b): in ISUP, SCCP, and TCAP documents.

7.3.1 Signal Unit Types

In addition to the message signal unit, SS7 includes two other SU types [6]. The *link status signal units* (LSSU) and *fill-in signal units* (FISU) originate in the MTP2 at one end of the signaling link, and are processed by the MTP2 at the other end (they are the SS7 equivalents of the "link" SUs described in Section 5.2). LSSUs are used for the control of a signaling link, and FISUs are sent when no MSUs or LSSUs are waiting to be sent out.

Figure 7.3-2 shows a MSU in some detail. It consists of a MTP3 message, surrounded by MTP2 data. Length indicator (LI) has a dual role. In the first place, it indicates number of octets, measured from octet 4 through octet $(n-2)$. In addition, the value of LI implies the SU type. In MSUs, LI exceeds 2, LSSUs have LI = 1 or 2, and FISUs have LI = 0.

The MTP2 data are added by the MTP2 of the signaling point where the message originates, and are processed, and then removed, by the MTP2 at the signaling point on the other end of the signaling link.

The *MTP3 message* consists of the *service information octet* (SIO) and the *signaling information field* (SIF). The maximum length of SIF is 272 octets.

SIO identifies the MTP-user (TUP, ISUP, or SCCP) that has originated the message, and is used by the MTP3 in the destination signaling point to deliver the message to the "peer" user.

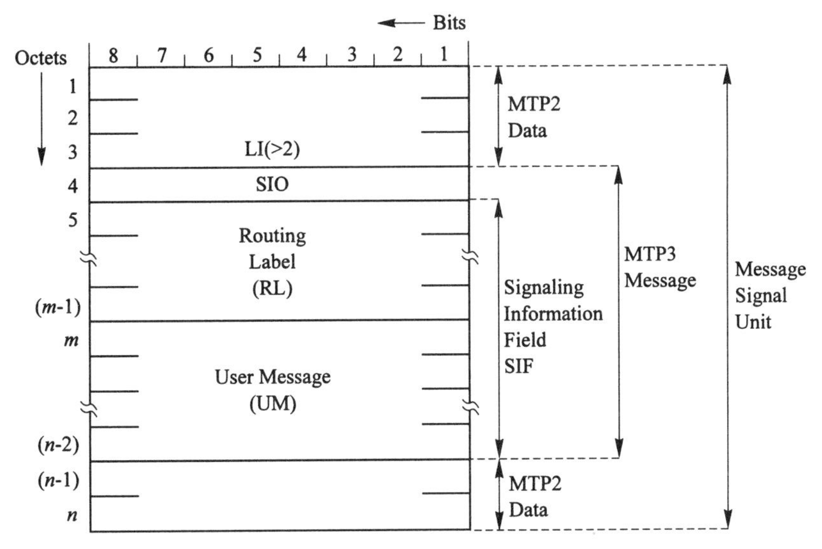

Figure 7.3-2 Message signal unit. (From Rec. Q.700. Courtesy of ITU-T.)

The SIF consists of the *routing label* (RL) and the *user message* (UM). The routing label contains parameters that are used by the MTP3s in the signaling points along the message path to route the MSU to its destination signaling point. The user message contains information for the MTP-user at the destination signaling point.

The user message, and the information in SIO and RL, are supplied by the originating MTP-user. The parameters in SIO and RL are interface parameters, and are used by the MTP3s, and by the MTP-user at the destination signaling point. The user message is passed transparently (not examined by MTP3).

7.3.2 Primitives

In the CCITT model of SS7, the messages between protocols in a signaling point are passed in standardized interface elements called *primitives* [7]. There are four types of primitives: *requests* and *responses* pass information from a higher level protocol to a lower level protocol, and *indications* and *confirmations* pass information in the opposite direction.

Primitives between two protocols are named after the lower level protocol. For example, the primitives for the message transfer between MTP3 and the MTP-users are known as *MTP-transfer* primitives.

CCITT has defined the information to be included in each primitive. Since primitives pass information inside a signaling point only, their (software) implementation is left to the equipment manufacturers.

Figure 7.3-3 shows the message-transfer primitives between MTP3 and a MTP-user at a signaling point. On outgoing messages, the user passes a MTP-transfer request that includes the user message, and the parameters in SIO and

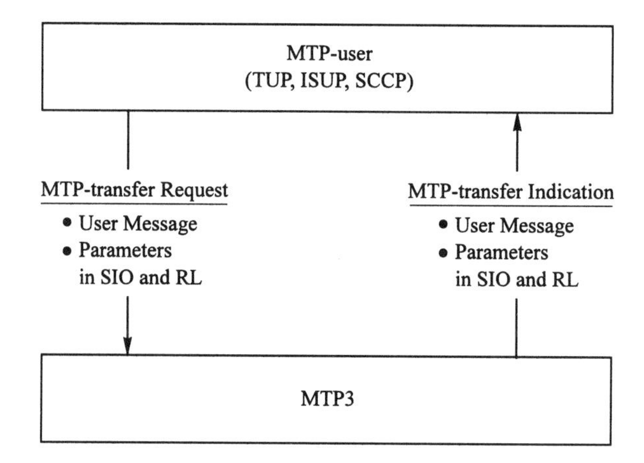

Figure 7.3-3 Message transfer primitives. (From Rec. Q.701. Courtesy of ITU-T.)

RL. On incoming messages, MTP3 delivers these data to the MTP-user, in a MTP-transfer indication that contains the same information.

7.4 ACRONYMS

ANSI	American National Standards Institute
CB	Check bits
CCITT	International Telegraph and Telephone Consultative Committee
CIC	Circuit identification code
F	Flag
FISU	Fill-in signal unit
ISC	International switching center
ISO	International Standards Organization
ISDN	Integrated Services Digital Network
ISUP	ISDN user part
ITU	International Telecommunication Union
kb/s	Kilobits per second
LI	Length indicator
LSSU	Link status signal unit
MSU	Message signal unit
MTP	Message transfer part
OSI	Open systems interconnection
PC	Point code
PCM	Pulse code modulation
RL	Routing label
SCCP	Signaling connection control part
SIF	Signaling information field
SIO	Service information octet

SL	Signaling link
SP	Signaling point
SS6	Signaling system No.6
SS7	Signaling system No.7
STP	Signal transfer point
TCAP	Transaction capabilities application part
TUP	Telephone user part
UM	User message

7.5 REFERENCES

1. A.R. Modaressi, R.A. Skoog, "Signaling System No.7: A Tutorial," *IEEE Comm. Mag.*, **28**, No.7, July 1990.
2. R. Manterfield, *Common Channel Signalling*, Peter Peregrinus Ltd, London, 1991.
3. K.G. Fretten, C.G. Davies, "CCITT Signalling System No.7: Overview," *Br. Telecomm. Eng.*, **7**, April 1988.
4. *Specifications of Signalling System No.7*, Rec. Q.700, CCITT Blue Book, **VI.7**, ITU, Geneva, 1989.
5. *Ibid.*, Rec. Q.708.
6. *Ibid.*, Rec. Q.703.
7. *Ibid.*, Rec. Q.701.

8

SS7 MESSAGE TRANSFER PART

The *message transfer part* (MTP) of SS7 has two main functions. In the first place, it handles the transfer of MTP-user messages across the SS7 signaling network. In the second place, it includes functions to keep the message traffic flowing when failures occur in the signaling network.

8.1 INTRODUCTION TO MTP

8.1.1 Structure of MTP

MTP is divided into three parts, located at levels 1, 2, and 3 of the SS7 hierarchy. The main functions of these parts are outlined below [1–5].

MTP Level 1 (MTP1) is the physical *signaling data link* (SDL), which consists of a pair of 64 kb/s digital transmission channels, and transports SS7 signal units between two signaling points. MTP1 is described in Section 8.2.

MTP Level 2 (MTP2). A *signaling link* (SL) between signaling points A and B consists of a SDL between the signaling points, and MTP2 functions located at both signaling points (Fig. 8.1-1). The MTP2 functions are similar to the functions of the signaling terminals in signaling system No.6 (see Section 6.1). They relate to individual signaling links, and include synchronization, and the detection and correction of errors in *message signal units* (MSU). MTP2 is discussed in Sections 8.3 through 8.6.

MTP Level 3. MTP3 is the interface between MTP and the MTP-users (level 4 protocols) at a signaling point. In addition to providing services for the

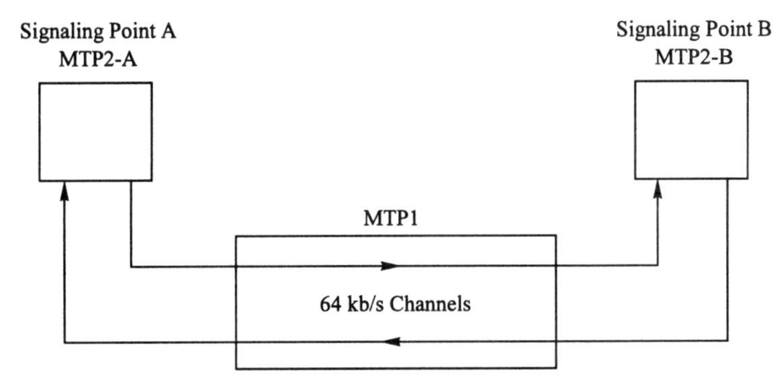

Figure 8.1-1 SS7 signaling link.

transfer of user messages, MTP3 includes procedures to reroute messages when a failure occurs in the SS7 signaling network. MTP3 is discussed in Sections 8.7 through 8.9.

At a signaling point, there is an MTP1 and an MTP2 for each signaling link, and a single MTP3 (Fig. 8.1-2). The signaling links carry message signal units, *link status signal units* (LSSU), and *fill-in signal units* (FISU). The LSSUs and FISUs originate in the MTP2 at one end of a signaling link, and terminate in the MTP2 at the other end.

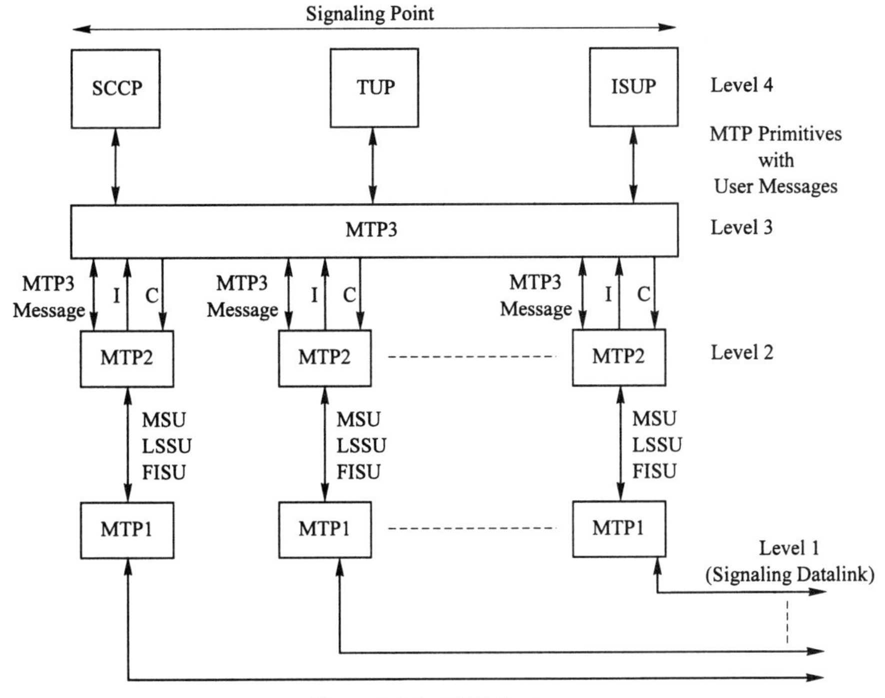

Figure 8.1-2 MTP structure.

8.1.2 Message Transfer

A MTP user passes its outgoing message to MTP3, in a MTP-transfer primitive. MTP3 expands the user message into a MTP3 message, selects the outgoing signaling link, and passes the MTP3 message to the MPT2 of that link. The MTP2 expands the MTP3 message into a MSU, which is sent out.

A MTP2 extracts the MTP3 message from a received MSU, and passes it to MTP3. MPT3 extracts the user message, and passes it to the appropriate user.

8.2 MTP LEVEL 1

MTP1 defines the physical aspects of SS7 signaling data links, which are carried by digital 64 kb/s time-division channels (time slots) of digital transmission systems [6].

In today's telecommunication networks, the number of digital trunks exceeds the number of SS7 signaling links by about two orders of magnitude. It is not economical to install digital transmission systems that are dedicated to signaling data links. Instead, the digital transmission systems are shared by trunks and signaling data links.

Figure 8.2-1 shows a building that houses an exchange that has SS7 trunks, and a signal transfer point. The digital transmission systems (a) that enter the building from other buildings of a telecommunication network can be first- or higher-order transmission systems. Some of these carry trunks only, and others carry trunks and signaling links. After demultiplexing the higher order systems, a number of first-order multiplexes (b) is obtained. These multiplexes have 24 or

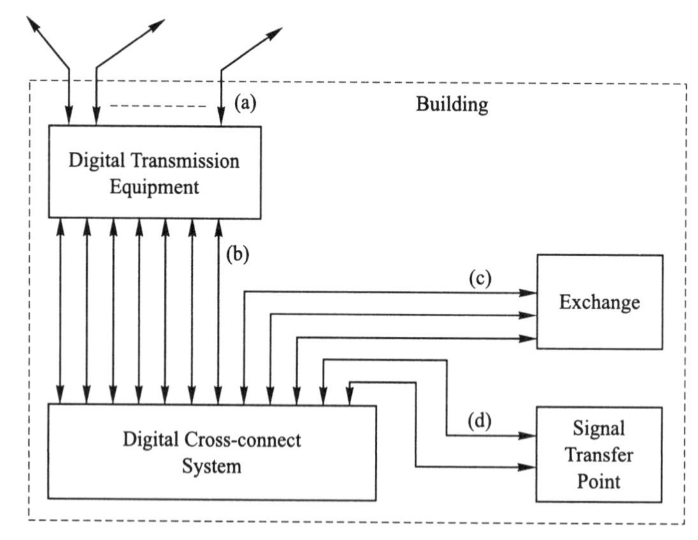

Figure 8.2-1 Attachment of signaling data links to exchanges and signal transfer points. (a): digital multiplex circuits of any order. (b),(c),(d): first-order digital multiplex circuits.

30 digital 64 kb/s channels (1.5.2). The multiplexes are attached to a *digital cross-connect system.*

Digital cross-connect systems are very similar to the digital switchblocks in exchanges (1.7.2). They have a number of ports to which first-order digital multiplexes are attached, and the paths in cross-connect systems are bidirectional 64 kb/s paths that connect a channel (time slot) of one multiplex to a channel of another multiplex. However, unlike the paths in exchange switchblocks, these paths are semi-permanent, and are established and released under command of telecom personnel.

In Fig. 8.2-1, each channel in the multiplexes (b) corresponds to a channel in one of the digital transmission systems (a). The cross-connect system segregates the channels attached to the exchange and the channels attached to the signal transfer point. This is done by setting up paths such that the channels in the multiplexes (c) are associated with the digital trunks and SS7 signaling data links of the exchange, and the channels in multiplexes (d) carry the signaling data links of the signal transfer point.

Transfer Rates of Signaling Data Links. The channels of North American first-order digital multiplex transmission systems (T_1 systems) nominally transfer bits at a rate of 64 kb/s—see Fig. 1.5-4(a). However, there is a limitation on bit patterns: at least one bit in every time slot should be a "1" (pulse). This is minor problem for time slots that carry the speech samples of PCM trunks, and is solved by substituting the bit pattern 0000 0000 by the pattern 0000 0001. This causes a small—but acceptable—distortion in the decoded analog waveform.

The sequences of 1s and 0s on a SS7 signaling data link are unpredictable. Therefore, in the time slots associated with signaling data links, bit 8 is permanently set to "1." This reduces the transfer rate of SS7 signaling links carried by these transmission systems to $8000 \times 7 = 56$ kb/s.

The first-order digital multiplexed transmission systems defined by CEPT, and used in most countries outside North America, use a form of transmission that can handle any sequence of 1s and 0s. SS7 signaling links on these transmission systems can therefore operate at 64 kb/s.

8.3 OVERVIEW OF MTP LEVEL 2

In a signaling point, each signaling data link is connected to a MTP2. The MTP2 functions are similar to those of the signaling terminals in SS6 signaling. The primary MTP2 responsibility is to transfer MSUs across the signaling link, including error detection and correction. MTP2 also monitors and controls the status of the link [5,7].

The reliability objectives for SS7 signaling links are [8]:

1. The probability that errors in a received MSU are not detected should be less than one in 10^{10}.

2. Failures should not cause the loss of more than one MSU in 10^7.
3. Less than one in 10^{10} messages should be delivered out-of-sequence to the user parts.

In SS6 signaling, error correction consists of retransmitting a message signal unit that has incurred transmission errors (6.1.3). The retransmitted MSU arrives after MSUs that were sent later and did not incur errors (out-of-sequence delivery).

Since the *bit error rate* on typical signaling links is in the range of 10^{-4} to 10^{-6}, objective 3 dictates an error-correction procedure that does not cause out-of-sequence message delivery. This is accomplished as follows: when a MTP2 receives a mutilated message signal unit, say MSU_m, it discards that MSU *and* all subsequently received MSUs, until it receives an error-free retransmitted copy of MSU_m. Conversely, when the distant MTP2 receives the negative acknowledgment of MSU_m, it retransmits that MSU *and* all subsequently sent MSUs.

SS7 error correction is discussed in Sections 8.4 and 8.5. By eliminating the major cause of out-of-service MSU delivery, SS7 allows simpler (and more drastic) procedures to cope with the receipt of an out-of-sequence MSU.

8.3.1 MTP2 Structure

The main parts of MTP2 are shown in Fig. 8.3-1. *Link control* (LC) controls the other functional units of MTP2. In the first place, it coordinates the transfer of signal units. LC also monitors the operation of the signaling link. It communicates with its MTP3, accepting link status commands (C), and reporting link status information with indications (I). Finally, LC communicates with the LC at the distant end of the signaling link, using link status signal units.

The MTP3 in a signaling point places its outgoing MTP3 messages in the *output buffer* (OB) of the signaling link. *Retransmission buffer* (RB) stores messages that have been sent, but have not yet been positively acknowledged by the distant MTP2.

Each message to be transmitted or retransmitted passes through *outgoing processing* (OP), and then enters the signaling data link as a MSU. A signal unit received from the signaling data link is processed by *incoming processing* (IP). The MTP3 messages in the MSUs that are accepted by IP are placed in *input buffer* (IB), and are retrieved by MTP3.

All buffer transfers are "first in, first out:" a MTP2 takes outgoing messages from its output buffer in the same order in which they were placed there by MTP3, and MTP3 takes received messages from the input buffer of a MTP2 in the same order in which they were entered by MTP2. This is one of the requirements for in-sequence MSU delivery.

8.3.2 Outgoing Processing

In this section, we examine the MTP2 actions in transmitting a MSU, with the aid of Fig. 8.3-1 and Fig. 8.3-2.

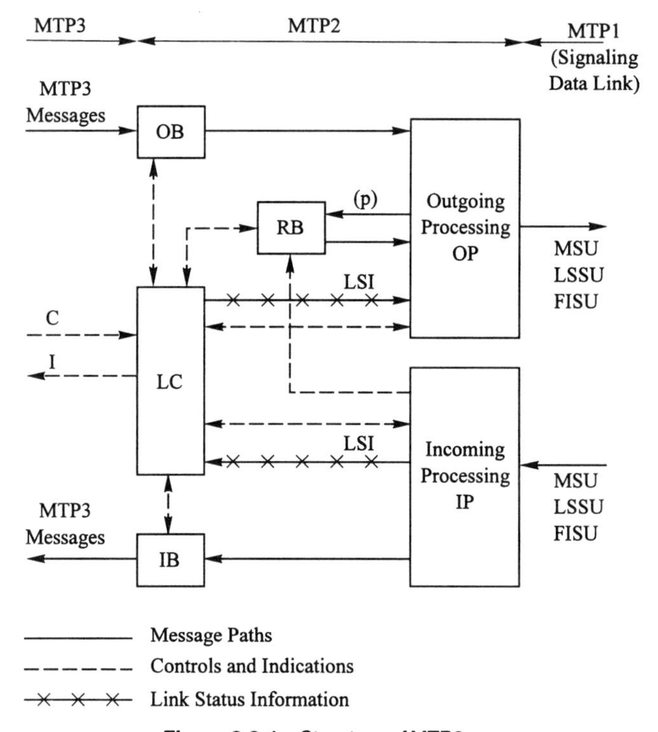

Figure 8.3-1 Structure of MTP2.

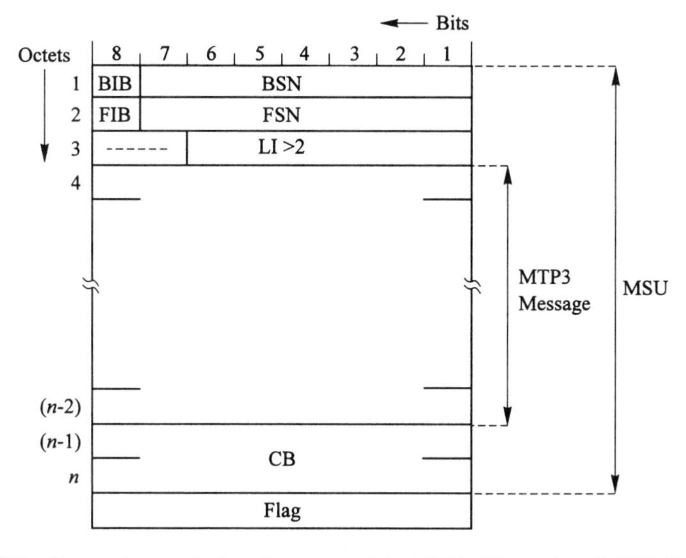

Figure 8.3-2 Parameters added and processed by MTP2. (From Rec. Q.703. Courtesy of ITU-T.)

A working signaling link transfers uninterrupted streams of signal units in both directions. When the transmission of a SU has been completed, LC selects the next SU to be sent out. LSSUs have the highest priority. When no LSSUs have to be sent, the oldest message in buffer OB or buffer RB is selected for transmission.

The rules for selecting a message from OB or RB depend on the type of error control, and are discussed in Sections 8.4 and 8.5. When no LSSU or message is awaiting transmission, fill-in signal Units (FISU) are sent.

MTP2 Parameters. Outgoing processing adds a number of MTP2 parameters to the message (Fig. 8.3-2):

Forward Sequence Number (FSN). A forward sequence number is assigned to a message when it is removed from the OB for its initial transmission. The FSN field has seven bits, and the sequence numbers are therefore confined to the range 0–127. The numbers are assigned cyclically to consecutive messages: 0, 1, 2, ... 126, 127, 0, 1, ... etc.

Backward Sequence Number (BSN). This number indicates the FSN of the most recently accepted incoming MSU.

Forward- and Backward Indicator Bits (FIB, BIB) are discussed in Section 8.4.

Length Indicator (LI) indicates the length (number of octets) of the message.

Check Bit Field (CB). As in SS6 signaling, error detection uses cyclic redundancy checking (see Section 5.2.3). The SS7 checking algorithm is a modulo 2 division of the contents in octets 1 through $(n-2)$ of the MSU by the 17-bit divisor 1 0001 0000 0010 0001. The remainder of the division is placed in the CB field.

Flag (F). Consecutive signal units on a signaling data link are separated by a flag octet F, coded 01111110.

We now consider the outgoing processing of a MTP3 message from buffer OB (Fig. 8.3-1):

1. The message is moved from OB to OP, where its length (LI) is determined, and a FSN is assigned.
2. A copy of the message, including its LI and FSN, is sent to OB (path P), where it remains until a positive acknowledgment is received.
3. The values of BSN, FIB, and BIB are determined.
4. The contents of CB are determined.
5. Zero insertion. All SUs are separated by flags. Since the 01111110 flag pattern may also occur in the MSU (octets 1 through n), the MSU is

scanned, and a "zero" is inserted behind each string of five consecutive "ones." In this way, the 01111110 pattern never occurs inside the MSU. The MSUs on a signaling data link thus are not always completely identical to MSU shown in Fig. 8.3-2, because of possible inserted zeros.
6. A flag is appended to the MSU, which then enters the signaling data link.

8.3.3 Incoming Processing of Signal Units

Incoming processing of received SUs (MSUs, LSSUs, and FISUs) is described below, again with the aid of Fig. 8.3-1 and Fig. 8.3-2.

1. Zero deletion. Having recognized the closing flag of the previous signal unit, IP scans the subsequent incoming bits, until it encounters the next flag. During the scan, it removes all zeros that follow five consecutive ones. After this, the SU consists again of an integral number of octets.
2. Error detection. This consists of modulo 2 division of the contents of octets 1 through $(n-2)$ by the divider polynomial. The result is compared against the received CBs. All SUs with errors are discarded. In SS6 signaling, all error-free SUs are accepted and processed (6.1.3). In SS7, error-free SUs are separated by type (LSSU, FISU, MSU—derived from the value of LI), and acceptance criteria depend on the SU type.
3. Error-free MSUs are "sequence screened." The basic criterion for acceptance of a MSU is that its FSN should exceed the FSN of the most recently accepted MSU by 1 (modulo 128), because this indicates that the MSU has been received in-sequence. This step is part of the procedure that maintains in-sequence delivery of MSUs when transmission errors occur.
4. IP processes the parameters BSN and BIB of accepted MSUs, which play a role in error correction (Sections 8.4 and 8.5).

8.3.4 Other Outgoing and Incoming Processing

So far, we have discussed the processing steps for the initial transmission and the acceptance of MSUs. We now examine the processing in other cases.

Retransmitted MSU. When a message that has already has been transmitted needs to be retransmitted, the message and parameters LI and FSN are copied from buffer RB. The retransmitted MSU thus has the same FSN value as the originally transmitted one. This is part of the procedure for in-sequence MSU delivery when transmission errors occur. Outgoing processing determines the values of FIB, BSN, and BIB, and then does steps 3 through 6 of Section 8.3.2.

Retransmitted MSUs remain in RB until they have been positively acknowledged.

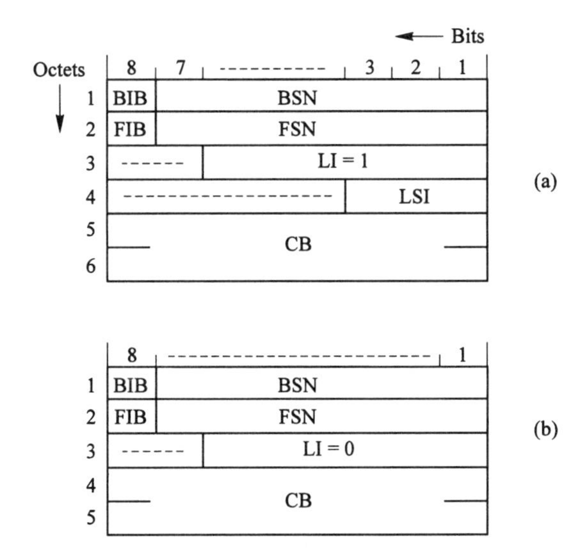

Figure 8.3-3 Link status signal unit (a) and fill-in signal unit (b). (From Rec. Q. 703. Courtesy of ITU-T.)

Transmission of LSSUs and FISUs. Outgoing LSSUs and FISUs originate at link control LC. They are processed by outgoing processing, and therefore include the parameters FSN, FIB, BSN, and BIB (Fig. 8.3-3). However, OP does not assign a "new" FSN value to these signal units; they receive the FSN value of the most recently transmitted MSU. Outgoing LSSUs and FISUs are never retransmitted, and link control periodically retransmits a LSSU until it receives a response from the LC in the distant MTP2. The parameters FSN, FIB, BSN and BIB are processed by incoming processing only if the LSSU or FISU is error-free and in-sequence (its FSN matches the FSN of the most recently accepted MSU). Link status information (parameter LSI) in error-free LSSUs is always passed to LC for processing.

8.4 BASIC ERROR CORRECTION

8.4.1 Introduction to Error Correction

Error correction of MSUs consists of screening received error-free MSUs for acceptance, positively acknowledging accepted MSUs, and retransmitting MSUs that have not been accepted by the distant MTP2.

CCITT has defined two error-correction procedures for SS7 [7]. Basic error correction, described in this section, is used on signaling links whose lengths do not exceed some 8000 km. Preventive cyclic retransmission, which is used on longer links, is discussed in Section 8.5.

We examine error correction of MSUs sent by MTP2-A on the signaling link of Fig. 8.1-1. The same procedure is used for MSUs sent in the opposite direction.

8.4.2 Actions at the MTP2s

Acknowledgments. Basic error correction includes *positive* and *negative* acknowledgments of received MSUs. The acknowledgment information for MTP2-A is in the BSN (backward sequence number) and BIB (backward indicator bits) of SUs (of any type) sent by MTP2-B. In both positive and negative acknowledgments, the BSN is equal to the FSN (forward sequence number) of the MSU that has been most recently accepted by MTP2-B. Positive and negative acknowledgments are indicated by the BIB bits in consecutive SUs. In Fig. 8.4-1, SU_2 has the same BIB value as SU_1. This signifies the positive acknowledgment of MSUs with FSN up through 26. SU_3 indicates a negative acknowledgment, because its BIB differs from the BIB of SU_2.

Response to Acknowledgments. MTP2-A can be in one of two transmission states: transmitting MSUs from its output buffer, or retransmitting previously sent MSUs from its retransmission buffer. When receiving a positive acknowledgment, MTP2-A removes the acknowledged MSUs from its RB, and remains it its current state. On receipt of a negative acknowledgment, it starts (or restarts) a retransmission cycle, beginning with the MSU in its RB whose FSN exceeds the FSN of the most recent positively acknowledged MSU by 1 (modulo 128).

In Fig. 8.4-1, after receipt of the negative acknowledgment in SU_3, MTP2-A starts retransmitting MSUs, beginning with the MSU which has FSN = 27.

The FIBs of the SUs sent by MTP2-A are copied from the BIB of the most recently received SU. For example, in the MSU sent after receipt of SU_3, FIB is set to 1.

Acceptance of Received MSUs. All SUs received at MTP2-B input processing are checked for errors, and SUs with errors are discarded. Input

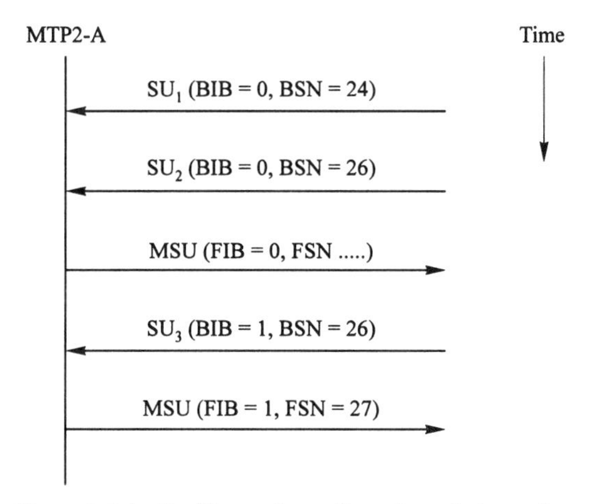

Figure 8.4-1 Positive and negative acknowledgments.

processing subjects error-free MSUs to two additional screening steps. It first compares FIB of the received MSU with the BIB it sent in its latest SU. If FIB is not equal to BIB, then MTP2-A has not yet received the latest negative acknowledgment from MTP2-B, and the MSU is discarded. If FIB = BIB, the MSU is sequence-checked. It is accepted if its FSN exceeds the FSN of the most recently accepted MSU by 1 (mod 128). If the check passes, MTP2-B accepts the MSU (places it in its input buffer), and positively acknowledges the MSU in the next sent SU. If the sequence check fails, it discards the MSU, and includes a negative acknowledgment in its next SU.

8.4.3 Error Correction Example

An example of basic error correction for MSUs sent by MTP2-A is shown in Fig. 8.4-2. The LSSUs and FISUs from MTP2-A are not shown, because they play no role in the error correction of the sent MSUs. The time required by the SUs to traverse the signaling link is indicated by the sloping lines.

The values of FSN and FIB in the MSUs from MTP2-A, and of BSN and BIB in SUs (MSUs, LSSUs, FISUs) from MTP2-B, are shown at both MTPs. This makes it easier to correlate the figure and the text below.

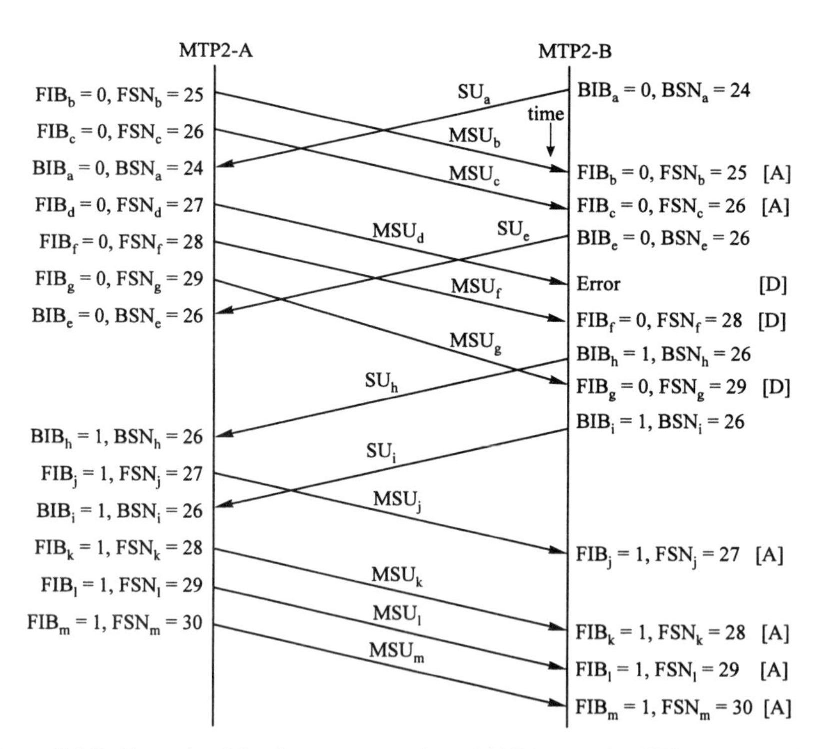

Figure 8.4-2 Example of basic error correction of MSUs sent by MTP2-A. [A]: MSU is accepted. [D]: MSU is discarded.

We assume that, prior to the transmission of SU_a, MTP2-B has been accepting MSUs, and has been sending SUs with BIB = 0. MTP2-A has been sending MSUs from its output buffer, with FIB = 0.

SU_a positively acknowledges MSUs with FSN up through 24. MTP2-A keeps sending MSUs. MSU_b and MSU_c pass the tests at MTP2-B, and are accepted, because FIB_b and FIB_c are equal to BIB_a, and the FSNs indicate the proper sequence. SU_e positively acknowledges the MSUs in SU_e ($BIB_e = BIB_a$, and $BSN_e = 26$).

MTP2-A keeps transmitting MSUs from its output buffer. MSU_d incurs a transmission error, and is discarded by MTP2-B incoming processing. MSU_f arrives without error, and with the proper FIB value, but fails the sequence test. It is therefore discarded, and negatively acknowledged by SU_h (BIB_h not equal to BIB_e, $BSN_e = 26$).

MSU_g is discarded because its FIB_g does not match BIB_h. In this situation, MTP2-B does not send another negative acknowledgment, but signals a positive acknowledgment in SU_i ($BIB_i = BIB_h$), in which BSN_i indicates that MSU_c (with FSN = 26) is still the latest accepted MSU.

MTP2-A receives the negative acknowledgment in SU_h. Since $BSN_h = 26$, it starts a retransmission cycle of MSUs in its RB, beginning with MSU_j (retransmission of MSU_d, which has $FSN_d = 27$), and setting FIB to 1. The receipt of the positive acknowledgment in SU_i does not change things at MTP2-A, which continues its retransmission cycle until all MSUs in its retransmission buffer have been sent again.

MTP2-B accepts MSU_j, MSU_k, and MSU_l, because their FIBs are equal to BIB_i, and their FSNs indicate in-sequence delivery.

When MSU_l (retransmission of MSU_g) has been sent, the retransmission cycle is complete, and MTP2-A resumes the initial transmission of MSUs in its output buffer.

8.4.4 Message Transfer Delays

We now consider the initial transmission of a MSU, sent by MTP2-A. It arrives at MTP2-B after a propagation delay (T_p) (Fig. 8.4-3). Suppose that the MSU is rejected by MTP2-B, and negatively acknowledged in a signal unit (SU). This signal unit arrives at MTP2-A after a delay of at least T_p.

MSU_r, the retransmitted copy of MSU, arrives at MTP2-B with a delay of at least $3T_p$ (measured from the time when the original MSU left MTP2-A).

Since SS7 call-control applications are real-time critical, basic error correction is used only on signaling links, which have propagation times (T_P) that do not exceed 40 ms [1]. This corresponds to signaling links with lengths below about 8000 km.

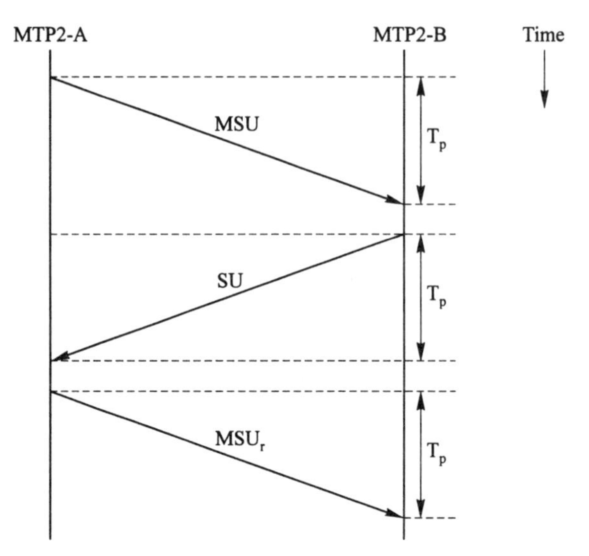

Figure 8.4-3 Transfer delay of a retransmitted MSU.

8.5 PREVENTIVE CYCLIC RETRANSMISSION

8.5.1 Introduction

Preventive cyclic retransmission (PCR) is designed for use on signaling links with large propagation times (T_P), for example, signaling links that are carried on satellite circuits (1.5.4). As in basic error correction, the FSN identifies the position of a MSU in its original sequence of transmission, and the BSN always identifies the most recently accepted message signal unit.

PCR uses positive acknowledgments only. Indicator bits FIB and BIB are ignored (they are permanently set to "1"), and incoming processing simply accepts or discards an error-free MSU based on the value of its FSN, which has to exceed the FSN of the most recently accepted MSU by one unit (modulo 128).

8.5.2 Preventive Retransmission Cycles

Whenever there is no LSSU or new MSU to be sent out, MTP2-A, in lieu of sending a FISU, starts a *preventive retransmission* cycle in which the MSUs in its retransmission buffer are retransmitted in-sequence and starting with the oldest one (lowest FSN). The retransmitted copies of MSUs that have already been accepted by MTP2-B now arrive out-of-sequence and are discarded. However, if any of the original MSUs were not accepted, their retransmission takes place much sooner than in basic error correction (where a time of at least $2T_P$ is required for the negative acknowledgment to reach MTP2-A). This is why PCR can be used on signaling data links with propagation times that make basic error correction impractical.

A preventive retransmission cycle ends in one of two ways. If all MSUs in RB have been retransmitted and no LSSU or new MSU has to be sent out, MTP2-A starts sending FISUs. A retransmission cycle also ends when a LSSU or new MSU has to be sent.

8.5.3 Forced Retransmission Cycles

Under normal traffic loads, about 20% of the octets on a signaling link with PCR carry MSUs that are transmitted for the first time. This provides ample opportunity to start and complete preventive retransmission cycles. However, during bursts of high MSU volume, preventive cyclic retransmissions may not occur as often as needed. Therefore, PCR also includes *forced* retransmission cycles.

Link control (LC) in a MTP2 (Fig. 8.3-1) constantly monitors the number of message signal units (N_1), and the number of octets (N_2), in its RB. If either of these numbers exceeds a predetermined threshold value, a *forced retransmission cycle* is initiated. This type of retransmission cycle ends only after all MSUs in the RB have been retransmitted.

8.5.4 Comparison with Basic Error Correction

As has been mentioned, PCR is used on signaling links with propagation times in excess of some 40 ms, because basic error correction on such links results in MSU queuing delays that are unacceptable for call control applications (TUP, ISUP).

On the other hand, basic error correction is preferred on signaling links with propagation times below 40 ms, because it allows higher MSU loads on the signaling links than PCR [7].

8.6 SIGNALING LINK MANAGEMENT

Signaling link management responsibilities at a signaling point are shared by the MTP2s of the individual signaling links and MTP3. This section describes the MTP2 signaling link management functions [7]. The MTP2 management functions monitor the status of the signaling link and, when necessary, pass status indications to MTP3.

We consider a signaling link between signaling points A and B (Fig. 8.6-1). The management functions are performed by link controls (LC-A and LC-B). Each LC communicates with the MTP3 at its signaling point, accepting controls (C), and sending status indications (I).

LC-A and LC-B also send link status information (LSI) to each other. LSI originated by LC-A is embedded in a link status signal unit (LSSU) by OP-A (outgoing processing at A). IP-B (incoming processing at B) extracts the LSI, and passes it to LC-B. Transfer of LSI in the other direction is done in the same manner.

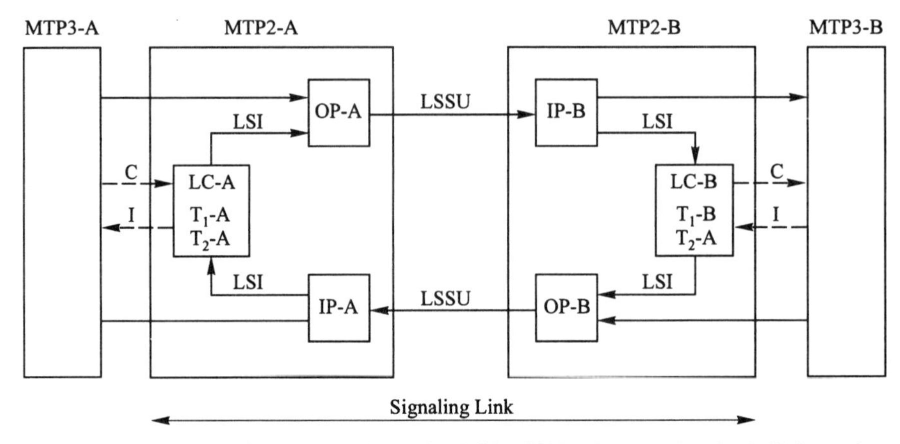

Figure 8.6-1 Transfer of link status information (LSI). LSSU: link status signal unit. IP: incoming processing. OP: outgoing processing. LC: link control. T_1,T_2: Timers. C: commands from MTP3. I: indications to MTP3.

The LSI is a parameter in a LSSU—see Fig. 8.3-3(a). The other LSSU parameters are the same as the MTP parameters in message signal units. Length indicator (LI) = 1 indicates that the signal unit is a LSSU. The link status information is coded as follows:

LSI Bit: 321	Acronym	Link Status
000	SIO	Out of alignment
001	SIN	Normal alignment
010	SIE	Emergency alignment
011	SIOS	Out of service
100	SIPO	Processor outage
101	SIB	Busy

In the link management examples that follow, a LSSU with a particular value of LSI is denoted by the corresponding acronym. For example, a "SIN" denotes a LSSU in which LSI = 001.

8.6.1 Initial Alignment

When a signaling link is turned on, both LCs start sending SIOs. When LC-A acquires alignment (i.e., recognizes the flags in the incoming bit stream), it starts sending SINs. When LC-A begins to receive SINs, indicating that LC-B also has achieved alignment, a proving period of a few seconds is started. If both LCs maintain their alignment during this period, they start sending FISUs, and indicate to their respective MTP3s that the link is *aligned and ready* for service.

If LC-A has acquired alignment but keeps receiving SIOs or SIOSs, it knows that LC-B is not in a working condition. LC-A then passes an *alignment failure* indication to MTP3-A.

8.6.2 Error Monitoring

When the signaling link is in service, each LC monitors the error rate of received signal units. When one of the following conditions occurs, the MTP3 in the signaling point is alerted with a *link failure* indication:

1. Sixty-four consecutive signal units have been received with errors.
2. The error rate of received signal units exceeds one error per 256 signal units.
3. An "impossible" bit pattern, consisting of at least seven consecutive 1s, has been received, and a flag has not been detected within 16 octets following that pattern.

8.6.3 Delayed Acknowledgments

Each LC expects that, under normal conditions, its outgoing MSUs are acknowledged within a short time. Excessive delay of acknowledgments is an indication of problems on the signaling link. Acknowledgment delays are checked by both LCs, using timers (T_1), which expire in 0.5–2 s (Fig. 8.6-1).

The procedure at MTP2-A is as follows. Timer T_1-A is restarted each time an acknowledgment with a new value of BSN (Section 8.3) is received from MTP2-B, and there is at least one MSU (waiting for acknowledgment) in the RB. If T_1-A expires, LC-A interprets the absence of received acknowledgments during the time-out interval as a signaling link problem, and passes a *link failure* indication to its MTP3-A.

8.6.4 Level 2 Flow Control

When a link control, say LC-A, detects that the number of received MSUs in its IB exceeds a particular value, because MTP3-A has fallen behind in taking out these MSUs, it starts sending SIB (link busy) status units to LC-B, at intervals of 80–120 ms. It continues sending outgoing MSUs and FISUs, but discards incoming MSUs, and "freezes" the value of BSN in the SUs it sends out. The delay in acknowledgments would normally cause timer (T_1-B) of LC-B to time out. However, timers (T_1) are also restarted each time a SIB is received from the distant MTP2. Therefore, T_1-B does not expire as long as SIBs are being received.

When LC-B receives the first SIB, it also starts a timer T_2-B, which expires in 3–6 s. If the congestion at MTP2-A abates, it again acknowledges received signal units, and MTP2-B starts to receive SUs with "new" values of BSN. If this happens before T_2-B has expired, LC-B stops the timer, and resumes normal operation. However, if T_2-B expires and LC-A is still sending SUs with the "frozen" BSN value, LC-B passes a *link failure* indication to its MTP3-B.

8.6.5 Processor Outage

With this procedure, the MTP3 at either end of the signaling link can temporarily suspend the operation of the link. Assume that LC-A in Fig. 8.6-1 has received a command (C) from MTP3-A to suspend incoming and outgoing message traffic on the link. LC-A then starts sending a continuous stream of SIPOs (processor outage), and discards received MSUs. On receipt of the SIPOs, link control (LC-B) starts sending fill-in signal units FISU—see Fig. 8.3-3(b)—and indicates to its MTP3-B that the signaling link is out of operation.

When LC-A receives a command from its MTP3-A to resume operation, it stops sending SIPOs, and resumes sending MSUs and FISUs. LC-B, which no longer receives SIPOs, indicates to its MTP3-B that the signaling link is back in operation, and resumes sending MSUs.

8.6.6 Outgoing Congestion

Link controls (LC) constantly monitor N, the total number of outgoing MSUs in their output and retransmit buffers, OB and RB (Fig. 8.3-1). When N exceeds a threshold value N_{on}, LC passes an *onset of congestion* indication to its MTP3, which then takes measures to reduce the outgoing MSU traffic temporarily. When N drops below a threshold value N_{off}, the link control passes an *end of congestion* indication, and MTP3 resumes normal outgoing message flow. To avoid rapid fluctuations in link status, N_{off} is set below N_{on}. This gives a certain amount of hysteresis in the status transitions.

8.7 OVERVIEW OF MTP LEVEL 3

The message transfer part–level 3 (MTP3) is divided into the following two groups of functions (Fig. 8.7-1).

Signaling Message Handling (SMH) handles the transfer of messages between "peer" MTP-users–telephone user part (TUP), integrated services digital network user part (ISUP), and signaling connection control part (SCCP). SMH is described in Section 8.8.

Signaling Network Management (SNM) keeps the message traffic flowing under abnormal conditions (congestion, failures) in the signaling network. SNM is described in Section 8.9.

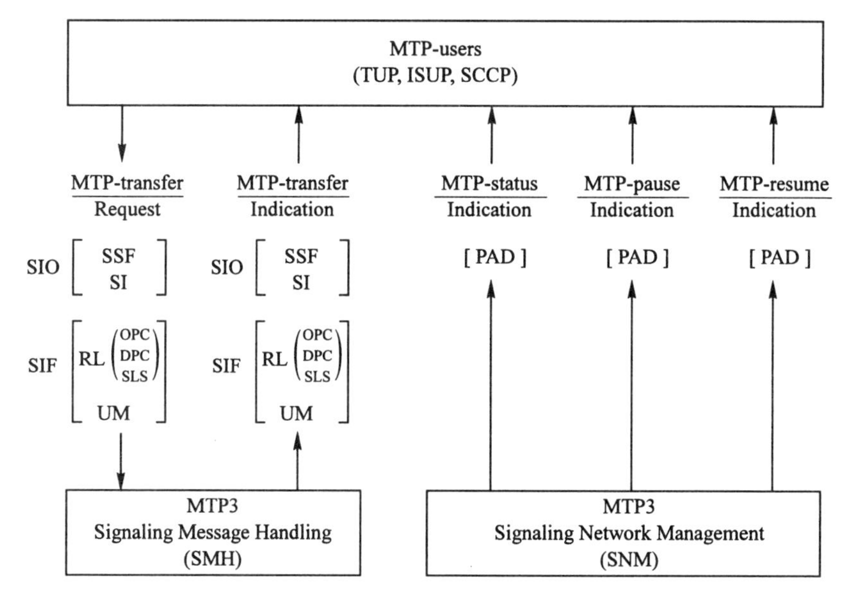

Figure 8.7-1 MTP3 primitives. S10: service information octet. SSF: sub-service field. SI: service indicator. SIF: signaling information field. RL: routing label. OPC: originating point code. DPC: destination point code. SLS: signaling link selector. UM: user message. PAD: point code of affected destination. (From Rec. Q.701 and Q.704. Courtesy of ITU-T.)

8.7.1 MTP-transfer Primitives

MTP-users pass their outgoing messages to SMH in MTP-transfer requests, and SMH delivers incoming messages to the MTP-users in MTP-transfer indications (7.3.2).

The primitives include the parameters in the service information octet (SIO) and the signaling information field (SIF) (Fig. 7.3-3). The parameters are discussed in Section 8.8.

8.7.2 MTP Status, Pause, and Resume Indications

SNM passes these indications to the MTP-users at its signaling point, to inform them about abnormal conditions (failures, congestion) in the signaling routes to particular destinations. The affected destinations are identified by their point codes.

8.8 MTP3 SIGNALING MESSAGE HANDLING

8.8.1 Message Format

The general format of MTP3 messages is shown in Fig. 8.8-1. We distinguish *service information octet* (SIO), and the *signaling information field* (SIF). In

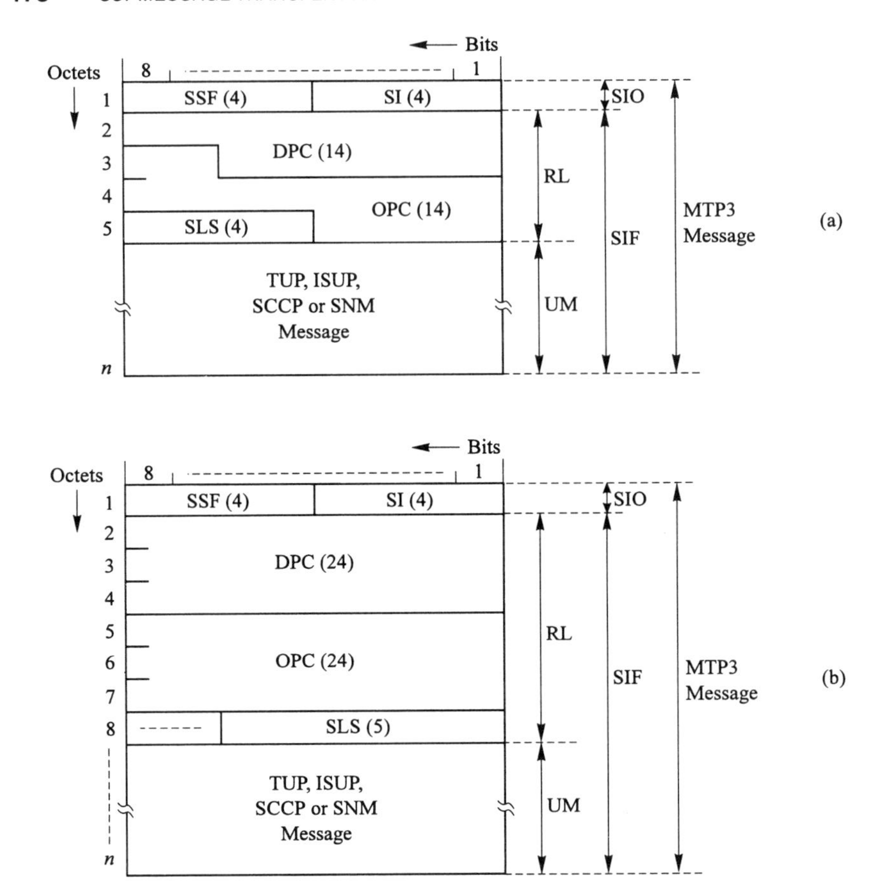

Figure 8.8-1 MTP3 message format. (a): CCITT No. 7. (b): ANSI No. 7. (From Rec. Q.704. Courtesy of ITU-T.)

turn, SIF is divided into the *routing label* (RL), and the *user message* (UM). All information originates at the MTP-user that sends the message, and is included in a MTP-transfer primitive (Fig. 8.7-1). MTP3 examines the data in SIO and RL. However, the user message is transferred transparently (without examination).

8.8.2 Parameters in Routing Label

The parameters in the RL are used by the MTP3s in the signaling points along the message path to determine the signaling route set to the message destination. The RL formats of CCITT No.7 and ANSI No.7 are slightly different—see Fig. 8.8-1, because of differences in parameter lengths [5,9].

Destination Point Code and Originating Point Code (DPC, OPC) identify the point code of the originating and destination signaling point respectively.

The point codes in CCITT No.7 and ANSI No.7 have 14 and 24 bits, respectively (Fig. 7.2-1).

Signaling Link Selector (SLS). This parameter divides the outgoing message loads of the MTP-users in a signaling point into 16 or 32 approximately equal parts. SLS is used for the selection of a particular signaling link in the signaling route set to the message destination (8.8.5).

8.8.3 Parameters in Service Information Octet

These parameters are used by the MTP3 in the destination signaling point to deliver the message to the appropriate MTP-user.

Service Indicator (SI). The code in of SI represents the MTP-user. For the international network, CCITT has specified the following SI codes:

SI	MTP-user
0000	SNM
0001	Signaling network testing and maintenance
0011	SCCP
0100	TUP
0101	ISUP

The MTP-users represented by SI = 0000 and 0001 have not been mentioned before, and are described in section 8.9.

Sub-service Field (SSF). Two SSF codes have been defined by CCITT:

SSF	
0000	International network
0010	National network

The SSF field is of importance in *international switching centers* (ISC), which belong to both the international network, and a national network. Since a MTP-user protocol, say TUP, for the international network is usually somewhat different from the same protocol in a national network, an ISC is equipped with two TUP versions, and SSF specifies the version to which the message is to be delivered.

8.8.4 SMH Functions

Signaling message handling functions are divided into three groups (Fig. 8.8-2).

SMH Message Discrimination uses the values of *SSF* and *DPC* to determine

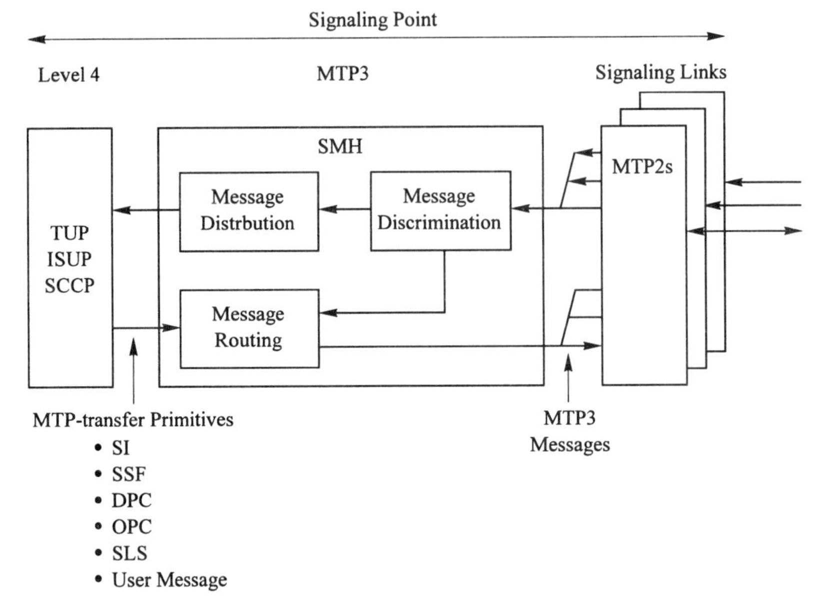

Figure 8.8-2 Structure and interfaces of signaling message handling (SMH). (From Rec. Q.704. Courtesy of ITU-T.)

whether a message received from a MTP2 is destined for its signaling point. The value of *SSF* indicates whether DPC contains a national or international point code.

When SMH message discrimination recognizes the DPC as the point code of its signaling point, it passes the message to SMH message distribution. When the message has another destination, the message handling depends on the type of signaling point. If the signaling point is a signal transfer point (STP), or a combined transfer and end point (STEP) (5.1), SMH message discrimination passes the message to SMH message routing. If the signaling point is a signaling end point (SEP), the message is discarded.

SMH Message Distribution delivers the messages received from SMH message discrimination to the MTP-user specified by SI and SSF.

SMH Message Routing selects the outgoing signaling links for outgoing messages (received from the MTP-users in the signaling point) and transits messages (received from SMH message discrimination).

8.8.5 Selection of Outgoing Signaling Links

SMH message routing first determines the signaling route set for the message, based on the values of *SSF* and *DPC*, which indicate the point code of a national or international destination.

The selection of a particular outgoing signaling link in a route set has two objectives. In the first place, a signaling point should distribute its messages to a particular destination equally among the signaling links in its route set to that destination.

In the second place, all messages pertaining to a particular trunk, or a particular transaction, should be routed on the same signaling route. This is necessary to avoid out-of-sequence delivery of messages pertaining to a trunk, or a transaction. We have seen that SS7 messages are delivered in-sequence from the MTP2 at one end of the link to the MTP2 at the other end (8.3). However, messages spend some "queueing delay" time in the output and input buffers of the MTP2s at the ends of a signaling link. These delays vary from moment to moment, and from link to link. If two messages, M_1 and M_2, which pertain to a trunk or transaction would traverse different signaling routes, inequalities in the delays on the signaling links along routes could result in out-of-sequence message delivery.

Therefore, the MTP-users assign a particular SLS value to all outgoing messages relating to a particular trunk or transaction, and SMH message routing determines the outgoing signaling link in a route set from the value of SLS.

The assignments of SLS can be made in several ways. For example, in TUP and ISUP, each trunk in a trunk group is identified by a circuit indication code (CIC)—see Section 7.2.5—and SLS values can be derived from CIC codes as follows:

CCITT No.7: SLS = CIC mod 16
ANSI No.7: SLS = CIC mod 32

Since each SLS value covers approximately the same number of trunks, the loads of outgoing messages with the various SLS values are approximately equal.

We shall see in Chapter 13 that SCCP can provide in-sequence delivery of messages relating to individual transactions, by assigning the same SLS value to all outgoing messages of a particular transaction.

A number of link-selection examples are outlined below.

Example 1. Figure 8.8-3(a) shows the signaling links in the route set between signaling points A and F in the original Bell System signaling network (see Section 5.1). We examine the signaling link selection for messages from A to F [10]. Under normal (non-failure) conditions in the signaling route set, signaling links SL_9 and SL_{10} are not used.

We assume a four-bit SLS field, and denote the bits in this field by:

SLS: s_4 s_3 s_2 s_1

The route sets to destination F at signaling points A, B, and C contain two signaling links each. We assume that SMH message routing at originating signaling points uses bit s_1 of SLS to select the outgoing signaling link. In this example, messages from A with $s_1 = 0$ and $s_1 = 1$ are sent on respectively SL_1 and SL_2.

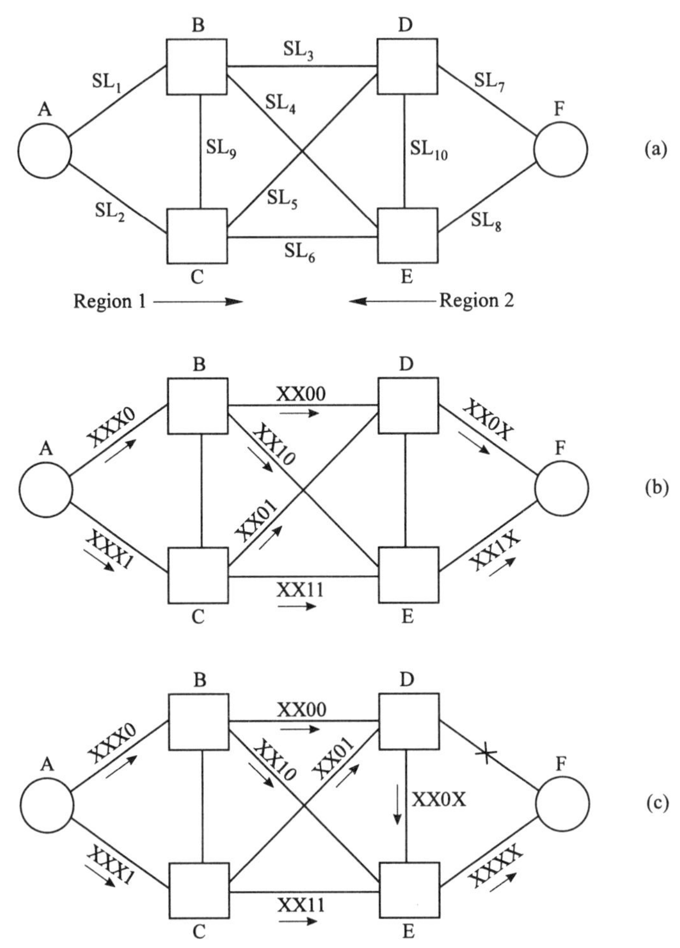

Figure 8.8-3 Load sharing examples. (a): signaling route set for messages from A to F. (b): load sharing under normal conditions. (c): load sharing on failure of SL_7. (From Rec. Q.705. Courtesy of ITU-T.)

Signal transfer points base their selections on bit s_2. In this example, we assume that signal transfer points B and C select respectively SL_3 and SL_5 when $s_2 = 0$, and SL_4 and SL_6 when $s_2 = 1$. Formally, signal transfer points D and E also use s_2 to make the outgoing link selection. However, D and E have only one normal route to F, which is used for both values of s_2.

Figure 8.8-3(b) displays the SLS codes in the messages carried by the individual signaling links (bits that can be O or 1 are marked X). The links SL_1, SL_2, SL_7, and SL_8 carry messages with eight of the 16 SLS codes each (about 50% of the messages). Also, each of the signaling links SL_3, SL_4, SL_5, and SL_6 carries messages with 4 SLS codes (about 25% of the messages).

Under this form of signaling link selection, the signaling route for messages with a particular value of SLS, and sent by signaling point A, can be different from the route for messages with the same SLS that are sent by signaling point F.

Example 2. Suppose that SL_7 in Fig. 8.8-3(a) has failed. Signal transfer point D thus has to divert the messages with destination F to SL_{10}. Link SL_{10} then carries the messages with $SLS = XX0X$, and SL_8 carries the entire message load—see Fig. 8.8-3(c).

Example 3. In some signaling networks, normal signaling routes pass through more than two signal transfer points (STP). Figure 8.8-4 shows an example that occurs in the post-divestiture North American signaling network [1]. Exchange A and signal transfer points B and C are owned by a local exchange carrier, while exchange H, and signal transfer points D, E, F, and G belong to the signaling network of an interexchange carrier (see 1.1.3). We examine the distribution of messages from exchange A to exchange H over the various signaling links.

The outgoing link selections at A, B, C, D, E, F and G are based on the values of bits in the five-bit (ANSI) signaling link selector SLS field. For equal message load division, the selections of outgoing signaling links at the signaling points (SP) along the route have to be based on different bits of SLS. In this example, signaling point A, which is the originating SP in the signaling route, uses bit s_1. STPs B and C are the second SPs in the routes from A to H, and therefore use bit s_2. STPs D and E are the third SPs in the routes, and thus use bit s_3. STPs F and G are the fourth signaling points in the routes, and formally base their selection on the value of s_4. However, F and G have only one normal signaling route to destination H, which is used for both values of s_4. Figure 8.8-4 shows the SLS codes on the individual signaling links, for messages sent by A.

Since the STPs cannot determine whether they are the second, third, or fourth SP in the signaling route of an incoming MSU, the link selection procedure is implemented as follows: All signaling points use the rightmost bit of SLS for outgoing link selection, and then *rotate* the contents of SLS one position to the right.

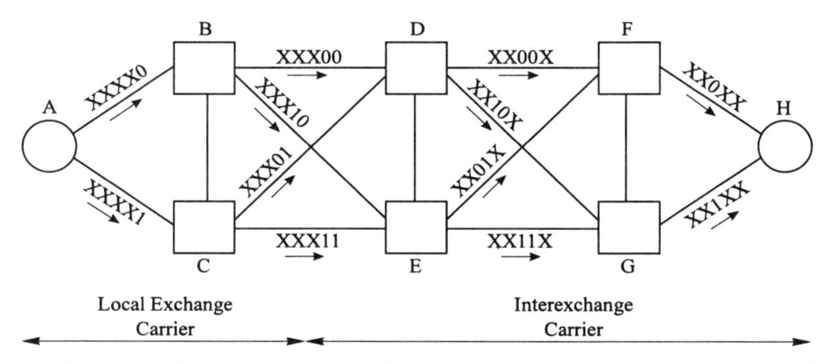

Figure 8.8-4 Load sharing for inter-network messages from A to H. (From *IEEE Comm. Mag.* **28.7**. Copyright © 1990 IEEE.)

In this way, signaling point A uses bit s_1, STPs B and C use s_2, STPs D and E use s_3, and F and G (formally) use s_4 of the *original* SLS code (assigned by the originating MTP-user at signaling point A). For simplicity, the SLS codes on the signaling links in Fig. 8.8-4 are shown "without rotation" (s_1 always appears in the rightmost position of SLS). The actual codes on the links can be derived easily. For example, the SLS code in the messages on the signaling link between B and D is shown as *XXX*00. The actual code has undergone two rotations, and is therefore 00*XXX*.

8.9 MTP3 SIGNALING NETWORK MANAGEMENT

The purpose of MTP3 signaling network management is to keep the signaling-message traffic flowing under abnormal (congestion, failure) conditions in the signaling network. Some of these conditions may require a temporary reduction—or suspension—of outgoing message traffic to certain destinations. In these cases, SNM alerts the level 4 protocols at its signaling point [8,10].

8.9.1 Structure and Interfaces of SNM

SNM consists of three parts—see Fig. 8.9-1.

SNM Link Management monitors and controls the status of the individual MTP2s (signaling links) of the signaling point.

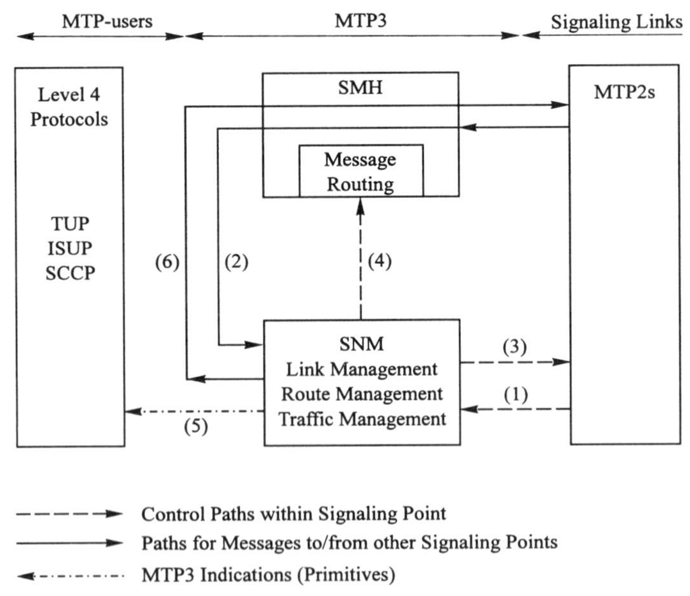

Figure 8.9-1 Structure and interfaces of signaling network management (SNM). (From Rec. Q.704. Courtesy of ITU-T.)

SNM Route Management communicates with its "peer" functions at other signaling points, sending and receiving information regarding the status of signaling routes to individual destinations.

SNM Traffic Management receives information from SNM link management on the status (available/unavailable) of the signaling links at its signaling point, and information from SNM route management about problems on the signaling route sets to particular destinations. When necessary, it informs SMH message routing and the MTP users at its signaling point.

Some SNM procedures involve more than one part of SNM, and require communications between these parts.

Interfaces of SNM. SNM has interfaces with the MTP-users (level-4 protocols), the MTP2s of the signaling links, and the SMH message routing function at its signaling point. It also communicates with SNMs at other signaling points, sending and receiving *SNM messages*.

Inputs. SNM bases its actions on the following inputs (the numbers below correspond to the numbers in Fig. 8.9-1):

1. Indications on the status of the signaling links at its signaling point, received from the MTP2s.
2. SNM messages received from the SNM functions at other signaling points.

Outputs. The actions of SNM result in the outputs listed below:

3. Commands to the MTP2s of the signaling links.
4. Commands to the SMH message routing function at its signaling point. For example, to divert signaling messages to a certain destination from their normal signaling link to an alternative link.
5. Indications (primitives) to the MTP-users at its signaling point, about the status of signaling route sets to individual destinations (Fig. 8.7-1). MTP-status, MTP-pause, and MTP-resume indicate that the signaling route set to a particluar destination has become congested, unavailable, and available again.
6. Messages to SNMs at other signaling points.

8.9.2 SNM Messages

To send and receive messages to/from other signaling points, SNM uses the services of SMH in the same manner as the MTP-users.

The format of an MTP3 message with a SNM message in its user message field (UM) is shown in Fig. 8.9-2. The contents of routing label (RL) are shown in the CCITT No.7 format.

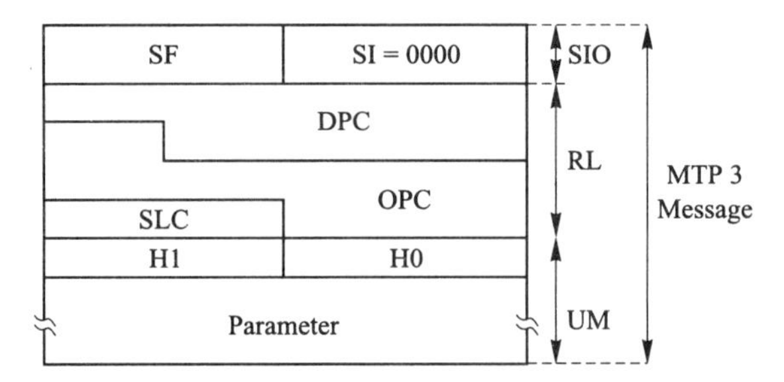

Figure 8.9-2 MTP3 message with a signaling network management message in the user message field. (From Rec. Q.704. Courtesy of ITU-T.)

SNM messages have service information code (SI) = 0000. RL contains OPC and DPC and, in lieu of the signaling link selection parameter (SLS), the parameter *signaling link code*.

Signaling Link Code (SLC). Some SNM messages have status information about a particular signaling link. In these messages, the signaling link is identified by the combination of the parameters: DPC, OPC, and SLC. In SNM messages that do not pertain to a particular signaling link, the link code is set to $SLC = 0$.

Since the MTP3 messages that carry SNM messages do not have a SLS parameter, SMH message routing has to use special routing rules. When $SLC = 0$, the message can be routed on any link in the route set to the message destination. When SLC is not equal to 0, it identifies a particular signaling link. In this case, the message has to be sent either on the identified link, or on an alternate of the identified link, depending on the SNM message type. We will see the reason for this in the sections that follow.

Headings and Parameters. SNM messages always contain a heading, and may include a parameter. The heading consists of the H0 and H1 fields. The code in H0 represents a group of functionally related messages, and the code in H1 indicates a particular message in the group—see Table 8.9-1. The table also lists the message acronyms, and—if the message includes a parameter—the parameter name and acronym. In the sections that follow, the messages and parameters are denoted by these acronyms.

When a procedure involves signaling network management at several signaling points, we shall denote the SNM at a particular signaling point—say point A—by SNM-A, etc.

8.9.3 Procedures for Congestion Control [11]

The MTP2s of each signaling link constantly monitor the number of waiting messages in their output- and retransmission buffers. When this number exceeds a preset threshold value, SNM is alerted with an *onset of congestion* indication (8.6.6). Outgoing signaling link congestion is determined independently at

Table 8.9-1 Messages for signaling network management.

Message Name	Acronym	HO	H1	Parameter
Changeover order	COO	0001	0001	FSNR (1)
Changeover acknowledgment	COA	0001	0010	FSNR (1)
Changeback declaration	CBD	0001	0101	CBC (2)
Changeback acknowledgment	CBA	0001	0110	CBC (2)
Emergency changeover	ECM	0010	0001	—
Emergency changeover acknowledgment	ECA	0010	0010	—
Transfer prohibited	TFP	0100	0001	PAD (3)
Transfer allowed	TFA	0100	0101	PAD (3)
Transfer controlled	TFC	0011	0010	PAD (3)

Notes: (1) Forward sequence number of last accepted MSU. (2) Change-back code. (3) Point code of affected destination.
Source: Rec. Q.701. Courtesy of ITU-T.

each end of the link. A signaling link between signaling points A and B can be thus congested at A, and not congested at B.

When a signaling link is congested, SNM considers all destinations whose signaling route sets include that link as "congested." In Fig. 8.9-3, signaling end point A has two signaling links (SL_1 and SL_2), to the signal transfer points B and C in its region. The route set at A for all destinations thus consists of SL1 and SL_2. When SNM-A receives an "onset of congestion" indication from the MTP2 of signaling link SL_1, it therefore concludes that its route sets to all destinations are congested.

It is important to keep congestion within limits, because otherwise the queuing delays for messages can become unacceptably large for call-control applications.

Congestion Control at Signaling End Points. In a signaling end point (SEP), the most likely cause of signaling link congestion is an excessive amount

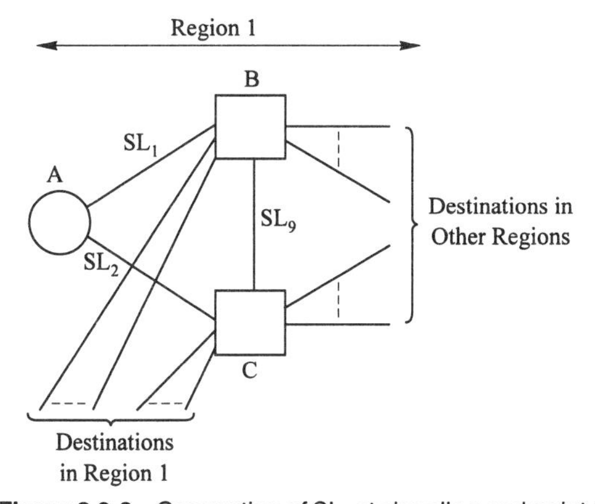

Figure 8.9-3 Congestion of SL_1 at signaling end point A.

of messages originated by the MTP-users. SNM then passes *MTP-status* indications to all users, identifying the affected destination by its point code PAD (Fig. 8.7-1). In CCITT No.7, separate indications have to be sent for each affected destination. In ANSI No.7, certain point codes represent a group of destinations.

In response, the MTP-users reduce their outgoing message traffic to the affected destinations for a certain amount of time. For example, TUP and ISUP suspend the seizure of SS7 trunks to these destinations.

SNM keeps monitoring the outgoing messages to the affected destinations and, when necessary, repeats the MTP-status indications. When SNM receives an *end of congestion* indication from the MTP2 of the previously congested signaling link, it stops passing the MTP-status indications, and the MTP-users then end their restrictions on outgoing messages.

Congestion Control at Signal Transfer Points. When the SNM at a signal transfer point (STP) receives a congestion indication from one of its signaling links, it informs the signaling points from which it receives messages to the affected destinations, by sending *transfer controlled* (TFC) messages. For example, when SNM-D of signal transfer point D (Fig. 8.9-4) receives an onset of congestion indication from the MTP2 of signaling link SL_7, it considers its route to F as congested. Then, on receipt of the first message to F from a signaling end point, say A, it sends a TFC message to SNM-A. Parameter PAD holds the point code of affected destination F. SNM-D repeats the TFC message each time it has received eight additional messages for F from A. The same procedure is carried out by SNM-D for all other SEPs from which it receives messages for F.

On receipt of each TFC for affected destination F, the SNMs in the signaling end points that have originated messages to F pass MTP-status indications for affected destination F to their MTP-users.

When SNM-D receives an indication that the congestion on SL_7 has ended, it stops sending TFC messages for affected destination F to the originating SEPs, which then end their restrictions on outgoing messages to F.

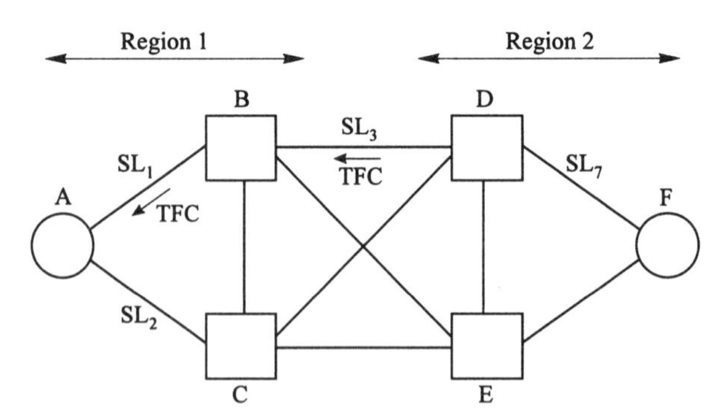

Figure 8.9-4 Congestion control by signal transfer point D.

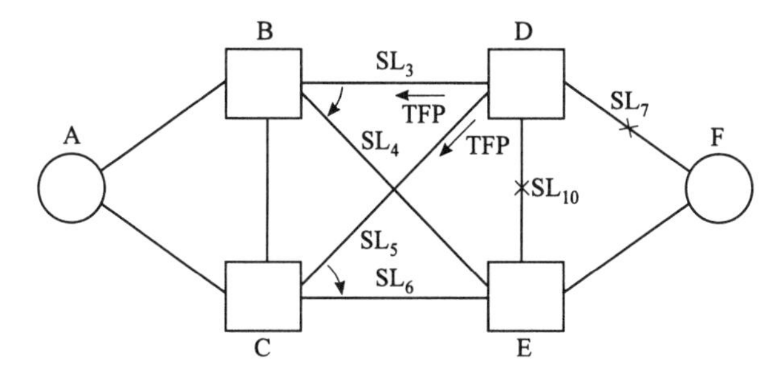

Figure 8.9-5 Rerouting of messages from A to F on failure of SL_7 and SL_{10}.

8.9.4 Rerouting of Messages

In Fig. 8.9-5, signaling links SL_7 and SL_{10} have failed simultaneously. Signal transfer point D is therefore no longer able to transfer messages to destination F.

SNM-D then alerts the SNMs in all directly connected signaling points (such as B and C) that it can no longer transfer messages to F, by sending *transfer prohibited* (TFP) messages. In these messages, PAD holds the point code of affected destination F.

We now examine the actions at signal transfer point B, in response to the TFP message. Suppose that SMH-B message routing normally routes messages for F in which SLS = *XX0X* and SLS = *XX1X* on respectively SL_3 and SL_4—see Fig. 8.8-3(b). After receipt of the TFP, SNM-B instructs SMH-B message routing to reroute messages to F with SLS = *XX0X* to link SL_4, starting with the messages to F that are waiting in the output buffer of SL_3. However, the messages waiting in the retransmission buffer of SL_3 are not retransmitted because SNM-B cannot determine whether they have reached their destination. The rerouting procedure thus may cause the loss of some messages.

Similar actions occur at signal transfer point C, and at all other signaling points that have received the TFP.

When SL_7 and/or SL_{10} become available again, SNM-D informs the previously alerted signaling points with *transfer allowed* (TFA) messages, indicating that it can again transfer messages with destination F. These SPs then resume their normal routings of messages to that destination.

8.9.5 Signaling Route Set Unavailable

Certain combinations of multiple failures in a signaling network can completely disable a route set. Consider two signaling end points, say A and F. When SNM-A determines that its route set to a destination F has become unavailable, it informs its MTP-users by passing *MTP-pause* indications (Fig. 8.7-1), which include the point code (PAD) of affected destination F. The users then take the necessary measures. For instance, if signaling point A has a TUP trunk group to F, TUP-A has to take this group out of service.

When SNM-A determines that the route set is no longer disabled, it informs its users with *MTP-resume* indications, again identifying the affected destination F by its point code.

8.9.6 Changeover and Changeback

When a signaling link fails, or is taken out of service for maintenance, it becomes unavailable for the transfer of messages. The messages normally sent out on the link have to be diverted to an alternative signaling link. This requires changeover actions by the SNMs at both ends of the link. When the link becomes available again, changeback actions by the SNMs terminate these diversions.

We examine these procedures for the case that link SL_1 in Fig. 8.9-6 has failed. Signaling end point A has to divert all messages that are normally sent on SL_1 to SL_2, and signal transfer point B, which normally sends all messages with destination A on SL_1, has to divert these messages to SL_9, and the messages arrive at A on signaling link SL_2.

Whenever possible, the changeovers and change-backs are executed in such a way that the affected messages are still delivered in-sequence, and without loss or duplication.

Changeover. When SNM-A decides that SL_1 has to be made *unavailable* for messages, it starts the changeover by ordering the MTP2-A of SL_1 to suspend sending messages, and to "freeze" the contents of its output and retransmission buffers.

SNM-A then sends a *changeover order* (COO) message to SNM-B. The message identifies the unavailable link SL_1 by the combination of the values of parameters DPC, OPC, and SLC. Also included in the message is FSNR, the forward sequence number of the last message received on SL_1 that has been accepted at A (see Table 8.9-1).

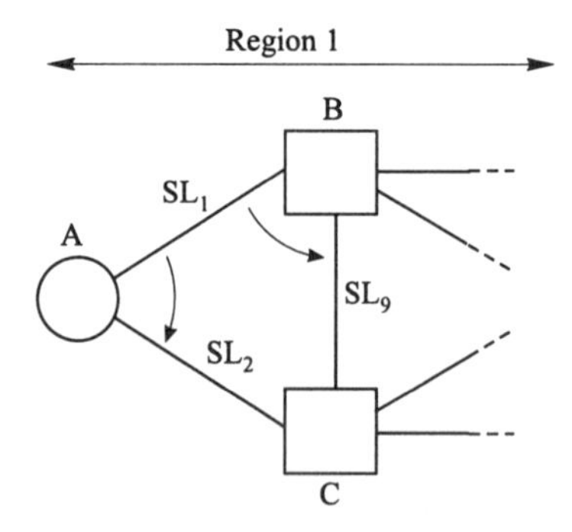

Figure 8.9-6 Changeover of signaling link SL_1.

For COO messages, SMH message routing selects an outgoing signaling link *other than* the link identified by SLC. In this example, the COO thus goes out on SL_2, and reaches B on SL_9.

When SNM-B receives the changeover order, it orders the MTP2-B of SL_1 to stop sending messages on SL_1. SNM-B then retrieves all messages not yet acknowledged by MTP2-A (that,is: messages with forward sequence numbers exceeding FSNR) from the retransmission buffer of SL_1, and moves them into the output buffer of alternative link SL_9. These messages are therefore sent out on that link.

In the next step, SNM-B retrieves the messages in the output buffer of SL_1, and moves them into the output buffer of SL_9. These messages are also sent out on that link. SNM-B then orders SMH-B message routing to route all subsequent messages that would normally be sent on SL_1, on SL_9.

As a result, all messages from B that are affected by the changeover are sent, in-sequence and without loss or duplication, on SL_9. They reach A on SL_2.

SNM-B now sends a *changeover acknowledgment* (COA) to SNM-A, in which the parameter FSNR represents the forward sequence number of the last message it has accepted from SL_1. The COA message is also sent on SL_9, and reaches signaling point A on SL_2.

When SNM-A receives the COA message, it executes a similar procedure, which has the result that all messages waiting in the buffers of SL_1 are sent out on SL_2, without loss or duplication, and in-sequence. They reach signaling point B on SL_9.

Emergency Changeover. The above "best case" procedure can not always be carried out, because failures in the MTP2-A of SL_1 may make the determination of FSNR impossible. In this case SNM-A sends an *emergency changeover message* (ECM), which does not include FSNR. When SNM-B receives the ECM, it ignores the contents of the retransmission buffer of SL_1, and only moves the contents of the output buffer to SL_9. This results in the loss of some messages.

A failure of SL_1 is likely to be noticed by both SNM-A and SNM-B. If SNM-A receives the changeover order from SNM-B after it has already started its changeover actions, it still acknowledges the order.

Changeback. When SL_1 becomes available again for the transfer of messages, it is necessary to end the diversion. This requires changeback actions by SNM-A and SNM-B.

SNM-A and SNM-B are informed by their respective MTP2s when SL_1 is again in working condition. Both SNMs may initiate the changeback. The procedure initiated by SNM-A is described below—the procedure at SNM-B is similar. The objective of the changeback procedures is to start using SL_1 again, preferably without causing message loss, duplication, or out-of-sequence delivery. The procedure is executed in two steps:

Step 1. SNM-A informs SMH-A message routing that messages should be no longer be diverted to SL_2, but stored in a *changeback buffer* (CBB) instead. SNM-A then sends a *changeback declaration* (CBD) to SNM-B, identifying SL_1 by the combination of the values in DPC, OPC, and SLC. Changeback declarations, like changeover messages, are routed on the alternate of the identified link SL_1. In our example, the CBD goes out on SL_2. The message reaches signaling point B on SL_9, and SNM-B acknowledges with a *changeback acknowledgment* (CBA).

Step 2. When SNM-A receives the CBA, it knows that all diverted (to SL_2) messages that were on their way to B when it sent the CBD have been accepted there, because signaling links SL_2 and SL_9 accept incoming messages *in-sequence*. Therefore, the messages that have accumulated in buffer CBB can now be sent out without causing out-of-sequence (premature) deliveries at their destinations. SNM-A thus orders SMH-A message routing to send out the messages in changeback buffer CBB on SL_1, and then to resume using that link for subsequent messages.

CBD and CBA messages include a parameter called *changeback code* (CBC) (see Table 8.9-1). This parameter is needed because a SNM may be involved in several changebacks simultaneously. The parameter value identifies a particular changeback action—and the associated CBB changeback buffer—at signaling point A. When SNM-B acknowledges the changeback declaration, it copies the value of CBC in its acknowledgment. This enables SNM-A to select the proper CBB buffer and signaling link for step 2.

8.10 ACRONYMS

ANSI	American National Standards Institute
BIB	Backward indicator bit
BSN	Backward sequence number
CB	Check bits
CBA	Changeback acknowledgment message
CBB	Changeback buffer
CBC	Changeback code
CBD	Changeback declaration message
CCITT	International Telegraph and Telephone Consultative Committee
CEPT	European Conference of Postal and Telephone Administrations
COA	Changeover acknowledgment message
COO	Changeover order message
DPC	Destination point code
ECA	Emergency changeover acknowledgment message
ECM	Emergency changeover message
FIB	Forward indicator bit
FISU	Fill-in signal unit

FSN	Forward sequence number
FSNR	Forward sequence number of the most recently accepted message
H0	Message heading field 0
H1	Message heading field 1
IB	Input buffer
IP	Incoming processing
ISC	International switching center
ISUP	ISDN user part of SS7
kb/s	Kilobits per second
LC	Link control
LI	Length indicator
MSU	Message signal unit
MTP	Message transfer part
MTP1	Message transfer part–level 1
MTP2	Message transfer part–level 2
MTP3	Message transfer part–level 3
OB	Output buffer
OP	Outgoing processing
OPC	Originating point code
PAD	Point code of affected destination
PCM	Pulse code modulation
PCR	Preventive cyclic retransmission
RB	Retransmission buffer
RL	Routing label
RST	Route-set test message
SCCP	Signaling connection control part of SS7
SDL	Signaling data link
SEP	Signaling end point
SI	Service indicator
SIB	Signaling link congested (LSSU)
SIE	Emergency alignment (LSSU)
SIF	Signaling information field
SIN	Normal alignment (LSSU)
SIO	Out of alignment (LSSU)
SIO	Service information octet
SIOS	Link out of service (LSSU)
SIPO	Processor outage (LSSU)
SL	Signaling link
SLC	Signaling link code
SLS	Signaling link selector
SMH	Signaling message handling
SNM	Signaling network management
SP	Signaling point
SSF	Sub-service field
SS6	Signaling system No.6

SS7	Signaling system No.7
STEP	Signal transfer and end point
STP	Signal transfer point
SU	Signal unit
TFA	Transfer allowed message
TFC	Transfer controlled message
TFP	Transfer prohibited message
TUP	Telephone user part of SS7
UM	User message

8.11 REFERENCES

1. A.R. Modaressi, R.A. Skoog, "Signaling System No.7: A Tutorial," *IEEE Comm. Mag.*, **28**, No.7, July 1990.
2. R. Manterfield, *Common Channel Signalling*, Chapter 5, Peter Peregrinus Ltd, London, 1991.
3. B. Law, C.A. Wadsworth, "CCITT Signalling System No.7: Message Transfer Part," *Br. Telecomm. Eng.,* **7**, April 1988.
4. *Specifications of Signalling System No.7*, Rec. Q.701, CCITT Blue Book, **VI.7**, ITU, Geneva, 1989.
5. *American National Standard for Telecommunications-Signaling System No.7 (SS7)-Message Transfer Part,* ANSI T.111, New York, 1992.
6. *Specifications of Signaling System No. 7*, Rec. Q.702, CCIT Blue Book, **VI.7**, ITU, Geneva, 1989.
7. *Ibid.*, Rec. Q.703.
8. *Ibid.*, Rec. Q.706.
9. *Ibid.*, R. Q.704.
10. *Ibid.*, Annex A to Rec. Q.705.
11. D.R. Manfield, G. Millsteed, M. Zukerman, "Congestion Controls in SS7 Signaling Networks," *IEEE Comm. Mag.,* **31,** No.6, June 1993

9

TELEPHONE USER PART

The *telephone user part* (TUP) is the first SS7 user part defined by CCITT. An early version appeared in the 1980 CCITT *Yellow Book*. The 1985 *Red Book* and the 1989 *Blue Book* include a number of additions and modifications [1].

The term "telephone" dates back to the beginnings of the TUP development, when all calls were "speech" calls. The calls in the present telecommunication networks can also be used for facsimile, and other data communications. TUP is primarily a link-by-link signaling system. It can be used on analog FDM trunks, and on 64 kb/s digital trunks (see Sections 1.4 and 1.5).

The CCITT specifications cover TUP applications in national networks, and in the international network. TUP has been designed to be backward compatible with R2 and SS6 signaling (Chapters 4 and 6), and includes all features of these systems. National versions of TUP coexist with R2 signaling (Chapter 4) in the networks of several countries. Since R2 provides end-to-end signaling (between the originating and terminating exchanges of a connection), TUP includes a similar procedure.

One important aspect of TUP is its support of *digital connectivity* (the provision of transparent end-to-end 64 kb/s digital connections). The demand for this service is growing steadily, because of new digital customer equipment, such as high-speed facsimile machines. Telecoms in several countries have installed subscriber lines that can transfer 64 kb/s digital information. These lines are the precursors of the digital subscriber lines for the *integrated digital services network* (ISDN) which are discussed in Chapter 10.

TUP was introduced in the international network in the mid-1980s. It is also used—with some modifications—in the national networks of several countries. Telecoms in other countries have chosen to bypass TUP signaling. For example, the telecoms in the United States have gone directly from common-channel interoffice signaling (CCIS—see Chapter 6) to the ANSI-defined version of the ISDN user part (ISUP). In Japan, a national version of ISUP is gradually replacing multi-frequency signaling.

Sections 1 through 5 of this chapter primarily describe TUP as specified in CCITT recommendations. Section 6 briefly discusses some aspects of national versions.

9.1 MESSAGES AND PRIMITIVES

This section outlines the structure of TUP messages, and reviews the primitives that involve TUP.

9.1.1 General Message Structure

The general TUP message format is shown in Fig 9.1-1 [2,3]. Octet (a) is the *service information octet* (SIO) (8.8.3), consisting of *service indicator* (SI), and *sub-service field* (SSF). The value SI = 0100 indicates a TUP message.

The *routing label* (RL) is in octets (b) through (e), and contains the *originating* and *destination point codes* (OPC, DPC), and the *signaling link selector* (SLS) (8.8.2). The trunk for which the message is intended is identified by the combination of OPC, DPC, and CIC (circuit identification code).

9.1.2 Signaling Link Selector and Circuit Identification Code

Bits 8–5 of octet (e) have a dual function: they represent both the SLS, and the four low-order bits (CIC_L) of the circuit identification code. The high-order bits

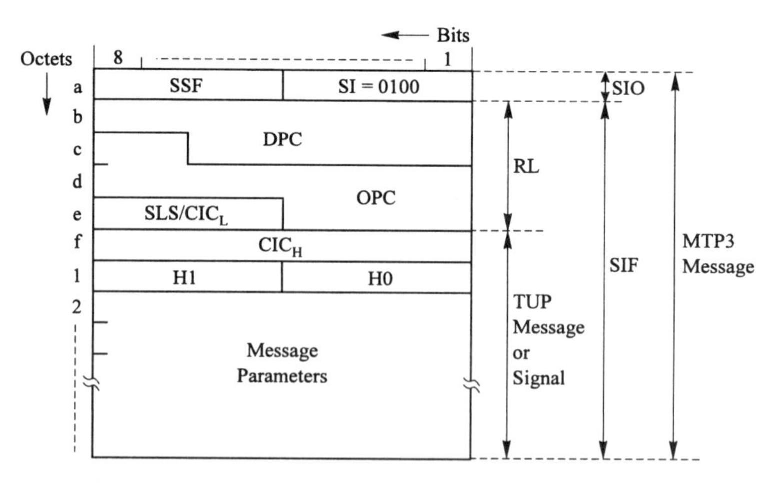

Figure 9.1-1 General format of TUP messages and signals. SIO: service information octet. SIF: signaling information field.

(CIC_H) are in octet (f). The value of SLS thus equals the value of CIC, modulo 16.

As a result, each trunk has an associated signaling route. This is one of the requirements for in-sequence message delivery (8.8.5).

9.1.3 TUP Messages and Signals

TUP literature makes a distinction between messages and signals. A TUP *signal* consists of octets (f) and (1) in Fig. 9.1-1. In a TUP *message*, one or more octets with message parameters follow octet (1).

9.1.4 Heading

The H1/H0 octet is known as the *heading*, and identifies a particular message or signal. H0 represents a group of functionally related messages/signals, and H1 identifies a particular message/signal within that group. This structure is the same as the heading structure of signaling network management messages (8.9.2).

9.1.5 Primitives

TUP communicates with the *message transfer part* (MTP) with the primitives shown in Fig. 8.7-1. MTP-transfer requests and indications pass information in the SIO and SIF fields of TUP messages/signals from TUP to MTP, and vice versa.

The MTP-status, MTP-pause, and MTP-resume indications alert TUP that the signaling route set to a destination is congested, has become unavailable, and has become available again. These primitives contain the parameter PAD (point code of the affected destination).

9.2 CALL-CONTROL MESSAGES AND SIGNALS

This section describes the most important messages, parameters, and signals for TUP call control. Most signals and messages are similar to those in SS6 signaling, and have the same three-character acronyms.

The octets shown in the figures of this section correspond to octets 1 through *n* of Fig. 9.1-1.

9.2.1 Initial Address Message

The *initial address message* (IAM) is the first forward message in a call set-up. It contains the calling party category, a number of message indicators, and digits of the called party number—see Fig. 9.2-1.

Calling Party Category (CPC). This parameter, which is used as in SS6 signaling, is coded as follows:

Bits 654321	
000000	Unknown calling party category
000001	French-speaking operator
000010	English-speaking operator
000011	German-speaking operator
000100	Russian-speaking operator
000101	Spanish-speaking operator
001010	Ordinary calling subscriber
001101	Test call

Message Indicators (A,...,K). An expanded (compared to SS6) set of indicators is used by the exchanges along the connection during the call set-up:

(BA)	Nature of (Called) Address Indicator
00	Subscriber number
10	National number (area code + subscriber number)
11	International number (country code + national number)

(DC)	Nature of circuit indicator
00	No satellite circuit in connection
01	One satellite circuit in connection

(FE)	Continuity Check Indicator
00	Continuity check not required
01	Continuity check required on this circuit
10	Check not required on this circuit, but has been performed on a previous circuit

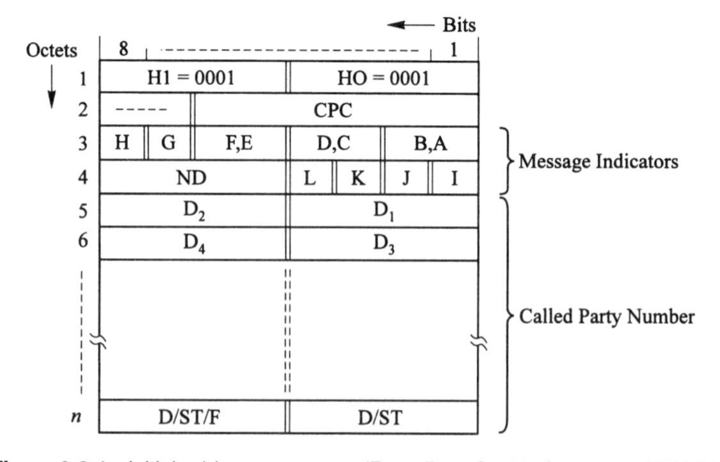

Figure 9.2-1 Initial address message. (From Rec. Q.723. Courtesy of ITU-T.)

(G)	Echo Control Indicator
0	Outgoing suppressor/canceler not included
1	Outgoing suppressor/canceler included

(H)	Incoming Call Indicator
0	Not an incoming international call
1	Incoming international call

(I)	Redirected (Forwarded) Call Indicator
0	Not a redirected call
1	Redirected call

(J)	Digital Path Indicator
0	All digital (64 kb/s) path not required
1	All digital (64 kb/s) path required

(K)	Signaling Indicator
0	Path with SS7 signaling not required
1	Path with SS7 signaling required

Number of Digits (ND). This indicates the number of digits of the called address that are included in the IAM.

Digits (D_i). Digit coding is different from SS6. Digit values 0 through 9 are coded 0000 through 1001; codes 11 and 12 are coded as 1011 and 1100, and *end of address* digit (ST) is coded as 1111.

SS7 uses 0000 as "digit 0," and as a "filler" code. If 0000 appears in bits 8–5 of octet *n*, it means "digit 0," if ND is *even*, and "filler," if ND is *odd*.

9.2.2 Initial Address Message with Additional Information (IAI)

This is an initial address message that includes one or more additional parameters. These parameters are *optional*: they are included only when necessary. Octets 2 through *n* in Fig. 9.2-2 are the same as in an IAM.

Octet ($n + 1$) contains indicator bits that show whether the optional parameters are present (1) or not (0). CCITT has defined three optional parameters and indicator bits:

Indicator Bit	Optional Parameter
E	Calling line identity
H	Original called address
B	Closed user group information

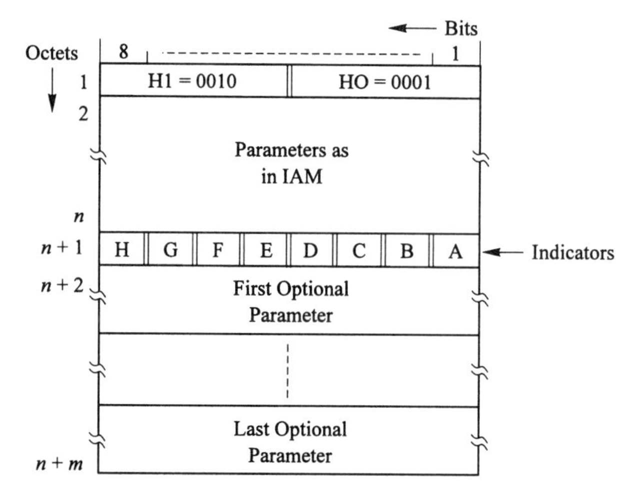

Figure 9.2-2 Initial address message with additional information (IAI). (From Rec. Q.723. Courtesy of ITU-T.)

When the IAI includes more than one optional parameter, they appear in the alphabetical order of their indicator bits.

Calling Line Identity and Original Called Address. The format of these parameters is shown in Fig. 9.2-3. ND indicates the number of digits; bits B and A indicate the nature of the address:

Bits BA	Nature of Address
00	Subscriber number
10	National number
11	International number

Indicator bit C is used in the calling line identity only. It indicates whether the calling party allows the presentation of his number to the called party:

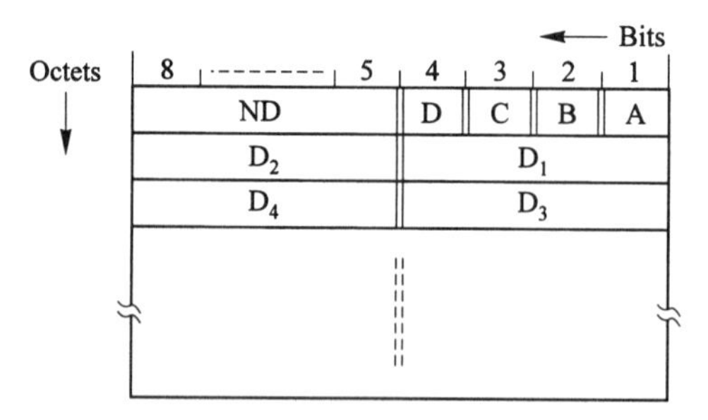

Figure 9.2-3 Format of calling line identity and original called address. (From Rec. Q.723. Courtesy of ITU-T.)

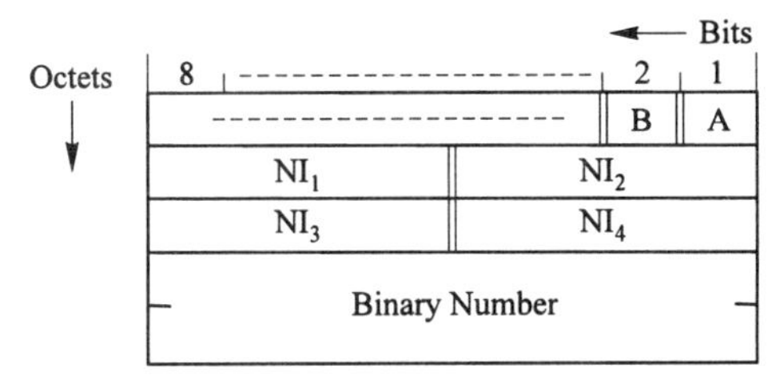

Figure 9.2-4 Coding of closed user group information. (From Rec. Q.763. Courtesy of ITU-T.)

Bit C	Calling Number Presentation
0	Allowed
1	Not allowed

Closed User Group Information. In some countries, subscriber lines and subscriber numbers can be marked at their local exchanges as belonging to a *closed user group* (CUG). Special call-control procedures apply to the members of these groups (9.4).

The CUG information parameter is shown in Fig. 9.2-4. Bits B and A indicate whether a calling CUG member has "outgoing access" (is allowed to make calls to subscriber numbers that are not members of the caller's CUG):

Bits BA	Outgoing Access
10	Allowed
01	Not allowed

The second and later octets hold the *CUG interlock code* which uniquely identifies a CUG world-wide. *Network identity* digits, NI_1,..,NI_4, identify the network of a particular telecom, and the binary number identifies a CUG that is administered by this telecom.

9.2.3 Subsequent Address Messages

TUP address signaling allows both *en-bloc* signaling, in which the complete called number is included in the IAM or IAI, and *overlap* signaling, in which IAM (IAI) includes only those digits that are needed for outgoing trunk selection at the next exchange.

Under overlap signaling, the later address digits are sent in one or more subsequent messages, of which there are two types.

The *subsequent address message* (SAM), identified by H1 = 0011, H0 = 0001, is an IAM (Fig. 9.2-1) without octets 2 and 3. The number of included digits is again indicated by the value of ND. Bits L, K, J, and I are set to zero.

The *subsequent one-digit address message* (SAO) has the heading H1 = 0100, H0 = 0001, and one octet that contains one digit, in bits 4–1. Bits 8–5 are coded 0000.

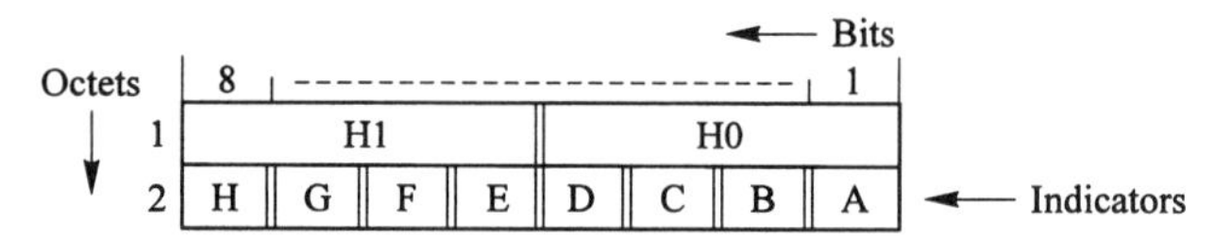

Figure 9.2-5 Address-complete message (ACM), automatic congestion control message (ACC) and general request message (GRQ). ACM: H1 = 0001, H0 = 0100. ACC: H1 = 0001, H0 = 1001. GRQ: H1 = 0001, H0 = 0011. (From Rec. Q.723. Courtesy of ITU-T.)

9.2.4 Address Complete Message (ACM)

This is a backward message, originated by the last exchange in a connection of TUP trunks. It includes one octet with indicators—see Fig. 9.2-5:

(BA)	Call Charging Indicator
00	No instructions on charging
10	Charge
11	Do not charge

(C)	Called Subscriber Status
0	No information available
1	Subscriber free

(D)	Incoming Echo Control Indicator
0	Incoming suppressor/canceler not included
1	Incoming suppressor/canceler included

(E)	Call Forwarding Indicator
0	Call not forwarded
1	Call forwarded

(F)	Signaling Path Indicator
0	Not a completely SS7 signaling path
1	Completely SS7 signaling path

9.2.5 General Request Message (GRQ)

This backward message requests actions and/or information from a preceding exchange in the connection. The message format (Fig. 9.2-5) consists of one octet of indicators, each of which represents a request for a particular action or information item (parameter):

Indicator	Requested Action or Parameter
A	Calling party category
B	Calling line identity
C	Original called address
D	Malicious call identification
E	Call hold
F	Inclusion of outgoing echo controller

To request a particular action or information item, the corresponding indicator bit is set to 1.

9.2.6 General Forward Set-up Information Message (GSM)

This forward message is sent in response to a received GRQ message. It includes an octet with indicator bits, and can also contain one or more of the requested parameters—see Fig. 9.2-6. Each indicator bit corresponds to a particular requested action, or parameter:

Indicator	Action or Parameter
A	Calling party category
B	Calling line identity
D	Original called address
E	Outgoing echo controller included
F	Malicious call identification
G	Call will be held under control of terminating exchange

The value 1 of an indicator bit signifies that the requested action has been (or will be) taken, or that the requested parameter is included in the message. Included parameters appear in alphabetical order of the corresponding indicator bits.

The coding of the calling party category is as in the IAM (9.2.1). The coding of the calling line identity and original called address is as in the IAI (9.2.2).

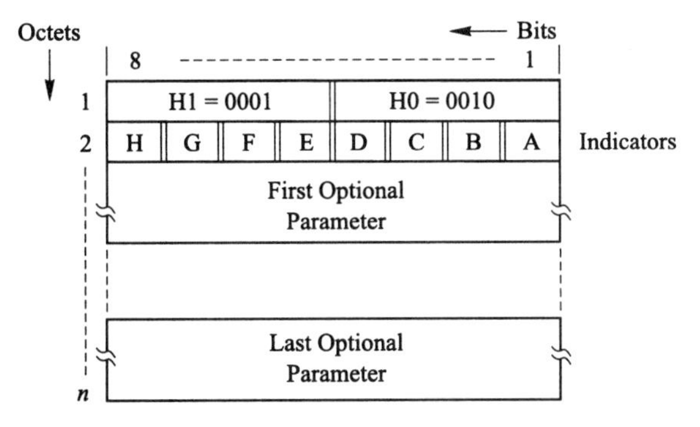

Figure 9.2-6 General forward set-up information message (GSM). (From Rec. Q.723. Courtesy of ITU-T.)

9.2.7 Automatic Congestion Control Message (ACC)

This message is sent by an exchange whose control equipment is congested (overloaded), and requests the directly connected exchanges to reduce the number of seizures of outgoing trunks to the exchange—see Fig. 9.2-5.

Indicator bits BA indicate the congestion level:

Bits AB	Congestion Level
01	Level 1 (moderate congestion)
10	Level 2 (severe congestion)

9.2.8 Call Control Signals

These signals consist of octets a through 1 of Fig. 9.1-1.

Continuity Signals. These forward signals report the success or failure of a continuity check on a trunk.

Signal	Meaning	H1	H0
COT	Continuity check success	0011	0010
CCF	Continuity check failure	0100	0010

Unsuccessful Backward Signals. These signals are sent backwards by the exchange that determines that the call set-up cannot be completed, and indicate the reason for the set-up failure (subscriber busy, address incomplete, etc.). The signal acronyms and heading codes are listed in Table 9.2-1.

Call Supervision Signals. This group consists of forward and backward that are sent on the occurrence of events in a connection that has been set up (answer, clear-forward, etc.). The signals and their heading codes are listed in Table 9.2-2.

Table 9.2-1 Unsuccessful backward set-up signals (H0 = 0101).

H1	Name	Acronym
0001	Switching-equipment congestion	SEC
0010	Circuit-group congestion	CGC
0011	National network congestion	NNC
0100	Address incomplete	ADI
0101	Call failure	CFL
0110	Called subscriber busy	SSB
0111	Unallocated number	UNN
1000	Called line out-of-service	LOS
1001	Send special information tone	SST
1010	Access barred	ACB
1011	Digital path can not be provided	DPN

Source: Rec. Q.723. Courtesy of ITU-T.

Table 9.2-2 Call supervision signals (H0 = 0110).

H1	Name	Acronym
0000	Answer	ANU
0001	Answer, charge	ANC
0010	Answer, no charge	ANN
0011	Clear-back	CBK
0100	Clear-forward	CLF
0101	Re-answer	RAN
0110	Forward-transfer	FOT

Source: Rec. Q.723. Courtesy of ITU-T.

9.3 BASIC SIGNALING SEQUENCES

This section illustrates the use of TUP call-control messages, parameters, and signals, with a number of basic signaling sequences [4]. References to the descriptions in Section 9.2 are included. It is helpful to look up these references while reading the examples that follow.

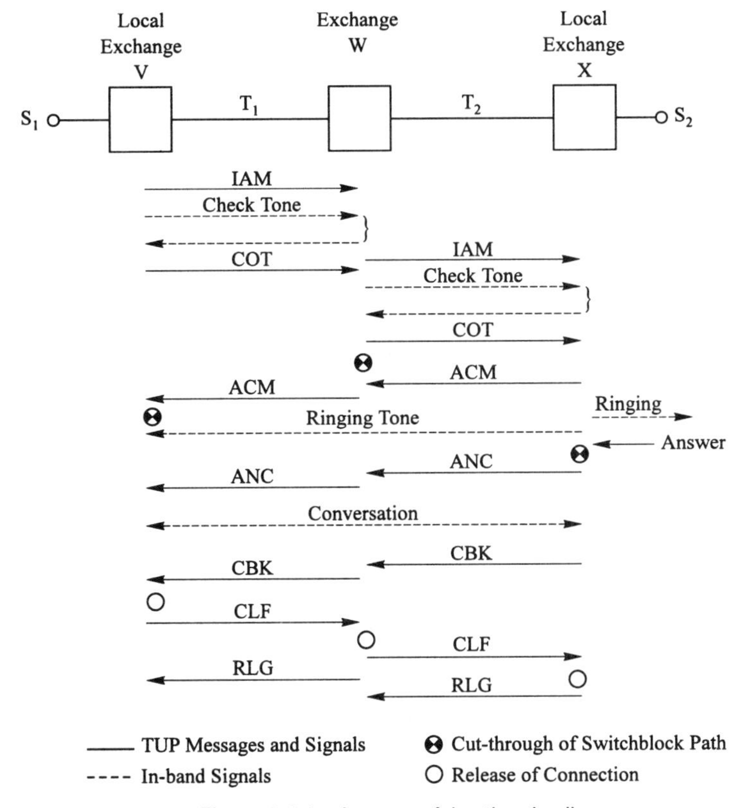

Figure 9.3-1 A successful national call.

9.3. Successful National Call

Figure 9.3-1 shows the signaling for a successful call in a national network from subscriber S_1 to subscriber S_2. Trunks T_1 and T_2 have TUP signaling. The address signaling is *en-bloc*. In this example, continuity checks are made on both trunks.

Set-up of Connection. Having received the called number from S_1, exchange V seizes trunk T_1, and sends an IAM, in which the message indicator bits (FE) are set to "continuity check required" (9.2.1). It also connects a continuity-test transceiver to T_1.

When exchange W has received the IAM, it knows that a continuity check will be made, and therefore connects the send and receive channels of T_1 to each other (loop-back). After analyzing the contents of IAM, exchange W seizes trunk T_2, sends an IAM that indicates "continuity check required," and attaches a continuity-test transceiver to T_2. On receipt of the IAM, exchange X establishes a loop-back on trunk T_2.

Assuming that the transmission of trunk T_1 is in working condition, the transceiver at exchange V receives the check tone. The exchange then disconnects the transceiver from T_1, and sends a COT (continuity signal, 9.2.8). On receipt of COT, exchange W ends the loop-back on trunk T_1.

The continuity check of T_2 is also successful, and exchange W detects the check tone on T_2. Since W has already received the COT signal for T_1, it now completes the call set-up, removing the transceiver from T_2, cutting through a switchblock path between T_1 and T_2, and sending a COT signal to exchange X. If W had received the check tone on T2 but no COT signal for T_1, it would have waited for the COT signal before completing the set-up.

Local exchange X examines the called number in the received IAM. We assume that the number is complete and valid, and that subscriber S_2 is free. From the IAM, exchange X knows that a continuity check will be made on T_2. It therefore awaits the COT signal for T_2 before proceeding. When the signal has been received, X sends an ACM (address complete message, 9.2.4), rings S_2. and sends ringing-tone on T_2. Exchange W repeats the ACM to exchange V, which then cuts through a switchblock path between S_1 and T_1. Subscriber S_1 now hears the ringing tone.

When S_2 answers, exchange X cuts through a path between T_2 and S_2, and sends an ANC (answer, charge signal, Table 9.2.2). The signal is repeated to exchange V, which now starts to charge for the call. The conversation begins.

Message Indicator Settings. We briefly examine the settings of some IAM message indicators (9.2.1) in the example. The call is national, and indicator BA is therefore set to "subscriber number" or "national number," and H is set to "not an international incoming call,"

In signaling system No.6, exchanges make continuity checks on all outgoing trunks. In TUP signaling, the exchange decides whether to perform the check, based on stored information about the trunk. For example, continuity checking is not required for digital trunks, which are carried on a 24- or 30-channel time-division multiplex transmission system. This is because the transmission system has no hardware units for individual trunks, and transmission problems can be detected at the multiplex port in the exchange. In this example, exchange V has set indicator FE to "continuity check required."

If V had decided not to make the check, it would have set FE to "continuity check not required." Exchange W then would not loop back trunk T_1, and would not wait for a COT signal before proceeding with the set-up.

Assuming that the subscriber line of S_1 is an analog line, bit J indicates "all digital path not required," and exchange W can then select an analog or digital outgoing trunk. Also, on calls from analog subscribers, exchange V usually indicates (bit K) that SS7 signaling is not required, and exchange W could therefore have selected an outgoing trunk with another signaling type. The settings of other message indicators are discussed in later examples.

Indicator Settings in ACM. The ACM sent by exchange X, and repeated by exchange W, includes indicators with information for originating exchange V (9.2.4). Bits BA indicate whether V has to charge S_1 for the call. Exchange X knows the status of called subscriber S_2, and sets indicator C to "subscriber free." The call has not been forwarded from S_2, and this is indicated by bit E. T_1 and T_2 have TUP signaling, which is indicated by bit F.

Release of Connection. In TUP signaling, the release of a connection takes place in the same way as in the signaling systems described earlier: it is initiated by originating exchange V. Connections are normally controlled by the calling party (2.1.2). When the calling party disconnects, V immediately initiates the release. In Fig. 9.3-1, the called party disconnects first. Exchange X sends a CBK (clear-back signal), which is repeated by exchange W. When exchange V receives the CBK, it starts a 30–60 s timer, and initiates the release when the calling party disconnects, or when the timer expires, whichever occurs first. The release of trunk T_1 then takes place as follows. Exchange V disconnects the trunk at its end, and sends a CLF (clear-forward signal) to exchange W. This exchange then clears T_1 at its end, and sends a RLG (release-guard signal), to indicate that the trunk is now available for new calls. Trunk T_2 is released in the same manner.

9.3.2 Successful International Call

Figure 9.3-2 shows the set-up of a successful international call. Exchanges V, W, and X are *international switching centers* (ISC) in the originating country, a transit country, and the destination country. Exchanges V and W have seized international trunks (T_1, T_2) with TUP signaling. In this example, address signaling is *en-bloc*, and no continuity checks are made.

Exchange X seizes a national trunk T_3 with channel-associated signaling (CAS)—see Chapter 4. Exchange X is thus the last exchange in the TUP

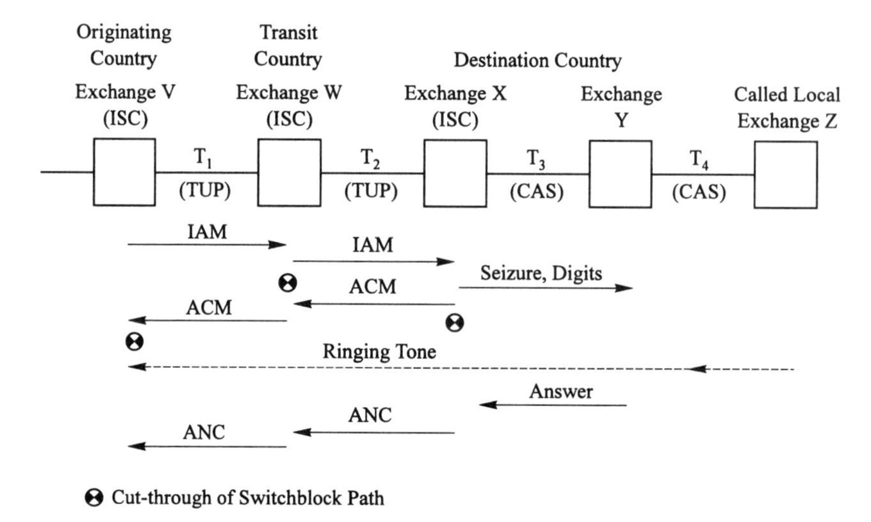

Figure 9.3-2 A successful international call.

signaling segment of the connection. We examine the information in the IAM- and ACM messages.

IAM Information. Exchange V has seized trunk T_1 to a transit country. The called number in its IAM (9.2.1) is therefore an international number, and indicator (BA) is set to "international number." Exchange W seizes trunk T_2 to the destination country. It therefore removes the country code from the received called number, places the national called number in its IAM, and sets BA to "national number." Indicator H is set to "incoming international call" by the originating ISC (exchange V).

Some international trunk groups are carried on satellite transmission systems. It is desirable to have at most one satellite trunk in a connection (1.4.6). If trunk T_1 a is satellite trunk, exchange V sets the IAM message indicator DC to "satellite in connection," and exchange W then selects a terrestrial outgoing trunk.

Address Complete Message. Since trunk T_3 has CAS, the ACM is originated by exchange X. It informs exchanges W and V that they can discard the called number, and other set-up data. In this example, ACM indicator F (9.2.4) is set to "not a completely SS7 signaling path."

On subscriber-dialed international calls, the exchanges in the originating country do not know whether a received called international number is complete, and the called number in the IAM sent by exchange V therefore does not include a ST (end of address digit).

Exchange X, the ISC in the destination country, has to determine whether the received national called number is complete. This is simple when the destination country has an uniform numbering plan (fixed-length national numbers), or when the country has variable-length national numbers and the received number has the maximum possible length. Otherwise, exchange X waits 4–6 s after the receipt of the IAM. If no subsequent address message (9.2.2) is received during this interval, it sends the ACM.

Echo Control. The lengths of the international trunks may make echo control necessary. It is desirable to include only one pair of echo control devices in a connection (1.7.2). In TUP signaling, this is accomplished in the following manner. Suppose that exchange V includes an outgoing echo-control device on trunk T_1. It sets IAM indicator (G) to "outgoing echo control included." Exchange W repeats this indication in its IAM. If exchange X includes an incoming echo-control device on trunk T_2, it sets ACM indicator (D) to "incoming echo control included" (9.2.4). In this case, exchange W merely repeats the ACM. However, if exchange X cannot include an incoming echo control on trunk T_2, it sets indicator (D) to "incoming echo control not included," and exchange W then includes an incoming echo-control device on trunk T_1.

9.3.3 Set-up Problems

Dual Seizures. When a bothway trunk, say trunk T_1 in Fig. 9.3-2, is seized simultaneously by exchanges V and W, the exchanges receive unexpected IAMs in response to their sent IAMs. In TUP signaling, the exchange with the higher point code (PC) controls the trunks with even CICs (circuit identification codes), and the exchange with the lower PC controls the trunks with odd CICs.

On a dual seizure, the controlling exchange continues its call set-up, and the non-controlling exchange "backs off" and attempts to seize another trunk for its call.

Response to Unsuccessful Backward Set-up Signals. An exchange unable to extend a call set-up informs the preceding exchange with one of the unsuccessful backward set-up signals listed in Table 9.2-1. Most of these signals are also available in SS6 signaling.

We first examine a national connection with SS7 signaling all the way (Fig. 9.3-3). When local exchange X determines that called subscriber S_2 is busy, it sends a subscriber buy (SSB) signal to exchange W. This exchange then initiates the release of trunk T_2, sends a clear-forward (CLF) signal, and also repeats the SSB. Exchange V then initiates the release of trunk T_1, and connects a busy-tone source to the line of S_1. Other problems detected at exchange X (called number incomplete or not allocated, called line out of service, etc.) are handled in the same manner.

In Fig. 9.3-2, trunks T_1 and T_2 have TUP signaling, but the signaling on T_3 and T_4 is channel-associated. Suppose now that the set-up fails at the called local exchange Z. If T_3 and T_4 have national R2 signaling (4.6), Z uses a group-B signal to inform exchange X that the called subscriber is busy, or that the called

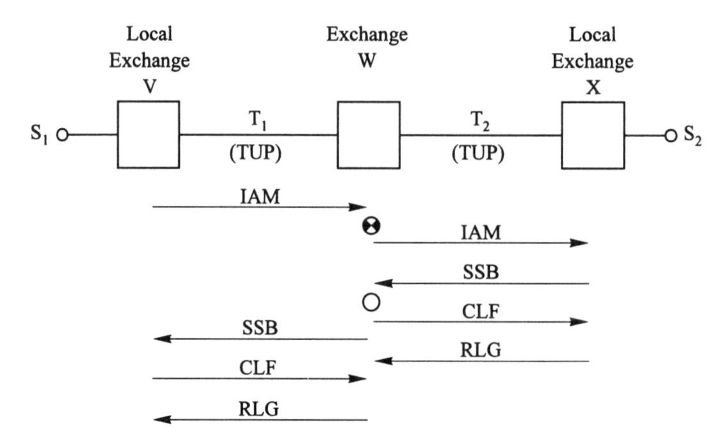

Figure 9.3-3 Release of a failed connection. Signaling path is TUP all the way.

number is not allocated. Exchange X then releases T_3, and sends a SSB or unallocated number (UNN) signal for T_2, and exchange W initiates the release of the trunk. If trunk T_3 and T_4 have a signaling system that cannot inform exchange X about set-up problems, local exchange Z connects its incoming trunk a busy-tone, reorder-tone, or a recorded announcement. The calling subscriber then disconnects, and this initiates the release of the connection.

Automatic Rerouting. Exchanges can be programmed to perform automatic rerouting (1.3.4). Suppose that in Fig. 9.3-2 international exchange V has seized trunk T_1 and receives a switching equipment congestion (SEC) or circuit group congestion (CGC) signal from exchange W. If V is arranged for automatic rerouting, it releases T_1, and attempts to set up the call on a trunk in a route to the destination that does no pass through exchange W.

Time-outs. TUP call set-up procedures include a number of time-outs that cause a call set-up to be abandoned. For example, when an exchange does not receive an ACM within 20–30 s after sending the final address message for the call, or does not receive an answer signal within 2–4 minutes after the receipt of the ACM, it clears the forward connection and sends a call failure (CFL) signal to the preceding exchange. Also, when an exchange receives an IAM with message indicator bits (FE) indicating that the preceding exchange will make a continuity check, it returns a CFL signal if it does not receive the COT signal within 10–15 s.

This concludes the overview of basic TUP signaling for call control.

9.4 TUP SUPPORT OF ADDITIONAL SERVICES

This section describes a number of services that do not fall in the category of "basic" call control, and the manner in which these services are supported in TUP signaling [4,5].

9.4.1 Services

The services to be described fall in one of the following classes.

Malicious Call Identification. Multi-frequency code (MFC) signaling in many national networks includes procedures to identify malicious callers (4.6). These are also included in TUP signaling.

Digital Connectivity. Telecoms in France [8] and several other countries have introduced digital subscriber lines, without waiting for the completion of the specifications for digital subscriber lines in the integrated services digital network (ISDN).

Normal subscriber signaling (Chapter 3) is used on these lines. When the connection has been set up, the subscribers communicate with 64 kb/s digital data.

Supplementary Services. CCITT has defined a number of supplementary services. These services require connections with SS7 (TUP or ISUP) signaling all the way:

Calling Line Identification. Subscribers marked for this service at their local exchanges have devices attached to their telephones that display the calling party number. In the U.S. this service is known as "caller ID" (3.7.2).

Call Forwarding Service. In this service, a subscriber can indicate to his local exchange that, until further notice, incoming calls should be forwarded to another subscriber.

Closed User Group Service [5]. In this service (not offered in the U.S.), a multi-location business customer can register his lines as members of a closed user group (CUG). A CUG is identified worldwide by four digits that represent a particular telecom, and an interlock code assigned by the telecom to the CUG (9.2.2). A CUG restricts the calls that can be made and received by its members. A CUG has three characteristics that can be specified by the business customer:

OUTGOING ACCESS. Members of a CUG with no outgoing access (NOA) can only make calls to other members of their CUG. Members of a CUG with outgoing access (OA) have no restrictions on their outgoing calls.

INCOMING ACCESS. Members of a CUG with no incoming access (NIA) can receive calls from other members of their CUG only. Members of a CUG with incoming access (IA) have no restrictions on their incoming calls, except as noted below.

INCOMING CALLS BARRED. Members of a CUG with incoming calls barred (ICB) are not allowed to receive calls from other members of their CUG. Members of a group without incoming calls barred (NICB) can receive calls from other members of their CUG.

The decision to allow or to reject a call involving one or two members of a CUG is made by the local exchange of the called party (see Section 9.4.7).

9.4.2 The GRQ-GSM Procedure

This procedure, which is used in some of the services outlined above, is the functional equivalent of end-to-end interregister signaling in the R2 signaling systems (4.4.4, 4.4.5). It allows the called exchange to make a request to the calling exchange, during the set-up of the call. In Fig. 9.4-1, called exchange X has received an IAM, and sends a *general request message* (GRQ). Each message indicator bit represents a particular request from the called exchange (9.2.5). When a request is made, the corresponding indicator bit is set to 1.

The requests are of two types: requests for additional information (calling line identity, original called address, etc.) and requests for special call-control actions (malicious call identification, holding the call under control of the called party, etc.). When a particular item is requested, the corresponding indicator is set to 1.

The GRQ message is transferred backwards, along the exchanges in the connection, and reaches calling exchange V. This exchange responds with a *general forward set-up information message* (GSM). Message indicator bits (9.2.6) confirm (1) or deny (0) the individual requests. The parameters included in the message appear in the alphabetical order of their indicator bits. The GSM message is transferred forwards, and reaches exchange X.

A GRQ-GSM message sequence has to be completed before the called exchange sends its ACM (address complete) message for the call.

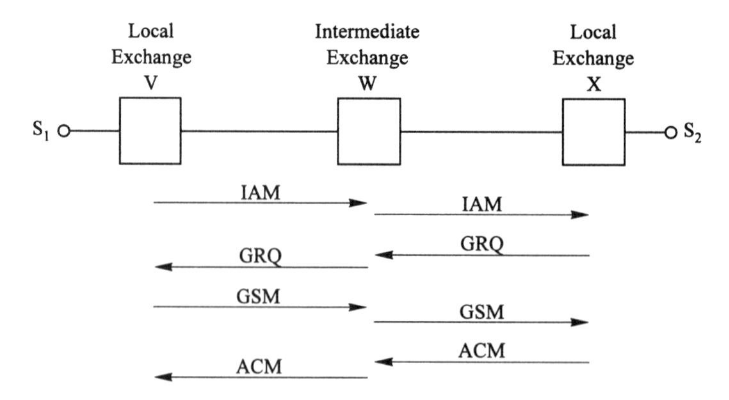

Figure 9.4-1 Request for information or action, and response.

9.4.3 TUP Support of Malicious Call Identification

The directory number of a subscriber who is subjected to malicious calls can be marked in his local exchange. Suppose that the directory number of a subscriber S is marked in this way.

When the local exchange of S receives an IAM or IAI for a call to S, it has to obtain the number of the calling party. If the initial address message is an IAI, this number may already be included. Otherwise, the exchange requests the calling number with a GRQ (9.2.5).

If the call turns out to be malicious, subscriber S alerts his exchange with a flash signal (3.3.2). The exchange then prints a record that includes the date and time-of-day of the call, and the calling and called numbers.

In some countries, the connections of malicious calls are also held under control of the called exchange, for tracing purposes. In these countries, the local exchange that receives an IAM or IAI for a call to subscriber S also requests "call holding." If the exchange receives a flash signal from S, it does not send a CBK (clear-back signal) when subscriber S disconnects. Moreover, when the calling party disconnects, the calling exchange will hold the connection until it receives a CBK from the called exchange. This enables personnel at the calling and called exchanges to trace the connection.

9.4.4 TUP Support of Digital Connectivity

For this service, the path between the calling and called digital subscriber lines has to be completely digital (64 kb/s in both directions). This requires that all exchanges in the connection have digital switchblocks, and that all trunks are digital trunks. These connections also require TUP signaling all the way. When a call originates on a (pre-ISDN) digital subscriber line, the originating exchange selects a digital outgoing TUP trunk to a next exchange that has a digital switchblock, and informs the succeeding exchanges that the call is digital, by setting the IAM message indicators J and K to "digital path required" and "path with SS7 signaling required" (9.2.1).

9.4.5 TUP Support of Calling Line Identification

This service requires the transfer of the calling line identity (calling party number) across the network. The service can be provided in two ways.

In some countries, the first address message of a call is always an IAI (initial address message with additional information) in which the calling number is included (9.2.2).

In other countries, the first message of a call is an IAM. The called exchange checks whether the called party is registered for the calling line presentation service. If so, the called exchange obtains the calling line identity with the GRQ-GSM procedure.

The format of the calling line identity parameter is shown in Fig. 9.2-3.

Subscribers can be marked at their local exchanges as allowing, or not allowing, the "presentation" of their number to called subscribers, and the originating exchange sets indicator bit C of the calling line identity parameter to "presentation allowed" or "presentation not allowed." The called exchange delivers the calling number only when the presentation is allowed.

9.4.6 TUP Support of Call Forwarding

A subscriber, say S_1, who is registered at his exchange for call-forwarding can activate and deactivate the service by dialing a service access code (3.7.1). On activation, S_1 also dials the number of subscriber S_2, to whom his incoming calls should be forwarded.

When a call for S_1 arrives while the service is active, his local exchange extends the call set-up to S_2, and sets indicator bit I in its outgoing IAM to "forwarded call" (9.2.1).

If the exchange serving S_2 receives the IAM of the forwarded call, and determines that S_2 also has call-forwarding and has activated his service, it does not forward the call again, but abandons the set-up. The call is thus forwarded only once. This avoids circular routing problems, for example when S_1 is forwarding his calls to S_2, and S_2 is forwarding his calls to S_1.

9.4.7 TUP Support of Closed User Groups

A local exchange has stored data that enable it to determine whether an attached subscriber line (or the directory number of the line) is a CUG member, and to obtain the interlock code and the characteristics of the CUG.

The decision to allow or reject a call involving one or two CUGs is made by the called exchange, and is based on the calling and/or called CUG. Information on the called CUG is stored at the called exchange. Information on the calling

Table 9.4-1 Acceptance and rejection of calls involving closed user groups (CUG).

Calling	CUG	Called CUG Type				
CUG Type	Match	NIA-NICB	NIA-ICB	IA-NICB	IA-ICB	Non-CUG
NOA	Y	A	R	A	R	R
	N	R	R	R	R	
OA	Y	A	R	A	R	A
	N	R	R	A	A	
Non-CUG	—	R	R	A	A	A

NOA: No outgoing access. OA: Outgoing access. Non-CUG: Calling or called party not a member of a CUG. NIA-NICB: No incoming access, incoming calls not barred. NIA-ICB: No incoming access, incoming calls barred. IA-NICB: Incoming access, incoming calls not barred. IA-ICB: Incoming access, incoming calls barred. A: Call accepted by called local exchange. R: Call rejected by called local exchange. Y: Yes. N: No.
Source: Rec. Q.723. Courtesy of ITU-T.

CUG is stored at the originating exchange. The called exchange knows whether the calling party is a CUG member because, on calls made by CUG members, the initial message sent by the originating exchange is an IAI in which indicator B of octet ($n + 2$) in Fig. 9.2-2 is set to "CUG information included," and includes the CUG information (9.2.2).

The criteria for accepting or rejecting the incoming call are listed in Table 9.4-1. If the calling and called parties are both CUG members, the decision depends on whether the members belong to the same CUG (indicated as CUG match in the table), and on the incoming and outgoing characteristics of the CUG (or CUGs).

If only the called party is a CUG member, the call is allowed if the called CUG has incoming access (IA). If only the calling party is a CUG member, the call is allowed if the calling CUG has outgoing access (OA).

9.5 OTHER TUP PROCEDURES, MESSAGES, AND SIGNALS

This section describes a number TUP procedures, messages, and signals that are not directly related to the control of calls [4,6]. We consider a trunk group (TG) between exchanges V and W—Fig. 9.5-1.

9.5.1 Overload at Exchange

During moments of high activity, the call processing equipment at an exchange may become overloaded. Suppose that this is the case for exchange W. The exchange then informs exchange V (and all other exchanges to which it is connected by a TUP trunk group) by sending *automatic congestion control* (ACC) messages (9.2.7). The ACC messages to exchange V are sent when exchange W receives a CLF for a trunk in group TG. Each ACC message is followed by a RLG signal, which is the normal response to the CLF.

ACC messages include one octet with indicators (Fig. 9.2-5). Bits B and A indicate the level of congestion:

(BA)	Meaning
01	Level 1 congestion (moderate)
10	Level 2 congestion (severe)

In response to the ACC message, exchange V reduces the number of seizures of trunks in trunk group TG. The degree of the reduction depends on the congestion level of exchange W. For example, exchange V may be programmed to abort every fourth seizure of a trunk in the group (25% reduction) for level 1 congestion, and all seizures (100% reduction) for level 2 congestion. If exchange V has other route choices for the call, it tries to set up the call on one of these routes. Otherwise, it abandons the set-up.

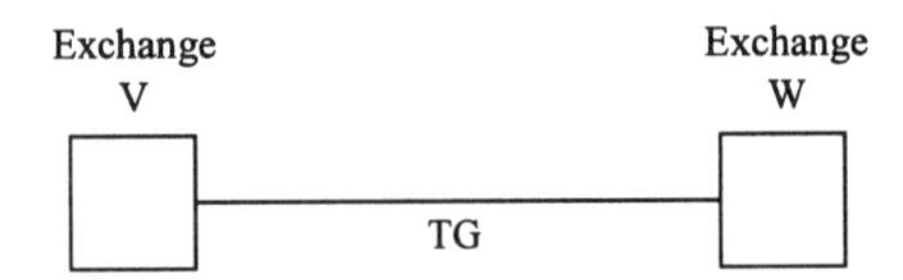

Figure 9.5-1 Exchanges V and W, interconnected by trunk group (TG).

Exchange V keeps reducing the number of seizures of trunks in group TG until some 5 s have elapsed without the receipt of a new ACC message from exchange W.

9.5.2 Trunk Blocking and Unblocking

Maintenance personnel (or an automatic process) in exchanges V and W can block a trunk. Suppose that exchange W needs to test a trunk T in group TG. It then sends a BLO (blocking) signal for the trunk to exchange V—see Table 9.5-1. The latter exchange acknowledges with a BLA (blocking acknowledgment) signal). The affected trunk T is identified by DPC, OPC, and CIC (Fig. 9.1-1). If trunk T is carrying a call, the call is not interrupted, and exchange W starts the test only after the call has ended. Exchange V will not seize T for new calls, but will process test calls from exchange W. In test calls, the CPC (calling party category) code in the IAM (9.2.1) is 001101. After the test, exchange W can unblock the trunk by sending an UBL (unblocking) signal to exchange V, which then sends an UBA (unblocking acknowledgment) signal, and resumes seizing trunk T for new calls. TUP trunks are usually bothway trunks, and both exchanges can therefore initiate the blocking of a trunk.

9.5.3 Resetting a Trunk

Memory devices of the processors in stored program controlled exchanges store records with status information for each attached trunk. When a memory malfunction causes exchange W to lose the status information of a trunk T of group TG, it sends a RSC (reset-circuit) signal—Table 9.5-1—for the trunk to

Table 9.5-1 Circuit supervision signals (H0 = 0111).

H1	Name	Acronym
0001	Release-guard	RLG
0010	Blocking	BLO
0011	Acknowledgment of BLO	BLA
0100	Unblocking	UBL
0101	Acknowledgment of UBL	UBA
0110	Continuity-check request	CCR
0111	Reset-circuit	RSC

Source: Rec. Q.730. Courtesy of ITU-T.

exchange V. This signal requests V to clear a possible connection involving T, and set the trunk to "idle."

If the memory at V indicates that T is currently involved on an incoming call (from W), the exchange regards the RSC as a clear-forward signal. It releases trunk T, sends a CLF (clear-forward) signal to the next exchange in the connection, and sends a RLG (release-guard) signal for trunk T to exchange W (Table 9.2-1). If the record at V shows that its call on T is an outgoing call (to W), it sets trunk T to "idle", sends a CLB signal to the previous exchange in the connection, and sends a CLF to W, which then responds with a RLG. If exchange V receives a RSC signal when the affected trunk T is idle, it responds with a RLG signal.

9.5.4 Circuit Group Blocking, Unblocking, and Resetting

TUP signaling includes a group of *circuit group supervision messages* that are used to block, unblock, and reset up to 256 trunks in a trunk group simultaneously. The message acronyms and heading codes are listed in Table 9.5-2.

The structure of a circuit group supervision message is shown in Fig. 9.5-2. The values of OPC and DPC (originating and destination point codes) identify the trunk group TG. The particular set of up to 256 trunks that are covered by the message is identified by the values of parameters CIC (circuit identification code) and R (range). For example, the combination (CIC $= c$, R $= r$) identifies the set of trunks in group TG that have the CIC values c, $c + 1$, $c + 2$,..., and $c + r$.

In Fig. 9.5-2, parameter R is followed by up to 32 octets of *status indicator* bits for the trunks in the set. Bits 1 through 8 in octet 3 represent the trunks with CIC values $(c + 0)$ through $(c + 7)$, bits 1 through 8 in the next octet represent trunks with CIC codes $(c + 8)$ through $(c + 15)$, etc.

In the request messages, the indicator value is 1 if the request is made for the

Table 9.5-2 Circuit group supervision messages (H0 = 1000).

H1	Name	Acronym
0001	Maintenance group blocking	MGB
0010	Acknowledgment of MGB	MGA
0011	Maintenance group unblocking	MGU
0100	Acknowledgment of MGU	MUA
0101	Hardware failure group blocking	HGB
0110	Acknowledgment of HGB	HBA
0111	Hardware failure group unblocking	HGU
1000	Acknowledgment of HGU	HUA
1001	Group reset	GRS
1010	Acknowledgment of GRS	GRA

Source: Rec. Q.723. Courtesy of ITU-T.

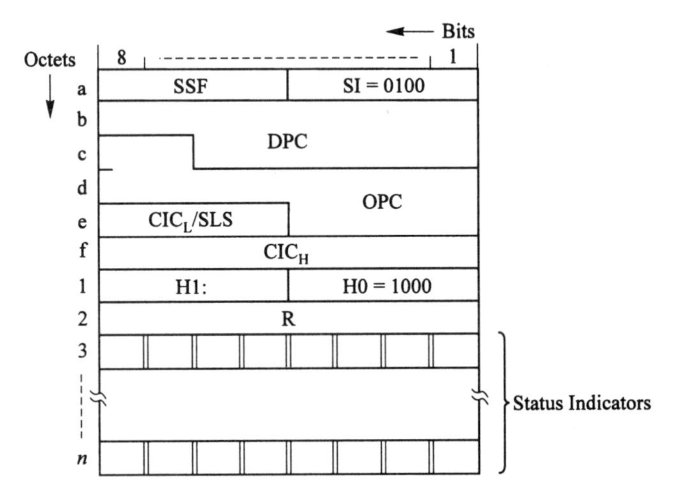

Figure 9.5-2 MTP3 message with circuit group supervision message. (From Rec. Q.723. Courtesy of ITU-T.)

particular trunk, and 0 otherwise. These indicator values are copied in the group blocking and unblocking acknowledgment messages.

In the group reset acknowledgment message (GRA), a status indicator value 1 or 0 indicates that the corresponding trunk is blocked, or not blocked, by the exchange that sends the GRA.

Maintenance and Failure Group Blocking. A maintenance group-blocking (MGB) message is sent by the exchange (V or W) that wants to test a number of trunks in trunk group TG. A *hardware failure group blocking* (HGB) message is sent by an exchange when it detects a failure on a number of trunks.

When an exchange receives a maintenance (MGB) or hardware failure (HGB) group-blocking message, it stops seizing the affected trunks for new calls. Existing calls are allowed to continue after receipt of a MGB, but are terminated immediately when a HGB is received.

A group-reset or group-blocking request can affect calls on a large number of trunks. Therefore, as a safety measure, the requesting exchange sends the message twice, and the other exchange takes action only if it receives two identical MGB or HGB messages in a 5-s interval.

9.5.5 Signaling Route Set Congested or Not Available

The TUP process at an exchange responds to MTP-pause, MTP-resume, and MTP-status indications from its MTP (8.9.3 and 8.9.5). We again consider the trunk group TG between exchanges V and W, and denote the TUP process at exchange V by TUP-V.

When TUP-V receives a MTP-pause indication for destination W, it knows that it can no longer send and receive signaling messages for the trunks in group

TG. It therefore clears all existing calls in TG, and suspends the seizure of trunks for new calls. When TUP-V receives a MTP-resume for destination W, it resumes seizing trunks in group TG.

When TUP-V receives a MTP-status indication for destination W, it reduces the number of seizures of trunks in group TG. The reduction remains in effect until a period of about 5 s has elapsed during which TUP-V has not received a new status indication for destination W.

9.6 VERSIONS OF TUP SIGNALING

CCITT Recommendations Q.721–Q.724 [1–4] are useful as a general framework for TUP signaling systems. However, many telecoms have found it necessary to tie down the "loose ends," select options, and make adaptations to accommodate the services and procedures that already existed in their respective national networks prior to the introduction of TUP. As a result, there are differences in national TUP versions. This situation is comparable to the national versions of national R2 signaling.

Moreover, in the 1984–1988 time frame, several telecoms began to introduce some ISDN services, which had not yet been standardized by CCITT (ISDN services and ISUP interexchange signaling only started to be adequately defined in the 1989 CCITT *Blue Books*). To handle these "pseudo-ISDN" services, some telecoms have defined and deployed "extended" versions of TUP.

This section briefly describes a number of TUP versions.

9.6.1 British Telecom National User Part

British Telecom has defined a national version of SS7 signaling called *CCITT No.7 (BT)*, for use in the United Kingdom. It consists of a slightly modified version of the CCITT-defined message transfer part, and a *national user part* (NUP), which supports, in addition to basic telephone call control, some supplementary services and some initial ISDN features [7].

NUP transcends TUP in that it handles calls between analog subscriber lines, and calls between subscribers who have digital access lines that use a signaling system called *digital access signaling system* (DASS), of which there are two versions.

NUP message coding resembles CCITT–TUP coding. However, headings H0 and H1 occupy one octet each, thus leaving room for many additional (future) messages.

Sending of Called Number. NUP includes three address messages. Initial address messages carry the leading digits of the called address only (for overlap address signaling). After receipt of an IAM, an exchange Y can request a specific number (N) of subsequent digits from preceding exchange X. This is an established signaling procedure in the U.K. The request is made with a *send-N-digits* message. As soon as X has received the requested digits, it sends them in a subsequent address message (SAM). The last subsequent address message is called a *final address message* (FAM). The *initial and final address message* (IFAM) carries a complete called number, and is used for *en-bloc* signaling.

The initial (IAM or IFAM) message of a call include a number of other parameters, some of which are not available in CCITT-defined TUP. Notably, *signaling handling protocol* (SHP) indicates whether the call originated from an analog- or digital subscriber line, and *call path indicator* (CPI) indicates whether an analog or 64 kb/s digital path is required for the call.

End-to-End Signaling. During the call set-up, the calling and called local exchanges can communicate to obtain additional information. A number of *service information messages* (SIM) have been defined for this purpose. Their functions are similar to those of the GRQ and GSM messages described in Section 9.4.2.

Swap Message. Two digital subscribers can change their mode of communication from speech to digital data and vice versa during a call. The subscriber originating the change signals his local exchange X with a DASS *swap* message. This information is transferred across the network to the distant local exchange Y by NUP *swap* messages, and delivered to the other subscriber as a DASS message.

9.6.2 France Telecom TUP-E

France Telecom recognizes three types of subscribers: conventional telephone subscribers, Transcom-ISDN subscribers, and ISDN subscribers [8]. Calls from telephone subscribers are handled on trunks with Socotel MF, basic (French) TUP, or *extended TUP* (TUP-E) signaling. Transcom-ISDN subscribers can make 64 kb/s calls (only). These calls are set up by "telephone" subscriber signaling, and have to be routed on digital trunks with TUP-E signaling. ISDN subscribers use a national form of ISDN access signaling, and can make speech and data calls, which are also routed on digital trunks with TUP-E signaling exclusively.

The purpose of TUP-E is similar that of British Telecom's NUP signaling. It has allowed the introduction of a number of ISDN services at a time that CCITT had not yet defined digital subscriber signaling (DSS) and ISUP (ISDN interexchange signaling) in sufficient detail.

9.6.3 TUP in the International Network

Basic International TUP. TUP signaling for basic call control, as described in Section 9.3, is used on a number of international (mostly transatlantic and transpacific) trunk groups [9]. This version of TUP also supports international digital connectivity (transparent transfer of 64 kb/s user communications)—see 9.4.4.

TUP+ has been introduced on international circuits between countries of the European Economic Community. It has been defined by the CEPT in its Recommendation 43-02 [10]. It includes signals and procedures to support a number of ISDN and supplementary services, such as closed user groups, calling line identification, and user-to-user signaling.

9.7 ACRONYMS

ACB	Access barred
ACC	Automatic congestion control
ACM	Address complete message
ADI	Address incomplete
ANC	Answer, charge
ANN	Answer, do not charge
ANSI	American National Standards Institute
ANU	Answer (no instructions on charging)
BLA	Blocking acknowledgment
BLO	Blocking request
CAS	Channel-associated signaling
CBK	Clear-back
CCITT	International Telegraph and Telephone Consultative Committee
CCF	Continuity check failure
CCIS	Common-channel interoffice signaling
CCR	Continuity check request
CEPT	European Conference of Postal and Telephone Administrations
CFL	Call failure
CGC	Circuit group congestion
CIC	Circuit identification code
CIC_L	Four low order bits of CIC
CIC_H	Eight high order bits of CIC
CLF	Clear-forward
COT	Continuity check successful
CPC	Calling party category
CUG	Closed user group
DPC	Destination point code
DPN	Digital path not provided
FDM	Frequency-division multiplex
FOT	Forward-transfer
GRA	Group reset acknowledgment
GRQ	General backward request message
GRS	Group reset request

GSM	General forward set-up information message
H0	Message heading field 0
H1	Message heading field 1
HBA	Hardware failure group blocking acknowledgment
HGB	Hardware failure group blocking request
HGU	Hardware failure group unblocking request
HUA	Hardware failure group unblocking acknowledgment
IA	Incoming access
IAI	Initial address message with additional information
IAM	Initial address message
ICB	Incoming calls barred
ISC	International switching center
ISDN	Integrated services digital network
ISUP	ISDN user part of SS7
LOS	Subscriber line out of service
MFC	Multi-frequency code signalling
MGA	Maintenance group blocking acknowledgment
MGB	Maintenance group blocking request
MGU	Maintenance group unblocking request
MSU	Message signal unit on signaling link
MTP	Message transfer part
MTP3	Message transfer part–level 3
MUA	Maintenance group unblocking acknowledgment
NCIB	No incoming calls barred
ND	Number of digits
NIA	No incoming access
NNC	National network congestion
NOA	No outgoing access
OA	Outgoing access
OPC	Originating point code
PAD	Point code of the affected destination
PCM	Pulse code modulation
RAN	Re-answer
RL	Routing label
RLG	Release-guard
RSC	Reset circuit
SAM	Subsequent address message
SAO	One-digit subsequent address message
SEC	Switching equipment congestion
SI	Service indicator
SIF	Signaling information field
SIO	Service information octet
SL	Signaling link
SSB	Subscriber busy
SSF	Sub-service field

SST	Send special information tone
ST	End of address signaling
SLS	Signaling link selector
SS6	Signaling system No.6
SS7	Signaling system No.7
TUP	Telephone user part of SS7
UBA	Unblocking acknowledgment
UBL	Unblocking request
UNN	Unassigned number

9.8 REFERENCES

1. *Specifications of Signalling System No.7*, Rec. Q.721, CCITT Blue Book, **VI.8**, ITU, Geneva, 1989.
2. *Ibid.*, Rec. Q.722.
3. *Ibid.*, Rec. Q.723.
4. *Ibid.*, Rec. Q.724.
5. *Ibid.*, Rec. Q.730.
6. D.R. Manfield, G. Millsteed, M. Zukerman, "Congestion Controls in SS7 Signaling Networks," *IEEE Comm. Mag.*, **31,** No.6, June 1993.
7. CCITT Signalling System No.7: National User Part, *British Telecomm. Eng.*, **7**, April 1988.
8. J. Craveur *et al*, *CCITT No.7 Common Channel Signalling in the French Telecommunication Network*, ISS Conf. Rec., Phoenix, 1987.
9. J.J. Lawser *et al.*, *Common Channel Signalling for International Services*, ISS Conf. Rec., Stockholm, 1990.
10. *Rec. 43-20,* CEPT/T/SPS, 1987.

10

DIGITAL SUBSCRIBER SIGNALING SYSTEM NO. 1

Digital subscriber signaling system No.1 (DSS1) is used for signaling between an ISDN (integrated services digital network) subscriber and his local exchange.

This chapter is a brief excursion from the description of signaling system No.7 (SS7), which has been the subject of Chapters 7 through 9, and will continue in Chapter 11, with the description of the *integrated services digital network user part* (ISUP). This order of presentation has been chosen because DSS1 can be discussed without references to ISUP, and because it is helpful to be familiar with DSS1 when exploring ISUP, which is the interexchange signaling system for the control of ISDN interexchange calls.

10.1 INTRODUCTION TO ISDN AND DSS1

This section presents a broad-brush introduction to ISDN, focusing primarily on its signaling aspects. More details on ISDN architecture and technology can be found in a number of texts [1–4].

10.1.1 Circuit-mode and Packet-mode Networks

So far, we have discussed (circuit-mode) telecommunication networks in which two subscribers communicate over a temporary and dedicated circuit, consisting of the subscriber lines and possibly one or more trunks. Connections are set up by exchanges at the start of a call, and released when the call has ended.

During the 1970s, a new network architecture was developed for data communications. These networks, which have been installed in a number of countries, consist of *nodes* that are interconnected by *data links*. The users of these network have *data terminals* (DTE) at their premises, which are attached to one of the nodes. Users communicate by sending *packets*, consisting of a *header* and a limited amount of data (say up 100 octets). The packets are transferred on the data links, and each node uses the address information in the packet header to select an outgoing data link to or towards the packet destination. A data link is not "dedicated" to the communications between a specific pair of users: it transfers packets of several users.

This type of communication is known as *packet-switched* (or *packet-mode*) communication. CCITT has defined standards for packet data networks in Recommendations of the "X" series. In particular, Rec. X.25 [5] defines the interface between the DTE and the data communication network.

We have already encountered an example of such a network: the signaling networks described in Chapter 5 are packet-mode networks for a particular application: the transfer of common-channel signaling messages between exchanges (and other entities in the telecommunication network). Even though the terminology is different, the concepts are the same:

Concept	Data Communication Network	Signaling Network
Node	Node	Signal transfer point
Link	Data link	Signaling data link
Data unit	Packet	Signal unit
User	Customer's DTE	Signaling end point

Circuit-mode and packet-mode data communications exist side by side. Packet-mode is the preferred choice for data traffic that consists of short bursts that are separated by comparatively long "silent" intervals. This is often the case for communications between data terminals and a centralized mainframe computer. For example, the verification of credit cards by stores, and the reservation of hotel rooms by travel agencies. Data communication networks are usually designed to handle packets with up to about 100 octets of information, which is sufficient for a query or a response.

In applications involving the transfer of large amounts of data, circuit-mode data communications is the better choice. In packet communications, the data would have to be segmented into packets at one end, and reassembled at the other end. Moreover, the data transfer would consist of a large number of packets, and the headers of each packet have to be processed (for packet routing) at each traversed node. On the other hand, in ciruit-mode communications, any amount of data can be sent in one stream, without processing by the network, once the connection has been set up.

10.1.2 The Objectives of ISDN

When discussing ISDN, it is customary to speak about *users* instead of subscribers. ISDN has two objectives. In the first place, it is an "integrated services network" that provides circuit-mode (speech and data) and packet-mode (data) communications for its users.

The second objective is *user-to-user digital connectivity*. This means that all components (lines, trunks, exchanges) in an ISDN connection transfer 64 kb/s digital data.

10.1.3 User Equipment and Access

Figure 10.1-1 shows the equipment at the premises of an ISDN user, and his access to the ISDN network. The user's *network termination* (NT) is connected to a number of *terminal equipments* (TE) of various types (telephones, facsimile machines, data processing equipment, etc.). The NT is attached to an *exchange termination* (ET) in his local exchange by a *digital subscriber line* (DSL). The line has a number of time-division multiplexed B-channels and one D-channel. The B-channels are used for circuit-mode communications (PCM encoded speech or voiceband modem data, or 64 kb/s digital data). The D-channel is used for DSS1 signaling messages, and for packet-mode data communications.

In most countries, the DSL, NT, and ET are considered to be internal parts of the telecommunication network. Therefore, the network-to-user interface specified in CCITT is the S/T interface (between NT and the TEs). In the U.S., the DSL and NT are considered to be outside the network, and the network-to-user interface is the U interface. We distinguish the following two DSL types.

Basic Access. A basic access DSL is intended for residential or small business use. In the U.S., it is a two-wire circuit with two bidirectional B-channels

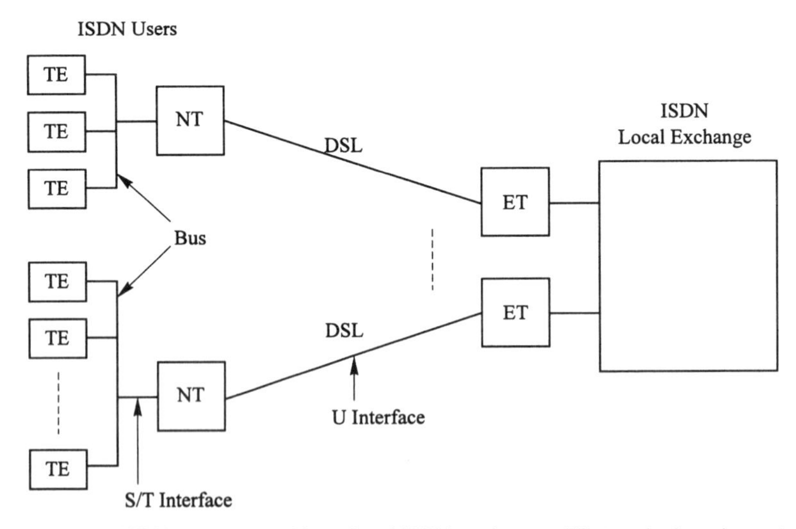

Figure 10.1-1 ISDN users served by a local ISDN exchange. TE: terminal equipment. NT: network terminal. DSL: digital subscriber line. ET: exchange terminal.

(64 kb/s), and one bidirectional D-channel that operates at 16 kb/s (2B + D). The bit rate in each direction is $2 \times 64 + 16 + 48$ (for overhead functions) = 192 kb/s. Physically, the DSL is an ordinary two-wire subscriber line. Circuitry in the ET and NT enables the line to transfer both bit streams simultaneously (full duplex operation). Since a call requires one B channel, this DSL allows two simultaneous calls, and up to eight TEs can be connected to the NT.

Other countries use a variety of basic access DSLs. They can be two- or four-wire circuits with one (B + D) or two (2B + D) B-channels.

At the user's premises, the TEs are attached to the NT by a bidirectional *passive bus*. Each terminal has access to the B-channels and the D-channel of the DSL.

Primary Rate Access. Primary rate access DSLs are intended for medium or large businesses, and resemble the first-order digital multiplex transmission systems that have been developed for PCM trunks (1.5.2). These DSLs consist of two amplified two-wire channels, one for each direction of transmission.

American primary rate DSLs have the channel format of the T1 digital transmission system. They operate at 1544 kb/s, and have twenty-four 64 kb/s channels. Channels 1–23 are used as B-channels, and channel 24 is the (64 kb/s) D-channel (23B + D). Most other countries use a digital transmission system that operates at 2048 kb/s. Channels 1–15 and 17–31 are used as B-channels and channel 16 is the D-channel (30B + D).

10.1.4 The ISDN Network

From the user's point of view, ISDN is an "integrated services" network. His DSL can be used for circuit and packet mode communications. Before the arrival of ISDN, a subscriber needed a line to his local exchange for circuit-mode communications, and a separate line to a node in a data communications network for packet-mode communications.

Actually, only the DSLs and the local ISDN exchanges are truly integrated. As shown in Fig. 10.1-2, an ISDN network is a group of the following three networks.

Telecommunication (Circuit-switched) Network. A circuit-switched "pure" ISDN network would consist of 64 kb/s digital trunks, and exchanges with digital switchblocks only. In practice, these networks are gradual evolutions of existing telecommunication networks. They include a mixture of analog (FDM) and digital trunks, and exchanges with analog or digital switchblocks. The analog equipment is gradually being replaced by digital equipment. The networks usually have interexchange signaling systems of several types, but are in the process of being converted to ISUP signaling. This system, to be described in Chapter 11, is the only system that meets the requirements of ISDN.

Since the number of ISDN users is still very small compared to the number of (analog) subscribers, only a small fraction of the local exchanges in a network is equipped to serve ISDN users, and these exchanges also serve analog subscribers.

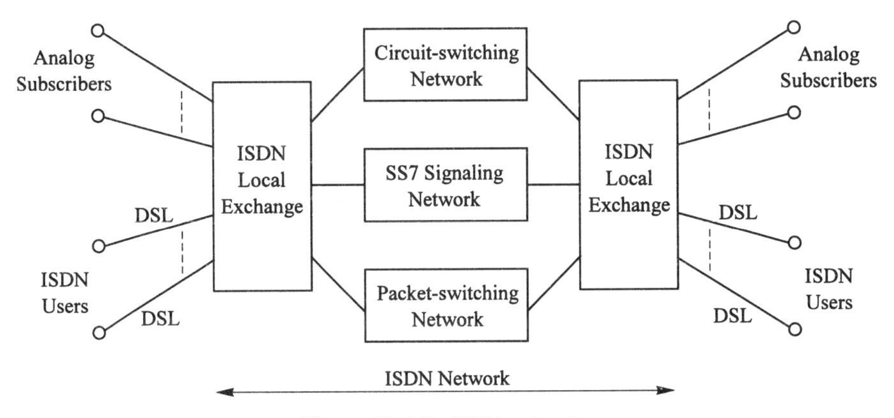

Figure 10.1-2 ISDN network.

SS7 Signaling Network. This may be an extension of an existing SS7 signaling network (if the telecommunication network is already using TUP signaling), or a newly installed network.

Packet-switching Network. This may be an existing data communication network, adapted with an interface to ISDN local exchanges, or a new network.

The local exchange segregates the DSS1 signaling messages and data packets that arrive on the D-channels of a DSL. The packets are transferred to the packet-switching network, and the incoming DSS1 signaling messages are processed by the exchange. In the reverse direction, the local exchange transfers the packets received from the packet-switching network, and its outgoing DSS1 signaling messages, to the D-channels of the DSLs.

10.1.5 ISDN Services for Circuit-mode Communications

This book covers signaling in circuit-switched networks. Therefore, this chapter describes the circuit-mode communications of ISDN only. Three groups of ISDN services for this mode of communications are listed below.

Bearer Service. This defines the type of communication service for a call. A calling ISDN user can request one of three bearer services: speech, 3.1 kHz audio (voiceband modem data), or 64 kb/s digital data. Information about the type of bearer service is transferred by the network to the called user, where it is one of the criteria to select an appropriate TE. For example, when the bearer service is "speech," the incoming call is connected to a telephone. We shall see in Chapter 11 that the bearer service is also taken into account by the exchanges in the network for the selection of outgoing trunks. For example, speech calls can be set up on analog or digital trunks, but 64 kb/s data calls require digital trunks.

Teleservice. The data in 3.1 kHz audio and 64 kb/s calls can pertain to various data services (facsimile, telex, teletext, etc.). The calling user specifies the type of teleservice for his call. The teleservice information is transferred transparently (i.e., without examination) by the network. It is processed by the called user equipment, to select the appropriate TE for the incoming call.

Supplementary Services. These services vary from country to country. In countries that use TUP signaling, the supplementary services supported by TUP (malicious call handling, calling line identification, call forwarding, and closed user group service (9.4) that are available to (analog) subscribers—are also available to ISDN users.

In addition, ISDN can provide *user-to-user signaling*. This allows two ISDN users to send signaling messages to each other during the set-up and clearing phases of a call. The network transfers this information transparently.

Closed user group service and user-to-user signaling are not offered in the U.S. On the other hand, the ISDN for the U.S. includes a number of services for multi-location businesses.

10.1.6 Introduction to DSS1

DSS1 is a message-oriented signaling system [1–5]. Literature on DSS1 is usually in terms of signaling between a *user* and the *network*. Actually, the signaling takes place between a TE of an ISDN user and the local exchange to which the user's DSL is attached. The DSS1 signaling messages are carried in the D-channel of the DSL, which is the common signaling channel for the TEs on a DSL.

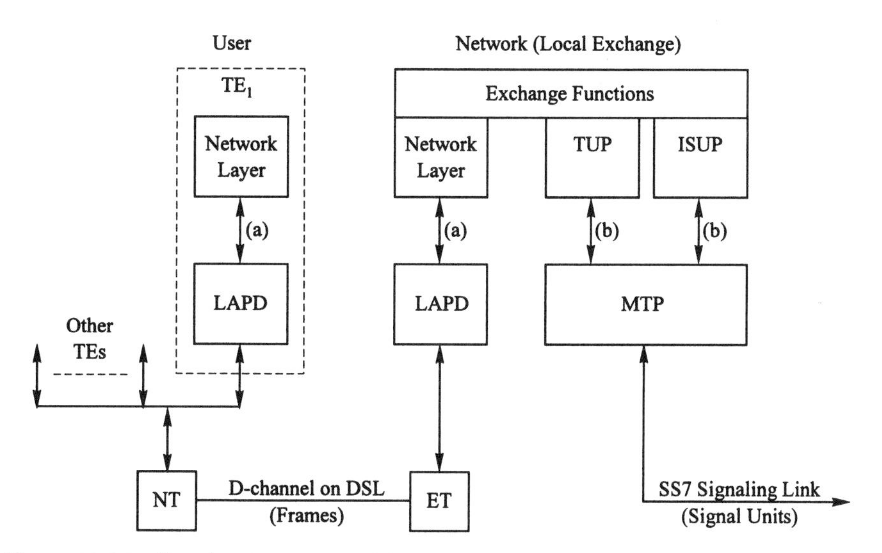

Figure 10.1-3 Functional entities in DSS1 and ISUP signaling. (a): DSS1 primitives. (b): SS7 primitives.

DSS1 and signaling system No. 7 (SS7) have been specified by different CCITT study groups, and use different "languages" (terms). However, many DSS1 concepts are similar to concepts of SS7.

Figure 10.1-3, shows a local exchange, a SS7 signaling link, and a DSL. The functions of the D-channel are comparable to those of the SS7 signaling link. The information units on the D-channel, which are known as *frames*, are similar to the signal units (SU) of SS7.

SS7 is organized as a hierarchy of protocols. The message transfer part (MTP) serves a number of SS7 User parts, such as the telephone user part (Chapter 9) and ISUP (Chapter 11). In a similar fashion, DSS1 is divided into the *data link layer*, also known as LAPD (*link access protocol for D-channels*), and the *network layer*.

The functions of LAPD are comparable to those of MTP. The network layer includes protocols comparable to those of ISUP. The network-layer protocols are usually referred to as "Q.931 protocols" because they have been specified in ITU-T recommendation Q.931.

As in SS7, LAPD and the network layer communicate by passing "primitives" (Section 7.3).

10.2 DATA LINK LAYER (LAPD)

The primary function of LAPD is the reliable transfer of frames between a TE and the local exchange [1–7]. It includes provisions for error detection and correction.

10.2.1 Data Link Connections

We start by describing the functional entities at the ISDN user and at the exchange in more detail. Figure 10.2-1 shows a user with two terminals at his premises.

Each TE on a DSL is identified by a *terminal end-point identifier* (TEI), which has a value in the range 0–126. The TEs in this example have TEI values 1 and 2.

A terminal has two LAPD functions. One is TE-specific, and identified by the TEI of the terminal. The other one is identified by TEI = 127 at all terminals. Each terminal LAPD has a "peer" LAPD at the exchange.

A TE has several network-layer functions. Each function is identified by a *service access point identifier* (SAPI). This chapter considers the following functions only:

SAPI	Function
0	ISDN call control
63	Management of data-link layer

These functions are present in each TE, and have peers in the exchange.

Frames are always transferred between a terminal LAPD and the peer LAPD

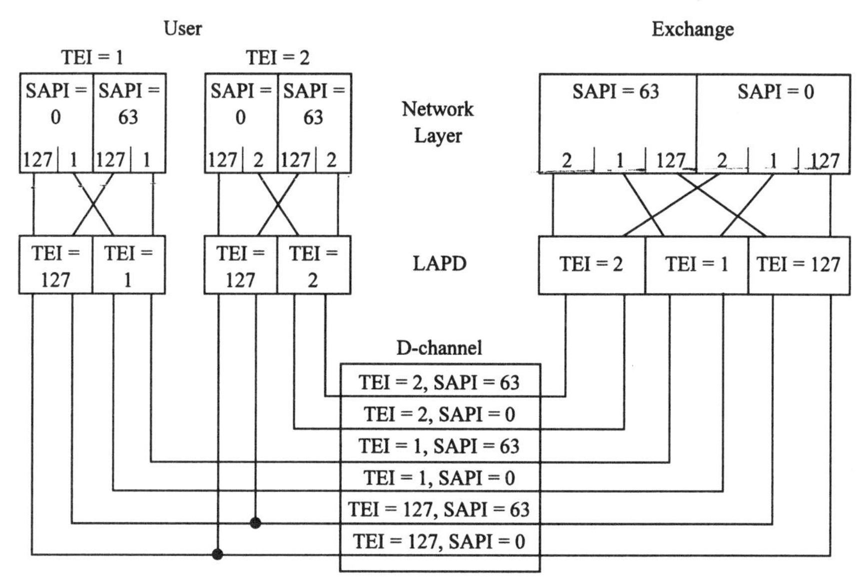

Figure 10.2-1 Data link connections on a D-channel.

at the exchange. Moreover, a frame that originates at a network-layer function at one end of the D-channel is delivered to the peer function at the other end.

There are a number of *data link connections* on a D-channel. Each connection is identified by a combination of TEI and SAPI values. The connections with TEI = 0–126 are bidirectional point-to-point connections, between a TE on a DSL and its "peer" function at the exchange. For example, the connection (TEI = 2, SAPI = 0) carries call-control frames to and from the terminal identified by TEI = 2.

The connections with TEI = 127 are "point-to-multipoint" in the direction from exchange to the user. All TEs on a DSL examine received frames with this TEI value. An exchange can broadcast a message to all TEs on a DSL, by sending a frame with TEI = 127.

10.2.2 Frames

The general format of LAPD frames is shown in Fig. 10.2-2. Frames are separated by one-octet flags. The flag pattern (0111 1110) is the same as in SS7.

The *address field* (octets 2 and 3) of a frame contains SAPI and TEI, and is used to route the frame to its destination. The *control field* starts in octet 4, and consists of one or two octets. The *information field* is present in some frame types only.

Octets $(n - 1)$ and n contain a 16-bit *frame checking sequence* field (FCS). It has the same function as the CB (check-bit) field in SS7 signal units (Section 8.3.2), and enables a LAPD to detect errors in a received frame.

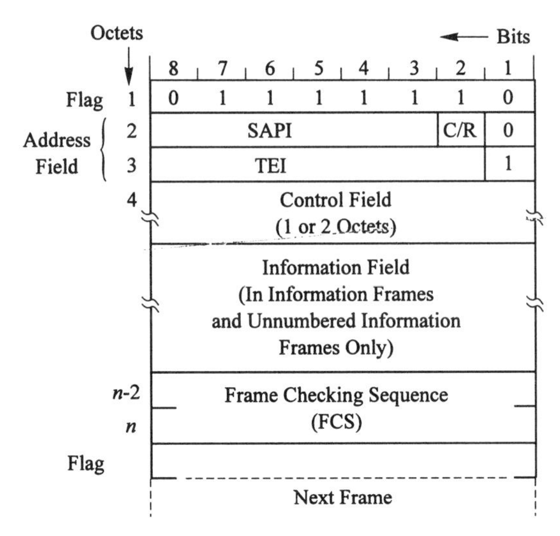

Figure 10.2-2 General LAPD frame format. (From Rec. Q.921. Courtesy of ITU-T.)

10.2.3 Frame Types

We distinguish the following frame types—see Table 10.2-1.

Information Frame (I). This frame is comparable with the message signal unit of SS7. It has an information field that carries a network-layer message, to/from a particular TE on a DSL. The TE is identified by the value of TEI (0 through 126).

Supervision Frames (Group S). These frames originate at the LAPD at one end of a data-link connection, are processed by the LAPD at the other end, and contain information on the status of the connection. These frames do not have an information field, and are comparable with the link status signal units of SS7 (7.3.1).

Unnumbered Frames (Group U) have no counterparts in SS7. The *unnumbered information* (UI) frame is the only frame in this group that has an information field, and carries a network-layer message. To broadcast a message to all TEs on a DSL, the exchange sends an UI frame with TEI = 127.

10.2.4 Control Fields, C/R, P, and F Bits

The control fields of the various frames consist of one or two octets (Table 10.2-1). Bits 1 and 2 of octet 4 identify a group of frames:

Table 10.2-1 Control fields.

Group	Frame type	C	R	Control Field (bits)								Octet
				8	7	6	5	4	3	2	1	
	I Information	X		N(S)							0	4
				N(R)							P	5
S	RR Receive ready	X	X	0	0	0	0	0	0	0	1	4
				N(R)							P/F	5
S	RNR Receive not ready	X	X	0	0	0	0	0	1	0	1	4
				N(R)							P/F	5
S	REJ Reject	X	X	0	0	0	0	1	0	0	1	4
				N(R)							P/F	5
U	UI Unnumbered information	X		0	0	0	P	0	0	1	1	4
U	SABME Set automatic balanced mode extended	X		0	1	1	P	1	1	1	1	4
U	DM Disconnected mode		X	0	0	0	F	1	1	1	1	4
U	DISC Disconnect	X		0	1	0	P	0	0	1	1	4
U	UA Unnumbered acknowledgment		X	0	1	1	F	0	0	1	1	4

S: supervision frames. U: unnumbered frames.
Source: Rec. Q.921. Courtesy of ITU-T.

	Octet 4, Bit 2	Octet 4, Bit 1
Information frame	0 or 1	0
Group S (supervision frames)	0	1
Group U (unnumbered frames)	1	1

In supervision and unnumbered frames, other bits in octet 4 identify a particular frame type within the group.

The C/R bit in the address field, and the P or F bits in the control field, have been carried over from the X.25 protocol, which has been specified by CCITT for data communication networks [5]. These bits are set by the LAPD at one end of a connection, and processed by its peer.

The value of C/R (Fig. 10.2-2) classifies each frame as a *command* or a *response* frame:

	Frames Sent by Exchange	Frames Sent by Terminal
Command frame	$C/R = 1$	$C/R = 0$
Response frame	$C/R = 0$	$C/R = 1$

The I, UI, SABME, and DISC frames are command frames, and the DM and UA frames are reponse frames. The supervision frames can be sent as command or response frames.

By setting the P bit in a command frame to $P = 1$, a LAPD orders its peer to respond with a supervision or unnumbered frame. A response frame with $F = 1$ indicates that it is sent in response to a received command frame in which $P = 1$.

10.2.5 Transfer Modes

LAPD transfers frames in one of the following two modes.

Multiframe Acknowledged Message Transfer. This mode is used only on point-to-point data-link connections, for the transfer of I frames. It includes error correction by retransmission, and in-sequence delivery of error-free messages. This mode is similar to the basic error correction of *message signal units* (MSU) in SS7 (Section 8.4).

The control field of an I frame has a *send-number* [N(S)] field and a *receive-number* [N(R)] field—see Table 10.2-1. These fields are comparable to the forward and backward sequence number fields (FSN, BSN) in MSUs. A LAPD assigns increasing "send" sequence numbers N(S) to consecutive transmitted I frames: N(S) = 0, 1, 2,...127, 0, 1,... etc.). It also stores the transmitted frames in a retransmission buffer where they are kept, and are available for retransmission, until positively acknowledged by the distant LAPD.

We examine the transfer of I frames on a data-link connection, sent from terminal to exchange—see Fig. 10.2-3. The procedure in the other direction is the same. LAPD-E and LAPD-T denote the data-link functions at the exchange and terminal respectively.

LAPD-E checks all received frames for errors. In addition, error-free I frames are "sequence checked." If the value of the N(S) is one higher (modulo 128) than the N(S) of the most recently accepted I frame, the new frame is "in-sequence" and therefore accepted, and its information field is passed to the specified network-layer function.

LAPD-E acknowledges accepted I frames with the receive number [N(R)] in its outgoing I and supervision frames. The value of N(R) is one higher (modulo 128) than the value of N(S) in the latest accepted I frame.

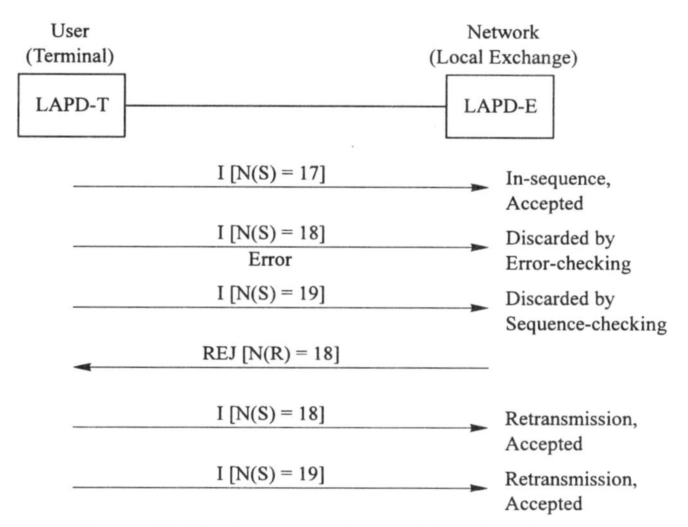

Figure 10.2-3 Error correction on information frame.

Suppose that the most recently accepted I frame had $N(S) = 17$, and that the I frame with $N(S) = 18$ incurred a transmission error, and was discarded by LAPD-E. The next I frame [with $N(S) = 19$] passes LAPD-E error-checking, but arrives out-of-sequence and is discarded by LAPD-E sequence-checking. LAPD-E now sends a reject (REJ) frame with $N(R) = 18$. This requests LAPD-T to retransmit the I frames in its retransmission buffer, starting with the frame with $N(S) = 18$. At LAPD-E, the sequence-checking continues to discard I frames until it receives the (retransmitted) frame with $N(S) = 18$.

The two message streams (terminal to network, and vice versa) in a point-to-point data-link connection are independent of each other, and independent of the message streams in the other point-to-point connections in the D-channel. On a D-channel with n point-to-point connections, there are $2n$ independent N(S)/N(R) sequences.

Unacknowledged Message Transfer. Supervision (S) and unnumbered (U) frames do not include a N(S) field. They are accepted when received without errors, and are not acknowledged. Supervision frames include a N(R) field to acknowledge received information frames.

Unnumbered information (UI) frames do not include N(S) and N(R) fields, because they are always sent with the "group" TEI (127), and it is not possible to coordinate send and receive sequence numbers for the "group" functions in the terminals on a DSL.

Primitives. A network-layer function requests the acknowledged mode for an outgoing message by passing it in a *DL-data* request primitive to its LAPD, which then transfers the message in an I frame. Unacknowledged transfer is requested by passing the message in a *DL-unitdata* request, and the message is then carried in a UI frame.

10.2.6 Supervision of Data Link Connections

The LAPDs at the ends of a point-to-point data link connection can send and receive the following supervision frames:

RR (receive ready) is sent by a LAPD to indicate that it is ready to receive I frames.

RNR (receive not ready) is sent by LAPD to indicate that it is not able to receive I frames, but will process received supervision frames.

REJ (reject) indicates that the sending LAPD has rejected a received I frame.

A supervision action can originate at either end of the connection. Figure 10.2-4 shows a few examples, and illustrates the use of the C/R, P, and F bits. LAPD-T and LAPD-E denote the data-link layer functions at the terminal and the network side of the connection.

In example (a), LAPD-E has received an out-of-sequence I frame, and rejects it with a REJ command frame in which P is set to 0 (no acknowledgment required). N(R) $= x$ indicates that the last accepted I frame had N(S) $= x - 1$. LAPD-T then retransmits the I frames in its retransmission buffer, starting with the frame whose N(S) equals x.

Example (b) deals with the same situation, except that LAPD-E has set P to 1 in its REJ command frame. This orders LAPD-T to acknowledge the frame.

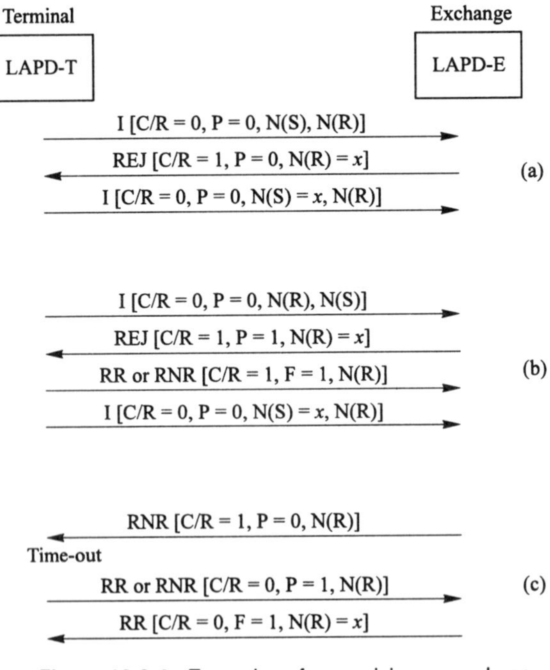

Figure 10.2-4 Examples of supervision procedures.

LAPD-T therefore first sends a RR or RNR response frame (C/R = 1, F = 1), and then starts the retransmission of I frames.

In example (c), LAPD-E indicates that it is unable to receive I frames, with a RNR command frame. LAPD-T suspends sending I frames and starts a timer. If it receives a RR frame before the timer expires, it resumes transmitting or retransmitting I frames.

If the timer expires and no RR frame has been received, LAPD-T sends a "command" supervision frame (C/R = 1), with P set to 1. This orders LAPD-E to send a supervision "command" frame. In the example, LAPD-E responds with a RR frame, indicating it is ready again to accept I frames, and that the last accepted I frame had N(S) = $x - 1$. LAPD-T then resumes transmitting or retransmitting I frames, starting with the frame with N(S) = x.

If the response of LAPD-E had been a RNR frame, LAPD-T would have restarted its timer, and waited again for a RR frame. If LAPD-E then remains receive-not-ready after several timeouts, LAPD-T informs its network-layer function.

10.2.7 SABME, DISC, DM, and UA Frames

These unnumbered frames (Table 10.2-1) are used to start and end multiframe acknowledged operation on a point-to-point data link connection.

For example, when the network-layer function identified by SAPI = 0 at an exchange needs to initiate multiframe acknowledged operation on a DSL connection identified by say, TEI = 5, SAPI = 0, it passes a *set automatic balanced mode extended* (SABME) frame to the LAPD-E for the connection, which then initializes its *N*(S) and *N*(R) counters to 0, and transfers the frame to the peer LAPD-T. That LAPD then initializes its counters, informs the SAPI = 0 network-layer function, and returns a unnumbered acknowledgment frame (UA) frame to the exchange. The LAPD-E then informs the requesting network-layer function that it can start sending I frames.

An existing multiframe acknowledgment operation on a connection is terminated the sequence of a disconnect (DISC) command, followed by a disconnect mode (DM) response.

10.2.8 TEI Management

The LAPD for point-to-point connections in a terminal (Fig. 10.2-1) stores a TEI, and checks the TEI in the address field of received frames to determine whether the frame is intended for it. It also places its TEI in the address fields of its outgoing frames. A terminal is "fixed", or "portable," depending on the manner in which its TEI value is entered.

A *fixed TE* is intended to be associated with a DSL on a long-term basis, say several years. These terminals have a number of switches whose positions determine the TEI value. The switches are set by the person who installs the TE, and the settings remain unchanged as long as the TE remains on the DSL. Fixed TEs can have TEI values in the range 0–63.

A *portable TE* is intended to be moved from DSL to DSL. For example, a user may have a DSL at home and another DSL in his office, and can take the TE along when he moves between these locations. It is inconvenient to change the TEI value manually on each move, and portable TEs are therefore arranged to receive a TEI value (in the range 64–126) that is assigned by the exchange. The assignment of a TEI value involves "TEI management" messages between the portable TE and the exchange.

TEI management messages are carried in UI frames with TEI = 127 (broadcast TEI), and SAPI = 63 (network-layer management entity). The messages are:

ID-request. This is sent by a portable TE, to request the assignment of a TEI value.

ID-assigned. The response by the network to an ID-request. It includes the assigned TEI value.

ID-denied. This is the response by the network, denying an ID-request.

ID-check-request. This is a command from the network to check out an assigned TEI value.

ID-check-response. This is the response by a portable TE to an ID-check-request.

ID-remove. This is a command sent from the network, to remove a portable TE with a specified TEI value from the DSL.

The information field of the UI frames is shown in Fig. 10.2-5. The code in octet 1 indicates a TEI-management message. The message type code is in octet 4 (see Table 10.2-2). The message includes the parameters *Ri* (reference number) and *Ai* (action indicator).

Table 10.2-2 Message type codes in TEI management messages.

Message Name	Message Type Code
ID-request	0000 0001
ID-assigned	0000 0010
ID-denied	0000 0011
ID-check-request	0000 0100
ID-check-response	0000 0101
ID-remove	0000 1010

Source: Rec. Q.921. Courtesy of ITU-T.

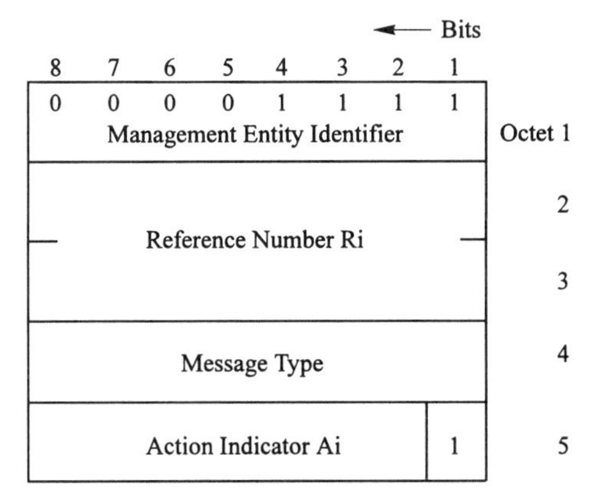

Figure 10.2-5 Information field of unnumbered information frame with a TEI management message. (From Rec. Q.921. Courtesy of ITU-T.)

We now describe the procedures for TEI assignment and TEI checking.

The *TEI assignment* procedure is shown in Fig. 10.2-6(a). When a portable TE is plugged into a DSL, it automatically sends an ID-request. Since the TE does not have a TEI, it generates a random reference number (Ri) to identify itself. A TE can request the assignment of a paricular TEI value by specifying it in the *Ai* field, or can leave the choice to the network by setting $Ai = 127$.

For each attached DSL, the network (local exchange) maintains a list of portable TEI values (range 64–126) that are currently assigned. When receiving an ID-request on a DSL, the exchange consults the list. When it can assign a TEI, it broadcasts an ID-assigned message in which the *Ri* value is copied from the ID-request, and the assigned TEI value is in *Ai*.

All TEs on the DSL examine the message, but only the TE that has sent the request, and recognizes its *Ri* value, accepts the assigned TEI. Therefore, two or more TEs on a DSL can make simultaneous ID-requests without causing problems.

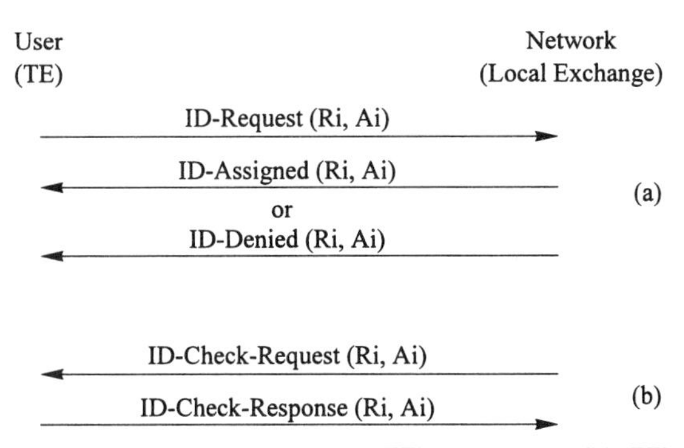

Figure 10.2-6 TEI management procedures. (a): TEI assignment. (b): TEI checking. All TEI management messages are sent in UI frames with TEI = 127 and SAPI = 63. (From Rec. Q.921. Courtesy of ITU-T.)

If the exchange cannot satisfy the ID-request because the requested TEI is already on the list of assigned TEIs for the DSL, or because all TEIs in the range 64–126 have already been assigned, it broadcasts an ID-denied message, again copying the *Ri* value from the received request. The TE then alerts its user that its request for a TEI assignment has been denied.

The *TEI checking* procedure allows the exchange to audit its list of assigned portable TEIs on a DSL—see Fig. 10.2-6(b). The exchange broadcasts an ID-check-request to the TEs on the DSL, in which *Ai* indicates the TEI value being checked, and the *Ri* field is set to 0. The network also starts a 200 ms timer. When a TE on the DSL has a TEI value that matches *Ai*, it responds with an ID-check-response that includes a randomly chosen number *Ri*, and the received *Ai* value.

Under normal conditions, the network receives one ID-check-response before the timer expires. This indicates that there is one TE with the particular TEI value. If the timer expires and no response has been received, the network repeats the ID-check-request and restarts the timer. If the timer expires again before a response has been received, the network assumes that the TEI is no longer in use, deletes it from the list of assigned TEIs for the DSL, and makes a report for the maintenance staff of the exchange.

When the network receives more than one response to an ID-check-request, it knows that the same TEI value has mistakenly been assigned to more than one TE on the DSL. In this case, it broadcasts an ID-remove command, with the TEI value in *Ai*. The TEs whose TEI values match *Ai* then stop sending and accepting frames on the DSL, and alert their user.

10.3 Q.931 CALL-CONTROL MESSAGES

This section describes the network-layer messages and parameters for the control of circuit-mode ISDN calls that have been specified in CCITT Recommendation Q.931 [8], and are generally known as Q.931 messages and parameters.

The CCITT specification is an "umbrella": the messages and parameters used in individual countries are subsets of those defined by CCITT. Some national Q.931 versions also include special parameter values and codes to support country-specific ISDN services and TE equipment characteristics. The U.S. version of Q.931 has been specified by Bellcore [9].

Q.931 messages are located in the information fields of I and UI frames (Fig. 10.2-2).

10.3.1 Message Direction and Scope

We can distinguish Q.931 messages by their direction and scope—see Fig. 10.3-1. *Network-to-user* messages (a,c) are sent from a local exchange to a TE, and *user-to-network* messages (b,d) are sent in the opposite direction. In

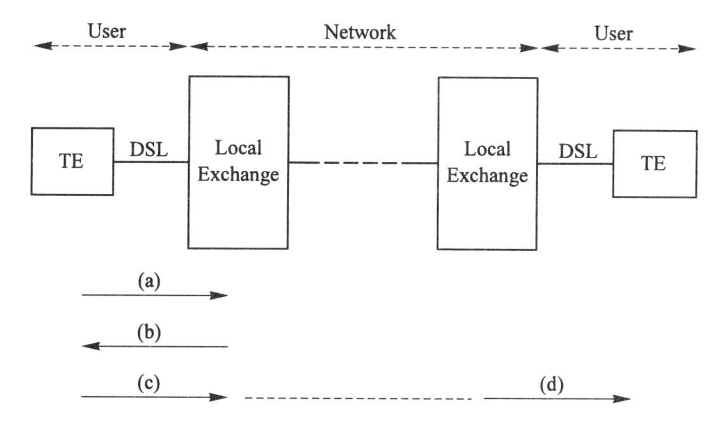

Figure 10.3-1 Classification of Q.931 messages by direction and scope.

addition, we speak of *local* messages and *global* messages. A local message (a,b) is of significance only to the TE that sends or receives the message, and its local exchange. A global message (c) is a message, sent by a TE, that has significance for its local exchange, and for the distant TE. Global messages are transferred across the network, and delivered (d) to the distant TE.

10.3.2 Q.931 Message Types

This section outlines the functions of the most important Q.931 messages, and introduces the message acronyms that will be used throughout this chapter. Most message types can be sent from the user to the network, and vice versa.

Set-up Message (SETUP). This is a global message that initiates a call. It is be sent from the calling user to the network, and from the network to the called user. It contains the called number, and other information for the call set-up.

Set-up Acknowledgment Message (SETACK). This is a local message from the network to the calling user. It indicates that more address information is needed to set up the call.

Call Proceeding Message (CALPRC). This is a local message, sent from the network to the calling user, or from the called user to the network. It confirms the receipt of a SETUP message, and indicates that the complete address information for the call has been received, and that the set-up is proceeding.

Progress Message (PROG). This is a local message, sent from the network to the calling user. It contains information about the progress of the call set-up.

Alerting Message (ALERT). This is a global message, sent from the called user to the network, and from the network to the calling user. It indicates that the called party is being informed (alerted) about the arrival of an incoming call.

Connect Message (CONN). This is a global message, sent from a called user to the network, and from the network to a calling user. It indicates that the called user has answered.

Connect Acknowledgment Message (CONACK). This is a local message, acknowledging the receipt of a connect message.

Disconnect Message (DISC). This is a global message, sent from the user to the network, and from the network to the user. It requests the release the connection.

Release Message (RLSE). This is a local message, acknowledging the receipt of a DISC message, and indicating that the sender has cleared the connection at its end.

Release Complete Message (RLCOM). This is a local message, acknowledging the receipt of a RLSE message, and indicating that the sender has released the connection at its end.

Information Message (INFO). This is a global message. It is sent by a calling user who enters called numbers from a keypad at his terminal, and contains one or more digits of the number. An INFO message can also be sent from the network to a calling user. In this case, it orders the user's TE to generate an audible progress tone (busy tone, etc.).

10.3.3 General Message Format

The general format of Q.931 messages is shown in Fig. 10.3-2. The bits are numbered from right to left, and the first bit sent is bit 1 of octet a. All messages begin with a standard header that consists of the following three parts.

Protocol Discriminator. The messages on the D-channel can belong to protocols other than Q.931 (for example, the X.25 protocol for packet-mode communications). The code shown indicates the Q.931 protocol.

Call Reference Value. The call reference value (CRV) is an integer that identifies the call to which the message relates. Q.931 messages are call-related instead of trunk-related because the Q.931 protocol covers both circuit-mode and packet-mode calls. In a packet-mode call, there are no dedicated circuits (trunks), and the Q.931 messages for these calls are therefore call-related. For uniformity, Q.931 also uses call-related messages for circuit-mode calls.

At a TE, a call is identified uniquely by a CRV. At a local exchange, the call is identified uniquely by the combination of a CRV, TEI, and the identity of the DSL.

Octet b indicates the length (number of octets) of the CRV. On basic rate DSLs, CRV values range from 1 to 127, and the CRV is located in bits 7–1 of

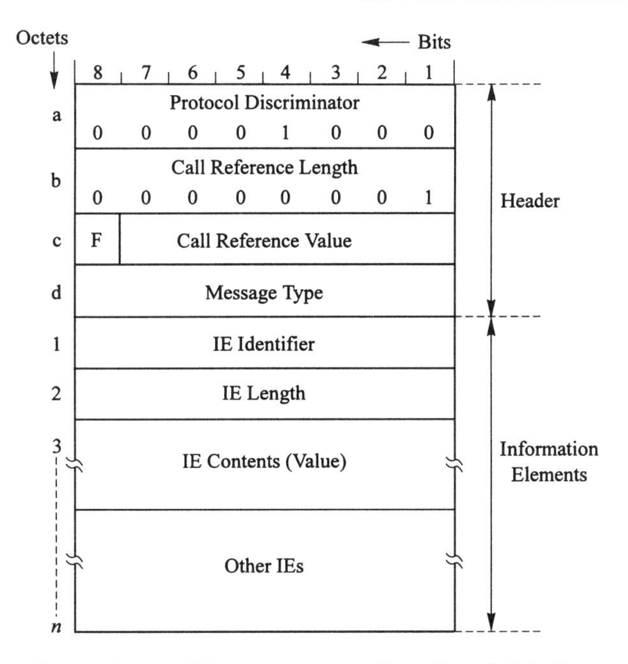

Figure 10.3-2 General format of Q.931 messages. (From Rec. Q.931. Courtesy of ITU-T.)

octet c—see Fig.10.3-2. On primary rate DSLs, CRV values range from 0 through 2^{15}–1, and the CRV occupies two octets.

Flag bit (F) indicates whether the CRV has been assigned by the sender or the recipient of the message.

Message Type. Octet d identifies the message type. The coding is shown in Table 10.3-1.

Table 10.3-1 Q.931 message type codes.

		← Bits							
Message Name	Acronym	8	7	6	5	4	3	2	1
Alerting	ALERT	0	0	0	0	0	0	0	1
Call proceeding	CALPRC	0	0	0	0	0	0	1	0
Connect	CONN	0	0	0	0	0	1	1	1
Connect acknowledgment	CONACK	0	0	0	0	1	1	1	1
Disconnect	DISC	0	1	0	0	0	1	0	1
Information	INFO	0	1	1	1	1	0	1	1
Progress	PROG	0	0	0	0	0	0	1	1
Release	RLSE	0	1	0	0	1	1	0	1
Release complete	RLCOM	0	1	0	1	1	0	1	0
Set-up	SETUP	0	0	0	0	0	1	0	1
Set-up acknowledgment	SETACK	0	0	0	0	1	1	0	1

Source: Rec. Q.931. Courtesy of ITU-T.

Table 10.3-2 Information element identifiers.

Reference Number	Information Element	← Bits 8	7	6	5	4	3	2	1
IE.1	Bearer capability	0	0	0	0	0	1	0	0
IE.2	Called party number	0	1	1	1	0	0	0	0
IE.3	Calling party number	0	1	1	0	1	1	0	0
IE.4	Called party subaddress	0	1	1	1	0	0	0	1
IE.5	Calling party subaddress	0	1	1	0	1	1	0	1
IE.6	Cause	0	0	0	0	1	0	0	0
IE.7	Channel ID	0	0	0	1	1	0	0	0
IE.8	High-layer compatibility	0	1	1	1	1	1	0	1
IE.9	Keypad	0	0	1	0	1	1	0	0
IE.10	Low-layer compatibility	0	1	1	1	1	1	0	0
IE.11	Progress indicator	0	0	0	1	1	1	1	0
IE.12	Signal	0	0	1	1	0	1	0	0
IE.13	Transit network selection	0	1	1	1	1	0	0	0
IE.14	User–user information	0	1	1	1	1	1	1	0

In addition, a message includes a number of parameters, which are known in Q.931 documents as *information elements* (IE).

10.3.4 Information Elements

An information element consists of three fields (Fig. 10.3-2):

IE Identifier: a one-octet field that holds the name of the IE—see Table 10.3-2.

IE Length: a one-octet field that indicates the length (number of octets) of the contents field.

IE Contents (or IE Value): a variable-length field that holds the actual information of the IE.

The identifier and length fields are in octets 1 and 2 of the IE. The contents field starts at octet 3.

In a message of a given type, the included IEs depend on the message direction. Moreover, an IE can be *mandatory* (always included in the message), or *optional* (included only when necessary).

The most important mandatory (M) and optional (O) IEs in user-to-network and network-to-user messages are listed in Tables 10.3-3 and 10.3-4 [8,9]. Each IE in the tables has a reference number (for example, IE.1). We shall use these numbers when referring to IEs in later sections of this chapter.

Table 10.3-3 Information elements in Q.931 call-control messages in the direction user → network.

Reference Number	Information Element	ALERT	CALPRC	CONN	CONACK	DISC	INFO	RLSE	RLCOM	SETUP
		Message Acronym								
IE.1	Bearer capability									M
IE.2	Called party number									M
IE.3	Calling party number									O
IE.4	Called party subaddress									O
IE.5	Calling party subaddress									O
IE.6	Cause					M		O	O	
IE.7	Channel identification	O	M	O						O
IE.8	High-layer compatibility									O
IE.9	Keypad						M			O
IE.10	Low-layer compatibility									O
IE.13	Transit network selection									O
IE.14	User–user information	O		O		O		O	O	O

M: mandatory. O: optional.

Table 10.3-4 Information elements in Q.931 call-control messages in the direction network → user.

Reference Number	Information Element	ALERT	CALPRC	CONN	CONACK	DISC	INFO	PROG	RLSE	RLCOM	SETUP	SETACK
		Message Acronym										
IE.1	Bearer capability										M	
IE.2	Called party number										M	
IE.3	Calling party number										O	
IE.4	Called party subaddress										O	
IE.5	Calling party subaddress										O	
IE.6	Cause					M		O	O	O		
IE.7	Channel identification		O								M	M
IE.8	High-layer compatibility										O	
IE.10	Low-layer compatibility										O	
IE.11	Progress indicator	O		O				M			O	O
IE.12	Signal	M		O	O	O	O	O	O	O	M	O
IE.14	User–user information	O		O		O			O	O	O	

M: mandatory. O: optional.

10.3.5 Information Element Contents

This section describes the contents fields of the IEs listed in Table 10.3-2. At this point, it is sufficient to read quickly through these descriptions, which are primarily intended as reference material for Sections 10.4 and 10.5.

General Remarks. Q.931 coding uses the concept of "extended" octets. In some octets of contents fields, bit 8 is an extension bit (ext). If *ext* = 1, the octet consists of one octet. An "extended" octet is an "octet" that is extended to a next octet. This is indicated by *ext* = 0. For example, when the data item in octet N has a length of up to seven bits, octet N is not extended (*ext* = 1). Next, suppose that the data item in octet N has a length of 18 bits. The initial seven bits are then in octet N, which is marked as extended (*ext* = 0). Bits 8–14 are in the next octet, which is numbered N_a (first extension of octet N) and is also marked as extended. Finally, bits 15–18 are in an octet, which is marked as "not extended" (*ext* = 1).

The parameter *coding standard* appears in the contents field of some IEs (for example, see Fig. 10.3-3). It indicates whether the field is coded according to CCITT standards:

0 0	CCITT standard
1 0	National standard
1 1	Network specific standard

This allows other standards organizations to define country-specific parameter codings. The descriptions that follow assume the CCITT standard.

Some contents fields contain a fairly large number of data items. Only the most essential ones are discussed here (for additional information, see [8] and [9]). In the figures that follow, the blank fields represent data items that are not discussed.

We now examine the IE contents.

IE.1 Bearer Capability (Fig. 10.3-3). This indicates the bearer service requested by the calling user.

Information transfer capability:
- 00000 speech
- 01000 unrestricted digital information
- 01001 restricted digital information
- 10000 3.1 kHz audio

Transfer mode:
- 00 circuit mode
- 10 packet mode

Information transfer rate:
- 00000 packet mode
- 10000 64 kb/s circuit mode

Bits 8	7	6	5	4	3	2	1	Octets
1 Ext	Coding Standard		Information Transfer Capability					3
0/1 Ext	Transfer mode		Information Transfer Rate					4
0/1 Ext			User Information Layer 1 Protocol					5
0/1 Ext			User Rate					5a

Figure 10.3-3 Format of IE.1: bearer capability. (From Rec. Q.931. Courtesy of ITU-T.)

User information layer 1 protocol:
- 00001 CCITT standardized rate adaptation
- 00010 mu-law coded analog signal (1.5.1)
- 00011 A-law coded analog signal

User rate (included only when the user information layer 1 protocol is 00001):
- 01111 rate adaptation between 56 and 64 kb/s

IE.2 Called Party Number (Fig.10.3-4). Octet 3a is not included.

Type of number:
- 001 international number
- 010 national number
- 100 subscriber (directory) number
- 000 special number

Numbering plan identification:
- 0001 ISDN/telephony numbering plan

Number digits: each digit is coded as a seven-bit IA5 (International Alphabet No.5) character, and occupies one octet. Bits 7,6,5 are set to: 011, and bits 4–1 represent the digit values, from 0000 (zero) through 1001 (nine).

Bits 8	7	6	5	4	3	2	1	Octets
0/1 Ext	Type of Number			Numbering Plan Identification				3
1 Ext	Calling Number Presentation			Calling Number Screening				3a
0	Number Digits (IA5 Characters)							4, etc

Figure 10.3-4 Format of: IE.2: called party number, IE.3: calling party number. (From Rec. Q.931. Courtesy of ITU-T.)

IE.3 Calling Party Number (Fig.10.3-4). The coding of this IE is the same as for the called number, except that octet 3a is included.

Calling number presentation:
- 000 presentation allowed (by calling party)
- 010 presentation not allowed

Calling number screening:
- 0000 user-provided number, not screened by network (local exchange)
- 0001 user-provided number, has passed screening
- 0010 user-provided number, failed screening
- 0011 number provided by the network

The screening status is important in calls to TE equipment that are not attended by humans (computers, facsimile machines). These TEs can be set up to accept only calls in which the calling number is as expected, and has passed screening.

IE.4 and IE.5 Called and Calling Party Subaddress (Fig. 10.3-5). The IE holds a subaddress associated with a specific TE on the called or calling user's DSL, and provides information for the selection of a compatible called TE. The IEs are transferred transparently (without inspection by the network) from the calling to the called user.

Type of subaddress:
- 000 coded per CCITT Recommendation X.213
- 010 coding specified by the user

Odd/even indicator. User-specified subaddress coding usually is a string of BCD coded digits (two digits per octet):
- 0 even number of digits
- 1 odd number of digits.

IE.6 Cause (Fig. 10.3-6). Indicates why a set-up has failed, or why a connection has to be released.

Location indicates where the message originated:
- 0000 (0) user
- 0010 (2) public network serving the local user

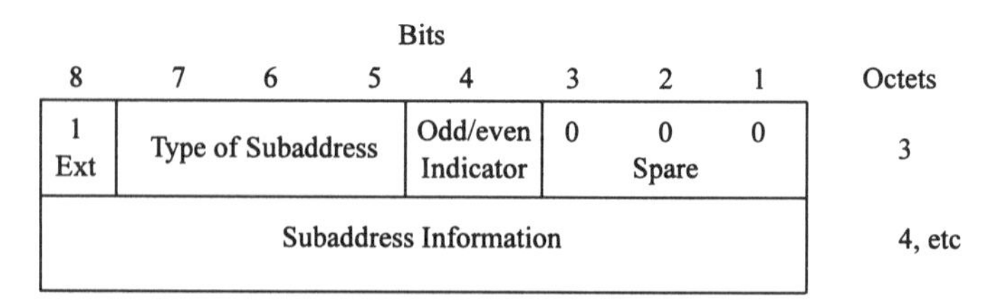

Figure 10.3-5 Format of: IE.4: called party subaddress, IE.5: calling party subaddress. (From Rec. Q.931. Courtesy of ITU-T.)

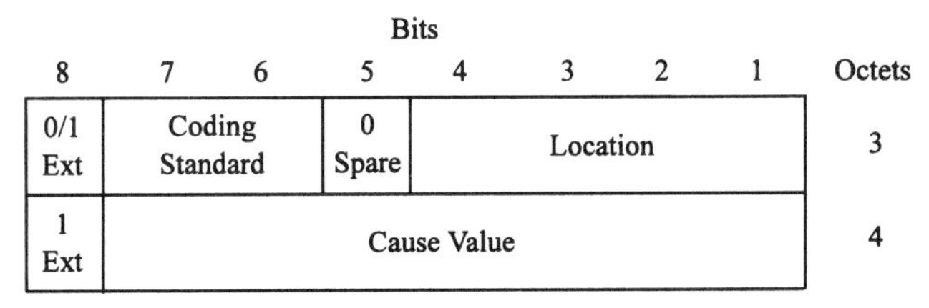

Figure 10.3-6 Format of IE.6: cause. (From Rec. Q.931. Courtesy of ITU-T.)

0011 (3) transit network
0100 (4) public network serving the remote user
0111 (7) international network

The terms *local* and *remote* are as perceived by the recipient of the message. For example, in a message to user U, code (2) indicates that the message was originated in U's local network.

Cause value identifies the reason for the termination of a call, or call set-up. CCITT has specified a large number of cause values [8], for example:

000 0001 (1) unassigned number
000 0010 (2) no route to requested transit network
000 0011 (3) no route to destination
001 0000 (16) normal call clearing
001 0001 (17) user busy
001 0010 (18) no user responding
001 0011 (19) user being alerted, no response
001 0101 (21) call rejected
001 1010 (26) clearing of non-selected user
001 1011 (27) destination out of order
001 1100 (28) invalid or incomplete called number
010 0010 (34) no trunk or B-channel available
010 1010 (42) switching equipment congestion
011 1001 (57) requested bearer capability not authorized
100 0001 (65) bearer capability not implemented
111 0110 (102) recovery action on expiration of a timer

Bellcore has specified a number of additional causes (coded as network specific) [9].

IE.7 Channel Identification (Fig. 10.3-7). This IE identifies the B-channel to be used for the call. Octets 3.1, 3.2, and 3.3 are included for primary rate DSLs only.

Interface type:

0 basic rate DSL
1 primary rate DSL.

Bits 8	7	6	5	4	3	2	1	Octets
	Channel Identification							
0	0	0	1	1	0	0	0	1
	Information Element Identifier							
Length of Channel Identification Contents								2
1 Ext		Int. Type	0 Spare			Information Channel Selection		3
1 Ext	Coding Standard							
Channel Number								

Figure 10.3-7 Format of IE.7: channel identification. (From Rec. Q.931. Courtesy of ITU-T.)

Information channel selection identifies the B-channel on a basic rate DSL:

01 channel B_1

10 channel B_2.

Channel number identifies the B-channel on a primary rate DSL.

IE.8 High-Layer Compatibility (Fig. 10.3-8). This IE is used for the selection of a compatible TE on the called DSL. It is transferred transparently across the network.

High layer characteristics identification

1000 0001 telephony

0000 0100 facsimile group 4

0110 0001 teletex

0110 0101 telex

IE.9 Keypad (Fig. 10.3-9). This IE holds one or more digits that are received from a keypad TE.

Keypad information. A string of digits coded as IA5 characters. Digits 0–9 are coded as in IE.2. Digits * and # are coded as 010 1010 and 010 0011.

Bits 8	7	6	5	4	3	2	1	Octets
1 Ext	Coding Standard		Interpretation			Presentation Method of Protocol Profile		3
0/1 Ext	High Layer Characteristics Identification							4

Figure 10.3-8 Format of IE.8: high-layer compatibility. (From Rec. Q.931. Courtesy of ITU-T.)

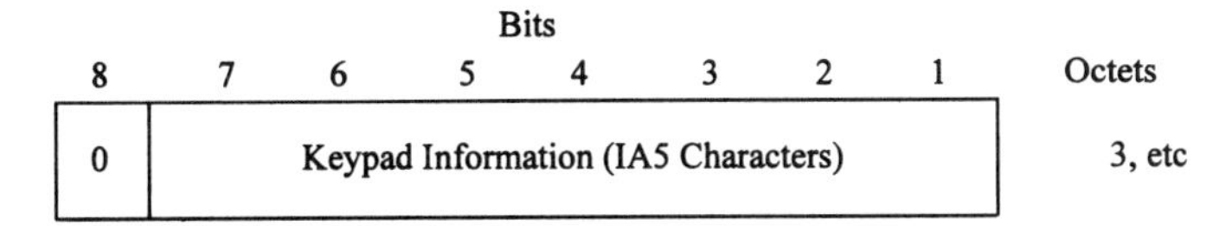

Figure 10.3-9 Format of IE9: keypad facility. (From Rec. Q.931. Courtesy of ITU-T.)

IE.10 Low-layer Compatibility. This IE is used for the selection of a compatible TE on the called DSL. It is transferred transparently across the network. The contents are the same as the contents of IE.1 bearer capability (IE.1).

IE.11 Progress Indicator (Fig. 10.3-10). This IE informs the user about certain characteristics of the connection, or about tones or patterns on the B-channel.

Location holds the location of the IE originator, and is coded as in IE.6 (Cause).

Progress description:

000 0001 (1) call is not end-to-end ISDN
000 0010 (2) called equipment is non-ISDN
000 0011 (3) calling equipment is non-ISDN
000 1000 (8) in-band tone or information now available on B-channel
000 1010 (10) delay in response of called user

IE.12 Signal (Fig. 10.3-11). This IE allows the network to inform the user about in-band tones.

Signal value:

000 0000 (0) dial-tone on
000 0001 (1) ringing-tone on
000 0011 (3) reorder/congestion-tone on
000 0100 (4) busy-tone on
100 1111 (79) tones off

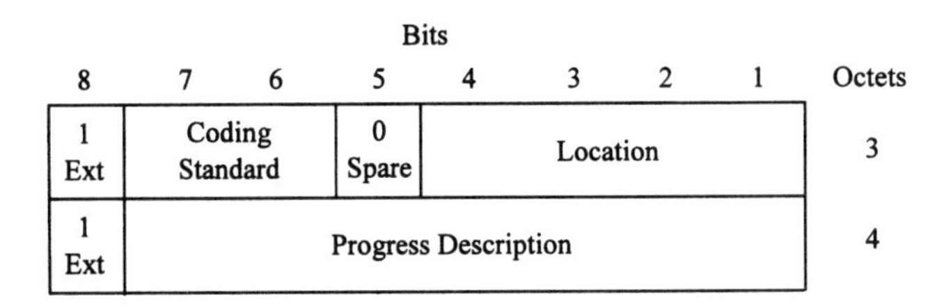

Figure 10.3-10 Format of IE.11: progress indicator. (From Rec. Q.931. Courtesy of ITU-T.)

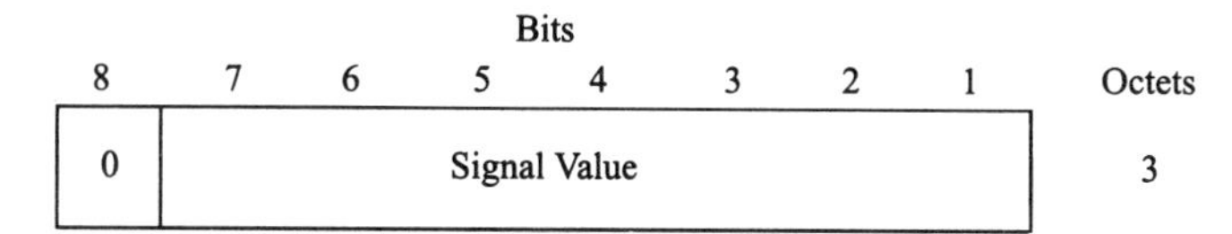

Figure 10.3-11 Format of IE.12: signal. (From Rec. Q.931. Courtesy of ITU-T.)

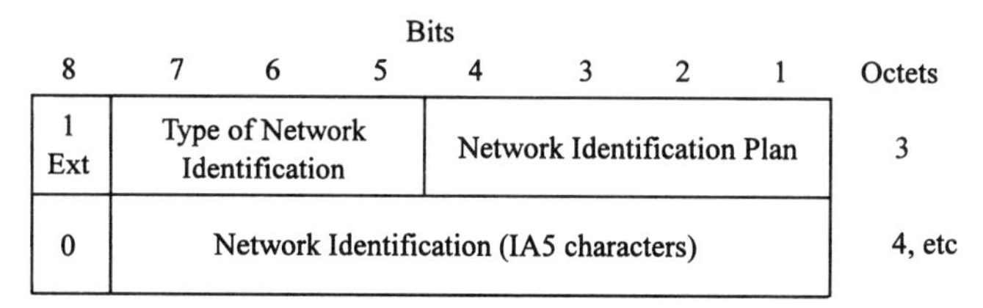

Figure 10.3-12 Format of IE.13: transit network selection. (From Rec. Q.931. Courtesy of ITU-T.)

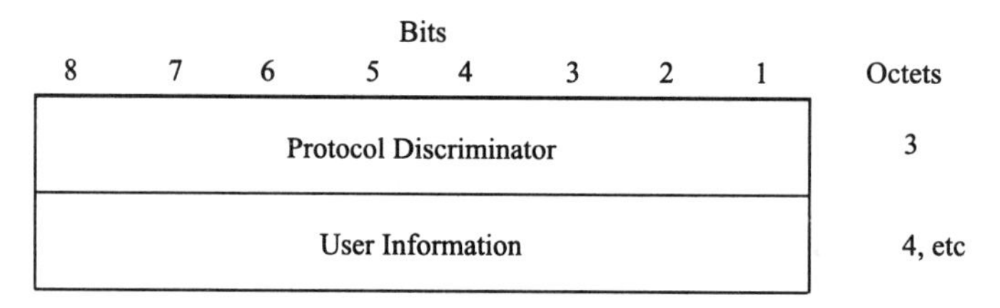

Figure 10.3-13 Format of IE.14: user–user information. (From Rec. Q.931. Courtesy of ITU-T.)

IE.13 Transit Network Selection (Fig. 10.3-12). Specifies the transit network selected for the call. If an ISDN user does not include this IE in the SETUP message for a long-distance or international call, the local exchange routes the call to the "normal" (default) interexchange carrier (IXC) of the user, which is stored at the exchange. A user can specify a different IXC for a call by including an IE.13 in the SETUP message.

Type of network identification:
- 010 national network identification
- 011 international network

Network identification plan:
- 0001 carrier identification code
- 0011 data network identification code

Network identification. A string of IA5 characters.

IE.14 User–user (Fig. 10.3-13). User-to-user signaling is a service offered in several countries, but not in the U.S. The IE, which can be included in SETUP, ALERT, CONN, DISC, RLSE, and/or RLCOM messages, allows the calling and called users to send up to 128 octets of information during the set-up and release of a connection. The IE is transferred transparently across the network.

Protocol discriminator:
  0000 0000 user-specific coding
  0000 0010 information in the form of IA5 characters

User information. This information is not standardized, and has to be prearranged by individual pairs of ISDN users.

10.4 INTRODUCTION TO CALL-CONTROL SIGNALING

This section begins by outlining a typical sequence of Q.931 messages for a call between two users who are served by the same local exchange. It then introduces some ISDN call-control concepts, and sets the stage for the additional examples of Section 10.5.

We shall use the message acronyms of Section 10.2.3, and refer to information elements by their reference numbers. When reading this section, it is helpful to look up the IE descriptions of Section 10.3.5.

10.4.1 Typical Signaling Sequence

We first examine the Q.931 messages in a typical signaling sequence for an intraexchange call from user P to user Q (see Fig. 10.4-1). In this section, the focus is primarily on the messages, and the IEs are not discussed in detail. The message sequence is indicated by letters in parentheses, (a), (b), etc.

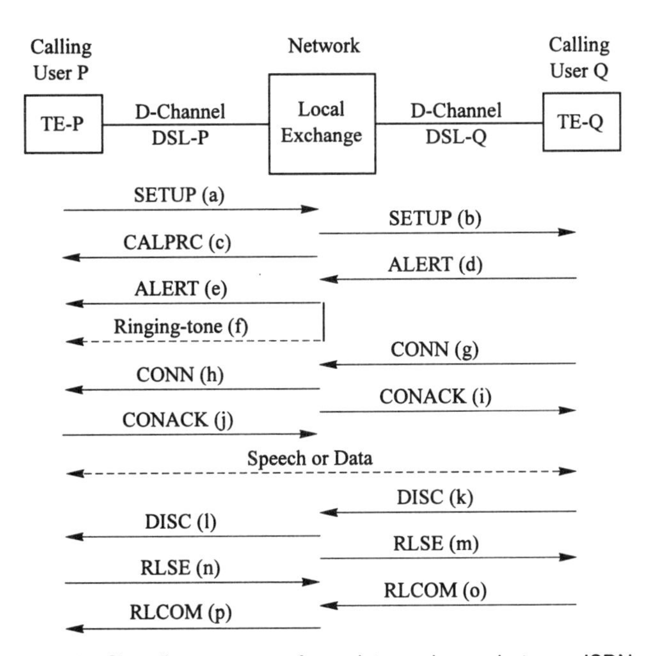

Figure 10.4-1 Signaling sequence for an intraexchange between ISDN users.

Call Set-up. User P starts the call by sending a SETUP message (a) from his terminal TE-P to the exchange. We assume that the message includes the complete called number (*en-bloc* signaling). The exchange then sends a SETUP message (b) on DSL-Q. It also assigns B-channels on DSL-P and DSL-Q, and informs TE-P that the call set-up has started, with a CALPRC message (c).

We also assume that TE-Q accepts the incoming call. It alerts the called user with a visual or audible signal, and returns an ALERT message (d) to the exchange. The exchange then sends an ALERT message (e) to TE-P. If the bearer service for the call is "speech" or "3.1 kHz audio," the exchange also connects a ringing-tone source to the B-channel of the calling user (f).

When user Q answers, TE-Q sends a CONN message (g) to the exchange, which then connects the B-channels, and sends a CONN message (h) to TE-P, indicating that the call has been answered. The exchange always acknowledges a received CONN message with a CONACK message (i). The figure also shows that TE-P acknowledges its CONN message with a CONACK message (j). This is not always the case: some TEs do not provide this message, and exchanges therefore do not require a CONACK response from a calling TE.

At this point, the users can begin to talk, or to send data.

Call Clearing. In the signaling systems discussed so far, calls are cleared by the calling party. Q.931 follows the procedure of data communication networks, in which calls can be cleared by either party. In Fig. 10.4-1, called user Q initiates the clearing, by ordering TE-Q to send a DISC message (k). The exchange then sends a DISC message (l) to TE-P, clears its end of the B-channel to TE-Q, and sends a RLSE message (m) to that TE. Terminal TE-Q then releases the B-channel at its end, and returns a RLCOM message (o). The B-channel is now available for a new call. The B-channel between TE-P and the exchange is released in a similar RLSE-RLCOM sequence.

10.4.2 Assignment and Release of Call Reference Values and B-Channels

We now explore how the exchange and the TEs inform each other about the assignment of *call reference values* (CRV) and B-channels at the start of a call, and the release of these items when the call ends. We use the signaling sequence of Fig. 10.4-1.

The exchange, and each TE, maintains a pool of available CRVs, and can assign a CRV to a call. The exchange knows the idle/busy status of the B-channels on its DSLs, and assigns the channels to the call.

Figure 10.4-2 shows only those messages in the sequence of Fig. 10.4-1 that play a role in informing the exchange and the TEs about assignments and releases of B-channels and CRVs.

Assignment of CRVs. The calling terminal TE-P assigns CRV_1 for its messages to and from the exchange, and includes it in the SETUP message sent to the exchange. The exchange copies CRV_1, and includes it in all messages to TE-P.

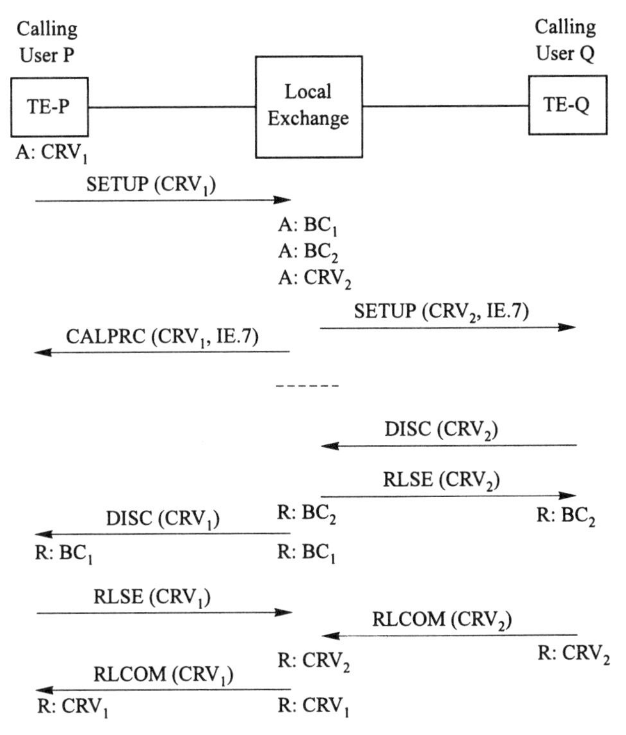

Figure 10.4-2 Assignment and release of call reference values and B-channels. IE.7: channel identification (IE). CRV: call reference value. BC: b-channel. A: assigned. R: released.

The exchange selects CRV_2 for messages to and from the called terminal, and includes it in its SETUP message to the called DSL-Q. If TE-Q accepts the call, it copies CRV_2, and uses it in its messages to the exchange.

Assignment of B-channels. B-Channels BC_1 and BC_2, on the calling and called DSLs, are assigned by the exchange when it receives the SETUP message from the calling TE. To identify the assigned channels, the exchange includes an IE.7 (channel identification) in its CALPRC message to TE-P, and in its SETUP message on the called DSL-Q.

Release of B-channels and CRVs. At the end of the call, CRVs and B-channels have to be released at both ends. The release BC_2 and CRV_2 is described first, because it occurs first in the signaling sequence of our example. When the exchange receives the DISC message from TE-Q, it releases BC_2 at its end, and sends a RLSE message to TE-Q. In response, TE-Q releases BC_2 at its end, sends a RLCOM message to TE-P, and then discards CRV_2. When the exchange receives the RLCOM, it returns CRV_2 to its pool of CRVs. BC_2 and CRV_2 are now available for new calls.

The release of BC_1 and CRV_1 is done in a similar manner. When TE-P receives the DISC message, it releases BC_1 at its end, and sends a RLSE message to the exchange. The exchange then releases the channel at its end, sends a RLCOM message to TE-P, and then discards CRV_1. On receipt of the RLCOM, TE-P returns CRV_1 to its pool of CRVs. BC_1 and CRV_1 are now available for new calls.

10.4.3 Called Terminal Selection

An ISDN user generally has TEs of several types on his DSL. When making a call, the calling user has to include one or more "TE-selection" IEs in his SETUP message, to indicate the type of TE to which he wants to be connected. The selection of a called TE is described below [3,10].

Each TE on a DSL needs to examine all received SETUP messages, to determine whether its type is appropriate for the incoming call. The local exchange therefore sends SETUP messages in a frame with $TEI = 127$, which is the "broadcast" address to all TEs on a DSL.

The SETUP message sent includes all TE-selection IEs received from the calling user. Each TE stores one or more of these IEs, which are entered when the TE is installed. A TE compares the values of a received IE.n with the value of the corresponding stored IE.n. If there is a mismatch in any of the received and stored IE values, the TE disqualifies itself, and does not respond to the SETUP message. The TE selection is thus an exclusion process, and becomes sharper as the number of compared IEs increases.

The SETUP message to a called DSL always includes the requested bearer capability (IE.1), which is a TE-selection IE. CCITT has specified three other selection criteria and IEs, which can be used singly, or in combination. These are listed below.

Multiple Subscriber Numbers. At the request of a user, the telecom assigns several subscriber numbers to a DSL, and each TE stores one or more of these numbers (IE.2). In a call to a DSL with multiple numbers, the local exchange includes IE.2 in it broadcast SETUP message, and a terminal does not respond if the received IE.2 does not match any of its stored IE.2 values.

Subaddresses. In this case, the TEs on a DSL store a subaddress (IE.4), consisting of say some 2–4 digits. The calling party includes an IE.4 in his SETUP message. Only the TE whose stored IE.4 matches the value of the received IE.4 responds to the received SETUP message.

When a user has a group of TEs of the same type, multiple subscriber numbers or subaddresses have to be used to request a connection to a particular TE in that group.

Compatibility Checking. The TEs on a DSL can store high-level and/or low-level compatibility data. The calling user can include an IE.8 (high-level compatibility) and/or IE.10 (low-level compatibility) in his SETUP message.

Again, only the TE or TEs whose stored data match the value(s) of the received IE(s) respond.

The subaddress, high-level compatibility and low- level compatibility IEs are optional parameters that are passed transparently (without examination by the exchange) from the calling to the called user.

10.4.4 Subscription Parameters

The local exchange stores a number of semi-permanent parameter groups. These parameters define the services to which the user has subscribed, and are used by the exchange when processing originating or terminating calls. This section examines the most important parameters that have been specified by Bellcore for local area (LATA) networks in the U.S. [9]. We can discern the following types of subscription-parameter groups.

DSL Subscription-parameter Groups. These parameter groups are associated with the individual DSLs that are attached to the exchange, and include the following parameters:

Type of DSL. This indicates the DSL type: basic access or primary rate access.

Bearer Capabilities. This is a list of bearer capabilities allowed on the DSL.

Directory Numbers (DN). This is a list of directory (subscriber) numbers assigned to the DSL.

Calling Party Number Provision. This indicates whether the local exchange should accept a SETUP message from the DSL that does not include IE.3 (calling party number).

Calling Party Number Screening. This indicates whether the local exchange should screen a user-provided calling party number (IE.3) against the list of DNs assigned to the calling DSL.

Default Directory Number. The directory number to be sent to the called party when the calling party does not include IE.3 in the SETUP message.

DN Subscription-parameter Groups. These groups are associated with the directory numbers (including the default DN) of the ISDN users served by the exchange. Examples of the parameters in these groups are:

Bearer Capabilities. This lists the bearer capabilities allowed for calls to and from this DN.

Interexchange Carrier Presubscription. This identifies the interexchange carrier, which should normally handle the user's long-distance calls. A calling user also can specify another carrier for a call, by including IE.13 (transit network selection) in his SETUP message.

Contention on Incoming Calls. This parameter has the values "allowed" and "not allowed," and is used by the terminating local exchange when several TEs on a DSL respond to a SETUP message—see Section 10.5.4

Calling Number Delivery. This indicates whether the DN of the called user has calling line identification (or caller ID) service. If yes, the terminating local exchange includes IE.3 (calling party number) and IE.5 (calling party subaddress) in its SETUP broadcast message to the DSL.

Network-provided In-band Tone/Announcement Service (NPIBS). On speech and 3.1 kHz audio calls, the calling users receive audible call-progress information (tones or announcements) when certain events occur during the set-up of a call. NPIBS indicates whether this information is to be provided by the network (exchange), or by the TE.

Some TEs used in the U.S. can generate dial-, busy-, and reorder tone, on command from the exchange. The directory numbers associated with these TEs are marked "not requiring NPIBS." Other TEs cannot generate these tones, and their directory numbers are marked as "requiring NPIBS."

Ringing-tone and recorded announcements are always provided by the network.

10.5 CALL-CONTROL EXAMPLES

We now examine a number of signaling sequences in more detail. When reading this section, it is helpful to look up the IE descriptions of Section 10.3.5.

10.5.1 Signaling Example

We start by revisiting the example of Fig. 10.4-1, this time including the most important information elements in the various messages (the IEs in Q.931 messages are listed in Tables 10.3-3 and 10.3-4). We also consider the influence of the subscription parameters (10.4.4) on the call-control actions.

(a) SETUP Message from the Calling User. IE.1 (bearer capability) is mandatory. Since *en-bloc* address signaling has been assumed, IE.2 is included, and contains the complete called number. IE.13 (transit network selection) is optional, and included only on long-distance calls if the calling user requests an interexchange carrier that is not his presubscribed carrier. IE.4, IE.5 (subaddresses), IE.8, and IE.10 (compatibility information), which play a role in the selection of the called terminal, are optional. IE.3 (calling party number) is mandatory or optional, depending on the contents of the "calling party number provision" subscription parameter of the calling DSL.

(b) SETUP Message to Called User. Except for the calling and called numbers, the IEs present in the caller's SETUP message (a) are always included. IE.7 (identification of the assigned B-channel on the called DSL) is mandatory. If the subscription information of the called DSL lists more than one directory number, the exchange also includes IE.2 (called number). This selects a particular called TE on the DSL. IE.3 (calling number) is included if the "calling number delivery" subscription parameter of the called DN indicates that the user has this service, *and* IE.3 indicates that the caller allows his number to be presented. The calling number is either the number provided by the caller, and included in his SETUP message (a), or provided by the local exchange, which uses the default DN associated with caller's DSL.

(c) CALPRC Message to Calling User. IE.7 (identification of assigned-B channel on calling DSL) is mandatory.

(d) ALERT Message from Called User. No optional IEs are necessary in this example.

(e) ALERT Message to Calling User. IE.12 (signal), with value "ringing-tone on," is mandatory. On speech and 3.1 kHz audio calls, the exchange also sends ringing-tone on the B-channel, and includes an IE.11 (progress indicator) in the message, to indicate that "in-band information is now available."

(g),(h) CONN Messages. Optional IE.12 (signal), indicating "tones off," is included in (h), because ALERT message (e) included an IE.12 that indicated "ringing-tone on."

(i),(j) CONACK Messages. Message (j) is sent only if the calling TE is able to provide it.

(k),(l) DISC Messages. IE.6 (cause) is mandatory. The general location is set to "user," and the cause value indicates "normal call clearing."

(m),(n) RLSE Messages. No optional IEs needed in this example.

(o),(p) RLCOM Messages. No optional IEs needed in this example.

10.5.2 Automatic Answering TEs

In the example of Fig. 10.4-1, the TE on the called DSL responds to a SETUP message with an ALERT message, and follows up with a CONN message when the TE user answers. This is not always the case. The ALERT message is sent by a TE (for example, a telephone) where a person is involved at the called end of the connection. *Automatic answering* TEs, such as facsimile machines, can receive calls without human intervention. When an automatic answering TE can accept an incoming call, it responds to the SETUP message with a CONN message.

10.5.3 Address Signaling

En-bloc Address Signaling. The example of Section 10.5.1 assumes *en-bloc* signaling of the called number. A TE arranged for *en-bloc* signaling waits until the user has entered the bearer service, the complete called number, and possibly other optional information elements, before sending the SETUP message. The procedures for entering the IEs into the TE are manufacturer-specific. When the local exchange receives the SETUP message, it assigns the B-channels to the call, sends a CALPRC message to the calling party, and starts the set-up.

Overlap Address Signaling. Some TEs have a keypad and a handset, and the user signals the called address in the same way as on a dual-tone multi-frequency (DTMF) telephone.

The SETUP messages from these TEs typically include only IE.1 (bearer capability), set to "speech" or "3.1 kHz audio." The local exchange assigns a B-channel, and sends a SETACK message, which is a request to send digits—see Fig. 10.5-1. The message includes IE.7 (channel identification), and IE.12 (signal) set to "dial-tone on." If the subscription parameters indicate that the TE does not have NPIBS, the IE.12 is a command to generate dial-tone. If the TE is marked as having NPIBS, the exchange also sends dial tone on the B-channel, and includes an IE.11 (progress indicator), indicating "in-band information now available," in the message.

The calling TE then sends a number of INFO messages that include one or more IE.9s (keypad), each of which represents a digit of the called number.

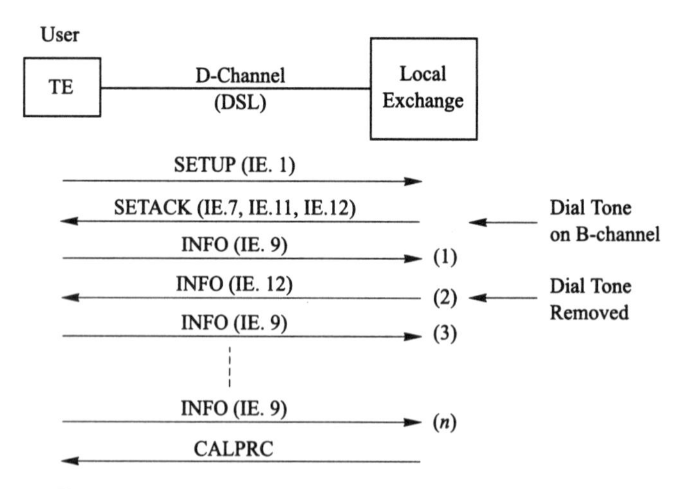

Figure 10.5-1 Overlap (keypad) signaling of called number.

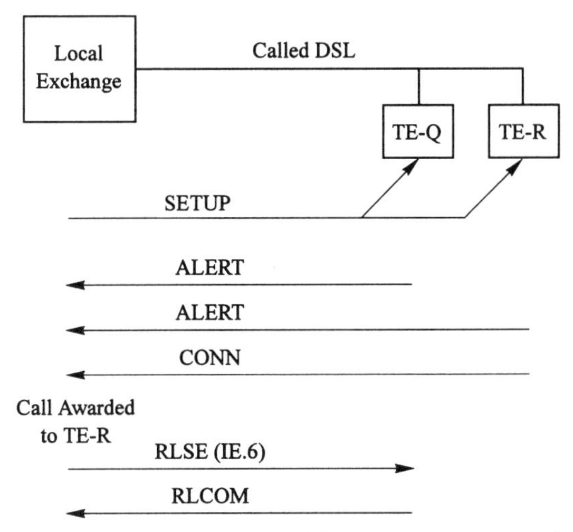

Figure 10.5-2 Multiple TE responding to SETUP message, contention allowed.

After receiving INFO message (1), the exchange returns INFO message (2), which has an IE.12 set to "tone off" and, if it has been sending dial tone on the B-channel, it stops the tone. The user then completes the called number, which is transferred to the exchange in INFO messages (3) through (*n*). The exchange then starts to set up the connection.

10.5.4 Multiple TE Responses

When more than one TE on a called DSL responds to a SETUP broadcast message with a CALPRC, ALERT, or CONN message, the exchange has to award to call to one of them. The decision depends on whether contention on incoming calls is allowed (see Section 10.4.4). If contention is allowed, the exchange awards the call to the first TE that sends a CONN message. If contention is not allowed, the call is awarded to the first TE that responds with a CALPRC, ALERT, or CONN message.

Figure 10.5-2 shows an example where contention is allowed, and TE-Q and TE-R have responded with an ALERT message. TE-R sends the first CONN message, and is therefore awarded the call. The exchange sends a RLSE message to TE-Q that includes an IE.6 (cause) with location "public network of the local user," and cause value "clearing of non-selected user."

10.6 FAILED ISDN SET-UPS

The subject of set-up failures in ISDN is more complicated than in telephony. In the first place, there are more failure types: in addition to the failures (subscriber busy, invalid called number, etc) that also occur in telephony, there are numerous failures that are specific to ISDN. In the second place, while a calling subscriber always receives an audible signal (tone or announcement) when a failure occurs, there are several ways in which a calling ISDN user can receive failure information.

This section explores a small number of failures that occur in the set-up of intraexchange ISDN calls. An exhaustive treatment of ISDN set-up failures in the U.S. networks can be found in [9].

10.6.1 Failure Causes

Failures can occur at several points in the set-up (Fig. 10.6-1). We first consider failures that occur immediately after the exchange has received the caller's

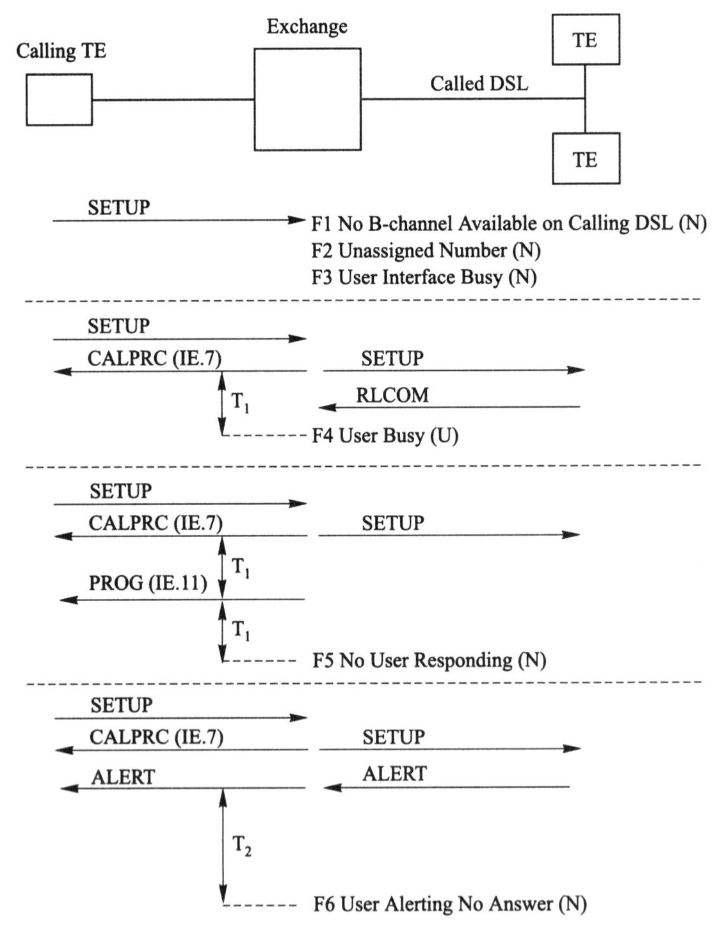

Figure 10.6-1 Set-up failure examples. N: failures determined by network (exchange). U: failures determined by user.

SETUP message. For example, the exchange can have determined that no B-channel is available on the calling DSL (F1), that the received called number is not assigned in the exchange (F2), or that no B-channel is available on the called DSL (F3). Failure F3 is known as "user interface busy."

Failures F4 and F5 occur after the exchange has sent its SETUP and CALPRC messages on the called and calling DSL. After sending the message, the exchange starts a 4-s timer T_1 and, on expiration of the timer, evaluates the received responses. In failure F4, the exchange has received RLCOM messages only. This means that all TEs that satisfy the TE-selection criteria are busy, or unable to accept the call for other reasons. Failure F4 is known as "user-determined busy."

Next, consider the case that no response has been received when T_1 expires. The SETUP broadcast message on the called DSL (10.4.3) is sent in an UI (unnumbered information) frame (10.2.5) that is not acknowledged by the TEs. It is possible that the message has incurred a transmission error, and has been discarded by the TEs. The exchange therefore broadcasts a second SETUP, restarts T_1, and sends a PROG message to the calling user that includes an IE.11 (progress indicator) with value "delay in response of called user." On speech and 3.1 kHz audio calls, the exchange also starts to send ringing-tone on the calling B-channel, and the PROG message includes a second IE.11, with value "in-band information now available" (not shown in Fig. 10.6-1). If no response has been received when T_1 expires again, the exchange concludes that the TEs on the DSL are not responding (F5).

Finally, suppose that the exchange has received one or more ALERT responses from the called DSL. It then sends an ALERT message to the calling TE. On speech and 3.1 kHz audio calls, the exchange also returns ringing tone on the calling B-channel, and the ALERT message includes an IE.11 set to "in-band information now available." The exchange then starts a 5-minute timer (T_2). If the timer expires and no CONN message has been received, the exchange concludes that the users at the TEs are not answering (F6).

10.6.2 Treatments of the Calling TE

On failed set-ups, the exchange has to inform the calling user, and to release B-channel connections that already have been established when the failure occurs.

The calling TE always receives a message that indicates the cause of the failure. On speech and 3.1 kHz audio calls, the user at the TE also receives in-band information (a tone or announcement) sent by the exchange on the B-channel, or generated by the TE in response to a "signal" command (IE.12) from the exchange.

The exchange gives one of the following failure treatments to the calling TE, depending on whether the failure occurs before or after it has sent a CALPRC message for the call (which identifies the allocated B-channel to the TE), and on whether it has to provide in-band information:

	In-band Information: Not Provided	Provided
Failure before CALPRC	FT1	FT1-IB
Failure after CALPRC	FT2	FT2-IB

The provision of in-band information depends on several factors. On 64 kb/s calls, the exchange never provides in-band information. On speech or 3.1 kHz audio calls, the exchange always provides ringing tone and announcements. However, busy-tone and reorder tone are sent only if the calling TE has network-provided in-band information service (NPIBS) (Section 10.4.4). Obviously, no in-band information is sent when no B-channel is available on the calling DSL (F1).

Table 10.6-1 lists the treatments for the failure cases described in the previous section. On speech and 3.1 kHz audio calls, the exchange is providing ringing-tone when failure F5 (no user responding) or F6 (user alerting, no answer) occurs. On failure F5, the exchange leaves the tone on, and gives the FT2-IB treatment. On failure F6, the exchange turns the tone off, and gives the FT2 treatment.

The treatments are outlined in Fig. 10.6-2. In FT1, the B-channel connection is not yet established, and the exchange ends the set-up with a RLCOM message. In FT2, the exchange initiates the release of the B-channel with a DISC message. In treatments FT1-IB and FT2-IB, the exchange sends a PROG message, returns a tone or announcement on the B-channel, and then waits for a DISC message from the calling TE. In FT1-IB, the PROG message is preceded by a CALPRC message, because the B-channel connection, which is needed for the transfer of the in-band information, has not yet been established.

Table 10.6-1 Treatments of calling TE on set-up failures.

			Bearer Service		
			64 kb/s	3.1 kHz	
				Calling TE	
Failure	Description	In-band Information		No NPIBS	NPIBS
F1	No B-channel available on calling DSL (N)	—	FT1	FT1	FT1
F2	Unassigned number (N)	Announcement	FT1	FT1-IB	FT1-IB
F3	User busy (N)	Busy-tone	FT1	FT1	FT1-IB
F4	User busy (U)	Busy-tone	FT2	FT2	FT2-IB
F5	No user responding (N)	Ringing-tone	FT2	FT2-IB	FT2-IB
F6	User alerting, no answer (N)	Ringing-tone Turned off	FT2	FT2	FT2

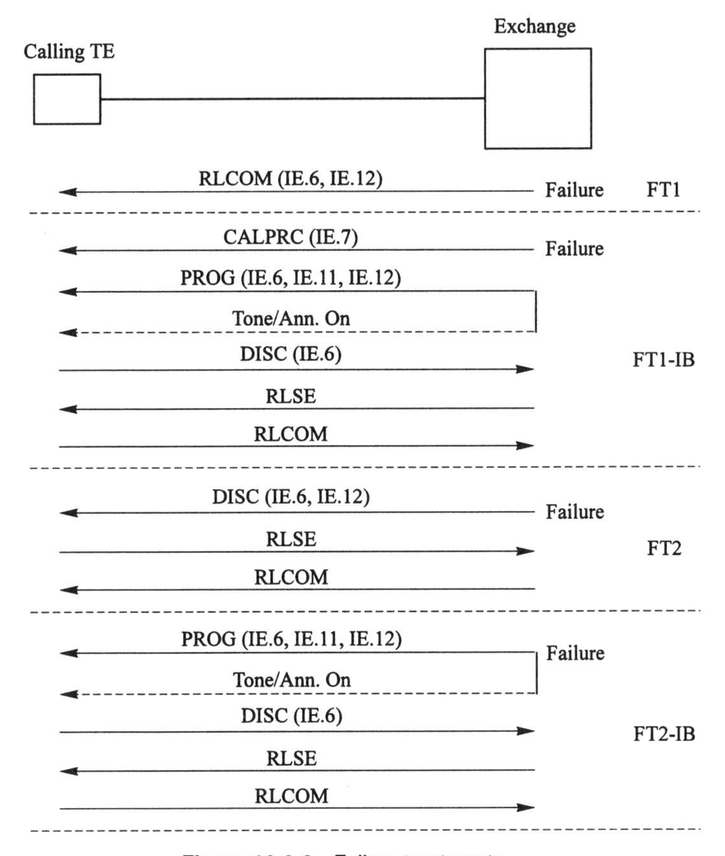

Figure 10.6-2 Failure treatments.

Table 10.6-2 shows the contents of IE.6 (cause) and IE.12 (signal) in the RLCOM, DISC, and PROG messages for the various failures. The contents generally do not depend on the bearer service, or on whether the calling TE has NPIBS. However, in failure F2 (unassigned number), IE.12 is not included on speech- or 3.1 kHz audio calls, and is set to "reorder-tone" on 64 kb/s calls.

When the exchange is giving an FT1-IB or FT2-IB treatment, the PROG message also includes an IE.11 (progress indicator) with value "in-band information now available."

10.6.3 Treatment of Called TEs

In failure F6 (no answer), the exchange initiates the release of the B-channel on the called DSL with RLSE messages to the called TEs that have sent ALERT messages. The IE.6 (cause) in the message has location "public network serving local user" and value "recovery action on expiration of timer."

Table 10.6-2 Coding of IE.6 (cause) and IE.12 (signal) in messages RLCOM, DISC, and PROG sent in set-up failures.

		IE.6		
Failure	Description	Location	Cause Value	IE.12
F1	No B-channel available on calling DSL (N)	2	34	3
F2	Unassigned number (N)	2	1	3 (note 1)
F3	User busy (N)	4	17	4
F4	User busy (U)	0	17	4
F5	No user responding (N)	4	18	1
F6	User alerting, no answer (N)	4	19	63

Notes: (1) IE.12 included only when bearer service is 64 kb/s. (2) Functional meanings of the listed decimal codes:

IE.6—location	
0	User
2	Public network serving the local user
4	Public network serving the remote user

IE.6—cause value	
1	Unassigned number
17	User busy
18	No user respondng
19	User alerting, no answer
34	Calling B-channel not available

IE.12	
1	Ringing-tone on
3	Reorder-tone on
4	Busy-tone on
63	Tones off

The exchange does not need to release the B-channel in the other failures, because it has not broadcast a SETUP message (failures F1, F2, and F3), or has sent a SETUP message and has received no response, or "busy" responses only (F4).

10.7 ACRONYMS

Ai	Action indicator
ALERT	Alerting message

B-Channel	64 kb/s channel for circuit-mode communications
BCD	Binary coded decimal digit
BSN	Backward sequence number
CALPRC	Call proceeding message
CB	Check-bit
CCITT	International Telegraph and Telephone Consultative Committee
CONACK	Connect acknowledgment message
CONN	Connect message
CRV	Call reference value
D-Channel	Signaling and packet channel on DSL
DISC	Disconnect message
DL	Data link
DM	Disconnect mode frame
DN	Directory (subscriber) number
DSL	Digital subscriber line
DSLI	DSL identifier
DSS1	Digital subscriber signaling system No.1
DTE	Data terminal equipment
ET	Exchange termination
Ext	Extension bit
FCS	Frame check sequence
FDM	Frequency-division multiplex
FSN	Forward sequence number
I	Information frame
IA5	International alphanumeric coding
ID	Identity
IE	Information element
INFO	Information message
ISDN	Integrated services digital network
ISUP	ISDN user part of SS7
IXC	Intertexchange carrier
LAPD	Link access protocol for D-channel
LATA	Local access and transport area
M	Mandatory
MSU	Message signal unit
MTP	Message transfer part
NPIBS	Network-provided in-band information service
N(R)	Receive sequence number
N(S)	Send sequence number
NT	Network termination
O	Optional
PCM	Pulse code modulation
PROG	Progress message
Q.931	ISDN network layer protocol for call control
REJ	Reject frame

Ri	Reference number
RLCOM	Release complete message
RLSE	Release message
RNR	Receive not ready frame
RR	Receive ready frame
SABME	Set automatic balanced mode, extended frame
SAPI	Service access point indicator
SETACK	Set-up acknowledgment message
SETUP	Set-up message
SS7	CCITT signaling system No.7
SU	Signal unit
TE	User terminal equipment
TEI	TE identifier
TUP	Telephone user part of SS7
UA	Unnumbered acknowledgment frame
UI	Unnumbered information frame

10.8 REFERENCES

1. W. Stallings, *ISDN, An Introduction*, MacMillan, New York, 1989.
2. P.K. Verma (editor), *ISDN Systems, Architecture, Technology and Applications*, Prentice Hall, Englewood Cliffs, N.J., 1990.
3. J.M. Griffiths, *ISDN Explained*, John Wiley, New York, 1990.
4. Roger L. Freeman, *Telecommunication System Engineering*, Wiley–Interscience, New York, 1996.
5. *Data Communication Networks: Services and Facilities, Interfaces*, Rec. X.1–X.32, CCITT Blue Book, **VIII.2**, ITU, Geneva, 1989.
6. R.J. Manterfield, *Common-Channel Signalling*, Peter Peregrinus Ltd., London, 1991.
7. *Digital Subscriber Signalling System No. 1 (DSS1), Data Link Layer*, Rec. Q.920–Q.921, CCITT Blue Book, **VI.10**, ITU, Geneva, 1989.
8. *Digital Subscriber Signalling System No. 1 (DSS1), Network layer, User-Network Management*, Rec. Q.930–Q.940, CCITT Blue Book, **VI.11**, ITU, Geneva, 1989.
9. *ISDN Access Call Control Switching and Signaling Requirements*, Report TR-TSY-000268, Issue 3 (1989), Bellcore, Piscataway, N.J.
10. D.R. Davies, R. Haslam, "Developments in ISDN: Signalling Standards and Terminals", *Br. Telecom. Technol. J.*, **9**, no. 2, 1991.

11

ISDN USER PART

This chapter resumes the discussion of SS7 signaling, and describes the ISDN (integrated digital services network) user part of SS7.

11.1 INTRODUCTION

The ISDN user part (ISUP) has been designed for use in the circuit-switched part of ISDN networks (see Fig. 10.1-2). It includes messages and procedures for the control of interexchange calls between two (analog) subscribers, two ISDN users, and between an ISDN user and a subscriber [1,2].

As in TUP signaling, ISUP call-control signaling is primarily link-by-link, but also includes procedures for end-to-end signaling.

Telecoms that have installed TUP signaling in their networks are gradually converting to ISUP signaling. The telecoms in the U.S. have chosen to bypass TUP, and are moving directly from multi-frequency and CCIS signaling (Chapters 4 and 6) to ISUP.

11.1.1 ISUP Call-control Requirements

The major difference between TUP and ISUP is that ISUP also supports signaling in calls in which one or both parties are ISDN users. ISUP has been designed to meet the signaling requirements for ISDN-related services.

Support of Bearer Services. The bearer service requested by a calling user can impose restrictions on the network equipment that can be used for the call.

Table 11.1-1 Trunk, transmission, exchange, and signaling types allowed in connections for individual bearer services.

	Speech	Bearer Service 3.1 kHz Audio	64 kb/s Data
Analog (FDM) trunk group	A	A	N
64 kb/s digital trunk group	A	A	A
Trunk group with ISUP signaling	A	A	A
Trunk group with other signaling	A	A	N
Circuit multiplication equipment	A	N	N
Echo control	A	N	N
Exchange with analog switchblock	A	A	N
Exchange with digital switchblock	A	A	A

A: allowed. N: not allowed.

These limitations are listed in Table 11.1-1. For example, speech calls can be carried on analog or digital (PCM) trunks that may be equipped with circuit multiplication equipment, and echo controls are usually enabled (see Section 1.7). Also, speech calls can be routed via exchanges with analog or digital (64 kb/s) switchblocks (Section 1.6).

On the other hand, 64 kb/s calls require digital trunks, and have to be routed via exchanges with digital switchblocks. Moreover, the bit patterns in the transmitted octets cannot be changed. Therefore, circuit multiplication equipment is not allowed, and echo cancelers have to be disabled.

Exchanges have to take these restrictions into account when selecting an outgoing trunk for an ISDN call.

Transfer of Q.931 Information. In interexchange calls, the calling and called users should be able to send and receive the same Q.931 messages and information elements as in intraexchange calls (Sections 10.3 and 10.6). ISUP signaling therefore includes messages and parameters to transfer the Q.931 information across the network. For example, the information elements included in a Q.931 SETUP message from the calling user for the selection of a particular called TE (Section 10.4.3) are transferred in ISUP messages to the terminating local exchange, which then includes them in its Q.931 SETUP message to the called DSL. Also, the "cause" information element of Q.931, which indicates the reason why a connection should be released, is transferred by ISUP messages to the calling or called local exchange, which then informs the ISDN user.

Services for Analog Subscribers. In present telecommunication networks, the number of digital subscriber lines is still a small fraction of the total number of lines. ISUP signaling includes messages and procedures to support the services already available to analog subscribers. In countries that are using TUP signaling, ISUP supports the services available with TUP (Section 9.4).

In the U.S., ISUP supports a number of services for business customers, and all services that require the transfer of the calling party number to the called party (Section 3.5.2).

11.1.2 ISUP Versions

CCITT/ITU-T has defined ISUP in Recommendations Q.761–Q.764 [3–6]. The descriptions of ISUP in Sections 11.2–11.7 are based on these recommendations.

ISUP is a very powerful and complex signaling system. Implemented versions usually include only a subset of the above recommendations. On the other hand, country-specific versions of ISUP—notably, the U.S. version—include signals and procedures that have not been defined by CCITT. Examples of implemented ISUP versions are given in Sections 11.8 and 11.9.

11.1.3 Interface with MTP

The interface between ISUP and the message transfer part (MTP) at a signaling point is as shown in Fig. 8.7-1. The MTP-transfer primitives include three groups of parameters:

Service Information Octet (SIO), consisting of parameters SI (service information) and SSF (sub-service field). In ISUP messages, $SI = 0101$.

Routing Label (RL), consisting of DPC (destination point code), OPC (originating point code), and SLS (signaling link selector).

User Message (UM), which holds the ISUP information that is passed transparently by MTP.

MTP-status, MTP-pause, and MTP-resume indications are sent by MTP to ISUP when a signaling route set to a particular destination becomes congested, unavailable, or available again.

11.1.4 Interface with SCCP

ISUP includes two methods for end-to-end signaling between the originating and terminating exchange of a call. One of these uses the services of the signaling connection control part (SCCP) of SS7.

11.2 ISUP MESSAGES, FORMATS, AND PARAMETERS

This section describes the most important ISUP call-control messages. Since we have returned to SS7, we speak again of *parameters*. ISUP messages have *mandatory* parameters (always present in messages of a given type), and *optional* parameters (included only when necessary). The number of optional parameters, and the number of messages that can include these parameters, are much larger than in TUP. This gives ISUP messages the flexibility to accommodate ISDN signaling requirements. This flexibility also has resulted in a rather small number of call-control messages (about 10 in ISUP versus over 30 in TUP).

11.2.1 Parameter Fields and Types [5]

Parameter Fields. ISUP parameters consist of one, two, or three fields—see Fig. 11.2-1. The "name" field (one octet) contains the parameter name, and thus indicates the meaning of the parameter; the "length" field (one octet) indicates the number of octets in the "value" (or "contents") field, and this latter field contains the actual information.

We distinguish three parameter types:

Mandatory, Fixed Parameters (MF Parameters). In a particular message type, these parameters are always present, and their "value" fields have fixed lengths.

Mandatory, Variable Parameters (MV Parameters). In a particular message type, these parameters are always present, but the lengths of their "value" fields are variable.

Optional Parameters (OP Parameters). These parameters, which may have fixed or variable length value fields, appear in a given message type only when necessary.

11.2.2 ISUP Message Format

This section describes the general format of ISUP messages (Fig. 11.2-2). The *circuit identification code* (CIC) has 14 bits, and is located in octets 1 and 2. In a

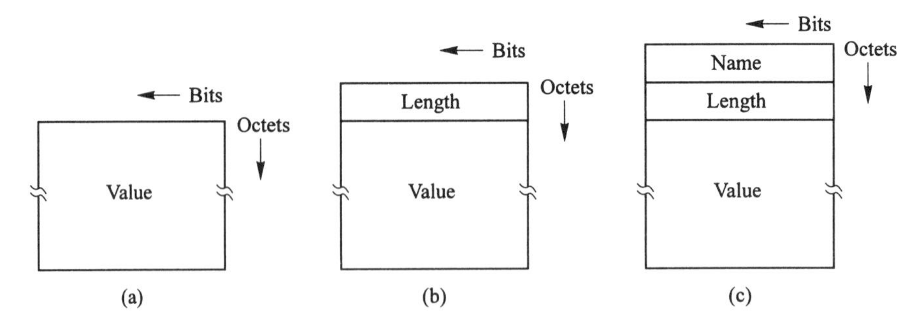

Figure 11.2-1 ISUP parameters. (a): mandatory, fixed-length (MF) parameters. (b): mandatory, variable-length (MV) parameters. (c): optional (OP) parameters.

network, an ISUP trunk is uniquely identified by the combination of DPC and OPC (located in RL—see Fig. 7.3-2), and CIC. In TUP signaling, the four low-order bits of CIC are also used as the signaling link selector (9.1.1). In ISUP signaling, SLS and CIC are separate parameters. This allows the rotation of the contents of SLS at the signaling points along the message route (see example 3 of Section 8.8.5).

The *message type code* (octet 3) identifies a particular ISUP message. CIC and the message type code are mandatory parameters (value fields only).

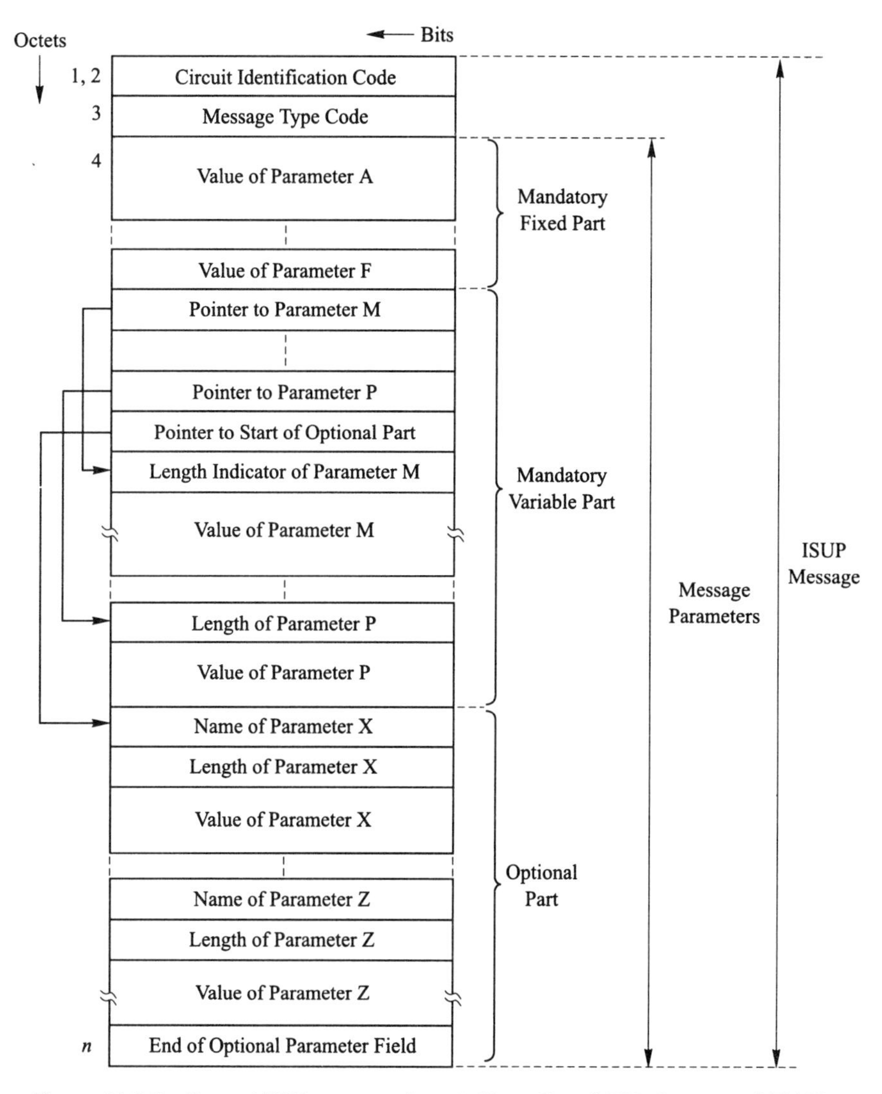

Figure 11.2-2 General ISUP message format. (From Rec. Q.763. Courtesy of ITU-T.)

Octets 4 through *n* hold the message parameters. In this example, A through F are mandatory fixed (MF) parameters. In a particular ISUP message type, these parameters always appear—in a predetermined order—starting at octet 4. Since their lengths are fixed, the locations of the MF parameters in the message are also fixed, and known by the software entity that reads the message. Therefore, "name" and "length" fields are not needed.

The mandatory variable (MV) part of the message-parameter field begins with a block of one-octet pointers to the locations of the "length" fields of the MV parameters (in this example, M through P). In messages of a particular type, this block always starts at the same location, and the pointers appear in a predetermined order. The software entity that reads the message can thus determine the locations of any MV parameter. "Name" fields are therefore not necessary.

The octet following the pointer of the last MV parameter holds the pointer to the first octet of the optional part of the message-parameter field, where the OP parameters X through Z are located. Optional parameters can appear in any combination, and in any order, and therefore appear with their "name," "length," and "value" fields.

When a message does not include optional parameters, the pointer to the optional part of the message is coded 0000 0000. When a message includes optional parameters, an octet coded 0000 0000 follows the value field of the last parameter.

The message structure described above is also used for messages of SCCP.

11.2.3 Call-control Messages

The most important ISUP call-control messages are outlined below. Most messages are the counterparts of TUP messages, but generally include more parameters.

Initial Address Message (IAM). As in No.6 and TUP signaling, the IAM is the first message in a call set-up. It includes the called number, and other parameters that are of importance for the set-up. In most countries, and in the international network, ISUP address signaling is *en-bloc* (the IAM includes the complete called number). However, CCITT has also specified a *subsequent address message* (SAM), for overlap address signaling. The IAM then holds the initial digits of the called address, and the remaining digits are transferred in one or more SAMs.

Continuity Message (COT). This message is sent forward during the call set-up, if the outgoing exchange has made a continuity test on an outgoing trunk. It indicates whether the test has been successful.

Address Complete Message (ACM). This is a backward message. When sent by the terminating local exchange, it indicates that the exchange is ringing the called subscriber, or has received a Q.931 ALERT message from a terminal on the called digital subscriber line (DSL). When sent by an intermediate exchange, it indicates that the exchange has seized an outgoing trunk that does not have ISUP signaling.

Call Progress Message (CPG). This is a backward message, sent by a terminating exchange that already has sent an ACM message, when it needs to report the occurrence of an "event" in the call set-up.

Answer Message (ANM). This is a backward message, sent when the called party answers.

Release Message (REL). This is a general message that requests the immediate release of a connection. It can be sent forward or backward, and includes—like the Q.931 DISC (disconnect) message—a "cause" parameter that indicates the reason for the release, and the originator of the release. In other interexchange signaling systems, the release of a connection is normally initiated by the calling subscriber. In ISUP signaling, the calling and called ISDN users can initiate a release.

A REL message is also sent by an intermediate exchange, or by the terminating local exchange, when the set-up of a call is not possible.

Release Complete Message (RLC). This is a forward or backward message, sent by an exchange in response to a received REL message for a trunk. It indicates that the exchange has completed the release of the trunk at its end. The RLC message has the same function as the "release guard" message of TUP signaling.

Suspend Message (SUS). This is a backward message that is similar to the "clear-back" message of TUP. It requests to suspend the call, but to leave the connection intact.

Resume Message (RES). This message requests to resume a suspended call.

Forward Transfer Message (FOT). An outgoing operator, who is assisting a subscriber or ISDN user in a call set-up, can request the help of an incoming operator, by sending a (forward) FOT message (see Section 4.3.5).

Information Request Message (INR). This is a request for additional call-related information, usually sent from the terminating exchange to the originating exchange of a connection (end-to-end signaling).

Information Message (INF). This is a response to an INR message, and containing the requested information.

Table 11.2-1 Message type codes of ISUP call-control messages.

Acronym	Name	Message Type Code
ACM	Address complete	0000 0110
ANM	Answer	0000 1001
CPG	Call progress	0010 1100
COT	Continuity	0000 0101
FOT	Forward transfer	0000 1000
IAM	Initial address	0000 0001
INF	Information	0000 0100
INR	Information request	0000 0011
PAM	Pass-along	0010 1000
REL	Release	0000 1100
RES	Resume	0000 1110
RLC	Release complete	0001 0000
SUS	Suspend	0000 1101

Source: Rec. Q.763. Courtesy of ITU-T.

Pass-along Message (PAM). This message holds another message that is transferred end-to-end with the pass-along method.

The message type codes are listed in Table 11.2-1.

11.2.4 Parameters in Call-control Messages [5]

This section describes the most important parameters of ISUP call-control messages. At this point, it is suggested that the descriptions are merely perused, and referred back to when reading the later sections of this chapter.

To allow the reader to quickly locate a parameter description, each parameter has a reference number (for example, Par.1). In the descriptions of this section, the parameters appear in the order of their reference numbers. In Sections 11.3–11.7, a parameter is always identified by its name and reference number.

The parameters, their reference numbers, and the messages in which they can appear are listed in Table 11.2-2.

In this section, the focus is on the parameter contents, and only their "value" fields are shown in the figures. The parameter "name" codes are listed in Table 11.2-3.

Par.1 Access Transport. This is an optional parameter in IAM that contains one or more Q.931 information elements. The IEs are transferred transparently, between the ISDN users. The parameter can include IE.4 (called party subaddress), IE.5 (calling party subaddress), IE.8 (high layer compatibility), IE.10 (low layer compatibility), and IE.11 (progress indicator)—see Section 10.3.5.

Table 11.2-2 Parameters in ISUP call-control messages.

Reference Number	Parameter Name	ACM	ANM	COT	CPG	FOT	IAM	INF	INR	REL	RES	RLC	SUS
Par.1	Access transport	O	O	O			O	O		O			
Par.2	Automatic congestion level									O			
Par.3	Backward call indicators	M	O		O								
Par.4	Call reference	O			O	O	O	O	O	O	O		O
Par.5	Called party number						M						
Par.6	Calling party category						M	O					
Par.7	Calling party number						O	O					
Par.8	Cause indicators	O			O					M			
Par.9	Closed user-group interlock code						O						
Par.10	Continuity indicator			M									
Par.11	Event information				M								
Par.12	Forward call indicators						M						
Par.13	Information indicators							M					
Par.14	Information request indicators								M				
Par.15	Nature of connection indicators						M						
Par.16	Optional backward call indicators	O	O		O								
Par.17	Optional forward call indicators						O						
Par.18	Original called party number						O						
Par.19	Redirecting number				O		O						
Par.20	Redirection information						O			O			
Par.21	Redirection number									O			
Par.22	Suspend/resume indicator										M		M
Par.23	Transmission medium requirements						M						
Par.24	User service information						O						
Par.25	User-to-user information	O	O	O			O			O			

Message Acronyms (column headings above).

M: Mandatory parameter. O: Optional parameter.
Source: Rec. Q.763. Courtesy of ITU-T.

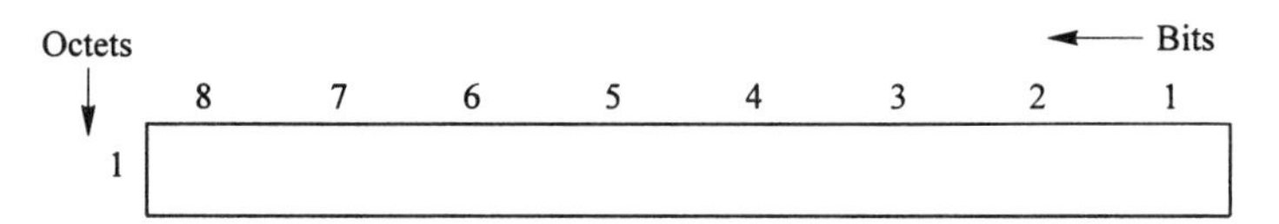

Figure 11.2-3 Format of parameters. Par.2: automatic congestion level. Par.6: calling party category. Par.11: event information. Par.23: transmission medium requirements.

Table 11.2-3 Parameter name codes.

Reference Number	Parameter Name	Name Code
Par.1	Access transport	0000 0011
Par.2	Automatic congestion level	0010 0111
Par.3	Backward call indicators	0001 0001
Par.4	Call reference	0000 0001
Par.5	Called party number	0000 0100
Par.6	Calling party category	0000 1001
Par.7	Calling party number	0000 1010
Par.8	Cause indicators	0001 0010
Par.9	Closed user group interlock code	0001 1010
Par.10	Continuity indicators	0001 0000
Par.11	Event information	0010 0100
Par.12	Forward call indicators	0000 0111
Par.13	Information indicators	0000 1111
Par.14	Information request indicators	0000 1110
Par.15	Nature of connection indicators	0000 0110
Par.16	Optional backward call indicators	0010 1001
Par.17	Optional forward call indicators	0000 1000
Par.18	Original called number	0010 1000
Par.19	Redirecting number	0000 1011
Par.20	Redirection information	0001 0011
Par.21	Redirection number	0000 1100
Par.22	Suspend/resume indicator	0010 0010
Par.23	Transmission medium requirements	0000 0010
Par.24	User service information	0001 1101
Par.25	User-to-user information	0010 0000

Note: Name codes are included only when the parameter is optional in the message.
Source: Rec. Q.763. Courtesy of ITU-T.

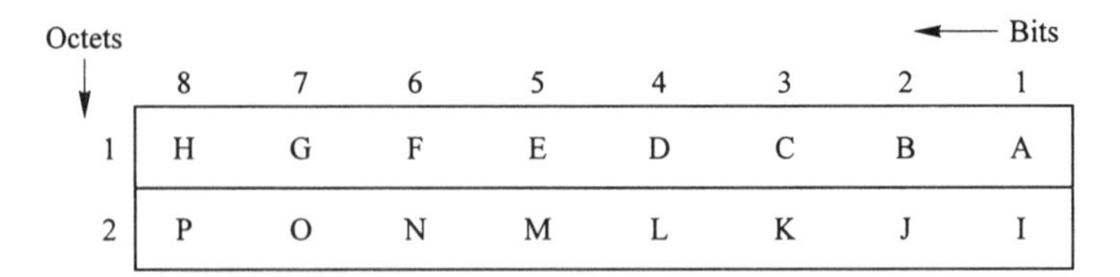

Figure 11.2-4 Format of parameters. Par.3: backward call indicators. Par.12: forward call indicators. Par.13: information indicators. Par.14: information request indicators.

Par.2 Automatic Congestion Level. This is an optional parameter in REL messages (Fig. 11.2-3). It indicates congestion at the exchange, and is coded as follows:

0000 0001	Congestion level 1 exceeded (moderate congestion)
0000 0010	Congestion level 2 exceeded (severe congestion)

Par.3 Backward Call Indicators. This is a mandatory parameter in ACM message, and optional in CPG and ANM messages (Fig. 11.2-4). It contains a number of indicator bits with information for the originating exchange:

Bits		Bits	
BA	*Charge Indicator*	*DC*	*Called Party Status*
00	No indication	00	No information
01	Do not charge	01	Subscriber free
10	Charge	11	Call set-up delay
FE	*Called Party Category*	*HG*	*End-to-end signaling*
00	Not known	00	Not available
01	Subscriber	01	Pass-along method available
10	Payphone	10	SCCP method available
		11	Pass-along and SCCP methods available
J	*Interworking Indicator*	*K*	*ISUP Indicator*
0	TUP or ISUP signaling all the way (no interworking)	0	Not ISUP signaling all the way
1	Interworking encountered	1	ISUP signaling all the way
L	*Holding Indicator*	*M*	*ISDN Access Indicator*
0	Holding not required	0	Called party has non-ISDN access (analog subscriber)
1	Holding required	1	Called party has ISDN access (ISDN user)
N	*Echo Control Indicator*		
0	Incoming half-echo control device not included		
1	Incoming half-echo control device included		

The combination of bits H, G, J, and K is known as the *protocol control indicator.*

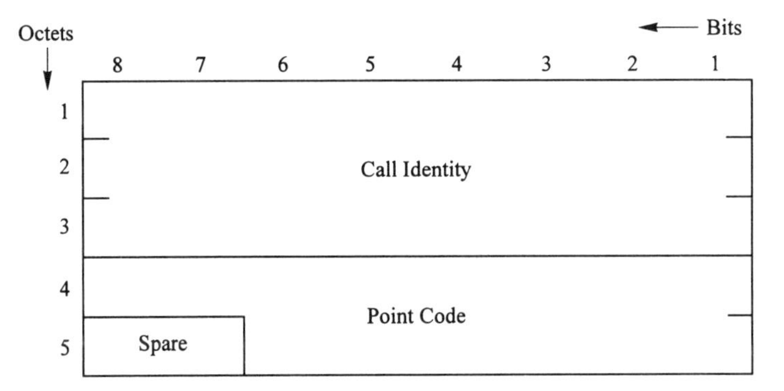

Figure 11.2-5 Format of Par.4: call reference. (From Rec. Q.763. Courtesy of ITU-T.)

Par.4 Call Reference. This is an optional parameter in IAM, ACM, CPG, REL, SUS, RES, FOT, INF, and INR messages (Fig. 11.2-5). The parameter identifies a particular ISUP call at an exchange, and consists of a *call identity* (assigned by the exchange that sends the message), and the point code of that exchange (two octets in CCITT-ISUP; three octets in ANSI-ISUP). It is used in one form of end-to-end signaling (Section 11.5.4).

Par.5 Called Party Number. This is a mandatory variable-length parameter in IAM. The parameter field consists of several subfields—see Fig. 11.2-6.

Odd/Even Indicator (O/E):

0 Even number of address digits
1 Odd number of address digits

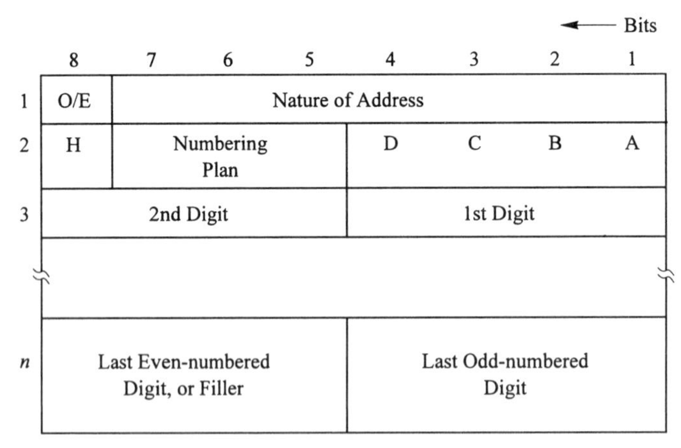

Figure 11.2-6 Format of parameters. Par.5: called party number. Par.7: calling party number. Par.18: original called number. Par.19: redirecting number. Par.21: redirection number. (From Rec. Q.763. Courtesy of ITU-T.)

Nature of Address:

000 0001	Subscriber (directory) number
000 0011	National number
000 0100	International number

Numbering Plan (only one value used):

001 ISDN/telephony numbering plan (see Section 1.3)

Bits H, and D through A, are not used.

Digits:

0000-1001	(Digit values 0-9)
1011	(Code 11)
1100	(Code 12)
1111	(End of address, ST)

If the number of address signals is odd, bits 8,...,5 of octet *n* are a filler coded 0000.

Par.6 Calling Party Category. This is a mandatory fixed-length parameter in IAM (Fig. 11.2-3), coded as follows:

0000 0001	Operator, French language
0000 0010	Operator, English language
0000 0011	Operator, German language
0000 0100	Operator, Russian language
0000 0101	Operator, Spanish language
0000 1010	Ordinary subscriber
0000 1100	Voiceband data call
0000 1101	Test call

Par.7 Calling Party Number. This is an optional parameter in IAM messages (Fig. 11.2-6). The coding is as in Par.5 (called party number), except that bits D,...,A are used:

Bits	
BA	*Screening Indicator (SI)*
01	Calling number provided by calling user, and verified by the network (originating local exchange)
11	Calling number provided by network

Bits	
DC	*Presentation Restriction Indicator (RI)*
00	Calling party allows the presentation of his number
01	Calling party does not allow the presentation of his number

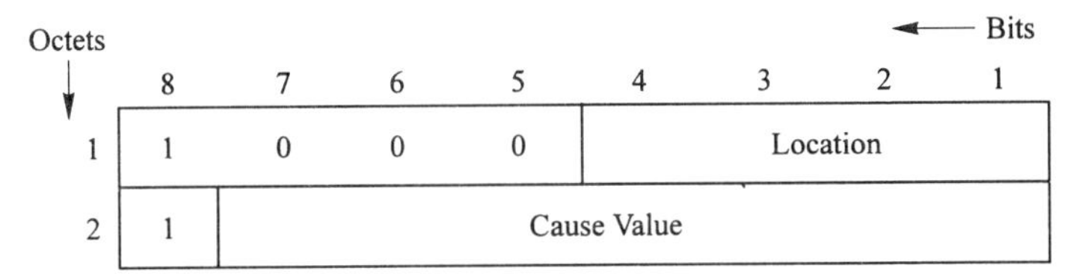

Figure 11.2-7 Format of Par.8: cause indicators. (Source: From Rec. Q.763. Courtesy of ITU-T.)

Par.8 Cause Indicator (Fig. 11.2-7). This is a mandatory (MF) in REL messages, and optional in ACM and CPG messages. The "location" field indicates where the message originated; the "cause" field indicates why a call cannot be set up, or why a connection has to be released. The coding is identical to the coding of octets 3 and 4 in the "cause" information element (IE.6) of Q.931—see Section 10.3.5.

Par.9 Closed User Group Interlock Code. This is an optional IAM parameter in countries that offer closed user group (CUG) service (Sections 9.2.2 and 9.4.1). It uniquely identifies a CUG. Octets 1 and 2 (Fig. 11.2-8) contain four digits that identify the telecom that administers the interlock code. The binary number in octets 3 and 4 represents a particular code assigned by that telecom.

Par.10 Continuity Indicator (Fig. 11.2-9). This is a mandatory (MF) parameter in COT messages. Only indicator bit A is used:

Bit A	
0	continuity check has failed
1	continuity check successful

Par.11 Event Information (Fig. 11.2-3). This is a mandatory (MF) parameter in CPG messages. It indicates an event that has occurred during the call set-up.

0000 0001 Called party has been alerted
0000 0010 Progress

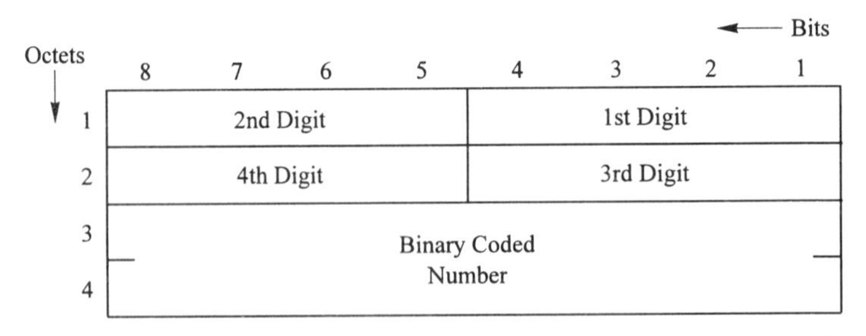

Figure 11.2-8 Format of Par.9: closed user group interlock code. (Source: From Rec. Q.763. Courtesy of ITU-T.)

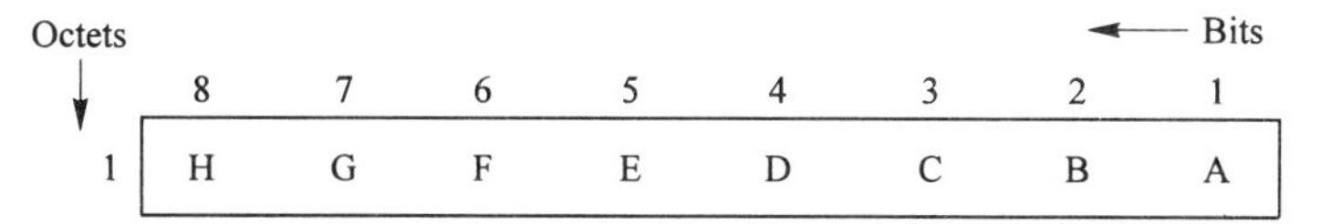

Figure 11.2-9 Format of parameters. Par.10: continuity indicators. Par.15: nature of connection indicators. Par.16: optional backward call indicators. Par.17: optional forward call indicators. Par.17: optional forward call indicators. Par.22: suspend/resume indicators.

0000 0011	In-band tone or announcement is now available
0000 0100	Call forwarded on busy
0000 0101	Call forwarded on no answer
0000 0110	Call forwarded unconditionally

Par.12 Forward Call Indicators (Fig. 11.2-4). This is a mandatory (MF) parameter in IAM messages. It contains indicator bits with information for the intermediate and terminating exchanges in a connection.

Bits		Bits	
A	*Call Type*	*CB*	*End-to-end Signaling*
0	Call to be treated as national call	00	Not available
1	Call to be treated as international call	01	Pass-along method available
		10	SCCP method available
		11	Pass-along and SCCP methods available
D	*Interworking Indicator*		
0	No interworking encountered (TUP or ISUP signaling all the way)		
1	Interworking encountered		
F	*ISUP Indicator*	*HG*	*ISUP Preference Indicator*
0	ISUP signaling not used all the way	00	ISUP signaling preferred all the way
1	ISUP signaling used all the way	01	ISUP signaling not required all the way
		11	ISUP signaling required all the way
I	*ISDN Access Indicator*		
0	Calling party has non-ISDN access (analog subscriber)		
1	Calling party has ISDN access (ISDN user)		

Note: the combination of bits C, B, D, F, H, and G constitutes the *protocol control indicator* (PCI).

Par.13 Information Indicators (Fig. 11.2-4). This is a mandatory parameter in INF (information) messages. It consists of bits representing the parameters or call-control functions that can be requested by an exchange. The bit values indicate whether the corresponding parameter is included in the message, or whether the corresponding service is being provided. For example:

Parameter or Function	Bit(s)	Value(s)	Meaning
Calling party address (Par.7)	BA	00	Not included
		01	Not available
		11	Included
Call holding	C	0	Not provided
		1	Provided
Calling party category (Par.6)	F	0	Not included
		1	Included

Par.14 Information Request Indicators (Fig. 11.2-4). This is a mandatory parameter in INR (information request) messages. It consists of indicator bits that represent parameters or call-control functions. The bit value indicates whether the corresponding parameter or function is requested.

Parameter or Function	Bit	Value	Meaning
Calling party address (Par.7)	A	0	Not requested
		1	Requested
Call holding	B	0	Not requested
		1	Requested
Calling party category (Par.6)	E	0	Not requested
		1	Requested

Par.15 Nature of Connection Indicators (Fig. 11.2-9). Mandatory (MF) parameter in IAM messages.

Bits	
BA	*Satellite Indicator*
00	No satellite circuit in connection
01	One satellite circuit in connection
10	Two satellite circuits in connection

Bits	
DC	*Continuity Check Indicator*
00	Continuity check not required
01	Continuity check required on this circuit

Bit	
E	*Echo-control Device Indicator*
0	No outgoing half-echo control device included
1	Outgoing half-echo control device included

Bits H, G, F are not used.

Par.16 Optional Backward Call Indicators (Fig. 11.2-9). Optional parameter in ACM, CPG, and ANM messages. Only bit A is used.

Bit	
A	*In-band Information Indicator*
0	No indication
1	In-band tone or announcement is now available

Par.17 Optional Forward Call Indicators. This is an optional parameter in IAM messages, which is used in countries that offer CUG service (Fig. 11.2-9). Only bits B and A contain information:

Bits BA	
00	Non-CUG call
10	CUG call; outgoing access allowed
11	CUG call; outgoing access not allowed

Par.18 Original Called Number. This optional parameter is included in IAM messages of calls that have been redirected to a new called number. It has the format of Par.7 (calling party number)—see Fig. 11.2-6. The presentation screening indicator (bits D,C) pertain to the original called number. If no presentation is allowed, octets 3–n are omitted. The screening indicator (bits B,A) is not used.

Par.19 Redirecting Number. This optional parameter is included in IAM messages of calls that are being forwarded. It is the number of the party that caused the call to be forwarded, and has format of Par.18.

Par.20 Redirection Information. This is an optional parameter, included in IAM and REL messages on calls that are being forwarded or rerouted—see Fig. 11.2-4.

Bits C,B,A indicate the type of redirection, for example:

0 0 1 call rerouted
0 1 0 call forwarded

Bits H,G,F,E indicate the original redirection reason:

0 0 0 0	unknown
0 0 0 1	user busy
0 0 1 0	no answer
0 0 0 0	unconditional call forwarding

Bits K,J,I hold a binary number between 1 and 5. It indicates the number or redirections the call has undergone.

Par.21 Redirection Number. Suppose that an exchange has received the IAM of an incoming call, and determines that the connection has to be rerouted. It then sends a (backward) REL message which may include a redirection number, which is the new called number. It has the format and coding of Par.6 (called party number).

Par.22 Suspend/Resume Indicator. This is a mandatory parameter in SUS (suspend) and RES (resume) messages. It indicates the originator of the message—see Fig. 11.2-9:

Bit A	Originator
0	ISDN user
1	Network (exchange)

Par.23 Transmission Medium Requirements. This is a mandatory IAM parameter, and indicates the nature of the information in the call. It is used by exchanges to select outgoing trunks of the appropriate type—see Fig. 11.2-3:

0000 0000	speech
0000 0010	64 kb/s data
0000 0011	3.1 kHz audio

Par.24 User Service Information. This is an optional IAM parameter, included only in calls originated by an ISDN user. It is coded as octets 3–5a of the bearer capability information element (IE.1) of Q.931—see Section 10.3.5.

Par.25 User–user Information. This is a variable-length optional IAM parameter, used in countries that offer user-to-user signaling to ISDN users. The parameter contents are not specified by CCITT, and are coded as agreed by individual user pairs.

11.3 SIGNALING FOR CALLS BETWEEN ISDN USERS

This section describes the Q.931 and ISUP signaling for a typical successful interexchange call between two ISDN users, with ISUP signaling all the way [6].

11.3.1 Connection Set-up

We first examine the set-up of the connection (Fig. 11.3-1). Trunks T_1 and T_2 have ISUP signaling. As in TUP signaling (9.3.1), an exchange decides whether to do a continuity check on the seized outgoing trunk, and informs the incoming exchange (9.3.1). In this example, no checks are made.

The focus is on the actions at the exchanges, and the interworking between Q.931 and ISUP signaling messages. We refer to the ISUP and Q.931 messages by their acronyms (Sections 11.2.3 and 10.3.2).

User U_1 originates the call at TE_1, with a SETUP message that includes the complete called address (*en-bloc* address signaling). Exchange P examines the message, seizes outgoing trunk T_1 to exchange Q, and sends an IAM. It also assigns a B-channel to TE_1, cuts through a unidirectional switchblock path (transmission in the backward direction only) between the B-channel and TE_1, and sends a CALPRC message to TE_1.

After receiving the IAM, exchange Q seizes outgoing trunk T_2 to exchange R, sends an IAM, and cuts through a bidirectional path between T_1 and T_2.

When exchange R receives the IAM, it assigns a B-channel on the DSL of U_2, and broadcasts a SETUP message to the TEs on the DSL. In this example, TE_2 responds with an ALERT message, and alerts user U_2 with an audible or visual signal. Exchange R then sends an ACM message to exchange Q. If Par.24 (user

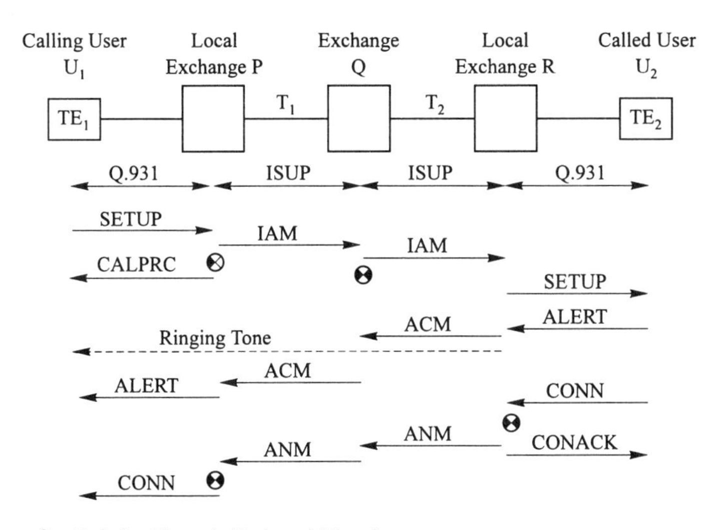

Figure 11.3-1 Set-up of a call between ISDN users.

service information) indicates a speech or 3.1 kHz audio call, exchange R also connects trunk T_2 to a ringing-tone source, and calling user U_1 then hears the tone. Exchange Q repeats the ACM to exchange P, which then sends an ALERT message to TE_1.

When U_2 answers, TE_2 sends a CONN message. Exchange R then cuts through a bidirectional path between T_2 and the B-channel allocated to TE_2, sends a CONACK (acknowledgment) to TE_2, and an ANM message to exchange Q. The message is repeated backwards, and reaches exchange P. This exchange then cuts through its forward transmission path, and sends a CONN message to TE_1. The TE may or may not acknowledge the CONN message; depending on its type and manufacturer.

The conversation or data transfer between U_1 and U_2 can now begin.

11.3.2 Release of the Connection

At the end of the call, the B-channels used by TE_1 and TE_2, and trunks T_1 and T_2 have to be released. In Q.931 and ISUP signaling, either user can initiate the release. In this example, U_2 initiates the release, and commands TE_2 to send a DISC message—Fig. 11.3-2. Exchange R then releases the B-channel to TE_2 at its end, and returns a RLSE message. TE_2 then releases the B-channel at its end, and sends a RLCOM message. This completes the release, and the B-channel is now available for a new call.

Exchange R also sends a backward REL message for trunk T_2 to exchange Q, and starts to clear the trunk at its end. In turn, exchange Q sends a REL message for trunk T_1 to exchange P, and starts to clear the trunk. After releasing respectively T_2 and T_1 at their ends, exchanges Q and P return RLC messages, indicating to exchanges R and Q that trunks T_2 and T_1 are available again for new calls.

In ISUP signaling, an exchange first sends a REL message for a trunk, and then clears the trunk. This results in a faster release than in TUP signaling,

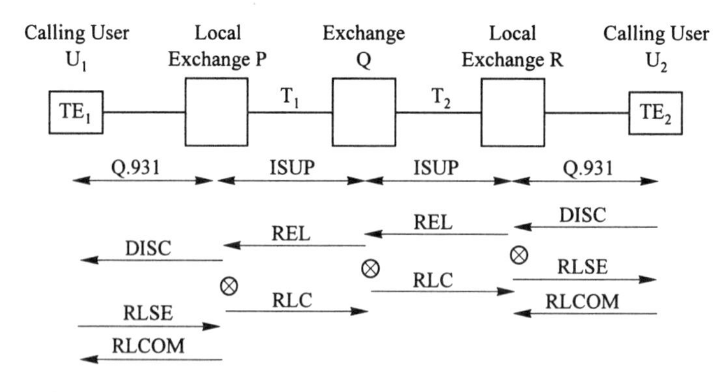

Figure 11.3-2 Release of the connection.

where an exchange first clears a trunk and then sends a clear-forward signal (Section 9.3.1).

When exchange P receives the REL message, it also requests TE_1 to release its B-channel, with a DISC message. TE_1 then clears the channel at its end, and returns a RLSE message. In response, exchange P releases the B-channel at its end, and sends a RLCOM message. The B-channel is now available for a new call.

11.3.3 Q.931 ↔ ISUP Interworking and Mapping

In the example of Figs. 11.3-1 and 11.3-2, interworking between Q.931 signaling and ISUP signaling takes place in local exchanges P and R. These exchanges have to "map" received Q.931 messages and information elements (IE) to outgoing ISUP messages and parameters (Par.)—and vice versa. The IEs and Pars. have been described in Sections 10.3.5 and 11.2.4. We now examine the mappings for the call [2].

Mapping of SETUP and IAM Messages (Fig. 11.3-3). Exchange P receives information elements IE.1 (mandatory), and IE.2, IE.3, IE.4, IE.5, IE.8, and IE.10. (optional) from calling ISDN user U_1.

The complete information in bearer capability (IE.1) is mapped into Par.24 (user service information), which is transferred transparently across the network. A part of the information in IE.1 is mapped into Par.23 (transmission medium requirements), which is used by the exchanges during the set-up of the connection.

IE.2 and IE.3 (called and calling party address) are mapped into Par.5 and Par.7. IE.2 is shown as optional, because U_1 may not include it in the SETUP message, and send the digits of the called party number in a series of INFO messages (Section 10.5.3). Our example assumes that U_1 has included IE.2 in the SETUP message. IE.3 is also optional. In countries where Par.7 is mandatory, and exchange P has not received IE.2, it enters the default directory number of U_1, which is a stored parameter at the exchange, into Par.7.

If included in the SETUP message, IE.4, IE.5, IE.8, and IE.10 are entered into Par.1 (access transport). This parameter is transferred transparently across the network.

The contents of parameters Par.6, Par.15, and Par.12 are set by exchange P.

We next explore the mapping—at terminating exchange R—of the IAM parameters into information elements that have to be included in the SETUP message to U_2. Par.24 is always mapped into IE.1. Par.5 (called party number) is mapped into IE.2 if necessary for terminal selection by U_2. Par.7 (calling party number) is mapped into IE.3 if U_2 has "calling line identity presentation service." The information elements present in Par.1 are retrieved, and included in the SETUP message.

Under conditions to be discussed later, some indicators of Par.12 (forward call indicators) are mapped to IE.11 (progress indicators) in the SETUP message to U_2.

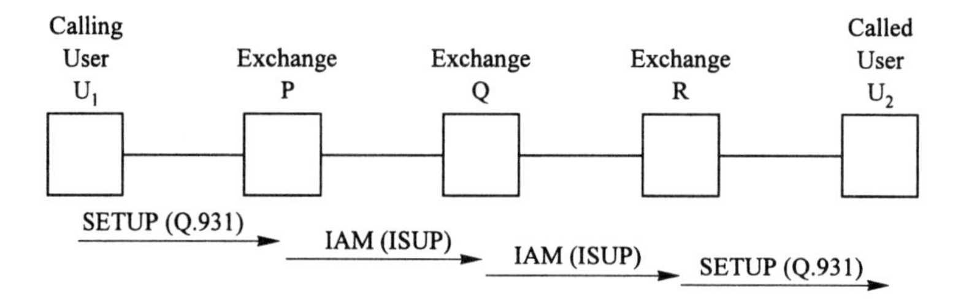

IE.I Bearer Capability (M)	Par. 23 Transmission Medium Requirements (M)	–
	Par. 24 User Service Information (O)	IE.1 (O)
IE.2 Called Party Number (O)	Par. 5 Called Party Number (M)	IE.2 (O)
IE.3 Calling Party Number (O)	Par. 7 Calling Party Number (O)	IE.3 (O)
IE.4 Called Party Subaddress (O)	Par. 1 Access Transport (O)	IE.4 (O)
IE.5 Calling Party Subaddress (O)		IE.5 (O)
IE.8 High Layer Compatibility (O)		IE.8 (O)
IE.10 Low Layer Compatibility (O)		IE.10 (O)
–	Par. 6 Calling Party Category (M)	–
–	Par. 15 Nature of Connection Indicators (M)	–
–	Par. 12 Forward Call Indicators (M)	IE.11 Progress Indicators (O)

Figure 11.3-3 Mappings of Q.931 SETUP messages and ISUP initial address message (IAM). Note: (M) mandatory parameter or IE. (O) optional parameter or IE.

Mapping of ALERT and ACM Messages (Fig. 11.3-4). The ALERT message received from user U_2 normally does not include IEs. The corresponding ACM messages include a Par.3 (backward call indicators), whose values are set by exchange R. Par.3 is processed by exchanges Q and P, and under certain conditions, an indicator is mapped to an IE.11 (progress indicators), which is included in the ALERT message to user U_1.

Mapping of DISC and REL Messages (Fig. 11.3-5). DISC and REL messages are general-purpose requests to release a B-channel or a trunk. They

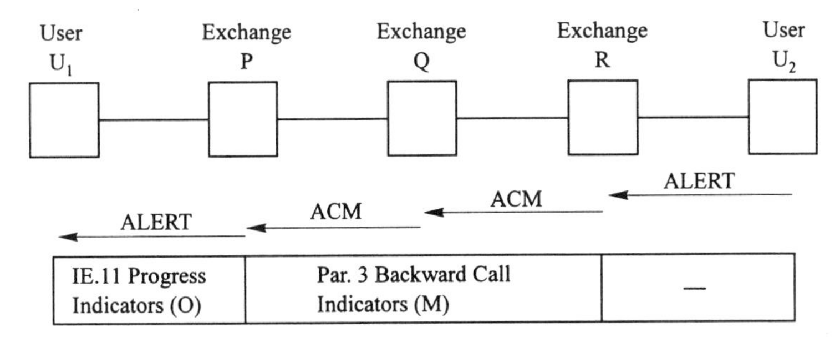

Figure 11.3-4 Mappings of Q.931 ALERT messages and ISUP address complete message (ACM). Note: (M) mandatory parameter or IE. (O) optional parameter or IE.

always include respectively an IE.6 (cause) and Par.8 (cause indicators), to indicate the reason for the release. Mappings between IE.6 and Par.8 take place at exchanges P and R.

The coding of the value fields of the information elements and parameters can be found in Sections 10.3.4 and 11.2.4. Some mappings require format changes. For example, the digits of the called and calling numbers have a seven-bit format in IE.2 and IE.3, and a four-bit format in Par.5 and Par.7. Other mappings, for example, between IE.6 (cause) and Par.8 (cause indicators), do not require reformatting.

11.3.4 Use of Message Parameters

This section revisits the call example described in Sections 11.3.1 and 11.3.2, focusing this time on the ISUP message parameters and their use. It is helpful to refer to the parameter descriptions of Section 11.2.4 when reading this material.

Routing Parameters in IAM. The selection of outgoing trunk groups by exchanges P, and Q is influenced by Par.5, Par.23, Par.15, and Par.12 of the IAM (Fig. 11.3-3).

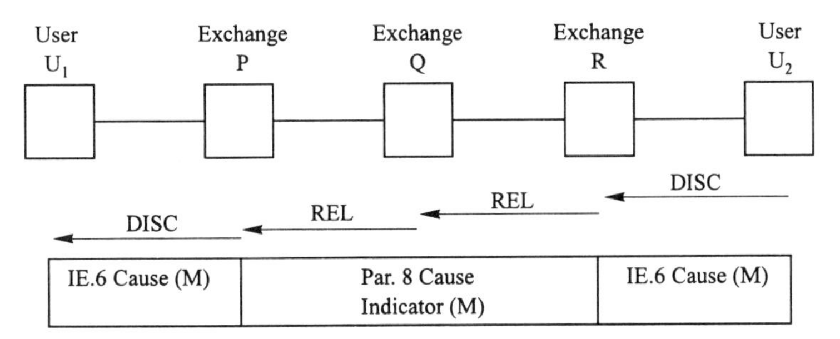

Figure 11.3-5 Mappings of Q.931 DISC messages and ISUP release message (REL). Note: (M) mandatory parameter or IE. (O) optional parameter or IE.

Par.5 (called address) is the primary routing parameter, from which the exchange derives the route set for the call. The other parameters can disqualify some outgoing trunk groups in the route set.

If Par.23 (transmission medium requirements) indicates that the transmission medium should accommodate a 64 kb/s data call, the connection is restricted to exchanges with digital switchblocks, and to digital trunk groups without circuit-multiplication equipment.

Par.15 (nature of connection indicators) is used to limit the number of satellite trunks in the connection, to control the inclusion (enabling) of echo control devices, and to indicate whether a continuity check will be made on the trunk.

In Par.12 (forward call indicators), bits CB, D, F, and HG constitute the signaling protocol indicator (PCI). Bits HG (ISUP preference indicators) specify whether ISUP signaling is required all the way. Originating exchanges are usually programmed to indicate that all-the-way ISUP signaling is required for calls originated by ISDN users, and preferred for calls originated by subscribers. Indicators HG are not modified by later exchanges along the connection. Indicators D (interworking indicator) and F (ISUP indicator) indicate the actual signaling on the connection. If exchange P in Fig. 11.3-1 has set HG, D, and F to "ISUP required all the way," "no interworking encountered," and "ISUP signaling used all the way," exchange Q selects an outgoing trunk with ISUP signaling, and does not change these indicators in its IAM to exchange R. The settings of these indicators in connections where interworking occurs between ISUP and other interexchange signaling systems are discussed in Section 11.4. The meaning of indicators CB is discussed in Section 11.5.

Call Processing Information at the Originating and Terminating Exchange. In the call of Fig. 11.3-1, the local exchanges know that the signaling is ISUP all the way. Terminating exchange R receives this information in bit F of Par.12 in the IAM, and originating exchange P is informed by bit K of Par.3 (backward call indicators) of the ACM message. We shall see later that, if the signaling is not ISUP all the way, both local exchanges are also made aware that this is the case. This information is sometimes needed for subsequent call-processing actions.

In connections with ISUP signaling all the way, both local exchanges also know whether the distant party in the connection is a ISDN user or a subscriber, from respectively bit I of Par.12 (in IAM), and bit M of Par.3 (in ACM). In the call of Fig. 11.3-1, both parties are ISDN users. In calls between an ISDN user and a subscriber, the ISDN user is informed by his local exchange that the distant party is non-ISDN. This is done by including an IE.11 (progress indicator) in a Q.931 message to the ISDN user (Section 11.4).

Information Included in DISC and REL Messages. The DISC and REL messages always include respectively an IE.6 (cause) and Par.8 (cause indicators). In the example of Fig. 11.3-2, the location and cause values indicate "public network of the remote user" and "normal call clearing" (10.3.5).

11.4 CALLS INVOLVING ANALOG SUBSCRIBERS

At present, the number of ISDN users is still very small, and the majority of calls carried on ISUP trunks are between two analog subscribers. This section explores ISUP signaling for calls in which one or both parties are subscribers.

11.4.1 Call between Two Subscribers

Figure 11.4-1 shows a call between subscribers, in which trunks T_1 and T_2 have ISUP signaling.

Set-up of Connection. Subscriber S_1 has originated a call. We first examine the IAM sent by originating exchange P—see Table 11.2-2. Par.5 holds the called party number dialed by S_1. Par.1 (access transport) and Par.24 (user service information) are absent because subscriber signaling does not provide bearer capability, subaddresses, and compatibility information.

The originating exchange cannot determine whether the call will be speech or

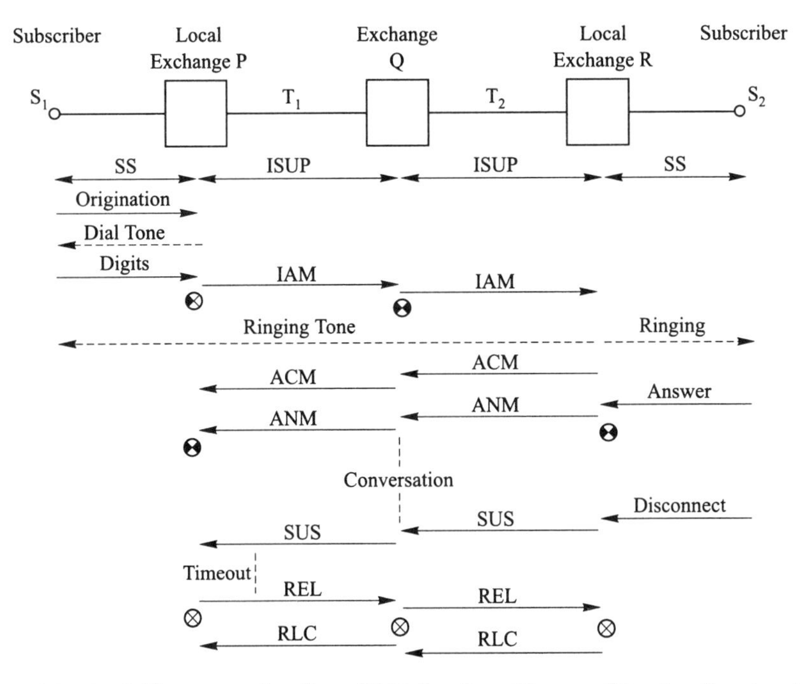

Figure 11.4-1 Call between subscribers. ISUP signaling all the way. SS: subscriber signaling.

3.1 kHz audio. Local exchanges are usually programmed to set Par.23 (transmission medium requirements) to "3.1 kHz audio."

The calling party category (stored in memory at the exchange) is entered into Par.6. We assume here that the local exchanges are programmed to also include Par.7 (calling party number).

In Par.12 (forward call indicators), bits D, F, and I are set to "no interworking encountered," "ISUP signaling used all the way," and "calling party has non-ISDN access." In calls originated by subscribers, bits HG are usually set to "ISUP signaling preferred all the way." In this example, exchange Q selects outgoing trunk T_2, which has ISUP signaling, and does not change the values of Par.12 in its IAM to exchange R.

When a call set-up arrives at its terminating exchange on an ISUP trunk, the exchange always checks whether the called party is compatible with the service indicated in Par.23 (transmission medium requirements). In this example, the called party is a subscriber, and is compatible with the "3.1 kHz audio service." Exchange R therefore proceeds with the call set-up.

Assuming that S_2 is idle, it rings the subscriber line, connects trunk T_2 to a ringing-tone source, and sends an ACM message to exchange Q, which repeats it to exchange P.

In the ACM, bits DC, FE, and M of Par.3 (backward call indicators) are set to "subscriber free," "subscriber," and "called party has non-ISDN access." Bits J and K are copied from bits D and F in Par.12 (forward call indicators) of the received IAM, and indicate "no interworking encountered" and "ISUP signaling all the way." Exchange P is therefore aware that signaling is ISUP all the way, and that the called party is non-ISDN (it may need this information for later call control actions).

Since exchange R receives the calling party number (Par.7) in the IAM, it can provide CLASS services to the called subscriber (Section 3.5.2).

When exchange R receives an answer signal from S_2, it cuts through the path between T_2 and the called subscriber line, and sends an ANM message. When the message reaches exchange P, it cuts through its forward transmission path, and the conversation or data transmission can start.

Release of the Connection. Figure 11.4-1 assumes that the called subscriber disconnects first. Called subscribers have a grace period that allows them to disconnect the telephone on which they answered the call, and to pick up another telephone, without causing the release of the connection (Section 3.1.1).

Therefore, when S_2 disconnects, exchange R sends a SUS message (a REL message would initiate the release of the network connection). On receipt of the message, exchange P starts a timer with a timeout of say 30 s. If S_2 reanswers before the timer expires, exchange R sends a RES (resume) message, exchange P stops the timer, and the call continues. In the example, exchange P has not received a RES message when the timer expires, and initiates the release of the connection.

11.4.2 Calls between a User and a Subscriber

We now consider calls between a user and a subscriber. The call configuration is as in Fig. 11.4-1, except that one party is a subscriber, and the other party is a user. Since the connection has ISUP signaling all the way, terminating exchange R always knows whether the remote (calling) party is a subscriber or ISDN user, from bit I of Par.12 (forward call indicators) in the IAM message. Likewise, originating exchange P knows the type of the remote (called) party from bit M of Par.3 (backward call indicators) in the ACM message.

Since some user actions depend on whether the distant party is a user or a subscriber, a local exchange always informs its user if the distant party is non-ISDN. This is done by including an IE.11 (progress indicator), in the CONN message to the TE of a calling user, or in the SETUP message to the DSL of the called user. The TE then presents this information to its user.

In calls from a subscriber to a user, the IAMs do not include a Par.1 (access transport), which contains IEs for called TE selection. If the called user has terminals of several types, the only method for the selection of a called TE is by multiple directory numbers (10.4.3).

11.4.3 Connections with Signaling Interworking

Figure 11.4-2 shows two connections in which one trunk has ISUP signaling, and the other trunk has multi-frequency (MF) signaling (Chapter 4). Connections involving MF-ISUP interworking can be used for calls originated by a subscriber, and for calls originated by a user who has requested a speech or 3.1 kHz audio call, and has not included information elements in his SETUP message that require ISUP signaling.

In these calls, a calling or called TE is always notified by its local exchange that the signaling on the connection is not ISUP all the way, and that therefore no information is available regarding the type (subscriber or user) of the distant party. The notification is in a message with an IE.11 (progress) set to "not ISUP signaling all the way."

In example (a), exchange P has indicated—in bits HG of Par.12 (forward call indicators)—that ISUP signaling all the way is preferred, or not required, and exchange Q has selected an outgoing MF trunk T_2. After seizing the trunk, Q returns an ACM for trunk T_1 in which bit K of Par.3 (backward call indicators) is set to "not ISUP signaling all the way." Exchange P then notifies TE_1 with a PROG message that includes the IE.11. Also, since the called party is an ISDN user and the call has arrived at exchange R on a MF trunk, the exchange includes the IE.11 in its SETUP broadcast message on the called DSL.

In example (b), exchange P has selected an outgoing MF trunk, and informs TE_1 with a PROG message that includes the IE.11. Moreover, exchange Q sets bit F of Par.12 (forward call indicators) in its IAM to "not ISUP signaling all the way", and exchange R then includes the IE.11 in its SETUP message.

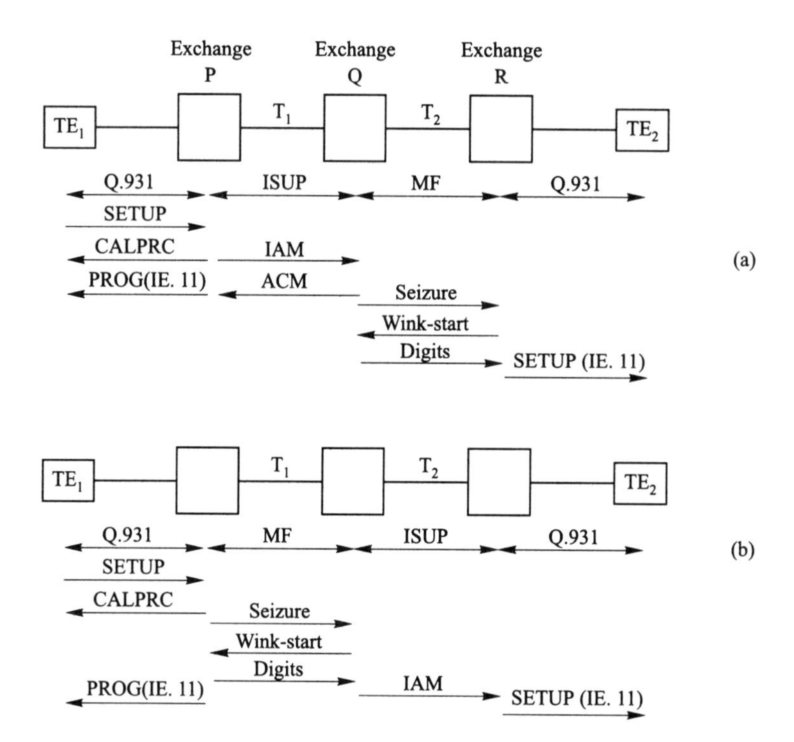

Figure 11.4-2 Connections with ISUP-MF interworking.

11.5 END-TO-END SIGNALING

ISUP signaling includes procedures for end-to-end signaling between the originating and terminating exchanges of a connection. The applications of ISUP end-to-end signaling are similar to those of CCITT R.2 and TUP (Sections 4.4.4 and 4.4.5). ISUP end-to-end signaling is possible only on connections in which all trunks have ISUP signaling. In this section, we explore this signaling in a connection from originating exchange A to terminating exchange C, and routed via intermediate exchange B.

ISUP has two methods for end-to-end signaling: the pass-along method, and the SCCP (signaling connection control part) method [2, 6]. In some networks, only one of these methods is implemented; in other networks, both methods exist side by side.

11.5.1 Information Request and Information Messages

In ISUP end-to-end signaling, the exchange at one end of the connection requests information, or an action, with an *information request* (INR) message. The exchange at the other end responds with an *information* (INF) message.

The INR messages include a mandatory Par.14 (information request indicators) whose bits represent individual information items and actions. The bits indicate that the corresponding information or action is being requested (1), or not requested (0). For example, the terminating local exchange can request the originating local exchange to hold the connection until the terminating exchange send a DISC message.

The INF messages include a mandatory Par.13 (information indicators) whose bits show whether a requested information item or action is provided. When an information item is provided, it is included as an optional parameter in the message.

11.5.2 Call Indicators

Early in the call set-up, the calling and called exchanges inform each other about their end-to-end signaling capabilities. The originating local exchange indicates its capabilities with bits C and B of Par.12 (forward call indicators) in its IAM, and the terminating local exchange does the same with bits H and G of Par.3 (backward call indicators) in its first backward message (CPG, ACM, or ANM) for the call.

11.5.3 Pass-along Method

In this method, the originating and terminating local exchanges place their outgoing INF and INR messages in a *pass-along message* (PAM), shown in Fig. 11.5-1. Octet x identifies the message as a PAM message, and octet y indicates the type of embedded message.

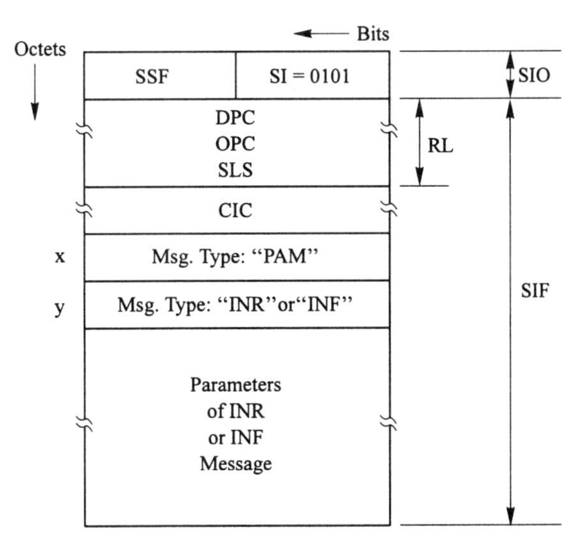

Figure 11.5-1 Pass-along message. SIO: service information octet. SIF: signaling information field. RL: routing label. CIC: circuit identification code.

Pass-along messages are transferred—forwards or backwards—by the exchanges along the connection that has been established for the call, just like other ISUP messages. Figure 11.5-2 shows the transfer of a PAM from originating exchange A to terminating exchange C, for a call routed on trunks T_1 and T_2.

The point codes of exchanges A, B, and C are a, b, and c. The circuit identification codes of the trunks are cic_1 and cic_2. The figure shows only the parameters DPC (destination point code), OPC (originating point code), and CIC (circuit identification code), in the MTP-transfer primitives.

ISUP-A passes the PAM to MTP-A, in a MTP-transfer request that includes the parameters DPC = b, OPC = a, and CIC = cic_1, which identify trunk T_1. The MTP at exchange B passes the received message to ISUP-B. A call record at the exchange shows that T_1 is connected to T_2. ISUP-B does not examine the contents of the embedded INF or INR message, and returns the PAM to MTP-B, with parameters DPC = c, OPC = b, and CIC = cic_2, which identify trunk T_2. MTP-C passes the message to ISUP-C, which then knows that the message relates to the call on trunk T_2.

11.5.4 End-to-end Signaling, SCCP Method

In the SCCP method, ISUP uses the services of SCCP, which is a part of signaling system No.7, and is described in Chapter 14. At this point, we only need to know that messages between ISUP and SCCP are passed in N-unitdata primitives, which include the point code of the message destination.

Figure 11.5-3 shows the transfer of a message from originating exchange A to terminating exchange C. ISUP-A passes its outgoing INF or INR message to the SCCP-A in a N-unitdata request that includes a *subsystem number* (SSN =

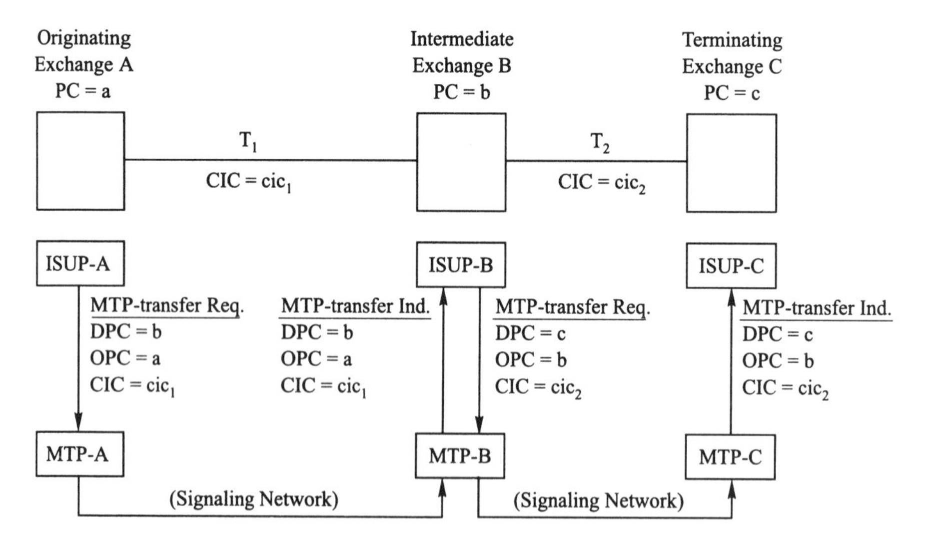

Figure 11.5-2 End-to-end signaling, pass-along method.

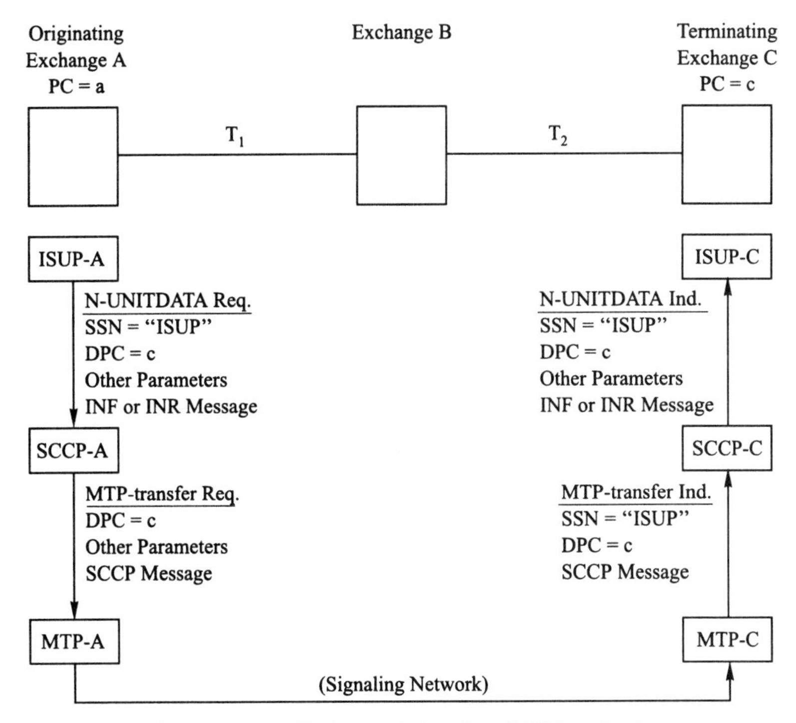

Figure 11.5-3 End-to-end signaling, SCCP method.

"ISUP"), which identifies the message as an ISUP message, the point code of the destination exchange (DPC = c), and the INF or INR message. SCCP-A embeds the message in a SCCP message, and passes it to MTP-A, in a MTP-transfer request that includes the message, and the point code of exchange C.

MTP-C passes the received SCCP message to SCCP-C, which extracts the embedded INF or INR message. Because SSN = "ISUP," SCCP passes the message to ISUP-C, with a N-unitdata indication.

Call References. In SCCP end-to-end signaling, exchanges A and C have to know each other's point codes, and need a mechanism to identify the call to which a INF or INR message pertains. This is done by including a Par.4 (call reference) in the message. The parameter consists of a point code (PC) and a call identity (CID).

When exchange A sends its IAM for the call, it assigns a call identity (CID = x), and includes a Par.4 with PC = a, and the assigned CID (Fig. 11.5-4).

When exchange C receives the IAM, it associates A's call reference with the call. It also assigns a call identity (CID = y) and, in its first backward message, includes a Par.4 with PC = c, and the assigned CID. On receipt of the message, exchange A associates C's call reference with the call.

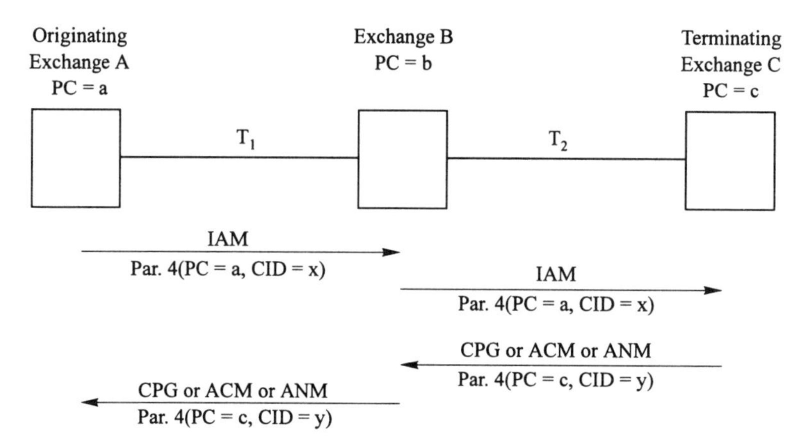

Figure 11.5-4 Establishing call references.

Exchanges A and C now know each other's call references. Then, if exchange A needs to send an INR message, it includes a Par.4 with C's call reference, and passes the message to its SCCP with a N-unitdata request, identifying the message destination by DPC = c. Likewise, exchange C responds with an INF message that includes a Par.4 with A's call reference, and passes it to its SCCP, identifying the message destination by DPC = a.

11.6 OTHER SIGNALING PROCEDURES

This section outlines several ISUP signaling procedures which have not yet been discussed. Some of these are the counterparts of the TUP procedures described in Sections 9.3 and 9.5.

11.6.1 Continuity Checking

As in TUP signaling, an exchange performs a continuity check on a seized outgoing ISUP trunk under certain conditions. When an exchange, say exchange A, seizes an outgoing trunk to exchange B, indicators in Par.15 (nature of connection indicators) of its IAM specify whether the continuity check will be made. If so, exchange B establishes a loopback on the incoming trunk.

When exchange A has made a continuity check, it sends a COT (continuity) message to exchange B. In the message, Par.10 (continuity indicator) indicates whether the check has been successful.

11.6.2 Calling Line Identity Presentation

This service, in which the calling party number is presented to the calling party, can be provided in two ways.

A telecom may program the exchanges in its network to always include Par.7 (calling party number) in their IAMs.

Alternatively, the local exchanges in a network can be programmed to obtain the calling party number on calls that terminate at subscribers who have the service. The terminating exchange then sends an information request (INR) message to the originating exchange, using end-to-end signaling. The message contains a Par.14 (information request indicators) which is set to request the calling party number. The originating exchange returns an information (INF) message that includes a Par.13 (information indicators) and a Par.7 (calling party number).

11.6.3 Closed User Group Service

In this service, the local exchange of the called subscriber allows or disallows the incoming call, based on closed user group (CUG) information about the calling and called parties (Section 9.4.7). The CUG information of the calling party is transferred by the IAM. Par.9 (closed user group interlock code) specifies the caller's GUG, and Par.17 (optional forward call indicators) indicates whether the calling party is allowed to access a party outside his GUG. The GUG information about the called party is available at the terminating local exchange.

11.6.4 User–user Signaling

In countries that offer this service, an ISDN user can exchange user-to-user information during the set-up and release of a connection, by including a user–user IE in a Q.931 SETUP, ALERT, CONN and/or DISC message (10.3.5).

A local exchange maps the IE.14 received in a Q.931 message into a Par.25 (user-to-user information, UUI), and includes the parameter in a sent ISUP message (Fig. 11.6-1). The other local exchange maps a received Par.25 into an IE.14, and includes it in a Q.931 message.

11.6.5 Call Forwarding

Figure 11.6-2 is an example of call-forwarding in a network with ISUP trunks. Customer C_1 (an analog subscriber or ISDN user) initiates a call to customer C_2, who is currently fowarding his incoming calls to customer C_3. In turn, this customer is forwarding his incoming calls to C_4. The numbers of customers C_2, C_3, and C_4 are N_2, N_3, and N_4.

We examine the parameters relating to call forwarding in the IAMs sent by exchanges A, B, and C. The called party number (Par.5) is the number as perceived by the exchange sending the IAM. The original called party number (Par.8) is the number dialed by S_1. The redirecting number (Par.19) is the number of the redirecting (forwarding) party. Finally, the number

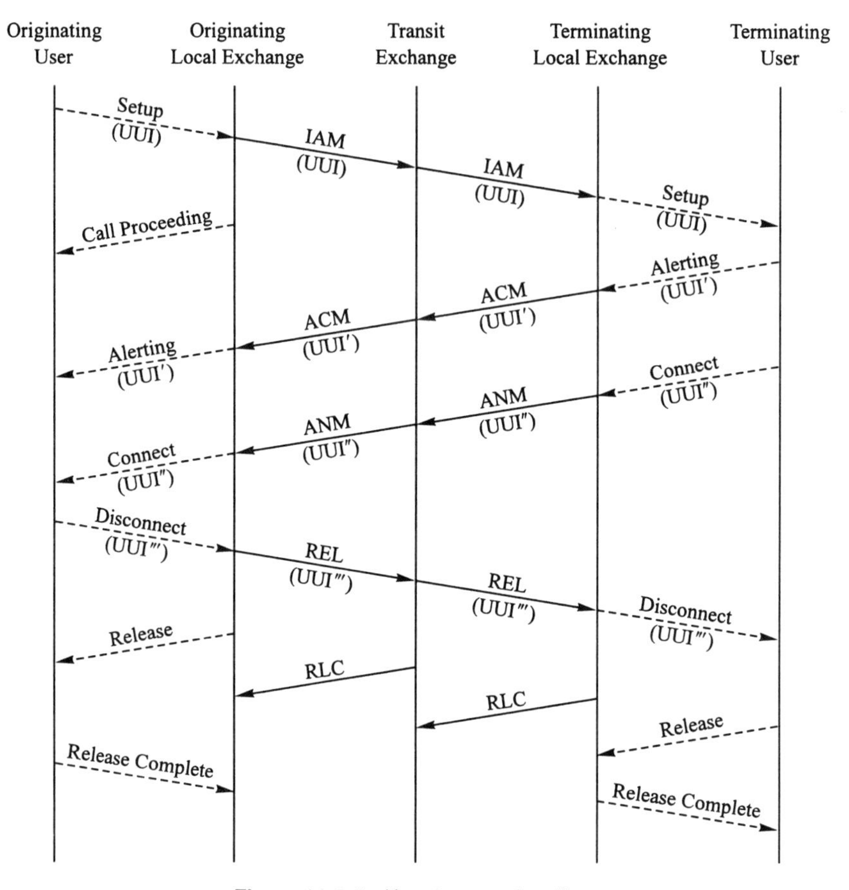

Figure 11.6-1 User-to-user signaling.

of redirections (an item in Par.20) shows how often the call has been redirected.

This information is used by the exchange that receives a forwarded call, to decide whether to forward the call again. In the first place, the call should not be forwarded to the original called number, or to the redirecting number. Also, the exchange may be programmed to not forward the call if it already has undergone a predetermined maximum number of redirections.

The forwarding exchange also sends a backward call progress (CPG) message, which is repeated to originating exchange A. The message includes a Par.11 (event information) which indicates the reason why the call has been forwarded, and a Par.19 (redirecting number) which represents the customer who is requesting the forwarding. If the calling customer is an ISDN user, the originating exchange sends this information to the customer, in a Q.931 message.

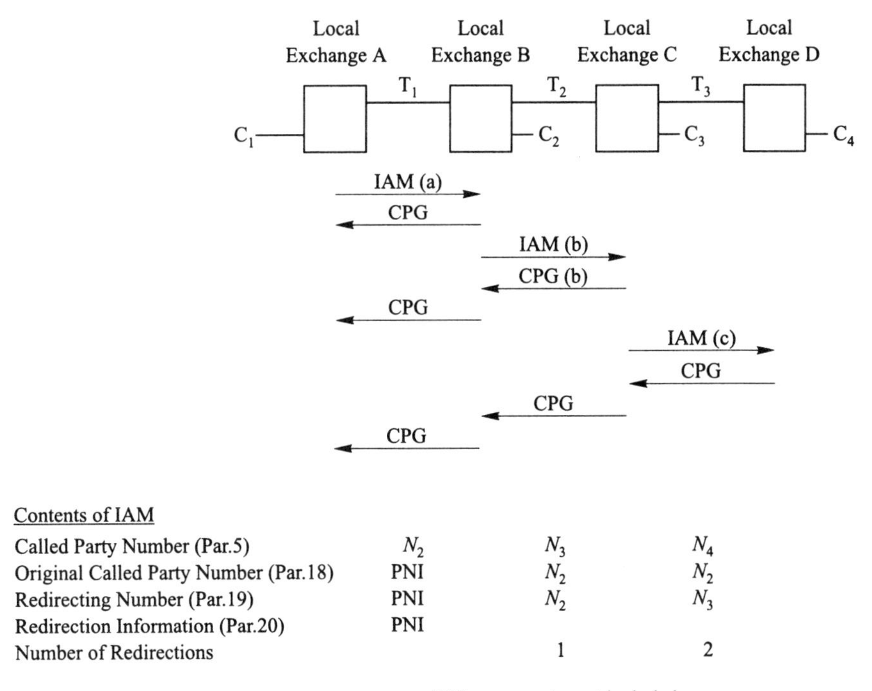

Figure 11.6-2 Call forwarding.

11.6.6 Congestion at Exchange

The procedure is the same as in TUP (Section 9.5.1). When an exchange A is overloaded, it informs the concerned exchanges (exchanges with ISUP trunk groups to A), by including a Par.2 (automatic congestion level) in the REL (release) messages sent for these trunks. Par.2 can indicate a moderate or a severe congestion.

On receipt of a congestion indication, the concerned exchanges temporarily reduce, or suspend, the seizures of trunks to exchange A.

11.6.7 Circuit Supervision

ISUP includes procedures and messages that request a circuit (trunk) to be blocked, unblocked, or reset, and messages that acknowledge these requests. They correspond to the TUP messages discussed in Section 9.5.2. The message acronyms and type codes are listed in Table 11.6-1. No parameters are included in these messages.

Table 11.6-1 Circuit supervision messages.

Acronym	Name	Message Type Code
BLA	Blocking acknowledgment	0001 0101
BLO	Blocking	0001 0011
CCR	Continuity check request	0001 0001
RSC	Reset circuit	0001 0010
UBA	Unblocking acknowledgment	0001 0110
UBL	Unblocking	0001 0100

Source: Rec. Q.763. Courtesy of ITU-T.

11.6.8 Circuit Group Supervision

The messages and procedures for blocking, unblocking, and resetting trunks correspond to those in TUP (Section 9.5.4). The names, acronyms, and type codes of the messages are listed in Table 11.6-2. The message parameters are outlined below:

Range and Status is a mandatory parameter in all circuit group supervision messages. It consists of two subfields (Fig. 11.6-3). The *range* subfield contains a number R which indicates the number of circuits in the affected group. The affected circuits are specified by R (range), and CIC (circuit identification code). If CIC = c and R = r, the potentially affected circuits are those whose CICs are in the range from c through c+r.

In the CCITT and U.S. versions of ISUP, the maximum values of R are 31 and 23, respectively. In this way, all trunks in a E1 (European) or T1 (North American) first order digital multiplex system (Section 1.5.2) can be blocked or unblocked simultaneously.

The *status* field has a variable length, and has a status bit for the individual circuits. Bits 1, 2, ..., 8 in octet 2 represent circuits with CIC = c, c+1, ..., c+7; bit 1 in octet 3 represents the circuit with CIC = c+8, and so on.

Table 11.6-2 Circuit group supervision messages.

Acronym	Name	Message Type Code
CGB	Circuit group blocking	0001 1000
CGBA	Circuit group blocking acknowledgment	0001 1010
CGU	Circuit group unblocking	0001 1001
CGUA	Circuit group unblocking acknowledgment	0001 1011
GRS	Circuit group reset	0001 0111
GRA	Circuit group reset acknowledgment	0010 1001

Source: Rec. Q.763. Courtesy of ITU-T.

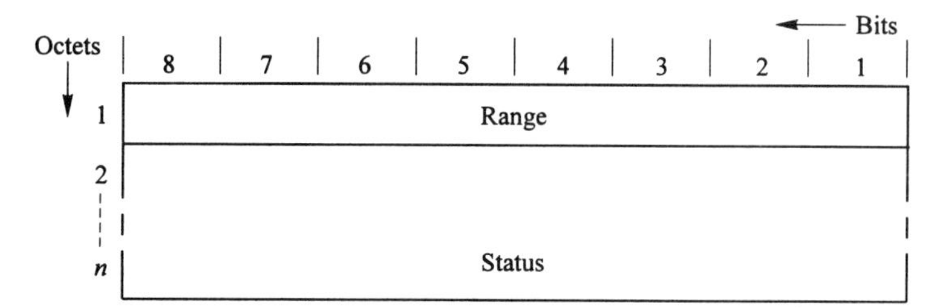

Figure 11.6-3 Range and status parameter.

In the circuit group blocking and unblocking messages (CGB and CGU), a status bit with value "1" indicates that the corresponding trunk should be respectively blocked or unblocked. In the circuit group blocking and unblocking acknowledgment messages (CGBA and CGUA), a status bit with value "1" indicates that the requested blocking or unblocking of the corresponding trunk is now in effect.

A circuit group reset request message (GRS) includes only the range octet of the range and status parameter. When exchange B receives a GRS message with CIC = c, and R = r, it releases all calls on circuits with CIC values c through c + r.

A group reset acknowledgment message (GRA) includes a complete range and status parameter; the bits in the status field again represent individual trunks in the group specified by the values of CIC and R. Exchange B sets the status bits of a trunk to "1" if it is currently blocking the trunk at its end. This informs exchange A which trunks cannot be seized for new calls.

Circuit Group Supervision Message Type. This is a parameter that appears in the messages CGB, CGU, CGBA, and CGUA (circuit group blocking and unblocking, and their acknowledgments) only. The parameter has a one-octet value field, in which bits 2 and 1 indicate the reason for the blocking or unblocking.

Bits 2,1 = 0,0: Blocking for maintenance reasons. No trunk should be seized for new calls, but existing calls are allowed to continue

Bits 2,1 = 0,1: Blocking because of a hardware failure. No new calls are allowed, and existing calls have to be released immediately

The "hardware failure" blocking and reset requests for a circuit group usually interrupt a number of calls. Therefore, as a safety measure against an undetected error in a CGB or GRS message, an exchange repeats a sent CGB or GRS message within 5 s. The other exchange only carries out the blocking or reset procedures if it has received two identical CGB or GRS messages within 5 s.

11.6.9 Signaling Route Set Congested or Unavailable

These procedures are the same as in TUP signaling (Section 9.5.5).

Signaling Route Set Congested. When the ISUP at an exchange receives a MTP-status indication for a destination (Section 8.9.3), it temporarily reduces the number of seizures of outgoing trunks in the routes to that destination.

Signaling Route Set Unavailable. On receipt of a MTP-pause indication for a destination (Section 8.9.5), ISUP clears all ISUP trunks to that destination, and suspends seizing these trunks for new calls until it receives a MTP-resume indication.

11.7 SIGNALING PROCEDURES FOR FAILED SET-UPS

The signaling procedures for interexchange calls that arrive on ISUP trunks at an exchange where a set-up failure occurs depend on whether the calling party is an analog subscriber or an ISDN user, and on whether the connection that has been set up so far has ISUP signaling all the way. This section explores a number of set-up failures.

11.7.1 Calls Originated by ISDN Users, ISUP Signaling All the Way

We first consider the case that the set-up of a call originated by a user at exchange P has reached the destination exchange R, and that one of the failures described in Section 10.6.1 occurs. The called party can be a subscriber or a user. From Par.12 (forward call indicators) of the IAM, exchange R knows that the interexchange signaling is ISUP all the way, and that the calling party is a user.

In most failures, terminating exchange R does not have to provide in-band information to the calling TE. It simply initiates the release of the connection (Fig. 11.7-1), and sends a REL message that includes Par.8 (cause). The coding of this parameter is the same as that of Q.931 information element IE.6—see Section 10.3.4. Originating exchange P bases its treatment of the calling TE on the value Par.8, which can indicate "user busy," "alerting, no answer," etc.

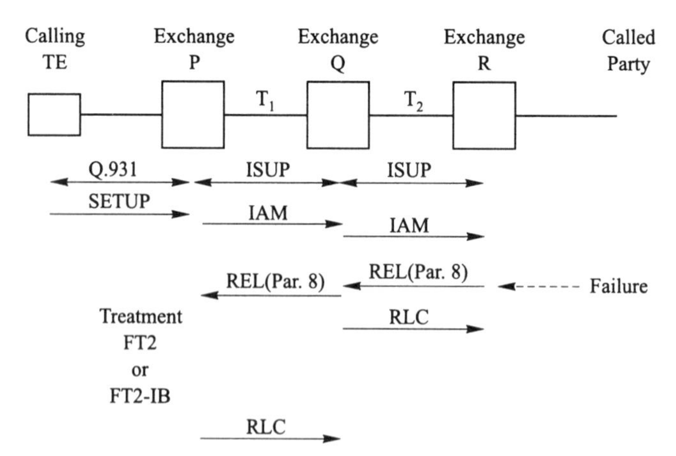

Figure 11.7-1 Failure procedure when exchange R does not provide in-band information.

On 64 kb/s calls, exchange P does not provide in-band information, and the calling TE receives failure treatment FT2 (10.6.1). On speech and 3.1 kHz audio calls, the treatment may require that exchange P provides in-band information (busy- or reorder-tone) to the calling TE (failure treatment FT2-IB).

On some failures of speech and 3.1 kHz audio calls, terminating exchange R provides ringing-tone, or an announcement, to the calling TE. Figure 11.7-2 shows the procedure for the case that exchange has determined that the called number is not assigned. The exchange connects trunk T_2 to an announcement source, and sends an ACM message that includes a Par.8 with location and cause values "public network serving the remote user" and "unassigned number." The message also includes a Par.16 (optional backward call indicators) with value "in-band tone or announcement now available."

Originating exchange P then gives treatment FT2-IB to the calling TE and, when the TE responds with a DISC message, starts the release of the network connection.

11.7.2 Calls Originated by Subscribers, ISUP Signaling All the Way

If the calling party is a subscriber, the procedures at the terminating exchange R are, with minor exceptions, as described above for speech or 3.1 kHz audio calls. On calls where the calling subscriber should hear busy-tone or reorder-tone, exchange R initiates the release of the connection. Exchange P provides the tone according to the information in Par.8 (cause) of the received REL message, releases trunk T_1 (as in Fig. 11.7-1), and awaits the disconnect signal from the calling subscriber. If the subscriber should hear ringing-tone or an announcement, it is provided by terminating exchange R (as in Fig. 11.7-2). When exchange P receives the subscriber's disconnect signal, it initiates the release of the connection.

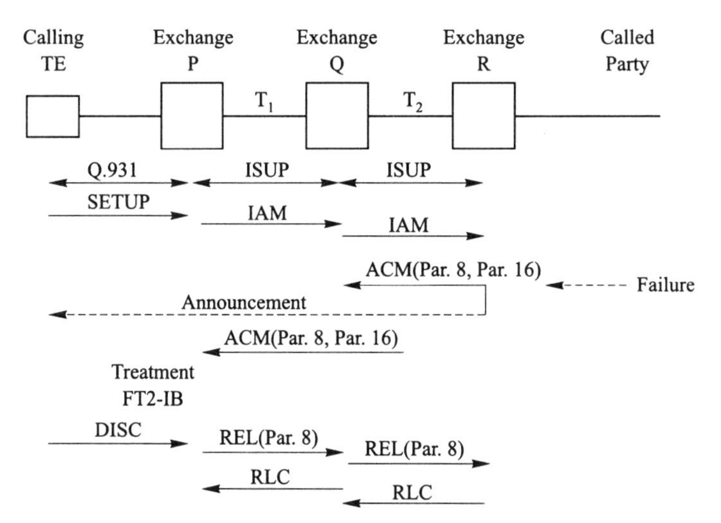

Figure 11.7-2 Unassigned number received by exchange R, on a speech or 3.1 kHz audio call.

11.7.3 Failure at Intermediate Exchange, ISUP Signaling All the Way

Suppose that the set-up of a call has arrived at intermediate exchange Q of Fig. 11.7-1, and that the set-up fails at that exchange. Trunk T_1 has ISUP signaling, and Par.12 (forward call indicators) in the IAM received by exchange Q thus indicates "ISUP signaling all the way."

On 64 kb/s calls, Q does not provide in-band information. It releases the connection, sending a REL message that includes Par.8 (cause indicators). Originating exchange P then sends a DISC message to the calling TE (treatment FT2), which includes an IE.6 (cause), in which the cause and general location information is copied from Par.8.

On speech and 3.1 kHz audio calls, exchange Q provides in-band information. It connects the incoming trunk to the appropriate in-band source, and returns an ACM message that includes Par.8, and a Par.16 (optional backward call indicators) that indicates "in-band information now available."

Typical cause values and locations for failures that occur at intermediate exchanges are (10.3.5):

	Cause Value
2	No route to specified transit network
3	No route to destination
34	No outgoing circuit (trunk) or channel available

	Location
2	Public network of local user
3	Transit network
4	Public network of remote user

11.7.4 Failures on Connections with Signaling Interworking

Finally, consider failures occurring at terminating exchange R, in the connection of Fig. 11.4-2(b). From Par.12 (forward call indicators) in the IAM, exchange R knows that signaling interworking has occurred, and that the call must therefore be a speech or 3.1 kHz audio call. Exchange R then connects trunk T_2 to an announcement or tone source. When exchange P receives a DISC message or a disconnect signal from the calling ISDN user or subscriber, it initiates the release of the network connection.

11.8 ISUP SIGNALING IN THE INTERNATIONAL NETWORK

11.8.1 Introduction

In the mid-1980s, representatives from AT&T, British Telecom International, and KDD (Japan), started to define a version of ISUP for use in the international network [7]. International ISUP signaling has now been installed in international switching centers (ISC) of several countries, and is being used for call control on a number of international trunk groups.

International ISUP, as documented in Recommendation Q.767 [8], is a subset of the CCITT Blue Book Recommendations Q.762–Q.764. Address signaling is *en-bloc* only (the IAMs always include the complete called party number). End-to-end signaling and call forwarding are not available.

11.8.2 Services Supported by International ISUP

In addition to the bearer services, and the transfer of parameters for the TE selection at the called ISDN user, international ISUP supports the following supplementary services: "calling line identity presentation," "user-to-user signaling," and "closed user group service." This support is limited to the transfer of parameters associated with these services across the international network.

11.8.3 Messages Not Provided

Of the messages listed in Sections 11.2.3, 11.6.7, and 11.6.8, international ISUP does not provide:

- Information (INF)
- Information request (INR)
- Pass along (PAM)

11.8.4 Parameters Not Provided

Of the parameters described in Sections 11.2.4 and 11.6.8, international ISUP does not provide:

- Call reference (Par.4)
- Information indicators (Par.13)
- Information request indicators (Par.14)
- Original called number (Par.18)
- Redirecting number (Par.19)
- Redirection information (Par.20)

11.8.5 Forward and Backward Call Indicators

Since international ISUP signaling does not support end-to-end signaling, bits H,G of Par.3 (backward call indicators), and bits C,B of Par.12 (forward call indicators) are always set to zero.

11.9 ISUP SIGNALING IN THE U.S.

This section outlines ISUP signaling in U.S. networks. The U.S. version of ISUP has been standardized by the American National Institute (ANSI) and Bellcore [9–12].

11.9.1 Services

The services supported by the U.S. version of ISUP are largely similar to those of CCITT-ISUP. They include call forwarding, and all services that require the transfer of the calling party number to the terminating exchange.

Not included are user-to-user signaling for ISDN users, and closed user group (CUG) service. However, U.S. ISUP supports CUG on international transit calls. For example, if an international switching center (ISC) on the west coast receives an IAM for a trunk from Japan (Fig. 11.9-1), the call destination is in the U.K., and the IAM includes a Par.9 and Par.17 (closed user group interlock code, and optional forward call indicators), these parameters are included in the IAMs for the trunks in the U.S., and in the IAM for the outgoing international trunk.

Some services and procedures provided in the U.S., but not in other countries, are outlined below.

11.9.2 Business Group Services [9]

In the U.S. a group of lines, or a private branch exchange (PBX), can be part of a *business group* (BG). Business groups can be established when requested by a business customer, a government agency, etc. Business groups have some similarities with closed user groups (CUG). At the request of individual

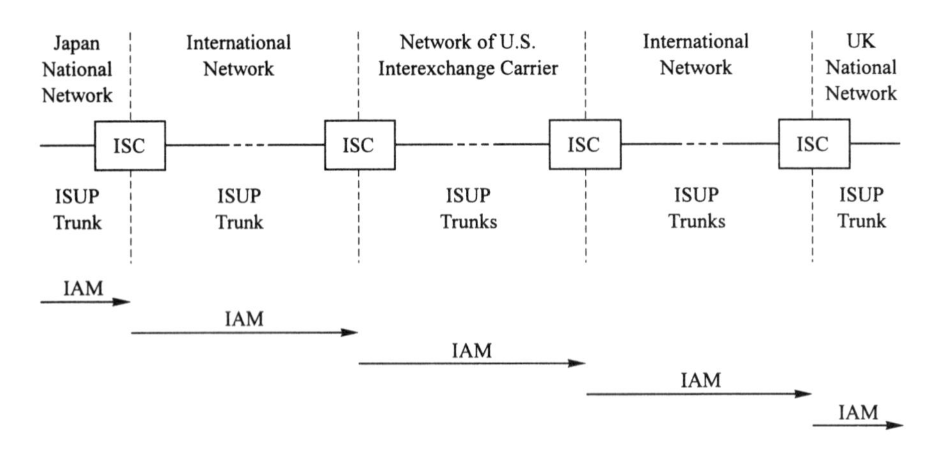

Figure 11.9-1 Call from Japan to the U.K.

business customers, the network can restrict calls to and/or from the lines of their BGs. A local exchange stores data about each attached BG line.

On calls originated by a BG line, these data are also needed at the terminating exchange. On these calls, the originating exchange therefore includes a "business group" parameter in its IAM—see Fig. 11.9.2.

The *attendant status* bit indicates whether the line is an attendant line (1) or not (0). The *BGID* bit is set to 0 for multi-location BGs, and to 1 if the BG is part of a private (corporate) network. The *party selector* indicates whether the parameter applies to the called, calling, original called, or redirecting number in the IAM. The *BG identifier* is a binary number that represents the BG. A business customer can also allocate *sub-group identifiers*, for example to individual locations of the BG.

The *LP* bit indicates whether the *privileges* of the line are network-defined (0), or defined by the BG customer (1). For network-defined privileges, bits 8–5 and 4–1 specify restrictions on originating and terminating privileges, respectively:

0000 Unrestricted
0001 Semi-restricted
0010 Fully restricted, intra-exchange
0100 Privileges denied

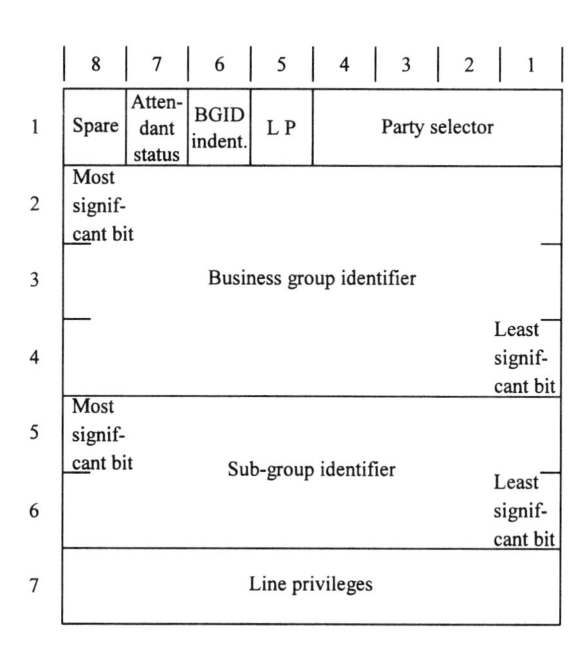

Figure 11.9-2 Business group parameter. (From ANSI T1.113–1992. Reproduced with permission of the Alliance for Telecommunications Industry Solutions, Inc.)

11.9.3 Transit Network Selection [9]

Long-distance calls in the U.S. are handled by an interexchange carrier (IXC) selected by the calling party (Section 3.7.1). If the caller's local exchange does not have a direct trunk group to an exchange in the IXC network, the IAM sent by the originating exchange includes a *transit network selection* parameter that identifies the IXC. The contents of the parameter are shown in Fig. 11.9-3 [9].

In the U.S., the *type of network identification* field is set to 010. This means that the identification is in accordance with a national network standard (as opposed to a CCITT standard). IXC networks have codes consisting of three or four digits. This is indicated by the *network identification plan* field:

0001 three-digit identification code
0010 four-digit identification code

The *circuit code* field shows whether the call is international, and whether operator assistance has been requested:

0000 national call
0001 international call, no operator requested
0010 international call, operator requested

8	7	6	5	4	3	2	1
Spare	Type of network identification			Network identification plan			
Digit 1				Digit 2			
Circuit code				Digit 3			

8	7	6	5	4	3	2	1
Spare	Type of network identification			Network identification plan			
Digit 2				Digit 1			
Digit 4				Digit 3			
Circuit code				Reserved			

Figure 11.9-3 Transit network selection parameter. (From ANSI T1.113–1992. Reproduced with permission of the Alliance for Telecommunications Industry Solutions, Inc.)

The parameter is removed from IAM by the exchange that seizes an outgoing trunk to an exchange in the network of the selected IXC.

11.9.4 Other Differences

In conclusion, we point out some messages and procedures that are particular to U.S. ISUP:

- All address signaling is *en-bloc*; subsequent address messages are not used.
- In IAMs of CCITT ISUP, Par.24 (user service information) is optional (included only on calls originated by ISDN users). It holds the bearer capability, and is transferred transparently by the network. Mandatory Par.23 (transmission medium requirements) is used by the network for the selection of outgoing trunks of the appropriate type (for example, digital trunks for 64 kb/s data calls).

  In the U.S., Par.23 is not used. Instead Par.24 is mandatory. In calls originated by subscribers, it is set to "3.1 kb/s audio." Par.24 is examined by the exchanges along the connection, to determine the type of outgoing trunk, and mapped into IE.1 (bearer capability) if the called party is an ISDN user.
- U.S. long-distance calls involve the local networks of the calling and called parties, and an IXC network. During the set-up of a call, the exchange in the originating local network that has seized an outgoing trunk to an exchange in the IXC network sends a backward *exit* message [9]. The message, which is repeated all the way to the originating exchange, confirms that the set-up in the originating network is complete. Likewise, when an IXC exchange has seized an outgoing trunk to an exchange in the terminating local network, it sends an exit message, which is repeated to all IXC exchanges along the connection.
- In addition to the Par.8 (cause indicator) values standardized by CCITT (Section 10.3.5), U.S. ISUP includes a number of values that have been specified by ANSI [9]. The coding standard (see Fig. 10.3.6) is set to 10. Some examples of cause values:

001	0111	unallocated destination number
001	1000	unknown business group
011	1110	call blocked by business group restrictions

11.10 ACRONYMS

ACM	Address complete message (ISUP)
ALERT	Alerting message (Q.931)
ANM	Answer message (ISUP)
ANSI	American National Standards Institute

Bellcore	Bell communications research
BG	Business group
CALPRC	Call proceeding message (Q.931)
CCITT	International Telegraph and Telephone Consultative Committee
CEPT	European Conference of Postal Administrations
CGB	Circuit group blocking
CGBA	Circuit group blocking acknowledgment
CGU	Circuit group unblocking
CGUA	Circuit group unblocking acknowledgment
CIC	Circuit identification code
CID	Call identity
CLASS	Custom local area subscriber services
CONACK	Connect acknowledgment message (Q.931)
CONN	Connect message (Q.931)
COT	Continuity message (ISUP)
CPG	Call progress message (ISUP)
CUG	Closed user group
DISC	Disconnect message (Q.931)
DPC	Destination point code
DSL	Digital subscriber line
EXM	Exit message (ISUP)
FDM	Frequency-division multiplex
FOT	Forward transfer message
GRA	Group reset acknowledgment
GRS	Circuit group reset request
IAM	Initial address message (ISUP)
IE	Information element (in Q.931 messages)
INF	Information message (ISUP)
INR	Information request message (ISUP)
ISC	International switching center
ISDN	Integrated services digital network
ISUP	ISDN user part of SS7
IXC	Interexchange carrier
LATA	Local access and transport area
MF	Multi-frequency signaling
MF	Mandatory fixed-length parameter
MTP	Message transfer part of SS7
MV	Mandatory variable-length parameter
NPIBS	Network-provided in-band information service
OP	Optional parameter
OPC	Originating point code
PAM	Pass-along message (ISUP)
Par.	Parameter (in ISUP messages)
PC	Point code
PCI	Protocol control indicator

PCM	Pulse code modulation
Q.931	Network-layer protocol of digital subscriber signaling system No.1
REL	Release message (ISUP)
RES	Resume message (ISUP)
RL	Routing label
RLC	Release complete message (ISUP)
RLCOM	Release complete message (Q.931)
RLSE	Release message (Q.931)
SAM	Subsequent address message
SCCP	Signaling connection control part of SS7
SETACK	Set-up acknowledgment message (Q.931)
SETUP	Set-up message (Q.931)
SI	Service indicator
SIO	Service information octet
SLS	Signaling link selector
SS7	Signaling system No.7
SSF	Sub-service field
SSN	Subsystem number
SUS	Suspend message (ISUP)
TE	Terminal equipment of ISDN user
TUP	Telephone user part of SS7
UM	User message
UUI	User-to-user information

11.11 REFERENCES

1. A.R. Modaressi, R.A. Skoog, "Signaling System No.7: A Tutorial," *IEEE Comm. Mag.*, **28**, no.7, July 1990.
2. R. Manterfield, *Common Channel Signalling*, Chapter 6, Peter Peregrinus Ltd, London, 1991.
3. *Rec. Q.761, Functional Description of the ISDN User Part of Signaling System No.7*, ITU-T, Geneva, March 1993.
4. *Rec. Q.762, General Function of Messages and Signals of the ISDN User Part of Signaling System No.7*, ITU-T, Geneva, March 1993.
5. *Rec. Q.763, Formats and Codes of the ISDN User Part of Signaling System No.7*, ITU-T, Geneva, May 1992.
6. *Rec. Q.764, Signalling System No.7 ISDN User Part Signalling Procedures*, ITU-T, Geneva, March 1993.
7. J.J. Lawser *et al.*, *Common Channel Signalling for International Services*, ISS Conf. Record, Stockholm, 1990.
8. *Rec. Q.767, Application of the ISDN User Part of Signalling System No.7 for International ISDN Interconnections*, ITU-T, Geneva, February 1992.

9. *American National Standard for Telecommunications—Integrated Services Digital Network (ISDN) User Part*, ANSI T1.113-1992, American National Standards Institute, N.Y.
10. *ISDN Access Call Control Switching and Signaling Requirements*, TR-TSY-000268, Issue 3, Bellcore, Morristown, N.J., 1989.
11. *Switching System Generic Requirements for Call Control Using the Integrated Services Digital Network User Part (ISDNUP)*, Tech. Ref. TR-NWT-000317, Issue 4, Bellcore, Morristown N.J., 1993.
12. *Switching System Generic Requirements Supporting ISDN Access Using the ISDN User Part*, Tech. Ref. TR-NWT-000444, Issue 3, Bellcore, Morristown N.J., 1993.

12

SIGNALING IN CELLULAR MOBILE TELECOMMUNICATIONS

Cellular mobile telecommunications is one of the most important telecommunication developments of the last decade. The technical concepts underlying this type of communications were developed by Bell Laboratories [1,2], and implemented in the *advanced mobile telephone service system* (AMPS). All current cellular systems in the U.S. are descendants of AMPS.

Early mobile stations (MS) were designed as car phones. They were too bulky to be carried around, and had to be powered by the battery of the car. Today, there are compact lightweight MSs with internal rechargeable batteries. They can be carried by hand, and are "personal" phones rather than car phones.

There are two groups of signaling procedures in cellular mobile telecommunications. This chapter describes the signaling between a MS and a cellular mobile network. The second group of signaling procedures involves various entities in a mobile network, and is discussed in Chapter 17.

Sections 12.1–12.6 of this chapter cover signaling in the AMPS system and its successors in the U.S. Sections 12.7–12.9 describe signaling in the *global system for mobile telecommunications* (GSM), which has emerged as the most important cellular system outside the U.S.

12.1 INTRODUCTION TO CELLULAR MOBILE NETWORKS

This section briefly describes some important aspects of cellular mobile networks [3–5].

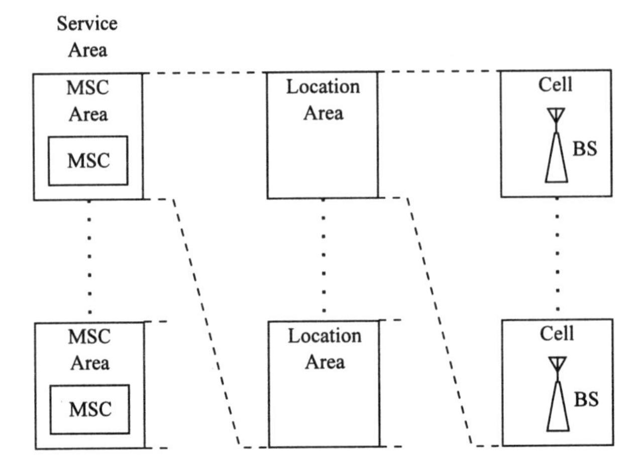

Figure 12.1-1 Cellular mobile network. (From EIA/TIA 553. Reproduced with permission of TIA.)

12.1.1 Definitions

A *cellular mobile network* (CMN) provides communication services for mobile stations that are operating in its *service area*. A service area typically covers a metropolis and its surrounding suburbs, or a number of medium-sized cities. The size of a service area is typically in the range of 100–4000 square miles.

The CMN service area is divided into a number of *MSC areas*, each of which contains an exchange known as a *mobile switching center* (MSC)—see Fig. 12.1-1. A MSC provides service to all MSs in its area. Some rural CMNs consist of just one MSC area. A MSC area is divided into a number of *location areas*, and each of these areas is divided into a number of *cells*. The cells are approximately circular, with radii that range from about 2 to 15 miles. Each cell has a *base station* (BS)—also known as *land station* and *cell site*—which houses radio-frequency (RF) transmitters and receivers.

A MSC has trunk groups (TG) to nearby exchanges in the public switched telecommunication network (PSTN) (also known as the "fixed" or "wire-line" network), and a base station trunk group (BSTG) to each base station—Fig. 12.1-2(a). When a CMN has several MSCs, there are also trunk groups (MSCTG) between these MSCs—Fig. 12.1-2(b).

A MS operating in a cell communicates on a RF channel with the BS of the cell. There are two channel types: voice channels and control channels.

Voice Channels. In a base station, each BS trunk is permanently wired to the transmitter and receiver of a RF voice channel. The combination of a BS trunk and its associated voice channel is the functional counterpart of a trunk in PSTN. A BS trunk and voice channel is assigned to a mobile at the start of a call, and released when the call ends. Figure 12.1-3 shows the connection for a call

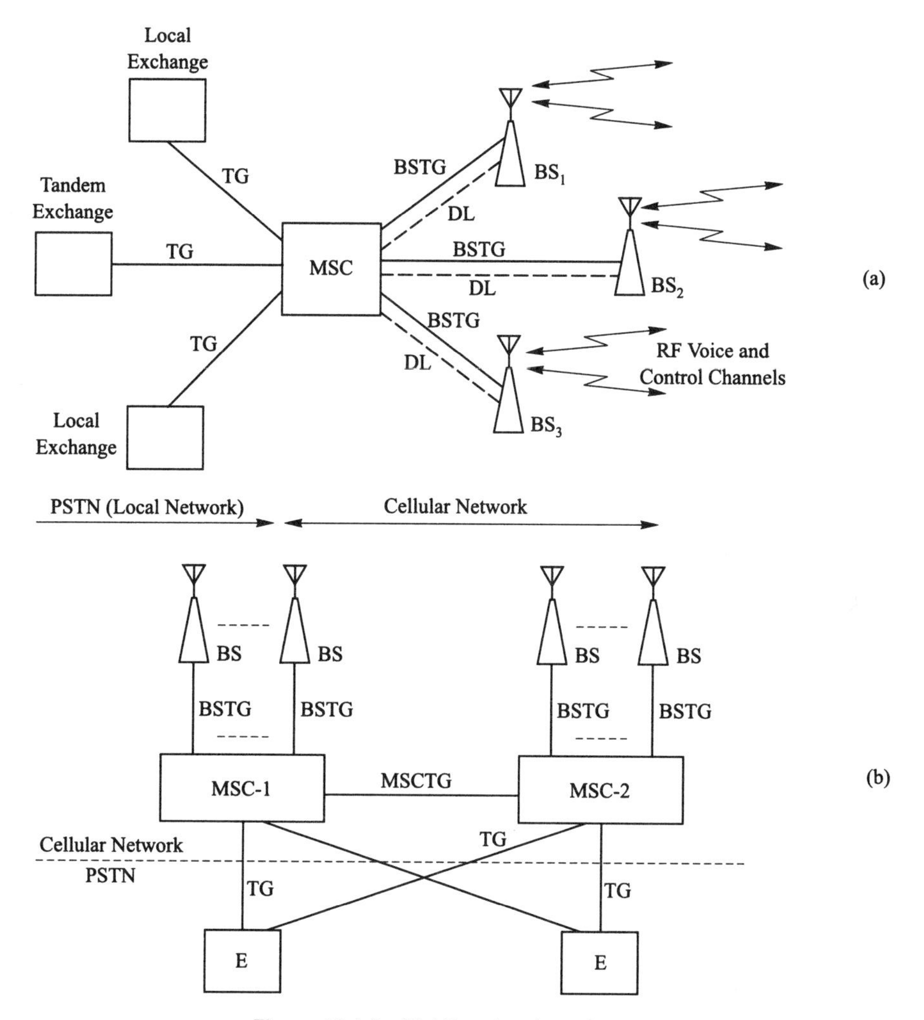

Figure 12.1-2 Mobile network equipment.

between a PSTN subscriber (S) and a mobile station (MS) operating in the cell of BS_1. The connection involves a path in the PSTN between S and exchange E, trunk T, a path—in the switchblock of MSC—between T and base station trunk (BST), trunk BST, and its associated RF voice channel (VC).

Control Channels. A base station has a pair (for reliability) of bidirectional data links (DL) to its MSC (Fig. 12.1-2(a)). In the BS, the data links are connected to the RF equipment of a group of RF channels, known as *control channels*. These channels carry signaling messages to and from mobiles that are operating in the cell when they are idle (turned "on," but not involved in a call). When a mobile is on a call, the voice channel that has been allocated to the call carries both speech and signaling.

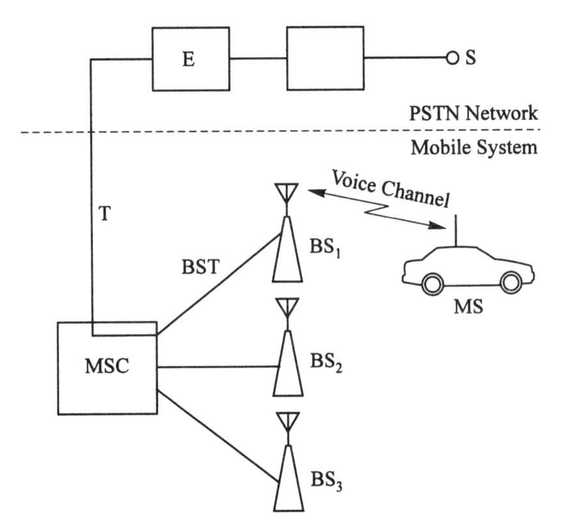

Figure 12.1-3 Connection between mobile (MS) and subscriber (S).

We distinguish three types of control channels. *Paging* channels carry "paging" messages that are sent from the MSC, to inform a mobile that it is being called. *Access* channels are used primarily by mobiles to originate a call, and to respond to a received paging message. *Combined* control channels are used for paging and accessing. A CMS is equipped either with separate paging and access channels, or with combined control channels.

12.1.2 AMPS Radio-frequency Channels

In mobile cellular literature, a channel is a bidirectional RF transmission facility, consisting of a *forward channel* for transmission from BS to MS, and a *reverse channel* for transmission in the opposite direction.

The forward and reverse channels of AMPS are analog channels, spaced at 30 kHz intervals. On the voice channels, speech is transmitted by frequency modulation. The digital messages on the voice and control channels are transmitted by frequency-shift keying, with a signaling speed of 10 kb/s.

The Federal Communication Committee, which controls the use of the RF spectrum in the U.S., originally allocated 40 MHz in the 850 MHz band to cellular mobile communications. This allowed 666 bidirectional channels in a service area. In 1987, the cellular spectrum was increased to 50 MHz, to accommodate 832 channels. Two competing cellular carriers—denoted as the "A" and the "B" carriers—are allowed to operate in a service area, using 416 channels each (395 speech channels and 21 control channels).

The 30 kHz channels are identified by *channel numbers*, which have been assigned as follows:

	"A" System	"B" System
Control channels	313–333	334–354
Voice channels	001–312 and 667–716 and 991–1023	355–666 and 717–799

The center frequency (f_c) of a forward channel can be determined from its channel number (N):

For N from 1 through 799:
$$f_c = (0.03\,N + 870)\ \text{MHz}$$

For N from 991 through 1023:
$$f_c = (0.03\,N + 839.31)\ \text{MHz}$$

The center frequencies of the reverse channels are 45 MHz below the center frequencies of their associated forward channels. When a MS is communicating on a particular channel, its receiver and transmitter are tuned to the center frequencies of respectively the forward and reverse channel.

12.1.3 Frequency Reuse

A key concept in cellular mobile systems is *frequency reuse*: the same RF frequency band (channel) is used in several cells of a cellular network. Without reuse, the maximum number of simultaneous calls in a CMN would be 395. Reuse greatly increases this number.

Frequency reuse is possible because, all other things being equal, the power of a received signal is roughly proportional to d^{-4}, where d is the distance between the transmitter and the receiver [3].

One can therefore allocate the same channels (frequencies) to cells that are sufficiently far apart from each other—say five cell radii (R). In Fig. 12.1-4, mobile MS_1 is operating in cell 1 and communicating on channel N with BS_1. MS_2 in cell 2, is communicating with BS_2 on the same channel. MS_1 receives a

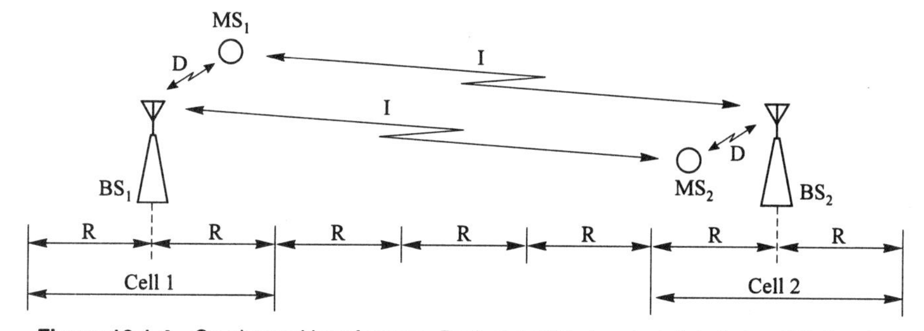

Figure 12.1-4 Co-channel interference. D: desired RF signals. I: interfering RF signals.

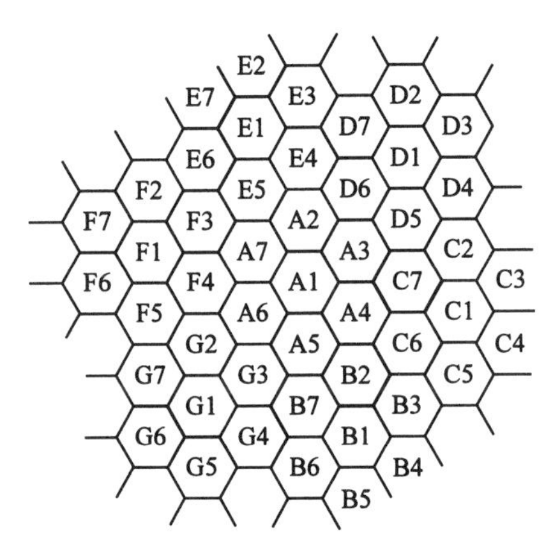

Figure 12.1-5 Seven-cell clusters.

desired signal from BS_1, and an interfering signal from BS_2. The minimum distance between MS_1 and BS_2 is $4R$, while the maximum distance between MS_1 and BS_1 is R. The transmit power at the base stations is equal. Therefore, the desired signal is approximately $4^4 = 256$ times stronger than the interfering signal. In frequency-modulated RF channels, this ratio of signal strengths is sufficient to suppress the effects of co-channel interference caused by BS_2. For the same reason, the co-channel interference caused by BS_1 does not affect MS_2. Mobiles MS_1 and MS_2 also transmit with equal power, and the effects of co-channel interference, from MS_1 to BS_2 and from MS_2 to BS_1, cause no problems at the base-station receivers.

A frequently used cell configuration is shown in Fig. 12.1-5. The cells are shown as hexagons for convenience only, and actual coverage areas of adjacent cells overlap slightly. Seven adjacent cells (for example, cells A_1 through A_7) form a cluster. The 395 voice channels and 21 control channels are divided into seven channel groups of about 56 voice channels and 3 control channels each. Cells A_1, B_1,...G_1 use channel group 1; cells A_2, B_2,...G_2 use channel group 2, and so on. In this way, there is no co-channel interference among cells of the same cluster.

Now consider cell A_1. The nearest cells that use the same channel group, and could cause co-channel interference, are B_1, C_1,...G_1. When the cell radius equals R, the minimum distance between two potentially interfering base stations in this configuration is $4.6R$, which is sufficient in actual systems, except during severe fading of the desired signal.

12.1.4 AMPS Color Codes and Supervisory Audio Tones

RF signals are subjected to short fades (decreases in received signal power), for example, when the MS moves through the "RF shadow" of a building, or because of multiple reflections. Suppose that mobile MS is in cell A_1 of Fig. 12.1-5, and is using voice channel N of the base station of that cell. During a fade, the strength of channel N received by MS from cell A_1 may fall below the strength of channel N received from cell B_1. The user of the mobile would then hear a short burst of speech from the call in cell B_1.

AMPS uses *color codes* and *supervisory audio tones* (SAT) to provide protection against problems of this nature. A color code is assigned to each cell. The codes can have three values: 00, 01, or 10, and each value is associated with a particular SAT frequency. Voice channels carry both speech and a SAT frequency.

Now consider the cells A_1, B_1,...G_1 of Fig. 12.1-5, all of which use the same channel group. The color codes and SAT frequencies could then be assigned as follows:

Color Code	SAT Frequency	Cells
00	5970 Hz	A_1
01	6000 Hz	B_1, D_1, F_1
10	6030 Hz	C_1, E_1, G_1

In this way, the color code of cell A_1 is different from the color codes of the cells in the adjacent clusters that use the same channel group, and could cause co-channel interference. Also, cells B_1 and D_1, who are nearby co-channel interferers for cell C_1, have color codes that are different from cell C_1, and so on.

All base stations transmit the SAT frequency of their respective cells on their forward voice channels. A mobile operating in A_1 expects to receive—and usually receives—the 5970 Hz SAT frequency of cell A_1.

However, during a fade it may receive a forward channel from an interfering BS in cell B_1,..., or G_1. When receiving an unexpected SAT frequency, the mobile mutes the received speech. The mobile user then experiences a silent interval, instead of a more disturbing burst of extraneous speech.

All mobiles transmit the received SAT frequency on their reverse voice channels. A base station therefore expects to receive the SAT frequency of its cell on its reverse voice channels. When a BS receives a different SAT frequency, it mutes the received speech, and the subscriber connected to a mobile is therefore also protected against extraneous speech.

The SAT frequencies are well above the highest transmitted speech frequency. The MS and BS receivers separate the speech and the SAT frequency with low-pass and high-pass filters, and no SAT tone is heard by the listener at the MS, or the party at the other end of the connection.

12.1.5 Cell Size

With seven-cell clusters, each cell can have a group of up to 56 voice channels. The load (the average number of simultaneous calls) on a group of 56 channels should not exceed 42 Erlang. Otherwise, the probability that all channels are busy when a new call has to be set up becomes unacceptably high. When the traffic density in a mobile system is T Erlang per square mile, and A is the coverage area of the cell (square miles), we have: $A < 42/T$. Assuming that the cell is approximately circular, its maximum radius R (miles) is:

$$R^2 < \frac{42}{(3.14)T}$$

For example, with $T = 2$, R should be < 2.6 miles, and therefore $A < 21$ miles^2. Call density T is usually high in metropolitan areas, and low in rural areas. Therefore, metropolitan systems usually have a large number (say, up to some 100) of small cells, and rural systems have a small number of large cells, which may be equipped with less than 56 voice channels.

12.1.6 AMPS Transmitter Power Levels

Consider a cell with a radius R, with a BS located at its center. The maximum distance between the BS and a mobile in the cell then equals R. In order to have a prescribed minimum signal power level at the receivers in the mobile and base stations when the distance between the BS and MS is at its maximum (R), the transmit power for a channel has to be approximately proportional to R^4. The power level of a BS transmitter can be set when it is installed, because cell radius R does not change. However, mobiles can operate in cells with different radii R, and their transmit power has to be adjusted accordingly.

This is done on command from the MSC that is serving the mobile. Each BS periodically measures the signal strength received on its reverse voice channels, and reports the results to its MSC. When MSC decides that a MS needs to change its transmit power, it sends a "change power" message that includes a *mobile attenuation code* (MAC) whose value ranges from 0 through 7, and represents the requested power level: MAC = 0 requests + 6 dBW, MAC = 1 requests + 2 dBW, and each next higher value reduces the requested power by another 4 dB.

Most of the power in a mobile station is consumed by its transmitter. The transmitters of mobile stations fall in power class I, II, or III, which have maximum power levels of +6, +2, and −2 dBW, respectively. Car-mounted mobile stations have class I transmitters. The power in hand-held stations has to be used sparingly to avoid frequent battery recharging, and these stations usually have class III transmitters. When a mobile receives a command for a power level that exceeds the capability of its transmitter, it simply transmits at its maximum power. Class III transmitters sometimes cause problems when the mobile operates near the edge of a large cell.

Some hand-held mobiles have a power-saving feature known as *discontinuous transmission* (DTX). When a mobile is in a call and its user is talking, the transmit power is as described above, but when the user is silent, the transmit power is reduced, usually by 8 dB.

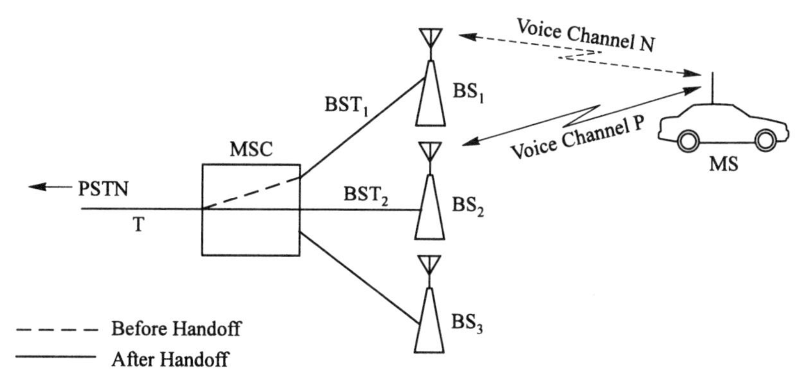

Figure 12.1.6 Handoff of a mobile.

12.1.7 Handoff

During a call, a mobile can move from one cell to an adjacent cell [3,6,7]. Suppose that in Fig. 12.1-6, mobile MS has started a call in the cell of BS_1. The connection involves trunk T to the PSTN, base-station trunk BST_1, and the associated voice channel N. When MS has moved to the cell of BS_2, it becomes necessary to hand it off to that base station.

The handoff decision is made by MSC, which periodically requests BS_1 to measure the strength of the signal received from the mobile (on reverse channel N). When the strength falls below a predetermined level, MSC requests the base stations in adjacent cells to measure the signal strength received on channel N. In this example, BS_2 reports the strongest signal, and is, therefore, in the best position to serve MS. MSC then seizes an idle voice channel P of BS_2, and the associated trunk BST_2. It then releases channel N and trunk BST_1, connects T with BST_2, and sends a "handoff" message that identifies the new channel P. The MS then tunes its transmitter and receiver to that channel.

12.1.8 Relationships of Mobile Station and Network

In the U.S., each cellular mobile service area is covered by two competing CMNs, known as the "A" and the "B" network. The owner of a MS selects one of these as his service provider. The selected network is known as the "home" CMN of the mobile.

When the MS operates in a location outside the service area of its "home" CMN, it can receive service from the A or B network that covers its current location. Also, when the mobile is in the service area of its "home" CMN and is unable to access the "home" CMN, it can access the competing CMN.

When a mobile is receiving service from a CMN other than its "home" CMN, we say that the mobile is "roaming," and is served by a "visited" CMN.

12.2 AMPS TONE SIGNALS AND MESSAGE WORDS

This section begins the description of the AMPS signaling protocol. The protocol is based on the original development by the Bell Laboratories [1,2], but includes a number of additions and modifications specified by the EIA/TIA (Electronics Industries Association/Telecommunications Industry Association)—see [6].

The signaling between mobile and base station is a combination of common-channel and channel-associated digital signaling messages, and a single-frequency signal.

The forward and reverse control channels (FOCC and RECC) carry signaling messages only. They are "common" channels, used for signaling between a base station and all mobiles in the cell that are active, but not involved in a call. The messages on these channels include a *mobile identification number* (MIN) that identifies a particular mobile.

A voice channel is allocated to a MS at the start of a call. Its forward and reverse channels (FVC and RVC) carry speech and channel-associated signaling messages (which do not include MIN).

12.2.1 Supervision Tone

This 10 kHz tone is sent on RVC only, and represents the on-hook (tone on) and off-hook (tone off) states of the mobile. When the mobile is in a conversation, on-hooks of 400 ms and 1.8 s indicate "flashes" and "disconnects," respectively.

12.2.2 Transmission of Messages

Messages consist of one or more words, transmitted at 10 kb/s. On forward channels (FOCC, FVC), the word length is 40 bits (28 information bits, followed by 12 parity bits for error checking). Words on reverse channels (RECC, RVC) have 48 bits (36 information bits, and 12 parity bits). Figure12.2-1(a), (b), and (c) show the general structure of transmitted word blocks for messages with two words, on respectively RVC, FVC, and RECC. In the figure, the fields in the message streams are denoted by acronyms, and the corresponding field lengths (number of bits) are shown below the acronyms.

Repeated Transmission. Each message word is repeated several times. This greatly reduces the probability that a word is completely missed by a receiver because of fading. Figure 12.2-1(a) shows that message words W1 and W2 on a RVC are repeated five times (W1-1,...W1-5; W2-1,...W2-5). Words on FOCC and RECC (c), (d), are also repeated five times, but words on FVC (b) are repeated 11 times, for reasons to be discussed later.

Error Checking. Error correction by retransmission (used on SS7 signaling links—see Sections 8.4 and 8.5) is not practical in mobile signaling. Errors in

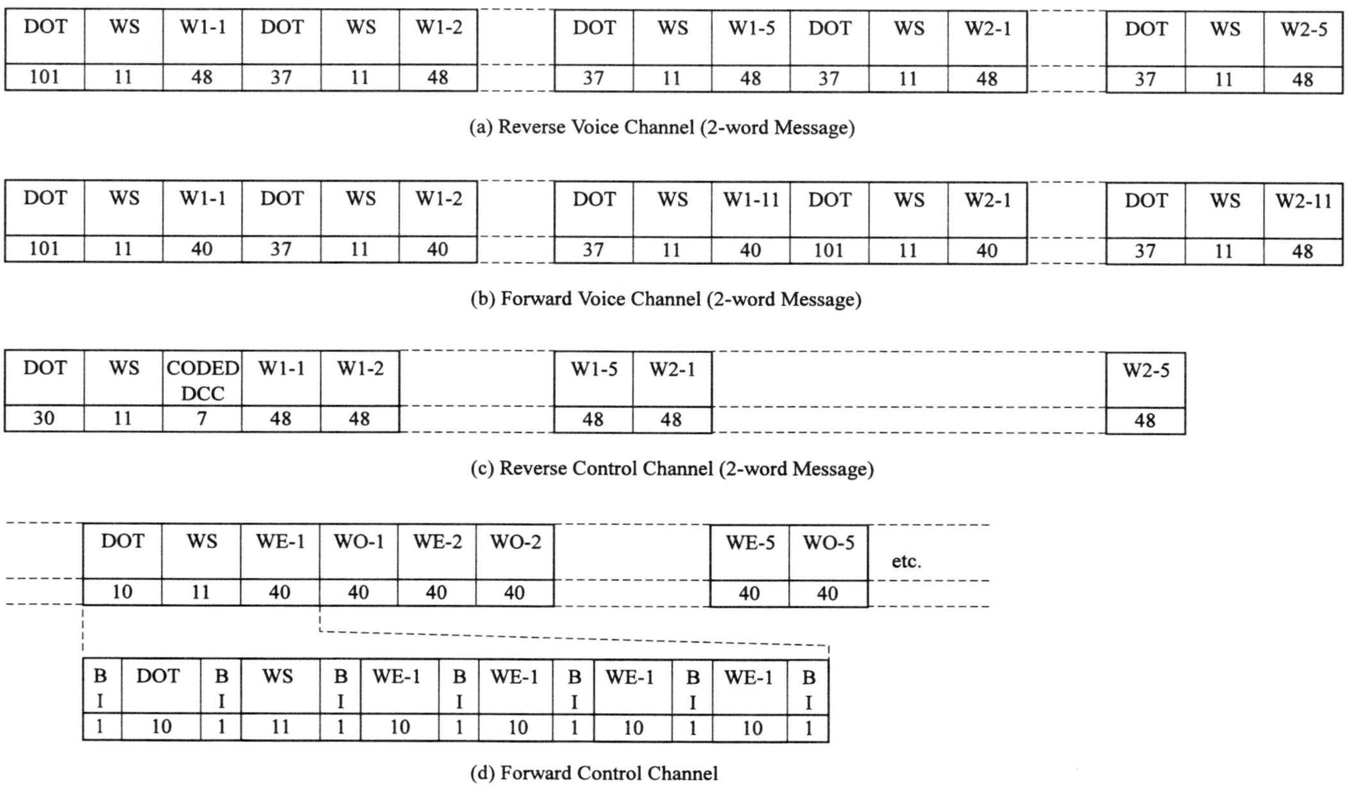

DOT	WS	W1-1	DOT	WS	W1-2	DOT	WS	W1-5	DOT	WS	W2-1	DOT	WS	W2-5
101	11	48	37	11	48	37	11	48	37	11	48	37	11	48

(a) Reverse Voice Channel (2-word Message)

DOT	WS	W1-1	DOT	WS	W1-2	DOT	WS	W1-11	DOT	WS	W2-1	DOT	WS	W2-11
101	11	40	37	11	40	37	11	40	101	11	40	37	11	48

(b) Forward Voice Channel (2-word Message)

DOT	WS	CODED DCC	W1-1	W1-2	W1-5	W2-1	W2-5
30	11	7	48	48	48	48	48

(c) Reverse Control Channel (2-word Message)

DOT	WS	WE-1	WO-1	WE-2	WO-2	WE-5	WO-5	etc.
10	11	40	40	40	40	40	40	

BI	DOT	BI	WS	BI	WE-1	BI	WE-1	BI	WE-1	BI	WE-1	BI
1	10	1	11	1	10	1	10	1	10	1	10	1

(d) Forward Control Channel

Figure 12.2-1 Transmission of message words.

received messages are minimized by taking advantage of the repeat transmissions of individual words. A MS or BS receiver first takes a majority vote of the corresponding bits in the N appearances of a word. It then checks the resulting "best guess" word for errors, using the parity (P) bits.

The sender of a message expects an acknowledgment in the form of a signal or a confirmation message. A failure to receive an acknowledgment within a specific time interval indicates that a problem has occurred.

Synchronization. Each message starts with a 10101... dotting sequence (DOT) that is used by the receiver for bit synchronization. The length of DOT depends on the channel type. DOT is followed by an 11-bit word synchronization pattern (WS): 11100010010. In voice-channel messages, DOT-WS sequences also appear between the message words.

Blank-and-burst. Messages on voice channels are sent as short data bursts (less than 0.1 s). While sending a message, the transmitter and receiver blank out the speech and tone signals.

Data Streams on Forward Control Channels—Fig. 12.2-1(d). FOCC channels continually transmit three interleaved data streams. The streams of message words WO and WE are read by mobiles with respectively odd and even MINs. Each word is again repeated five times (WO-1,...WO-5,...,etc.). The third stream consists of busy-idle (BI) bits that appear at the start of the DOT and WS sequences, and before the 1st, 11th, 21th, and 31th bit of each word. They indicate the status of the associated reverse control channel:

BI	RECC Status
0	Busy (receiving a message from a mobile)
1	Idle

A RECC is a common channel for messages sent by idle mobiles within a certain cell, or group of cells. To avoid "collisions" (simultaneous messages from two or more mobiles on the channel), a mobile that needs to send a message on a RECC first examines the BI bits on the associated FOCC. When the channel is idle, the mobile seizes the associated RECC, and starts to transmit its message. A mobile that finds the RECC busy (receiving a message from another MS) waits a random interval (0–200 ms) before trying again.

12.3 INTRODUCTION TO AMPS SIGNALING

This section describes a number of basic signaling procedures between the mobile and the cellular mobile network (CMN). The focus is on the messages in the radio-frequency channels between a mobile (MS) and a base station (BS).

The division of functions between the mobile switching center (MSC) and its base stations is implementation-dependent. We assume here that all "logic" of the CMN resides in the MSC, and that the BS merely transmits messages as directed by the MSC, and reports all received messages to the MSC. At this point, only a small number of message parameters are discussed.

Channel Number (CHAN). This identifies a voice channel.

Voice-channel Mobile Attenuation Code (VMAC). This indicates the power level at which the mobile should transmit on the voice channel.

System Identification (SID). This identifies a particular mobile system. By convention, the "A" and "B" cellular systems have odd and even SIDs, respectively.

Mobile Identification Number (MIN). A 10-digit national number that identifies a mobile. In the U.S., the numbering plan for mobile networks is integrated into the PSTN numbering plan. A MIN consists of a three-digit area and exchange codes AC–EC, followed by a four-digit "line number" LN (1.2.1). The AC–EC of a MIN identify the "home" MSC of the MS. Calls to a MS are routed by the PSTN to its home MSC.

Mobile Serial Number (MSN) or Electronic Serial Number (ESN). Uniquely identifies a mobile station.

A mobile station has variable, semi-permanent, and non-alterable memory devices. The SID of the "home" cellular system, and the MIN assigned to the mobile, are entered into semi-permanent memory by an agent of the "home" cellular system. The MSN of a mobile station is assigned by the manufacturer, and is stored in non-alterable memory.

12.3.1 Initialization

When a mobile is turned on, it has to establish contact with a cellular network. A mobile whose home is an "A" system first tries to establish contact with the "A" system that serves the area where the MS is located. If this fails, it tries the "B" system. Mobiles with "B" home systems first try to contact the "B" system.

In the example of Fig. 12.3-1 we assume that the mobile has an "A" home system. It starts by scanning the 21 dedicated forward control channels of "A" systems, and tunes to the strongest one (i.e., a control channel transmitted by the nearest BS).

All forward control channels of a CMN broadcast *overhead parameter* messages, at intervals of 0.8 s. These messages contain system-specific parameters. A mobile has to acquire this information before it can access (send messages to) a cellular system. Examples of system-specific information are: the

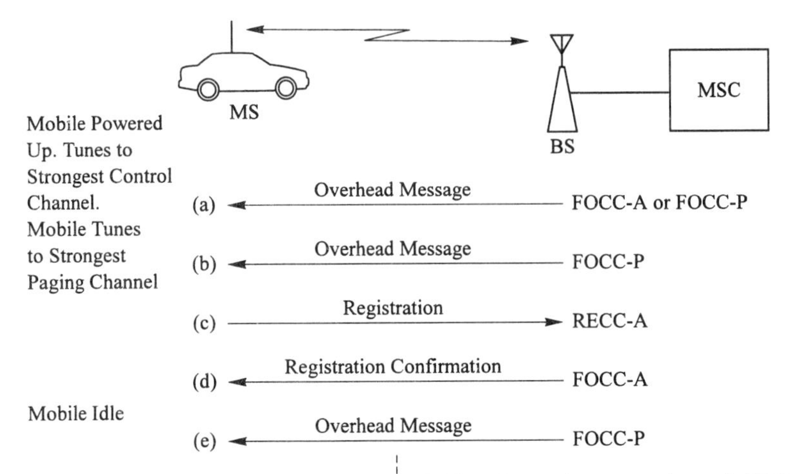

Figure 12.3-1 Initialization and registration. FOCC-A: forward access channel. FOCC-P: forward paging channel. RECC-A: reverse access channel.

SID of the CMN, data about its access and paging channels, and indicators that represent the characteristics and capabilities of the system.

When the mobile receives its first overhead message (a), it stores the received SID, and information about paging channels, in its variable memory. By comparing the received SID with the stored SID of its home system, it determines whether it is "at home" or roaming, and indicates this to the user. It also determines the first and last paging and access channels of the system.

The mobile then scans the "A" paging channels, tunes to the strongest one, again reads an overhead message (b), and completes its initialization by storing all parameters in the received message.

12.3.2 Registration

When the mobile has initialized, it makes the system aware of its presence. It scans the access channels of the system, tunes to the strongest one, and then sends a *registration* message (c). The message includes the MIN, MSN, and other information about the mobile. The MSC checks the validity of MIN and MSN. When MIN and MSN are valid and do not indicate a problem (for example, a MSN marked as a stolen mobile station), the MSC returns a *registration confirmation* message (d) on the associated forward channel. At this point, the mobile and MSC have sufficient information to handle calls from and to the mobile.

The mobile now enters the "idle" state, tunes to the strongest paging channel, and keeps reading overhead messages (e), updating its variable-memory when a parameter in an overhead message changes. The idle state ends when the mobile user originates a call, or is called.

12.3.3 Originating Call

The user originates a call by keying the called number, and then depressing the "send" button of the mobile station—Fig. 12.3-2. The mobile then tunes to the strongest access channel, and sends an *origination* message (a). The message includes MIN, MSN, and the called number. The MSC then seizes an available trunk T to an exchange in the PSTN exchange, and an available trunk and associated voice channel in the BS that received the origination. It then signals the called party number on trunk T, and sends an *initial voice channel designation* message in which the channel and attenuation code are specified by parameters CHAN and VMAC (b). The MSC also turns on the SAT tone on the voice channel (not shown).

The mobile then tunes its transmitter and receiver to the voice channel, sets its output power, returns the received SAT tone (not shown), and indicates off-hook (c). When the set-up of the call reaches the called exchange, the mobile user hears ringing tone (or busy tone). Assuming that the calling party answers, the conversation begins.

In this example, the mobile user disconnects first, by depressing the "end" button. The mobile sends an on-hook pulse (signaling tone "on") for 1.8 s, and then turns off its transmitter (d). The MSC recognizes the disconnect request, turns off the BS transmitter of the voice channel, releases the voice channel and associated BS trunk, and the trunk to the PSTN.

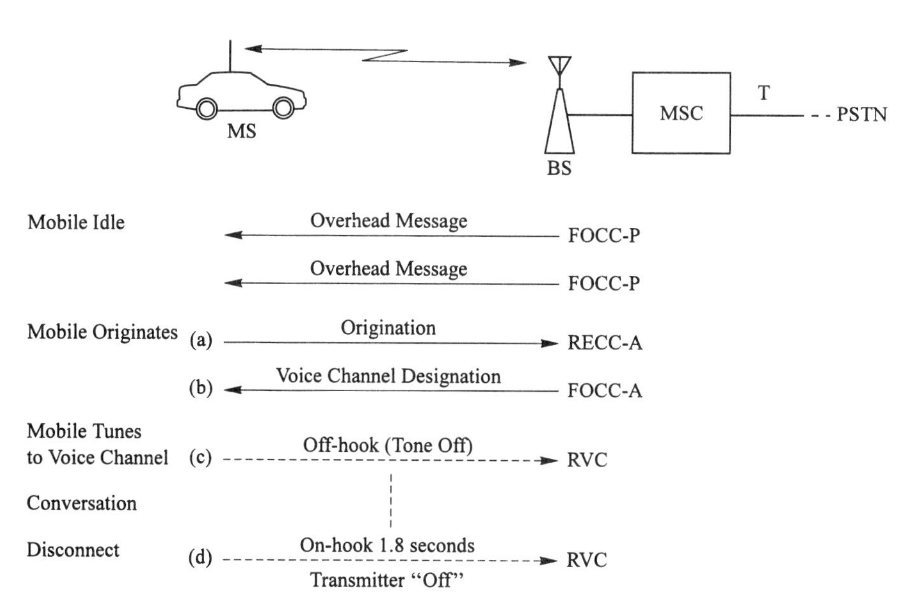

Figure 12.3-2 Call originated by mobile. FOCC-A: forward access channel. FOCCP: forward paging channel. RECC-A: reverse access channel. RVC: reverse voice channel.

12.3.4 Terminating Call

Idle mobiles are tuned to the strongest paging channel (Fig. 12.3-3). In addition to the overhead messages (a), the forward paging channels FOCC-P carry *page messages*, which inform the mobiles about incoming calls (b). These messages include the MIN of the called mobile. When a mobile reads a page message and recognizes its own MIN, it determines the strongest access channel, and sends a *page response* (c).

The MSC then seizes an available voice channel in the BS where the page response was received, starts to transmit SAT tone (not shown), and sends an *initial voice channel designation* message (d) on the access channel. Message parameters CHAN and VMAC specify the channel and the transmit power. The mobile tunes to the channel, sets its transmit power, and returns SAT tone and off-hook (supervision tone off) on the RVC (e). The MSC then sends an *alert* message (f). In response, the mobile generates ringing-tone for its user, and changes its state to on-hook (g). When the user answers, the mobile changes back to off-hook (h). In response, the MSC sends a *stop alert* message (i), and connects the BS trunk to the selected voice channel. The conversation can begin.

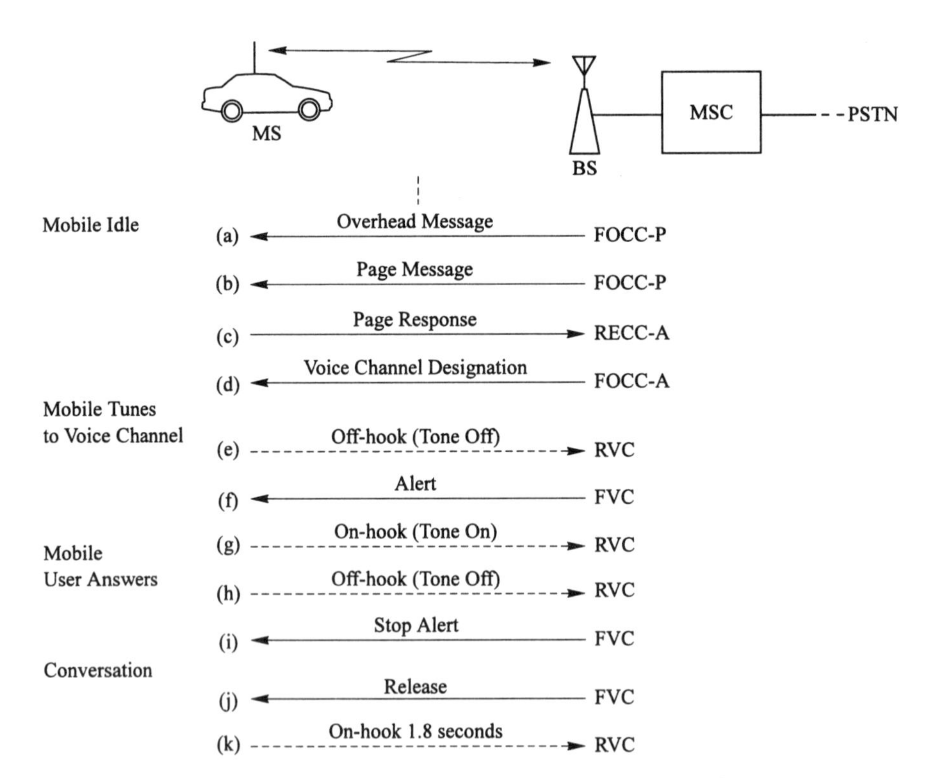

Figure 12.3-3 Call to mobile. FOCC-A: forward access channel. FOCCP: forward paging channel. RECC-A: reverse access channel. RVC: reverse voice channel. FVC: forward voice channel.

Assuming that the calling party disconnects first, the MSC sends a *release* message (j), and the mobile acknowledges with a disconnect signal (going on-hook for 1.8 s), and then turns off its transmitter (k). Finally, the MSC releases the voice channel and associated BS trunk, and the trunk to the PSTN exchange.

12.3.5 Power Change and Handoff

When on a call, a mobile can receive a command on its voice channel to change its transmitter power, or to tune to a different voice channel—see Fig. 12.3-4.

Power Change. A MSC monitors the mobile signal strengths, received by the base stations on their reverse voice channels. When it decides that a mobile should change its transmit power, it sends a "change power" order (a) that includes a new VMAC value. The mobile then adjusts its transmit power, and returns a "change power" confirmation message (b).

Handoff. In a handoff (see Section 12.1.7), the MSC sends a *handoff* message (c). The message includes the new channel number CHAN, and attenuation code VMAC. The mobile acknowledges with a 50 ms on-hook signal (d), and turns off its transmitter. It then tunes to the new channel, sets the new power level, and turns the transmitter on again.

12.4 AMPS MESSAGE FORMATS AND PARAMETERS

AMPS distinguishes two groups of messages, *overhead parameter messages* and *mobile control messages*. Overhead parameter messages are broadcast to

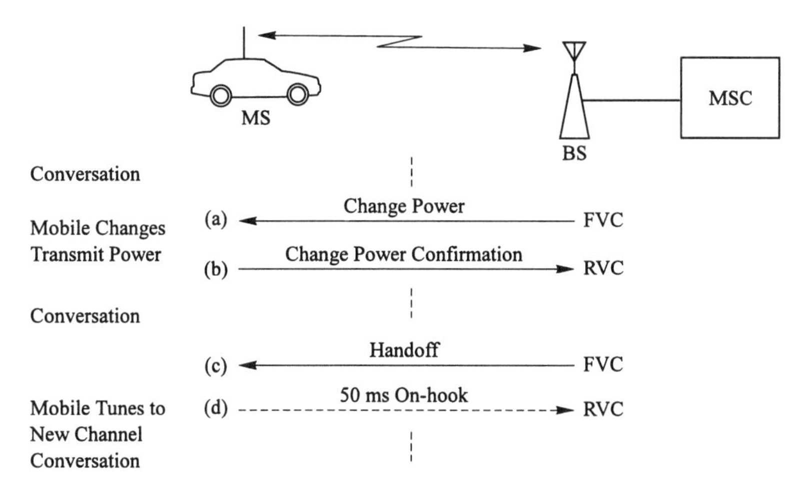

Figure 12.3-4 Power change and handoff. RVC: reverse voice channel. FVC: forward voice channel.

T1T2 = 11	DCC	SID1	RSVD = 000	NAWC	OHD = 110	P	PW1
2	2	14	3	4	3	12	

T1T2 = 11	DCC	S	E	REGH	REGR	DTX	N-1	RCF	CPA	CMAX-1	END	OHD = 111	P	PW2
2	2	1	1	1	1	2	5	1	1	7	1	3	12	

T1T2 = 11	DCC	ACT = 0010	REGINCR	RSVD = 0000	END	OHD = 100	P	GAW2
2	2	4	12	4	1	3	12	

T1T2 = 11	DCC	ACT = 1001	BIS	RSVD = 00......0	END	OHD = 100	P	GAW9
2	2	4	1	15	1	3	12	

T1T2 = 11	DCC	ACT = 1010	MAXBUSY - PGR	MAXSZTR - PGR	MAXBUSY - OTHER	MAXSZTR - OTHER	END	OHD = 100	P	GAW10
2	2	4	4	4	4	4	1	3	12	

T1T2 = 11	DCC	REGID	END	OHD = 100	P	RIDW
2	2	20	1	3	12	

T1T2 = 11	DCC	= 010111	CMAC	RSVD = 00	= 11	RSVD = 00	= 1	WFOM	= 1111	END	OHD = 001	P	CFW
2	2	6	3	2	2	2	1	1	4	1	3	12	

Figure 12.4-1 Overhead parameter message words. (From EIA/TIA 553. Reproduced with permission of TIA.)

all mobiles on the forward control channels (FOCC). Mobile control messages contain call-control messages for individual mobiles, and—depending on the message—are sent on forward or backward control, or voice channels (FOCC, RECC, FVC, and RVC).

Before exploring the individual messages, two general aspects of AMPS message words need to be mentioned—see Fig. 12.4-1. In the first place, each word has a 12-bit P (parity) field, which is used for error control (12.2.2). In the second place, the developers of AMPS, anticipating future additions to signaling in mobile systems, included a number of *reserved* (RSVD) fields in message words. In AMPS, a mobile or MSC sets the RSVD fields in its outgoing messages to 00,...0, and does not examine these fields in received messages. We shall see later how these fields are used in post-AMPS mobile systems in the U.S.

12.4.1 Overhead Parameter Message Format

An AMPS overhead parameter message (OPM) consists of several words.

Figure 12.4-1 shows the most important OPM words [6]. All messages include parameter words PW1 and PW2. The other words are included only when necessary.

In all OPM words, the *type* field (T1T2) is set to 11. This distinguishes these words from words of mobile-control messages on FOCC. The overhead (OHD) field identifies the word type:

OPM Word Type	Acronym	OHD
First parameter word	PW1	110
Second parameter word	PW2	111
Global action words	GAW	100
Registration ID word	RIDW	000
Control filler word	CFW	001

The global action words contain system parameters that were added at a later stage of AMPS development. The various GAW words are distinguishable by the codes in their action (ACT) fields. We limit the discussion to the GAW words shown in Fig. 12.4-1.

In word PW1, the NAWC (number of additional words coming) field indicates the number of subsequent message words. The latter words have END indicators that are set to 0 in all words, except the last one.

The control-filler word (CFW) also has T1T2 = 11, but is not part of an overhead message. CFWs are sent when no other messages have to be transmitted. CFW is similar to the fill-in signal unit (FISU) of signaling system No.7 (7.3), but—unlike the FISU—also holds a number of parameters.

12.4.2 Overhead Message Parameters

This section describes the parameters shown in Fig. 12.4-1. They are denoted by acronyms, and are listed in alphabetical order. At this point it is suggested that the descriptions are briefly perused, and referred to when reading the sections that deal with signaling procedures.

BIS: Busy-idle Status Indicator (in word GAW 9, which is the GAW with *ACT* = 1001). If BIS = 1, a mobile sending a message on a reverse control channel must check for an idle-to-busy transition on the associated forward control channel.

CMAC: Control-channel Mobile Attenuation Code (in CFW). This indicates the mobile transmit power level to be used on reverse control channels.

CMAX-1 (in PW2) . This is the number of access channels in a system, minus 1 (in a system with 10 access channels, CMAX-1 = 9).

CPA: Combined Paging and Access Indicator (in PW2). When CPA = 1, the control channels in the system are combined paging and access channels. When CPA = 0, the system has separate paging and access channels.

DCC: Digital Color Code (in all OPM words). This is the color code of the transmitting base station (values: 00, 01, or 10; see Section 12.1.4).

DTX: Discontinuous Transmission Indicator (in PW2). This indicates whether a mobile with discontinuous transmission (12.1.6), and transmitting on a voice channel, is allowed to reduce its transmit power when the user is silent:

$DTX = 00$ Power reduction not allowed
$DTX = 10$ Limited (8 dB) power reduction allowed
$DTX = 11$ Any power reduction allowed

When the mobile user speaks, or a message is sent, the transmitter should return immediately to its normal power level.

E: Extended Address Indicator (in PW2). If E = 1, all mobiles have to include their complete mobile identification number MIN in messages sent on a reverse control channel. If E = 0, only roaming mobiles need to include the complete MIN. Resident mobiles send a partial MIN (without area code).

MAXBUSY-PGR, MAXBUSY-OTHER, MAXSZTR-PGR, and MAXSZTR-OTHER (in GAW 10). When a mobile needs to send a message on a reverse control channel, these parameters indicate the maximum number of times the mobile can try to seize a reverse control channel.

N–1 (in PW2). This is the number of paging channels in the system minus one (in a system with five paging channels, N–1 = 4).

RCF: Read Control Filler Indicator (in PW2). If set to 1, the mobile must read and copy the parameters in a control-filler word (CFW) before accessing a reverse control channel.

REGH and REGR (in PW2). These are home and roamer registration indicators. When set to 1, the system allows registrations by resident and roaming mobiles respectively.

REGID and REGINCR (in respectively RIDW and GAW 2). These are the registration identification and registration increment, which are integers used by mobiles for autonomous registration.

S: Serial Number Indicator (in PW2). If S = 1, mobiles have to include their MSN (mobile serial number) when sending a message on a reverse control channel.

SID1: System Identifier, Part 1 (in PW1). The leading 14 bits of the 15-bit system identifier (SID). The least significant bit is omitted because, prior to sending a message, the mobile has decided to communicate with the "A" or "B" system, and therefore knows whether SID is odd or even (12.3).

WFOM: Wait-for-overhead-message Indicator (in CFW). If set to 1, mobiles have await an overhead message before seizing a reverse control channel.

12.4.4 Mobile Control Messages

We now examine the most important mobile-control messages. Some of these have already been mentioned in Section 12.3. Mobile-control messages on the forward and reverse control channels (FOCC, RECC) are used for signaling between the mobile system and a mobile that has been turned "on" and is not involved in a call. When a mobile is on a call, the signaling is on the forward and reverse part of the voice channel (FVC, RVC) that has been allocated to the call.

12.4.5 Mobile Control Messages on FOCC

The messages shown in Fig. 12.4-2 consist of word 1 (T1T2 = 01) and word 2 or word 2* (T1T2 = 10).

T1T2 = 01	DCC	MIN1	P	Word 1
2	2	24	12	

T1T2 = 10	SCC = 11	MIN2	RSVD = 0 - - - 0	ORDQ	ORD	P	Word 2
2	2	10	6	3	5	12	

T1T2 = 10	SCC ≠ 11	MIN2	VMAC	CHAN	P	Word 2*
2	2	10	3	11	12	

Message	Words
Page	1, 2
Registration Confirmation	1, 2
Release	1, 2
Reorder	1, 2
Initial Voice Channel Designation	1, 2*

Figure 12.4-2 Mobile control messages on forward control channel. (From EIA/TIA 553. Reproduced with permission of TIA.)

Page Message. This alerts the mobile about an incoming call.

Registration Confirmation Message. This confirms the registration of a mobile.

Release Message. This is sent when the system rejects the registration of a mobile.

Reorder Message. This is sent when the system cannot process an origination, for example, when no voice channel is available.

Initial Voice Channel Designation Message. This specifies the initial voice channel assigned to the call, the color code of the channel, and the mobile attenuation code.

12.4.6 Mobile Control Messages on RECC

In these messages, indicator bit F is 1 in the first message word, and 0 in succeeding words—Fig. 12.4-3. Messages always include words 1 and 2. Word 3 holds the MSN of the mobile, and is included only if indicator S in the overhead parameter message is set to 1 (12.4.2). Words 4 (and, when necessary, words 5, 6, etc.) appear in originations only, and hold up to eight digits of the called number.

F = 1	NAWC	T	S	E	RSVD = 0	SCM	MIN1	P	Word 1
1	3	1	1	1	1	4	24	12	

F = 0	NAWC	RSVD = 0 - - - 0	ORDQ	ORD	LT	RSVD = 0 - - - 0	MIN2	P	2
1	3	5	3	5	1	8	10	12	

F = 0	NAWC	MSN	P	3
1	3	32	12	

F = 0	NAWC	DIGIT	DIGIT	DIGIT	DIGIT	DIGIT	DIGIT	DIGIT	DIGIT	P	4, 5...
1	3	4	4	4	4	4	4	4	4	12	

Message	Words
Registration	1, 2, 3
Page Response	1, 2, 3
Origination	1, 2, 3, 4...

Figure 12.4-3 Mobile control messages on reverse control channel. (From EIA/TIA 553. Reproduced with permission of TIA.)

Page Response. This is sent by a mobile, to acknowledge a page message.

Registration. This is sent by a mobile, to announce its presence to the system.

Origination. This is sent by a mobile, to request the set-up of a call.

12.4.7 Mobile Control Messages on FVC

The type field in the messages is set to T1T2 = 10 (Fig. 12.4-4). All message types except "handoff" consist of word 1.

Alert. This is a request by the system to generate ringing-tone, to alert the user about an incoming call.

Stop Alert. This is a request to turn off ringing-tone.

Release. This orders the mobile to release the voice channel.

Change Power. This orders the mobile to change its transmit power.

Send Called Address. This orders the mobile to send digits. For example, a called party number, or a service access code (3.5.1).

Handoff. This alerts the mobile that it is being handed off to a new cell, and specifies the new voice channel, its color code, and the new transmit power. This message consists of word 1*.

T1T2 = 10	SCC = 11	PSCC	RSVD = 0 - - - 0	ORDQ	ORD	P	Word
2	2	2	14	3	5	12	1

T1T2 = 10	SCC ≠ 11	PSCC	RSVD = 0 - - - 0	VMAC	CHAN	P	Word
2	2	2	8	3	11	12	1*

Message	Word
Alert	1
Stop Alert	1
Release	1
Change Power	1
Send Called Address	1
Handoff	1*

Figure 12.4-4 Mobile control messages on forward voice channel. (From EIA/TIA 553. Reproduced with permission of TIA.)

F = 1	NAWC	T	RSVD = 0---0	ORDQ	ORD	RSVD = 0---0	P	Word
1	2	1	5	3	5	19	12	1

F = 0	NAWC	T	DIGIT	DIGIT	DIGIT	DIGIT	DIGIT	DIGIT	DIGIT	P	Word
1	2	1	4	4	4	4	4	4	4	12	2, 3...

Message	Words
Change Power Confirmation	1
Called Address	1, 2, ...

Figure 12.4-5 Mobile control messages on reverse voice channel. ((From EIA/TIA 553. Reproduced with permission of TIA.)

12.4.8 Mobile Control Messages on RVC

$F = 1$ in the first message word, and $F = 0$ in succeeding words (Fig. 12.4-5).

Change Power Confirmation (word 1). The mobile confirms the receipt of a change power order.

Called Address (word 1, word 2, and succeeding words when necessary). This is sent by a mobile, and represents a called party number (for three-way calling) or a service access code.

12.4.9 Mobile Control Message Parameters

The parameters in mobile control messages are described below in alphabetical order of their acronyms [6].

CHAN: Channel. This identifies an initial or new voice channel (Figs. 12.4-2 and 12.4-4).

DCC: Digital Color Code. This is a binary number that identifies the color code (see Section 12.1.4) of the FOCC in overhead parameter and mobile control messages (Figs. 12.4-1 and 12.4-2).

DIGIT. This is a digit of the called address or service access code (Figs. 12.4-3 and 12.4-5).

E: Extended Address Indicator. $E = 1/0$ in a RECC message indicates that the message includes/does not include a word 2, which holds the area code (MIN2) of the mobile identification number (Fig. 12.4-3).

LT: Last Try Indicator. LT = 1 indicates a final attempt by a mobile to send a message on RECC (Fig. 12.4-3).

MIN1: Mobile Identification Number 1. This is the seven-digit directory number (exchange code + line number) of the mobile identification number (MIN), coded as a 24-bit binary number (Figs. 12.4-2 and 12.4-3).

MIN2: Mobile Identification Number 2. This is the three-digit area code of the MIN, coded as a 10-bit binary number (Figs. 12.4-2 and 12.4-3).

MSN: Mobile Serial Number. This is a number assigned by the manufacturer, and stored in non-alterable memory of the mobile. It is included in RECC messages if requested by the system (indicator bit S = 1 in overhead parameter messages)—see Fig. 12.4.3.

NAWC: Number of Additional Words Coming. This indicates the number of subsequent words in the message.

ORD: Order Code. This identifies the type of an order or confirmation message. Most mobile control messages contain an ORD (Figs. 12.4-2 through 12.4-5). In these messages, SCC = 11 (a non-existent color code). The *initial channel designation* and *handoff* messages do not include an ORD, and SCC represents the color code of the initial or new voice channel (00, 01, 10—see Section 12.1.4). The order codes for the messages described in the previous section are listed in Table 12.4-1.

Table 12.4-1 Color codes (SCC) and order codes (ORD) in AMPS mobile control messages.

		On Channel Type:				
SCC	ORD	FOCC	RECC	FVC	RVC	Message
11	00000	x				Page message
11	00000		x			Origination, page response
11	00001			x		Alert
11	00011	x		x		Release
11	00100	x				Reorder
11	00110			x		Stop alert
11	01000			x		Send called address
11	01000				x	Called address message
11	01011			x		Change power order
11	01011				x	Change power confirmation
11	01101	x				Registration confirmation
11	01101		x			Registration
00, 01, 10	—	x				Initial voice channel designation
00, 01, 10	—			x		Handoff message

Source: EIA/TIA 553. Reproduced with permission of TIA.

ORDQ: Order Qualifier. This field is included in all messages that have an ORD. In most messages, ORDQ is not used and coded 000. In the change power order and its confirmation, ORDQ holds the mobile attenuation code that specifies the transmit power (Section 12.1.6). In registration messages, it differentiates autonomous and non-autonomous registrations.

PSCC: Present SCC. This is the color code of the present voice channel in FVC messages (Fig. 12.4-4). If PSCC does not agree with the SCC code specified in a channel-assignment or handoff message, the mobile ignores the message.

S: S Indicator. This is included in RECC messages. When $S = 1$, the message includes a word 3 (Fig. 12.4-3).

SCM: Station Class Mark. This provides information on the transmitter of mobile station (power class, and capability for discontinuous transmission—see Section 12.1.6).

SCC: SAT Color Code. This appears in messages on FVC (Fig. 12.4-4). The values SCC = 00, 01, or 10 indicate an initial voice channel designation or a handoff message, and represent the color code of the designated voice channel. SCC = 11 indicates messages of other types, specified by ORD.

T: T Indicator. This indicator differentiates orders and confirmations in reverse messages. In RECC messages, $T = 1$ and 0 indicate respectively orders, and order confirmations (Fig. 12.4-3). In RVC messages, $T = 1$ and 0 indicate respectively an order confirmation, and a called-address message (Fig. 12.4-5).

VMAC: Voice Mobile Attenuation Code. This specifies the mobile transmit power on the voice channel (Figs. 12.4-2 and 12.4-4).

Coded DCC. All messages on FOCC include the parameter DCC that indicates the color code of the transmitting base station. A mobile sending a message on a RECC transmits a seven-bit coded DCC that represents the received DCC:

Received DCC	Coded DCC
00	000 0000
01	001 1111
10	110 0011

In contrast with the other parameters described here, the coded DCC is not contained in a message word, but follows the dotting (DOT) and word synchronization (WS) sequences—see Fig. 12.2-1(c).

12.4.11 Color Codes and Message Acceptance

Message Acceptance on Control Channels. The DCC and coded DCC on control channels are the digital counterparts of the SAT tones on voice channels (12.1.4), and have the same purpose. A mobile discards a received mobile control message in which DCC is different from the DCC being received in overhead messages (Figs. 12.4-1 and 12.4-2). Likewise, a base station ignores a message received with a coded DCC that does not represent its DCC (Fig. 12.4-3), because it must have been sent by a mobile that is signaling to another BS.

Message Acceptance on Voice Channels. The "initial voice channel designation" and "handoff" messages include the color code of the allocated channel. A mobile discards a FVC message in which PSCC does not have the expected value (Fig. 12.4-4). There is no comparable color-code checking for messages on RVC.

12.5 AMPS SIGNALING PROCEDURES

This section discusses a number of AMPS signaling procedures, adding details that were omitted in Section 12.3. While reading this material, it is helpful to look up the messages in Figs. 12.4-1 through 12.4-5, and the parameter descriptions in Sections 12.4.2 (overhead message parameters) and 12.4.9 (mobile control message parameters).

12.5.1 Mobile Initialization

The general procedure has been described in Section 12.3.1. We focus here on the retrieval by the MS of the system-specific data from overhead parameter messages.

Step 1. When the MS is turned "on," it scans the dedicated forward control channels (FOCC) of the "A" system (channels 333–313), if its home system is an "A" system, or the "B" system (channels 334–354), if its home system is a "B" system. It tunes its receiver to the strongest one. After receipt of the first overhead message (Fig. 12.4-1), MS determines the first and last paging channels of the system from the value of $N-1$

	"A" Systems	"B" Systems
First paging channel	333	334
Last paging channel	$(333-N+1)$	$(334+N-1)$

and stores this information in its temporary memory.

Step 2. The mobile then tunes to the strongest paging channel, waits for another overhead message, and compares the received system identifier (SID1) with the SID of its "home" system, which is stored in its semi-permanent memory. If the identifiers match, the MS is "at home." If not, MS informs its user by turning "on" its "roaming" light.

Step 3. The MS determines the first and last access channels from parameters CPA and CMAX-1. If CPA = 1, the system has combined access and paging channels, and the first and last access channels are the same as the first and last paging channels. If CPA = 0, the system has separate access and paging channels, and the first and last access channels are:

	"A" Systems	"B" Systems
First channel	$333 - N$	$334 + N$
Last channel	$(333 - N - \text{CMAX} + 1)$	$(334 + N + \text{CMAX} - 1)$

The MS then stores the first and last access channel in its temporary memory.

Step 4. The MS awaits another overhead parameter message, and stores the other overhead message parameters and indicators (12.3.2).

The MS is now initialized and in the "idle" state. It remains tuned to the strongest paging channel. While in this state it keeps monitoring the overhead messages, and updates its memory when an overhead parameter changes. The idle state changes when the MS user originates, and when the MS receives a paging message, indicating a terminating call.

If the initialization procedure fails, the MS tries to contact the other system (the "B" system if the mobile homes on an "A" system, and vice versa).

12.5.2 Seizing a Reverse Access Channel

When an idle mobile MS needs to access the system (send a message), it has to seize a reverse access channel. This is done in the following series of steps.

Step 1. The mobile scans the access channels, locks on to the strongest one, reads an overhead message, and examines parameter RCF. If RCF = 0, the mobile sets its transmitter to maximum power, and starts step 3. If RCF = 1, it goes to step 2.

Step 2. The MS reads a control filler word CFW (Fig. 12.4-1), sets its transmit power to the value in CMAC, and, if WFOM = 1, also reads another overhead message, updates its parameters, and then goes to step 3.

Step 3. Any idle MS in a cell can seize an access channel. In order to minimize "collisions" (simultaneous seizure of an access channel by more than one mobile), the mobile first examines the BI (busy-idle) bits on the forward access channel, which indicate whether the associated reverse channel is idle (12.2.2).

If the channel is busy, the mobile waits for a random time (0–200 ms), and repeats this step. Up to NBUSY-PGR or NBUSY-OTHER busy occurrences are allowed, for respectively page response and other messages. If the channel is idle, the MS goes to step 4.

Step 4. The MS seizes the channel, and starts its transmission. In systems with BIS = 0, the mobile transmits the entire message. If BIS = 1, the system has a second defense against "collisions." The mobile then has to keep monitoring the BI bits on FOCC. BI turning to busy before 56 message bits have been sent indicates a collision, and BI not changing to busy after 104 bits have been sent indicates that the message is not being received. In either case, the mobile stops transmitting. The number of allowed RECC seizures is MAXSZTR-PGR or MAXSZTR-OTHER. As long as this limit is not exceeded, the mobile waits for a random time, and returns to step 3.

12.5.3 Registration

A mobile always tries to register when it has completed its initialization (12.3.2). Before sending a registration message, the resident or roaming MS first examines REGH or REGR, and attempts to seize an access channel only when allowed by the system. The mobile system acknowledges a received registration with a registration confirmation message (Fig. 12.3-1).

12.5.4 Determining the MS Location

The service area of an MSC is divided into a number of *location areas*, which are clusters of adjacent cells (12.1.1). When accessing the system with a registration or origination message, the MS transmits on the strongest access channel (that is, the channel of the nearest base station). MSC derives the current location area of MS from the identity of the BS that has received the message, and enters it into its record for the MS.

Autonomous Registration. Mobiles with "autonomous registration" capability also reregister periodically, on command from the MSC. Autonomous registration is governed by overhead parameters REGID and REGINCR, which are copied by the mobiles. The MSC increases REGID after a certain number of received registrations, by including a RIDW word with the new REGID value in its overhead parameter messages (Fig. 12.4-1).

On each registration, the MS copies the current values of REGID and REGINCR, say 34567 and 500, and then calculates the value of an internal parameter: NXTREG = 500 + 34567 = 35067. When the system has increased REGID to a value that exceeds NXTREG, the mobile reregisters, determines a new NXTREG value, and so on. On receipt of each reregistration, the MSC determines and stores the new location area, thus keeping track of the MS location.

12.5.5 Paging the MS

Consider a MSC area that contains *N* cells. If a MSC receives a call for an MS that has registered in its service area, it could transmit a page message on all *N* paging channels (FOCC-P). However, this is very ineffective if *N* is large (which is the case in metropolitan MSC areas), because only one of the *N* messages is received by the MS. The large number of ineffective page messages tends to overload the FOCC-P channels.

This is the reason for autonomous MS registration, which enables the MSC to keep track of the location areas of the individual MSs in its service area. A MSC starts by transmitting page messages on only the FOCC-Ps in the last known location area of the MS. If MS is still in this area, it responds with a page response message. If the MSC does not receive a page response within a few seconds, it assumes that the MS has moved to another location area, and repeats the paging message, this time on all FOCC-Ps.

12.5.6 Supplementary Services

Most cellular systems in the U.S. can provide call-waiting, call-forwarding, and three-way calling to mobiles that have subscribed to these services [7]. These services are supported by the signals and messages described so far.

To activate or deactivate call forwarding, the mobile user sends an origination message in which the digits represent a service access code. All access codes start with * (asterisk). In this way, the MSC can distinguish feature activations/deactivations and originations.

Mobile users who are on a call and have call-waiting service are informed by a tone that another call has arrived. The user can then switch back and forth between the original and the new call, by sending "flash" signals (12.2.1).

Mobile users who are on a call and have three-way calling service can initiate a call to a third party by sending a flash. The MSC responds with a "send called address" message, the mobile then sends the "called address" message, and the second call is added to the connection.

12.5.7 Protection against Cloning

A mobile identifies itself to a MSC by including its mobile identification number (MIN) and serial number (MSN) in its registration, origination and page-response messages, and the MSC serves the mobile only when it has verified these parameters. This gives some protection against customer fraud. For example, data bases in cellular systems maintain lists with the MSNs of stolen mobile stations, and do not give service to these stations. However, this does not protect against persons with the required equipment and technical knowledge who pick up messages on a reverse control channel, and thereby acquire valid combinations of MIN and MSN. These people are also capable of entering one of these combinations in a stolen MS. This activity is known as "cloning."

Mobile systems accept originations from these mobiles, and charge the calls to the owner of the MS whose MIN-MSN combination has been cloned. The fraud is detected when this owner complains about charges for unknown calls on his monthly invoice, but this happens on average one half-month after the illegally altered MS has started making calls. In the U.S., the annual losses resulting from cloning are estimated at about $500,000,000.

To protect against cloning, the operators of some mobile systems are issuing four-digit personal identification numbers (PIN) to MS users. When the user originates, or responds to a page message, the MSC allocates a voice channel, and sends an initial voice channel designation message. It then sends a short tone burst to the mobile, which prompts the user to send his PIN number. Only after the PIN has been validated, the MSC cuts through the connection between the trunk to the PSTN network and the BST and its associated voice channel (Fig. 12.1-3).

The protection offered by this arrangement is not complete. It is possible to construct equipment that can tune to a reverse access channel (RECC-A) to capture a MIN+MSN combination, then tune to the associated forward access channel (FOCC-A) to obtain the identity of assigned voice channel (VC), and finally tune to RVC, and capture the PIN code.

A more powerful anti-cloning arrangement is described in Section 12.6.7.

12.6 SIGNALING IN IS-54 CELLULAR SYSTEMS

Beginning in the early 1990s, a number of second-generation cellular systems are being deployed in the U.S and abroad [8–11].

Two second-generation mobile cellular systems have been defined for use in the U.S. These systems are known as the IS-54 and IS-95 systems—their names refer to the EIA/TIA Interim Standards in which they have been specified [12, 13]. Both have digital multiplex voice channels. IS-54 uses *time division multiple access* (TDMA), and IS-95 is based on *code division multiple access*. This section discusses the signaling in IS-54.

12.6.1 Dual-mode Systems

Because of the large installed base of AMPS mobiles and AMPS cellular networks, IS-54 and IS-95 are *dual-mode* systems. The networks are implemented as additions to—or partial replacements of—existing AMPS networks, and include both digital and analog (AMPS) RF channels. Likewise, the IS-54 and IS-95 mobile stations are dual-mode, and can operate with analog and digital channels. In addition, IS-54 and IS-95 networks and mobiles are capable of using their respective signaling protocols, and the AMPS protocol. This means that any MS can be served by any network. For example, an IS-54 mobile operating in an IS-95 network uses AMPS signaling and, if the MS originates a call, the network assigns an AMPS voice channel.

12.6.2 The IS-54 System

The digital channels of IS-54 use differential quaternary phase-shift keying (DQPSK), in which the phase of a signal with frequency f_c is changed 24,300 times per second. Each phase change has one of four values that represent the values of two consecutive bits. The bit rate in the channel is thus 48.6 kb/s. The power spectrum of the digital signal fits within 30 kHz, which is also the bandwidth of AMPS channels.

TDMA Traffic Channels. The 48.6 kb/s bit stream on the digital channels is divided into frames of 1994 bits (Fig. 12.6-1). The frame duration is 40 ms, giving a repetition rate of 25 frames/s.

Each frame is time-divided into six 6.67 ms time slots. Communications between the mobiles and a base station occur in TDMA bursts, during predetermined time slots of each frame. At present, the TDMA traffic channels are "full-rate," using two time slots in each frame (slots 1 and 4, or 2 and 5, or 3 and 6). A digital channel thus accommodates three full-rate TDMA traffic channels. "Half-rate" traffic channels, using only one time slot, are under development.

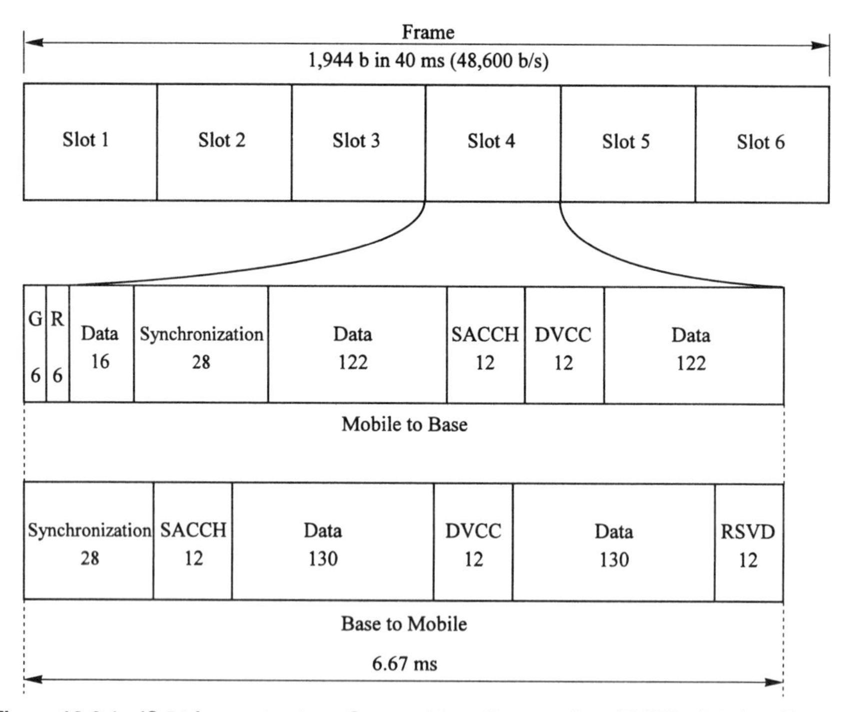

Figure 12.6-1 IS-54 frame structure. G: guard time. R: ramp time. DVCC: digital verification color code. RSVD: reserved for future use. (From IEEE *Comm. Mag.* **29.6.** Copyright © 1991 IEEE.)

A time slot has 260 bits of digital speech (or other user information). The user information rate on full-rate channels is thus 13 kb/s. The A/D and D/A converters for speech in the mobiles and base stations operate at 8 kb/s. Before transmission, five forward error-correction bits are added to each eight-bit speech sample, yielding the bit rate of 13 kb/s.

Each time slot includes a specific 28-bit synchronization sequence. This enables the receiver in a mobile to lock on to its assigned time slots.

IS-54 has 255 *digital verification color codes* (DVCC). They appear as 12-bit groups (eight information bits and four forward error correction bits) in each time slot. DVCC is the digital counterpart of the SAT tones on analog voice channels (12.1.4). Mobiles and base stations ignore the received information in time slots with incorrect DVCC codes.

System Capacity. Compared with AMPS, IS-54 considerably improves the RF system capacity, because the 30 kHz bandwidth required for one AMPS voice channel can also accommodate three IS-54 traffic channels. AMPS cells in seven-cell clusters have up to 56 voice channels, and a traffic capacity of 42 Erlang (12.1.5). Suppose that 28 AMPS voice channels in a cell are replaced by 3×28 digital traffic channels. This doubles the number of channels, and the traffic capacity, of the cell.

This is important in mobile systems that serve areas with high traffic densities (Erlangs per square mile). In AMPS, this would require a large number of small cells, and this has two disadvantages. In the first place, high traffic densities often occur in metropolitan areas, where suitable locations for base stations are hard to find. In the second place, in small cells, the probability that a call has to be handed off is rather high. Replacing a number of AMPS by digital traffic channels allows the use of fewer, and larger, cells.

12.6.3 IS-54 Dual-mode Signaling

In IS-54 systems, the dedicated AMPS control channels (channels 313–333 and 334–354, in the "A" and "B" systems—see Section 12.1.2) are used by AMPS and IS-54 mobiles. Some high-traffic IS-54 systems may require additional control channels. Therefore, channels 688–708 (in "A" systems), and 737–757 (in "B" systems), have been designated as secondary dedicated control channels. These channels can be used by IS-54 mobiles only, because AMPS mobiles are not designed to use them.

Messages on Control Channels. The messages on IS-54 control channels must be compatible with AMPS and IS-54 mobiles, and all AMPS messages described in Section 12.4 are therefore included without change.

To accommodate the additional information in AMPS overhead-parameter and mobile-control messages for IS-54 mobiles, use is made of the RSVD (reserved) message fields, which are ignored by AMPS mobiles and networks.

Messages on Voice and Traffic Channels. The signaling on the analog voice channels remains as in AMPS. Each IS-54 traffic channel has two associated control channels, which carry signaling messages during the times that the channel has been allocated to a mobile. The *slow associated control channel* (SACCH) uses 12 bits in each time slot—see Fig. 12.6-1. On SACCH, a signaling message word consists of 132 bits (66 for information; 66 for forward error correction). This gives a rate of $2 \times 12/2 = 12$ information bits per frame, or $12 \times 25 = 300$ information bits/s, which is slower than in AMPS. Most mobile-control messages are sent on the SACCH.

The *fast associated control channels* (FACCH) messages are sent as short bursts in the user information field. This is comparable with blank-and-burst technique for messages in AMPS voice channels (see Section 12.2.2). A FACCH message word consists of 65 information bits and 65 forward error-correction bits.

The FACCH channel is used primarily for handoff messages, where speed is important.

12.6.4 Determination of the System and Mobile Type

In the initial contact between a mobile and a mobile system, an IS-54 mobile needs to determine the type of the mobile network (AMPS or IS-54), and an IS-54 system needs to determine the type of the mobile.

An IS-54 system indicates its type—and other capabilities—in the three-bit RSVD field of word PW1 in the overhead parameter messages on FOCC—see Fig. 12.6-2. If *protocol control indicator* (PCI) = 1, the control channel belongs to an IS-54 system, and can assign digital traffic channels.

An IS-54 mobile uses the RSVD fields of word 2, in the messages it sends on RECC—see Fig. 12.6-3. In these messages, an IS-54 mobile sets the *mobile protocol capability indicator* (MPCI) to 01. Messages with this MPCI can include words that are not used in AMPS.

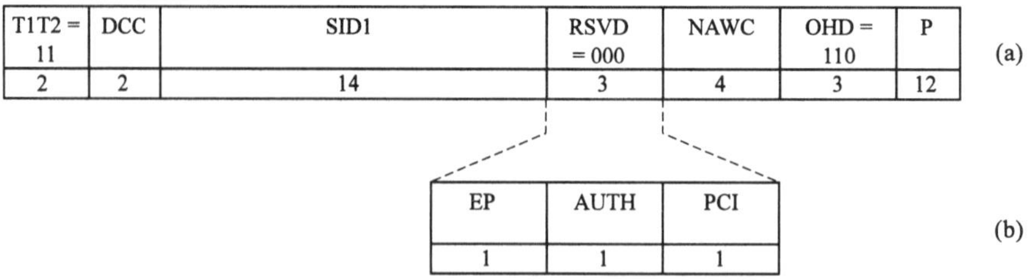

(a)

T1T2 = 11	DCC	SID1	RSVD = 000	NAWC	OHD = 110	P
2	2	14	3	4	3	12

(b)

EP	AUTH	PCI
1	1	1

Figure 12.6-2 Word PW1 of overhead parameter message. (a): AMPS (see Fig. 12.4-1). (b): use of RSVD field in IS-54. (From IS-54-B. Reproduced with permission of TIA.)

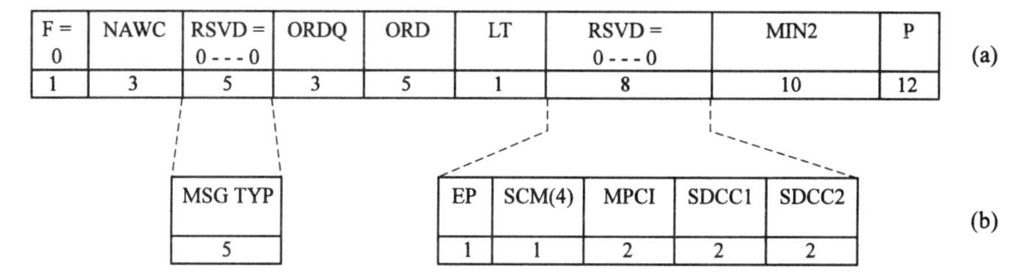

Figure 12.6-3 Mobile control messages on reverse control channel. (a): AMPS (see Fig. 12.4-3). (b): use of RSVD fields in IS-54. (From IS-54-B. Reproduced with permission of TIA.)

12.6.5 Order Qualifiers

In AMPS call-control messages, the ORD field identifies an order (or the acknowledgment of an order), and the ORDQ field qualifies the order—see Figs. 12.4-2 through 12.4-5. In IS-54 messages on the FOCC (forward) and RECC (reverse) control channels, the order is qualified by the contents of ORDQ in combination with the contents of a five-bit *MSG TYP* field.

In FOCC messages, MSG TYP occupies five bits in the RSVD field of word 2 (Fig. 12.4-2). In RECC messages, MSG TYP is in the five-bit RSVD field of word 2 (Fig. 12.4-3).

12.6.6 Assignment of Digital Traffic Channels

When setting up a connection, the MSC needs to assign a channel of a type that is compatible with the types of available channels in the cell, and with the channel types that can be handled by the mobile.

When an IS-54 mobile originates a call, the ORDQ and MSG TYP fields in its origination message indicate, among other things, whether the mobile can handle analog voice channels, and/or full-rate, and/or half-rate digital traffic channels. MSC uses this information to allocate a channel of the proper type.

When paging an IS-54 mobile, the MSC uses the ORDQ and MSG TYP fields in its paging message to indicate the types of available channels. The mobile includes the channel types it can handle in its page response message, and MSC then selects a channel of a compatible type.

To conclude this section, we examine a number of IS-54 features, and the messages and parameters that support them.

12.6.7 Authentication

This feature is a defense against the cloning of mobiles (12.5.7). It is based on the CAVE (cellular authentication and voice encryption) algorithm, which produces an 18-bit result AUTHR (authentication response), from the values of four inputs:

	Input	Acronym	Length (bits)
1	Random number generated by MSC	RAND	32
2a	In registrations and page responses: the first part of the mobile identification number	MIN1	24
2b	In originations: the last six digits of the called number (BCD coded)	6DIG	24
3	Mobile serial number	MSN	32
4	Shared secret data for authentication	SSD_A	64

In a MS with authentication capability, its SSD_A(M) is stored in semi-permanent memory, along with MIN, etc., and the MS can execute CAVE.

Mobile networks have (or will have) *authentication centers* (AUC). An AUC stores the parameters MSN(A) and SSD_A(A) associated with individual MINs of mobiles with authentication capability, and can also execute CAVE.

In the authentication procedure, the MS and AUC execute CAVE. The MS uses its stored SSD_A(M), and the AUC uses the SSD_A(A) which is associated with the MIN of the mobile. The other inputs to the algorithm are the same at MS and AUC.

Suppose that the mobile and the AUC produce the results AUTHR(M) and AUTHR(A). The AUC receives a copy of AUTHR(M), and compares it with its result. If the AUTHRs match, AUC concludes that the mobile stores the proper SSD_A(M) value, namely SSD_A(A), and must be "authentic." Only authentic mobiles will receive service from the network.

Authentication can take place when a mobile registers, originates a call, or responds to a page message. Figure 12.6-4 shows the procedure for originating mobiles. The message parameters that are significant for authentication are shown inside square brackets. The MSC indicates its authentication capability in its parameter overhead messages, by setting the authentication indicator in word PW1 to AUTH = 1 (see Fig. 12.6-2), and including two global action words each of which holds 16 bits of RAND. The MSC frequently generates new RAND values, say every minute.

The originating MS notes that AUTH = 1, and therefore executes CAVE, using the most recently received RAND, the final digits of the called number (6DIG), and its internally stored parameters MSN(M) and SSD_A(M). It then sends an origination message that includes, in addition to MIN, MSN(M), and the called number, a word that holds RANDC and AUTHR(M). RANDC contains the eight high-order bits of the RAND that was used in the calculation of AUTHR(M).

MSC may have changed its RAND after MS has executed CAVE. Therefore, on receipt of the origination message, it compares the eight high-order bits of its current and previous RANDs with RANDC, to determine the RAND used by MS.

MSC then sends an "authentication request" message to the authentication

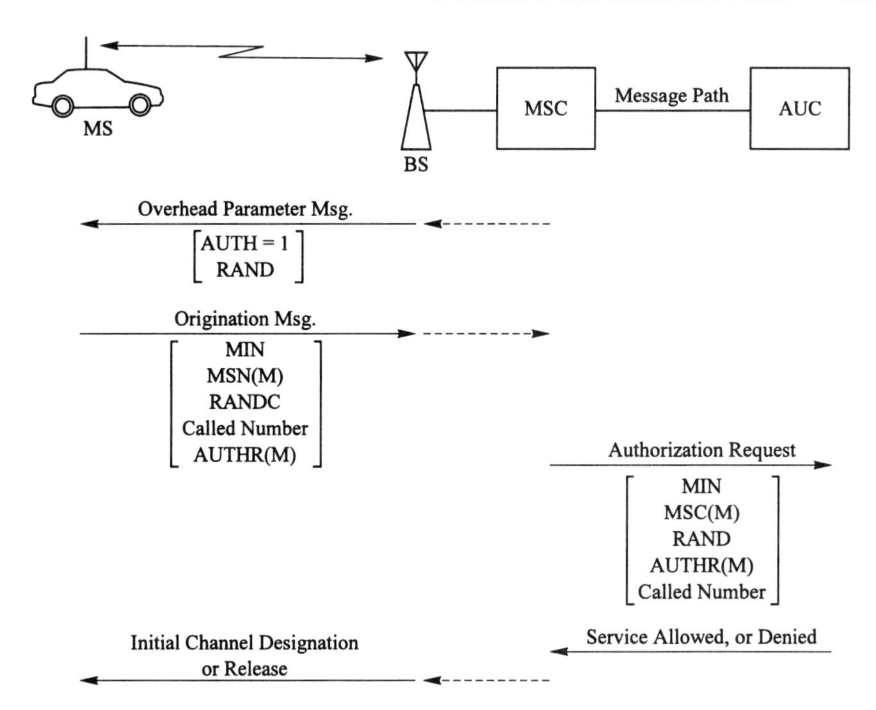

Figure 12.6-4 Authentication of originating mobile. (From TIA/EA TSB-51. Reproduced with permission of TIA.)

center. AUC then executes CAVE, using MIN, MSN(M), RAND received in the message, and its stored SSD_(A). It then compares its result AUTHR(A) with AUTHR(M). If the AUTHRs match, AUC sends a "allow service" message to MSC. If the AUTHRs do not match, a "deny service" response is sent. The signaling messages between MSC and AUC are discussed in Chapter 17.

When a MS registers, or sends a page response, it uses MIN1 instead of 6DIG for the execution of CAVE, and does not include a called number in its message. If AUC receives an authorization request that does not include a called number, it also uses MIN1 for its CAVE.

It is next to impossible to defeat the authentication procedure, because SSD_A is not transmitted on RF channels. It is possible to pick up, on a reverse control channel, combinations of parameters sent by mobiles in their origination messages, but AUTHR(M) is of no use for cloning, because its value is different for each call of the mobile, since it depends on RAND and the called number.

12.6.8 Voice Privacy

This feature protects the mobile user against unauthorized "listening in" on his calls. This is done by scrambling the speech on digital traffic channels before it is fed into the RF transmitter, and unscrambling it at the receiving side of the channel. The voice privacy procedure is an extension of the CAVE procedure.

A mobile with voice privacy capability stores a second 64-bit "shared secret data" parameter SSD_B in its memory. SSD_B is also stored at the authentication center.

During the set-up of a call, the MS and the AUC execute CAVE, using the inputs RAND, MIN1, MSN, and SSD_B, which produces two "voice privacy masks" (one for each direction of transmission). The AUC includes the masks in its response message to MSC.

A mask consists of 260 bits, each of which corresponds to a bit in the 260-bit user data fields shown in Fig. 12.6-1. Before transmitting a time slot, the MSC scrambles the user data fields that contain the digitally coded speech, by forming the "exclusive-or" of each data bit and its corresponding bit in the MSC-to-MS voice-privacy mask.

The mobile does the same operation on the received data, using its copy of the mask. In the same way, before transmitting a time slot, the MS scrambles the data with the MS-to-MSC mask, and the MSC performs the same operation on the received data.

If the inputs to the algorithm at the MS match those at the MSC, the voice-privacy masks at MS and MSC match also, and the original data are restored. Both masks depend on SSD_B which is not transmitted over the air. It is therefore practically impossible for outsiders to generate the masks.

12.6.9 Mobile-assisted Handoff

The AMPS handoff procedure has been described in Sections 12.1.7 and 12.3.5. In this procedure, the signal strength of the reverse voice channel used by the mobile is measured—on command from the MSC—by the base station of the cell that currently serves the mobile, and by a number of base stations in adjacent cells. As a result of the measurements, the MSC may decide to handoff the mobile to the cell that receives the strongest signal. This requires—in each cell—one or more measurement receivers that can be tuned to reverse channels that are not used in the cell itself.

In mobile-assisted handoff, an IS-54 mobile that is in conversation on a traffic channel can—on command from the MSC—make signal strength measurements on forward channels, and report its results. This eliminates the need for measurement receivers in base stations, and also reduces the handoff work at the MSC, which is important in high-traffic cellular systems. Measurements by an IS-54 mobile that is in conversation, and served by a traffic channel, are possible because the mobile receiver needs to be tuned to the assigned traffic channel only during two of the six time slots in a frame, and can be used for measurements during the other time slots.

The signaling for a handoff measurement, which requires several new messages, is shown in Fig. 12.6-5. The MSC orders the MS to start measure-

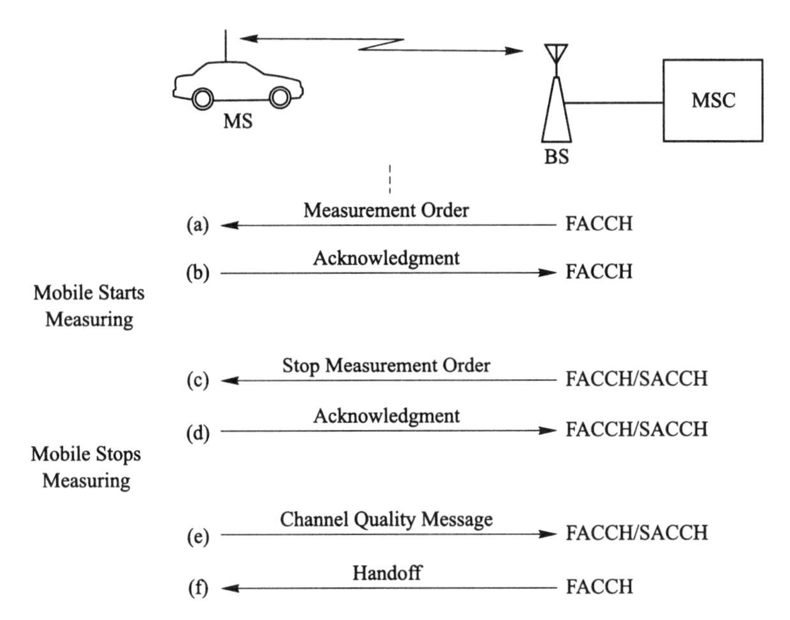

Figure 12.6-5 Mobile handoff signaling. FACCH: fast associated control channel. SACCH: slow associated control channel. (From IS-54-B. Reproduced with permission of TIA.)

ments with a measurement order (a) that identifies up to 12 channels to be measured in cells adjacent to the cell that is serving the MS. The MS responds with an acknowledgment message, and starts the measurements. The measurements continue until MS receives a stop measurements order (c). The MS acknowledges the order (d) and stops its measurements. It then sends a channel-quality message (e) that includes the received signal strength indicators (RSSI) of the currently used channel and of the channels specified by MSC. The value of RSSI ranges from 0 (very weak signal) to 31 (very strong signal). The MSC then decides whether to handoff the mobile to another cell. The measurement and handoff orders are sent on FACCH. The other messages can be sent on FACCH or SACCH.

IS-54 mobiles can be handed off to an analog or digital channel in an adjacent cell, regardless the type of the current channel.

12.7 INTRODUCTION TO THE GSM CELLULAR SYSTEM

The first-generation cellular systems developed in Europe were patterned after the AMPS system. However, each system has its own national standards, and a MS designed to work with the cellular system of a particular country cannot roam in another country.

In 1982, the Conférence Européenne des Postes et Télécommunications (CEPT) established a "Groupe Spéciale Mobile" (GSM) to start the definition of a Pan-European standard for a digital second-generation cellular system [8,9]. The resulting system is known today as GSM (Global System for Mobile

Communications). In 1988, the responsibility for GSM standards was transferred to the European Telecommunication Standards Institute (ETSI).

GSM was first deployed in 1993, and has gained wide acceptance, in part because GSM mobiles can roam in any country with a GSM network. As of November 1994, it was serving two million subscribers in 26 European countries, and had been adopted by another 26 countries outside Europe.

GSM radio channels are in the 900 MHz band. In the U.K. and Germany, this band was already allocated to other services, and these countries use the DCS1800 mobile system, which is very similar to GSM, but operates in the 1.8 GHz band.

GSM is a rather complicated system. This section presents a brief overview [8–11,14]. Signaling between a MS and a GSM network is discussed in Sections 12.8 and 12.9.

Most of the AMPS/IS-54 terminology also applies to GSM, but there are some differences. For example, in GSM documentation, cellular networks are known as public land mobile networks (PLMN), base stations (BS) are sometimes denoted as base station subsystems (BSS), and *location updating* is the equivalent of MS registration.

12.7.1 GSM Overview

GSM is a single-mode system, and has digital radio channels only. GSM mobile stations can receive service from GSM networks only.

GSM has several similarities with IS-54 (more correctly: IS-54, whose definition started in 1987, is similar in several respects to GSM). Both use a combination of frequency-division multiple access FDMA (RF carrier channels at different frequencies) and time-division multiple access TDMA (frames and time slots on the RF channels).

Like IS-54, GSM includes provisions for mobile-assisted handover (the European term for handoff), MS authentication, and *ciphering* (voice privacy). Moreover, it has been designed taking into account ISDN services, such as circuit- and packet-switched data communications.

12.7.2 GSM Radio Channels

A GSM network has 124 bidirectional RF channels, consisting of a *downlink* (forward, BS to MS) channel, and an *uplink* (reverse, MS to BS) channel [10, 14]. The GSM radio channels are spaced at 200 kHz intervals. The uplink and downlink channels occupy the frequency bands of 890–915 and 935–960 MHz. The center frequency of an uplink channel is 45 MHz below that of the downlink channel. The bit rate on RF channels is 270 kb/s in each direction.

Each RF (physical) channel contains a number of time-division multiplexed *logical* channels. We distinguish *dedicated* and *common control* channels:

Dedicated Channels. These are point-to-point (between a MS and a BSS) bidirectional channels. There are two groups of dedicated channels: *traffic channels* (TCH) and *stand-alone dedicated control channels* (SDCCH). A *slow* and a *fast associated control channel* (SACCH and FACCH) is associated with each TCH and SDCCH. The combination of a TCH and its SACCH and FACCH is known as a *traffic and associated control channel* (TACH).

A BSS has a pool of SDCCHs and a pool of TACHs. A SDCCH is allocated to a MS for call set-up signaling, and released when this signaling is complete. When the call set-up is successful, a TACH is allocated to the MS for the duration of the call. The TCH carries the user's speech or data, and the SACCH carries the "slow" signaling. When "fast" signaling is needed during a call (for example, when the MS has to be handed over), the user communication is suppressed for a short time, and the TCH functions as a FACCH (like the "blank and burst" technique used in AMPS—see Section 12.2).

Figure 12.7-1 shows the time slots, frames, and multiframes of TACHs. Each frame contains eight time slots, which have a duration of 0.557 ms and contain 148 bits, namely 114 information bits, 2 flag bits (F), and 32 bits for RF transmission functions. A frame consists of eight time slots, and has a duration of 4.615 ms. Twenty-six frames form a *multiframe* that has a duration of 120 ms.

A RF carrier can be used for eight TACHs, each of which uses a particular time slot. When a TACH has been allocated, the MS and the BS transmit in *burst periods* (BP) which occur in the time slot of the TACH. The TCH is sent in 24 BPs of the multiframe (frames 0–11 and 13–24). The SACCH is sent in frame 12, and no BP is transmitted in frame 25.

The information bit rate on a TCH or FACCH is therefore $24 \times 114/(0.120) = 22.8$ kb/s. The GSM speech coder produces 13 kb/s, and forward error-

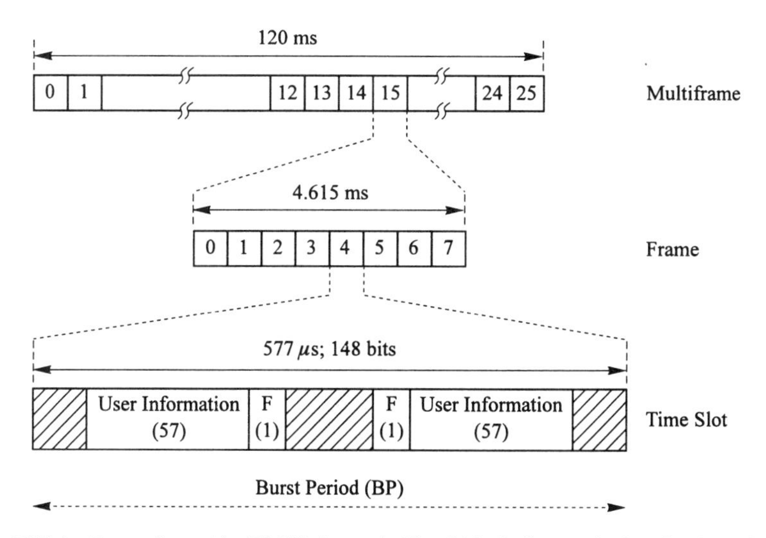

Figure 12.7-1 Frame format for TACH channels. The F bits indicate whether the time slot is a TCH or FACCH slot. (From *IEEE Comm. Mag.* **29.6**. Copyright © 1991 IEEE.)

correction of the speech (or user data) bits adds 9.8 kb/s. The information bit rate on a SACCH channel is 114/(0.120) = 0.95 kb/s.

Eight SDCCHs share one time slot on a RF carrier. Their information bit rate is thus one-eighth of the TCH bit rate (2.85 kb/s). The association of time slots and SDCCHs is rather complicated, and can be found in Ref. [14].

Common Control Channels (CCCH). One RF carrier in each cell contains a CCCH. This channel is always in time slot 0, and can be extended—if necessary—to time slots 2 and 4. The other time slots on the carrier can be used for dedicated channels.

Frames with CCCH are organized in multiframes that are different from the TACH multiframes. They contain 51 frames (numbered 0–50), and have a duration of 235 ms. A CCCH is time-divided into a number of common (point-to-multipoint, unidirectional) channels, for signaling between a BSS and all mobiles in the cell that are active, but not involved in a call:

BCCH (Broadcast Control Channel). This is a downlink channel that broadcasts information about the network (similar to the overhead parameter messages of AMPS).

SCH (Synchronization Channel). This is a downlink channel with information that enables the mobiles to acquire frame and time-slot synchronization with the BSS.

FCCH (Frequency Correction Channel). A downlink channel with information for mobiles concerning frequency synchronization with the RF carrier.

PAGCH (Paging and Access Grant Channel). This is a downlink channel that broadcasts paging messages. Also, when the network (MSC) allocates a SDCCH to a MS, it informs the MS with a message on PAGCH.

RACH (Random Access Channel). This is the only uplink common control channel, and is used by mobiles to request a SDCCH from the network.

On the downlink of CCCH, FCCH is always in frames 0, 10, 20, 30, and 40; SCH is always in frames 1, 11, 21, 31, and 41; and BCCH is always in frames 2–5.

The sizes (number of frames) of PAGCH and RACH depend on the traffic in the cell. They are determined by the operator of the PLMN, and allocated to respectively the remaining downlink frames and the uplink frames.

12.7.3 Signaling Interfaces and Protocols

The interfaces and protocols for signaling between a MS and the PLMN are shown in Fig. 12.7-2 [10]. ETSI has defined two interfaces: the Um (radio)

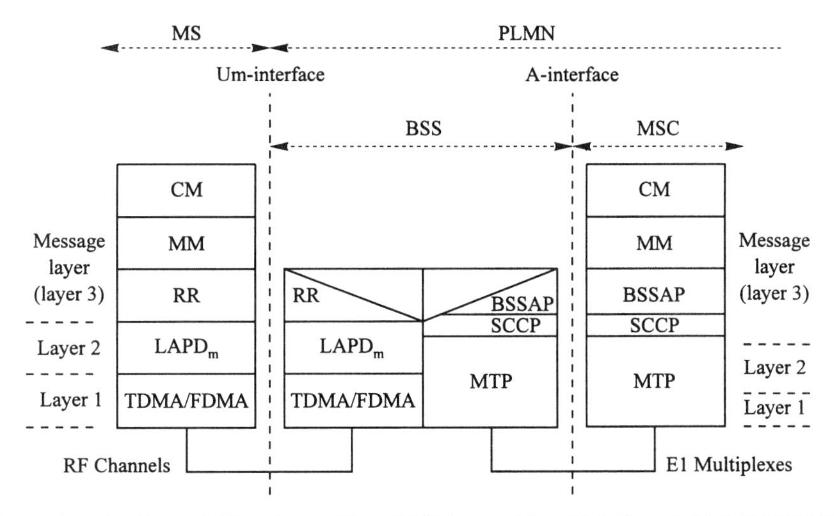

Figure 12.7-2 Um and A interfaces. (From *IEEE Comm. Mag.* **31.4**. Copyright © 1993 IEEE.)

interface between a MS and the BSS, and the A (cable) interface between BSS and MSC. This is different from the U.S. cellular systems (AMPS, IS-54) where it is assumed that an MSC and its associated base stations (BS) are supplied by the same manufacturer, and that the MSC-to-BSS interface is an internal manufacturer-specific interface.

The A-interface specifications of GSM enable the operator of a PLMN to purchase MSC and BS systems from different suppliers.

UM Interface. The signaling protocol on this interface has three layers.

Physical Layer (Layer 1). This consists of those parts of the RF channels that contain signaling channels (SACCH, FACCH, BCCH, SCH, FCCH, PAGCH, RACH, and SDCCH).

Data Link Layer (Layer 2). The protocol is known as LAPDm, and is a modified version of the link access protocol for D-channels of DSS1 (10.2). For details, the reader is referred to Ref. [15].

Message Layer (Layer 3). In the MS, this layer consists of three parts. The *radio resource management* (RR) sublayer at a MS communicates with its peer in the BSS. For example, when RR at BSS allocates a TACH or a SDCCH channel to MS, it informs the MS with a RR message.

The *mobility management* (MM) and *connection management* (CM) sublayers at a MS communicate with their peers at the MSC. MM and CM messages traverse the Um and A interfaces, and are transferred transparently across BSS.

The MM sublayer messages support MS *location updating* (the GSM counterpart of "registration") and *authentication.*

The CM sublayer has three parts: *call control* (CC), *supplementary services* (SS), and *short message service* (SMS). CC contains the messages for the set-up and release of connections to the MS. These messages are patterned after the Q.931 messages of DSS1 (Section 10.3). SMS is a service by which subscribers can send short (text) messages to a MS.

A Interface. The signaling protocol on this interface [16] consists of three layers that are similar to those in signaling system No.7.

Physical Layer (Layer 1). The BSS is connected to its MSC by digital E1 multiplexes (1.5.2). The majority of the 64 kb/s multiplexed channels are digital trunks. During a call, a digital trunk, in tandem with a TCH channel, conveys the MS user's speech or data between the MS and the MSC [17]. Other E1 channels are signaling data links. Layers 2 and 3 of the protocol are outlined below.

Data Link Layer (Layer 2). This consists of the message transfer part level 2 (MTP2) of signaling system No.7 (SS7)—see Chapter 8. MTP2 is responsible for the reliable transfer of signaling messages between MSC and BSS. Details can be found in Ref. [18].

Message Layer (Layer 3). The *base station subsystem application part* (BSSAP) is present at the MSC and BSS [19]. It is a user of the signaling connection control part (SCCP) of SS7 (Chapter 13). This is one of the few applications of connection-oriented SCCP. A signaling connection is established whenever a dedicated channel (SDCCH, FACCH, or FACCH) has been assigned to a MS.

BSSAP consists of two parts: the *direct transfer application part* (DTAP) and the *BSS management application process* (BSSMAP).

We start with BSSMAP at BSS. The RR at a BSS is involved in the allocation, encryption, and release of dedicated radio channels, and in the transmission and reception of RR messages on the common-control radio channels. The BSSMAP at BSS and MSC handles the transfer of RR-related BSSMAP messages. At a BSS, RR and BSSMAP communicate with each other. Figure 12.7-3 illustrates a few RR–BSSMAP interactions. In example (a), a MS sends a RR message—on the RACH channel—to request a dedicated radio channel. RR at BSS then allocates a channel, returns an RR message on PAGCH that includes the identity of the channel, and informs its BSSMAP, which then composes a message that informs MSC, and sends it to BSSMAP at the MSC. In example (b), MSC sends a BSSMAP message requesting the transmission of an RR message on a common-control radio channel. The requested RR message may be a paging message, which is then transmitted by BSS on the PAGCH channel, or a broadcast message, which is transmitted on the BCCH.

The DTAP at a BSS transparently (without processing by BSS), transfers MM and CM messages, received on dedicated radio channels, to a SS7 data link,

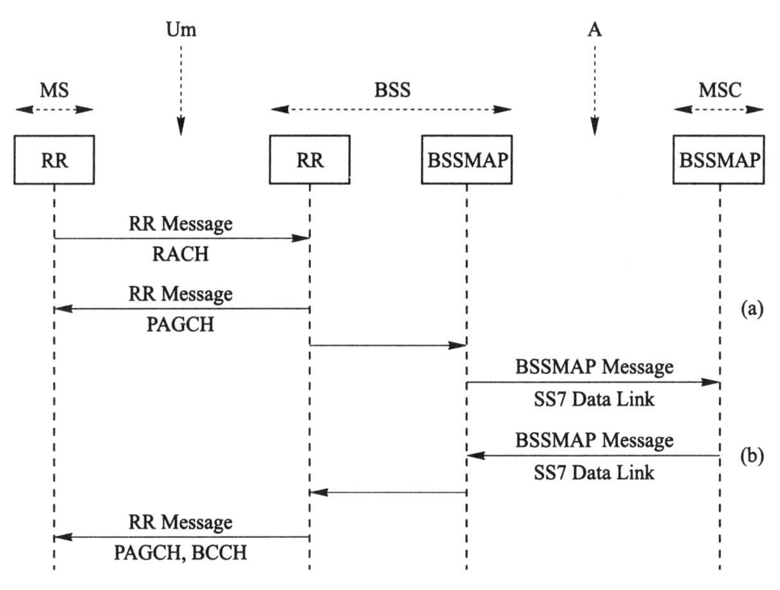

Figure 12.7-3 Transfer of RR messages. (From GSM 04.07 version 4.10.0. Courtesy of ETSI.)

which transports them to the MSC—see Fig. 12.7-4. Also, messages for a MS received from DTAP at the MSC are transferred to the radio channel that is currently dedicated to the MS.

The transfer of BSSMAP- and DTAP-messages involves the SCCP and MTP in the MS and BSS [12]. The SCCP at the sending end adds a *discrimination parameter* to the messages, which indicates whether the message belongs to BSSMAP or DTAP, and which is used by SCCP at the receiving end to deliver

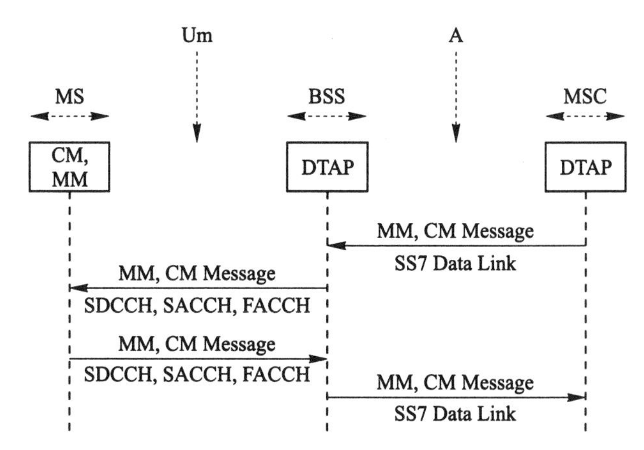

Figure 12.7-4 Transfer of MM and CM messages. (From GSM 04.07 version 4.10.0. Courtesy of ETSI.)

the message to the proper entity.

12.7.4 Identification of GSM Entities

Since GSM systems are deployed in many countries, CCITT and ETSI have standardized the identification of GSM entities according to a numbering plan specified in CCITT Rec. E.212 [20]. This plan is different from the CCITT Rec. 163/164 numbering plan for fixed networks.

PLMN Identity. A PLMN is uniquely identified by its *mobile country code* (MCC) and *mobile network code* (MNC). MCC consists of three digits, of which the first one indicates a world zone:

2 Europe
3 North America
4 The Mid-East and Western Asia
5 Eastern Asia and Australia
6 Africa
7 Latin America

The second and third digits represent individual nations in these zones. For example, the MCCs of the U.K. and Malaysia are 234 and 502.

The MNC code identifies a PLMN in a country. It consists of two digits, and is allocated by individual national organizations for mobile telecommunication standards.

Location Area Identity (LAI). This uniquely identifies a GSM location area. The service area of a PLMN is divided into a number of MSC service areas, and each of these is subdivided into location areas that consist of a number of adjacent cells. A MSC keeps track of the location areas of mobiles currently registered in its service area. When a MS has to be paged, paging messages are sent out in all cells of the mobile's present location area. The LAI format is:

LAI = MCC–MNC–LAC

where the location area code (LAC) identifies a location area within a PLMN. The code consists of up to four hexadecimal digits. LACs are allocated by the operators of individual PLMNs.

Mobile Station Identity. We begin by pointing out a significant difference between U.S. mobiles and GSM mobiles. A GSM mobile consists of two parts: the *mobile equipment* (ME) and the *subscriber identity module* (SIM). The ME is an "almost complete" mobile station, containing the RF equipment, keypad, mouthpiece, earphone, etc. The SIM is a small package (smart card) with semiconductor chips that store permanent and temporary information about the ME user. A user can insert his SIM into—and extract it from—any ME. The ME is operable only when a SIM has been inserted.

Splitting a MS into two parts allows a ME to be used—at different times—by different people. For example, a business may have a number of employees, each of whom has a SIM. The business also owns a smaller number of MEs which are used by the employees on a "when needed" basis.

Mobile station identities can be in two forms:

International Mobile Station Identity (IMSI). This uniquely identifies a mobile station in any GSM network. Its format is:

$$\text{IMSI} = \text{MCC–MNC–MSIN}$$

where MCC–MNC identifies the PLMN selected by the MS owner for mobile services, and MSIN (mobile station identity number) identifies a MS in that PLMN. The maximum length of MSIN is nine BCD digits.

An IMSI is allocated by the operator of the selected PLMN, and entered into permanent memory of the SIM.

The second MS identity format is a combination of LAI and TMSI.

Temporary Mobile Station Identity (TMSI). This is a 32-bit binary number which uniquely identifies the MS within one location area, or a group of adjacent location areas, of a PLMN. TMSI is a temporary identification, and is usually changed by the network when the MS enters a new location area. LAI and TMSI are stored in temporary SIM memory.

Most messages on the Um (radio) interface identify a MS by TMSI and LAI. IMSIs are used only in exceptional cases. TMSI gives protection against cloning (obtaining MS identifications for fraudulent use), because an intercepted TMSI no longer identifies the mobile after it has left the location area.

International Mobile Equipment Identity (IMEI). This is the counterpart of the mobile serial number (MSN) in AMPS. It uniquely identifies a ME. IMEI is a 15-digit number, entered into permanent ME memory by the manufacturer.

12.8 SIGNALING BETWEEN MOBILE AND NETWORK

12.8.1 Introduction

This section explores the most important signaling procedures on the Um (radio) interface between a MS and a GSM network [21]. As has been mentioned in the previous section, some layer 3 messages on this interface are between the MS and the BSS of the cell that serves the mobile, and others are between the MS and the MSC that controls the BSS. This section describes the signaling as seen by the MS, and considers the BSS and MSC as "the network."

Each message mentioned in this section points to the subsection of section 12.9 that contains the message description. It will be helpful to read Section 12.8 twice, first ignoring the message descriptions, and looking them up the second time around.

12.8.2 MS Initialization

When a MS is powered up by its user, it needs to find the RF carrier in its cell that carries the CCCH, and then to achieve synchronization with the broadcast channel (BCCH), the paging channel (PAGCH), and the random access channel (RACH), which enable the MS to listen to broadcast and paging messages, and to access the network, for location updating and originating calls.

In the AMPS/IS-54 system, a predetermined group of RF carriers has been designated by TIA as control channels, and a MS initializes by tuning to the strongest one of these channels (12.1.2). GSM does not designate the RF carriers that carry common-control channels, and a GSM mobile has to scan all RF carriers operating in the cell, and to find the carrier whose time slots 0 contain CCCH.

As a first step, MS scans the downlink RF channels, and looks for the frequency correction channel (FCCH), whose bursts have a distinguishable pattern (148 consecutive zeros), and occur in time slots 0 of frames 0, 10, 20, 30, and 40 of the 51-frame multiframe. When MS recognizes an FCCH burst, it knows that it is tuned to the RF carrier with CCCH, that the current time slot is slot 0, and that the current frame is one of the five listed above. The MS now initializes its (eight-step) time slot counter, and obtains time slot synchronization.

MS then reads the synchronization channel (SCH), whose bursts occur 8 BP after the FCCH bursts. SCH information indicates which one of the above frames is the current CCCH frame. The MS then initializes its (51-step) CCCH frame counter, and is now synchronized with CCCH frames. Since BCCH appears in predetermined frames (2–5) of CCCH, the MS is now also synchronized with BCCH, and begins to read the *system information* messages that are transmitted periodically on BCCH. There are messages that indicate the frames used for PAGCH and RACH. MS stores this information and is now synchronized with all common control channels in the cell.

SCH information also indicates the current frame number (0–25) in the multiframe for TACH channels. This enables the MS to initialize its (26-step) frame counter, for synchronization with the TACH channels.

12.8.3 Idle MS

Idle mobiles can move from cell to cell. In order to remain in contact with the network, a MS periodically compares the signal strength of the RF carrier with the CCCH channel in its cell to RF "CCCH" carriers in adjacent cells. The MS knows the channel numbers of these carriers from the system information messages.

If one or more of these RF carriers is stronger than the current carrier, MS retunes to the strongest one, and then reinitializes itself as described above.

12.8.4 Location Updating

GSM "location updating" is the counterpart of "registration" in AMPS and IS-54 systems. GSM networks keep track of the location area where the MS is operating [22]. When receiving an incoming call, the MS is paged in all cells of its current location area. GSM mobiles do a location update when entering a new location area, and at periodic intervals. In addition, some MSs are capable to do a location update when being activated or deactivated by their users.

Updating on Entering a New Location Area. The location area identity (LAI) is broadcast in system information messages on BCCH, and stored in MS memory. When a new received LAI does not match the previously stored LAI, the MS does a location update. This happens when an idle MS has tuned to a new BCCH carrier in a different location area.

Periodic Updating. Whenever a MS does a location update it resets a timer T. The timer has a time-out value of several hours. When the timer expires, the MS does a location update.

Updating on Deactivation and Activation. Mobiles equipped to do these updates send an IMSI DETACH message when being deactivated. The network then marks the MS as deactivated, and does not send paging messages for the MS until it is activated again. A MS that has sent an IMSI DETACH message does a location update when it has been activated again.

Figure 12.8-1 shows the messages and channels for a location updating procedure. Whenever a MS has to access the network, it first requests a SDCCH, by sending a CHANNEL REQUEST message (12.9.1) on RACH. The network then allocates a SDCCH channel, and returns an IMMEDIATE ASSIGNMENT message (12.9.1) on PAGCH, identifying the SDCCH. From this point on, all messages are on SDCCH.

MS sends a LOCATION UPDATING REQUEST message (12.9.2), which includes information about its transmission characteristics, its identity, and indicating whether it is equipped for enciphering (data encryption).

If the SIM in the MS has been used before, its memory stores the most recent location area identity (LAI) and TMSI, and MS identifies itself with these parameters. Otherwise, MS uses its IMSI (stored in semipermanent SIM memory). The network checks the MS identification and—if valid—starts the MS authentication.

The GSM authentication algorithm has two inputs: a random number RAND (128 bits) and an *individual subscriber authentication key* Ki (128 bits). The key is stored in the SIM of the mobile, and in the network. It is never transmitted on the UM (radio) interface. The MS and the network execute the algorithm, using the same RAND, and their respective Ki values. The results of the algorithm are the *signed result* SRES (32 bits), and the *cipher key* Kc (64 bits).

The network starts the authentication by sending an AUTHENTICATION REQUEST (12.9.2) which includes a RAND value. The mobile executes the algorithm, and returns an AUTHENTICATION RESPONSE (19.9.2) which includes SRES. The network compares the received SRES with the SRES it has

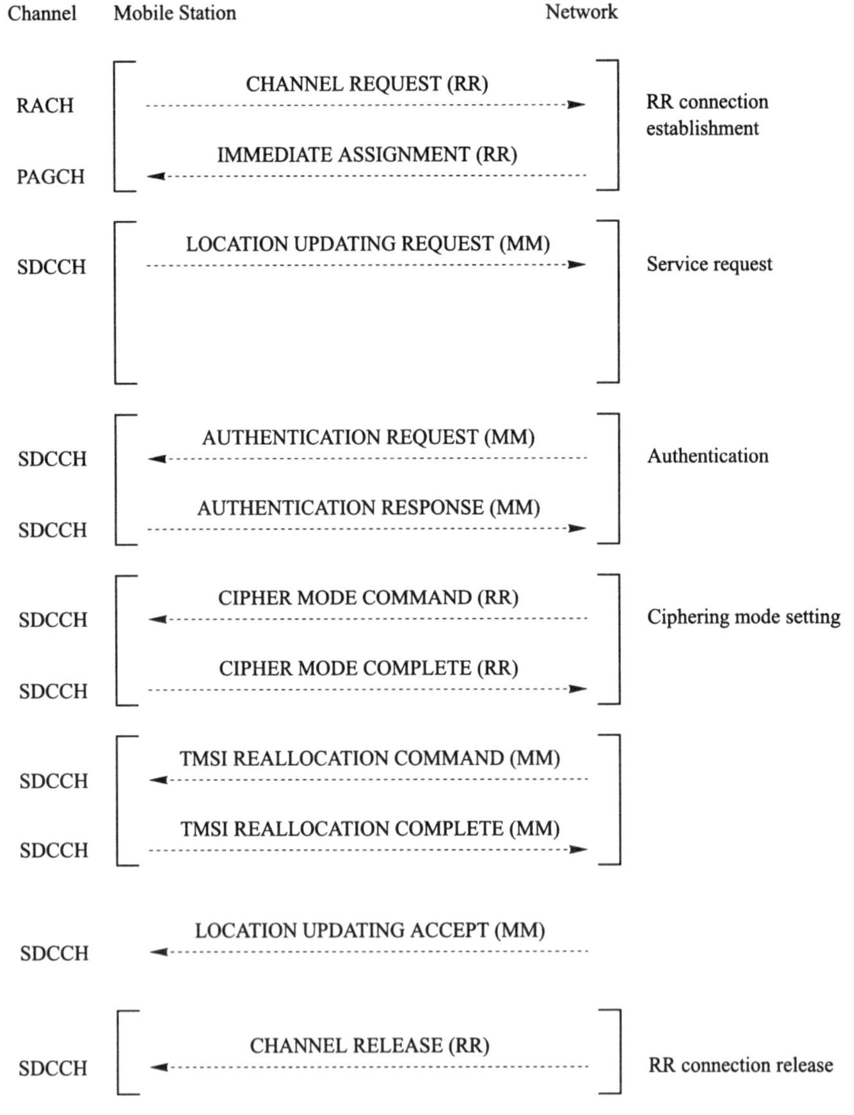

Figure 12.8-1 Location updating. (From GSM 04.08 version 4.10.0. Courtesy of ETSI.)

calculated. If they match, the network considers the MS as authenticated, and continues the location updating procedure.

If the MS has indicated that it can handle ciphered (encrypted) information, the network then sends a CIPHER MODE COMMAND message (12.9.1), and MS responds with CIPHER MODE COMPLETE (12.9.1) message. From this point on, MS and BSS encrypt their information on SDCCH, using the value of Kc to form an encryption/decryption mask. If the MS has indicated that it cannot handle ciphering, this step is omitted.

The network may decide to allocate a new TMSI to the mobile. In this case, it sends a TMSI REALLOCATION COMMAND (12.9.2) that includes the new TMSI. After entering the TMSI in its SIM, the MS sends a TMSI REALLOCATION COMPLETE message (12.9.2), and the network sends a CHANNEL RELEASE (12.9.1). It then releases SDCCH, and MS also releases the channel.

If neither the authentication nor TMSI reallocation is done, the network sends a LOCATION UPDATING ACCEPT message (12.9.2) to inform MS, and then sends the CHANNEL RELEASE.

12.8.5 Set-up of a Call Originated by MS

When the user of a MS originates a call, he first enters the called number and possibly additional information with the MS keypad, and then depresses the "send" button. Figure 12.8-2 shows the signaling for the set-up of the call.

The initial signaling is very similar to the initial signaling for location updating. MS first requests a SDCCH channel, with a CHANNEL REQUEST message (12.9.1). The network then assigns a SDCCH and returns an IMMEDIATE ASSIGNMENT message (12.9.1). At this point, the MS and BSS tune their transmitter and receiver to the SDCCH.

MS next sends a CM SERVICE REQUEST (12.9.2), indicating that it wishes to originate a call. The network then decides whether to execute or skip the MS authentication. If the MS has ciphering capability, the ciphering mode is started by command from the network.

The call-control messages that follow are similar to the corresponding Q.931 messages of Digital Subscriber Signaling System No.1 (10.3.2).

The MS sends a SETUP message (12.9.3) that includes the called party number, bearer capability, and other information. The network acknowledges with a CALL PROCEEDING message (12.9.3). It then allocates a TACH, sends an ASSIGNMENT COMMAND (12.9.1), and then releases SDCCH. On receipt of the command, MS sends an ASSIGNMENT COMPLETE (12.9.1) message, and tunes to TACH. From this point, the signaling is on the SACCH of TACH.

When the network receives an indication from the fixed network that the called party is being alerted, it sends an ALERTING message (12.9.3), and the MS notifies its user.

When the called party answers, the network sends a CONNECT message

(12.9.3), and the MS returns a CONNECT ACKNOWLEDGE (12.9.3). The conversation can now begin.

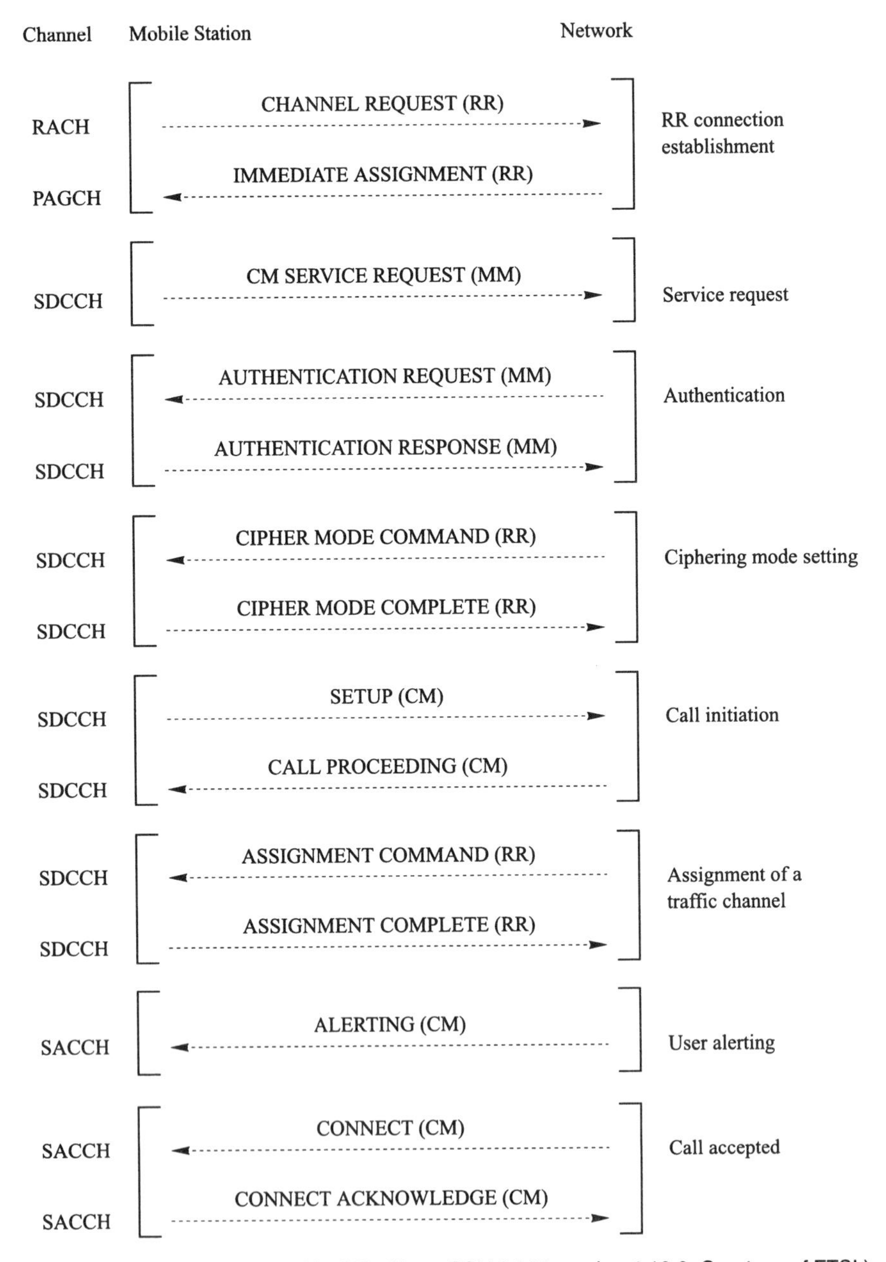

Figure 12.8-2 Call originated by MS. (From GSM 04.08 version 4.10.0. Courtesy of ETSI.)

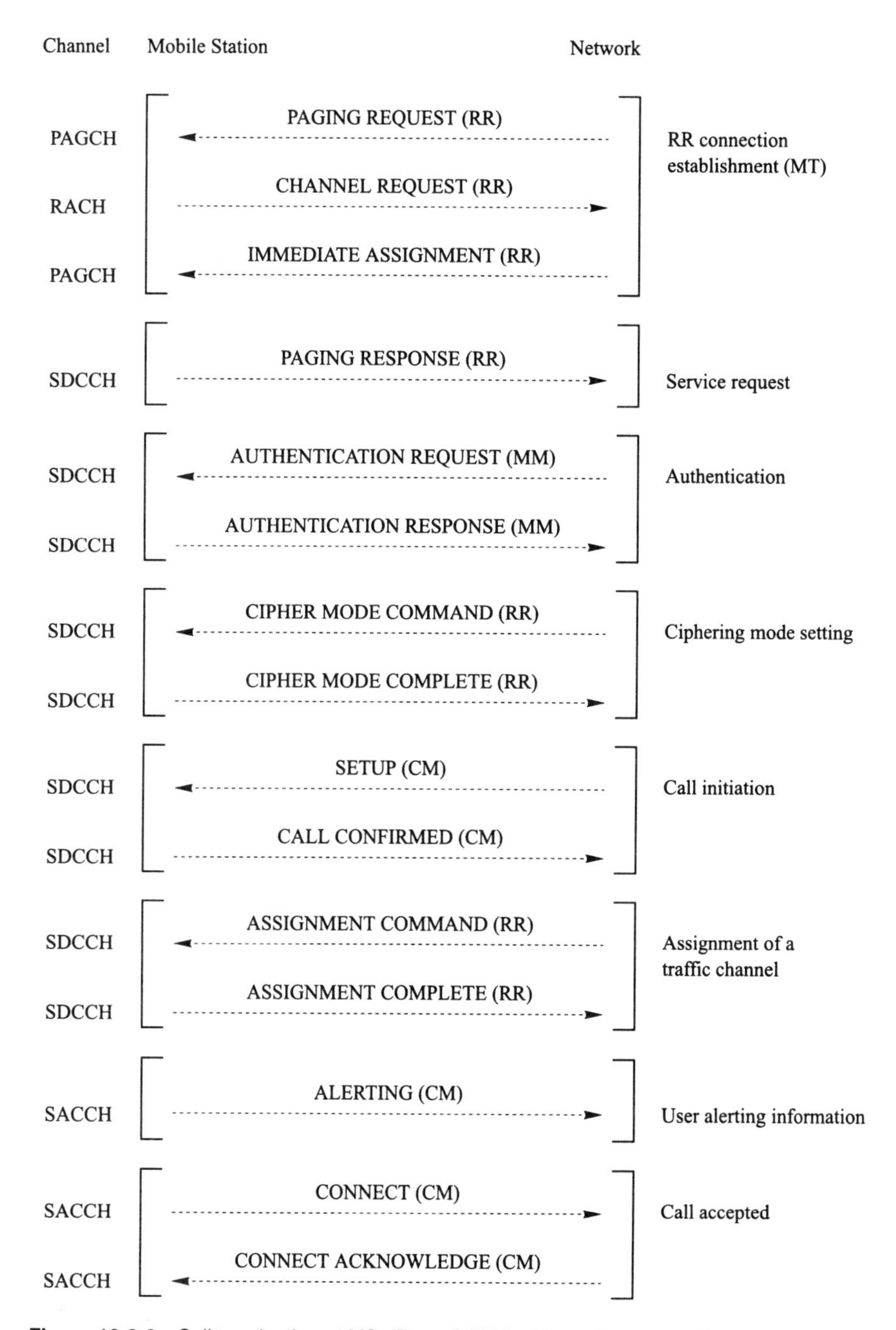

Figure 12.8-3 Call terminating at MS. (From GSM 04.08 version 4.10.0. Courtesy of ETSI.)

12.8.6 Connection Set-up for Call Terminating at MS

In Fig. 12.8-3, a MS is active and idle, and listening to the PAGCH of its cell. When the mobile network receives a call for the MS, it transmits a PAGING REQUEST in all cells of the MS location area (12.9.1). The called MS is identified by TMSI or IMSI. When recognizing that it is being paged, MS sends a CHANNEL REQUEST (12.9.1) message. The network then allocates an SDCCH, and returns an IMMEDIATE ASSIGNMENT message (12.9.1).

MS acknowledges the paging request with a PAGING RESPONSE message (12.9.1). The network then decides whether to execute or skip the MS authentication. If the MS has ciphering capability, the network initiates the ciphered mode on SDCCH.

The message sequence that follows is similar to that of 12.8.5, except that for some messages the roles of MS and the network are reversed. The network starts with a SETUP message (12.9.3), and MS acknowledges with a CALL CONFIRMED message (12.9.3).

At this point, the network allocates a TACH, and sends an ASSIGNMENT COMMAND (12.9.1). MS acknowledges with an ASSIGNMENT COMPLETE (12.9.3.1) message, and then tunes to TACH. The network then releases SDCCH, and signaling for the rest of the call is on the SACCH or FACCH of TACH.

MS sends an ALERTING message (12.9.3) when it starts to alert its user. When the user answers, MS sends a CONNECT message (12.9.3), the network returns a CONNECT ACKNOWLEDGE (12.9.3), and the conversation starts.

12.8.7 Release of Connection

The release of a connection can be initiated by a disconnect message from MS, or (via the network) by a disconnect message or signal from the other party. Figure 12.8-4 shows the message sequences for both cases. The DISCONNECT message (12.9.3) indicates that message originator has disconnected. The RELEASE message (12.9.3) acknowledges the receipt of the DISCONNECT, and indicates that its originator will stop sending and receiving on TACH. The RELEASE COMPLETE message (12.9.3) indicates that the other party also will stop sending and receiving information.

After the network has sent or received the RELEASE COMPLETE, it sends a CHANNEL RELEASE message (12.8.1), and then releases TACH. On receipt of the message, MS also releases TACH.

12.9 LAYER 3 MESSAGES ON THE Um INTERFACE

ETSI has defined some seventy GSM layer 3 messages on the Um interface. They are grouped according to the sublayers at the MS (Fig. 12.7-2). This section describes the messages mentioned in the previous section.

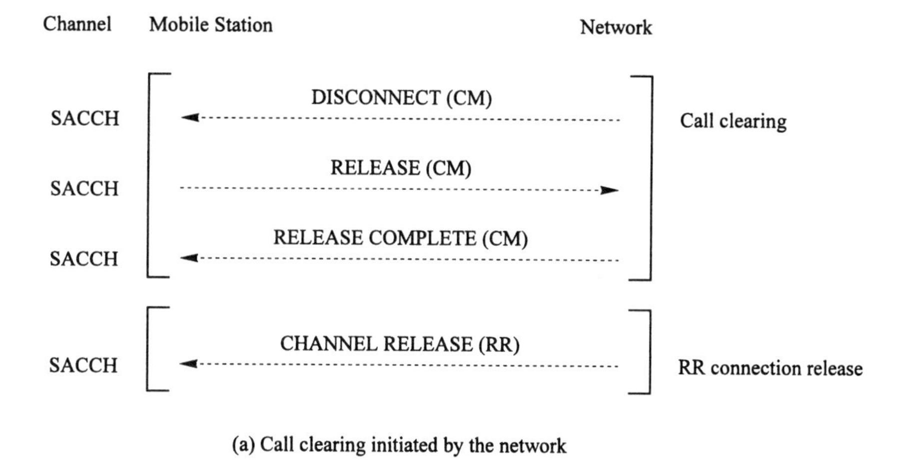

(a) Call clearing initiated by the network

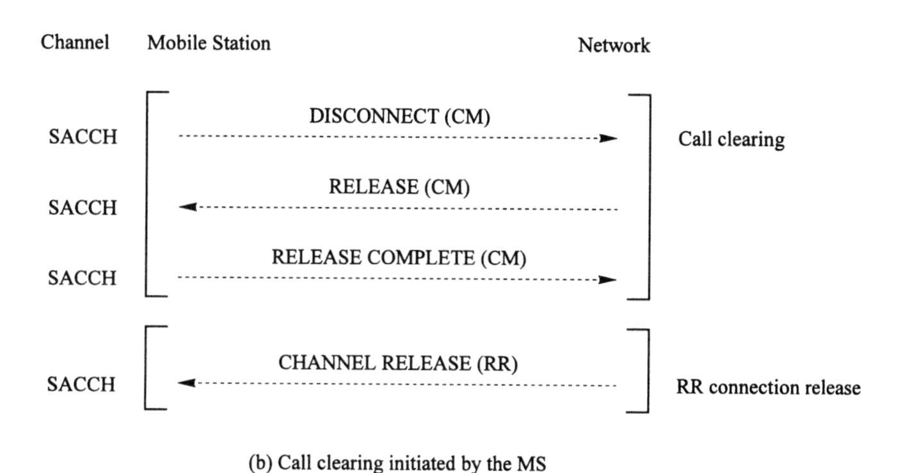

(b) Call clearing initiated by the MS

Figure 12.8-4 Call clearing. (From GSM 04.08 version 4.10.0. Courtesy of ETSI.)

Figure 12.9.1 shows the general message layout. The *protocol discriminator* specifies the message as an RR, MM or CM message.Within each protocol, the particular message is indicated by the *message type*. Messages can include mandatory and optional *information elements* IE. The descriptions include the most important IEs in the messages. For other IEs and details of IE coding, see Ref. [21].

12.9.1 Messages for Radio Resource Management (RR)

Assignment Command. Sent (on SDCCH) from the network to the MS, and indicating that a TACH has been assigned. Included is the *channel description* IE, which specifies the RF carrier channel number (ARFCN), and time slot of

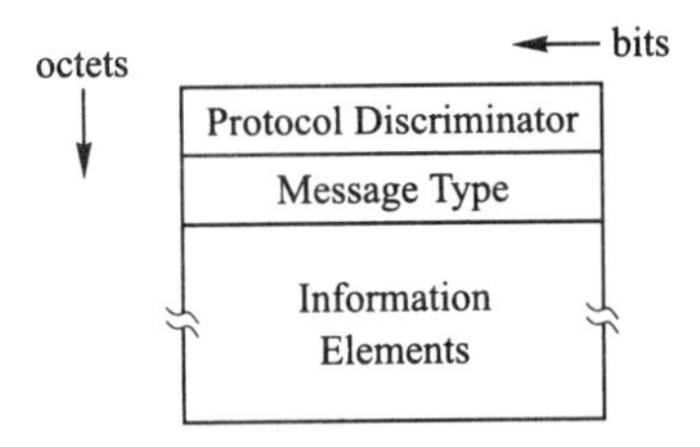

Figure 12.9-1 Layout of layer 3 messages on the Um interface. (From GSM 04.08 version 4.10.0. Courtesy of ETSI.)

TACH. Also included is the *channel mode* IE, which indicates the communication mode on the channel (signaling, speech, or data).

Assignment Complete. Response to the assignment command, sent by MS (on SDCCH). It indicates that the mobile will tune to the assigned channel.

Channel Release. Sent by the network—on the SDCCH or TACH that is currently assigned to the MS. It indicates that the network has released the channel. It includes the *release cause* IE, which gives the reason for the release (normal release, release caused by failure, etc.).

Channel Request. Sent (on RACH) by a MS which currently has no dedicated channel (SDCCH or TACH), to request a SDCCH. For security, the MS does not directly identify itself, but generates a random *reference number* IE, and includes it in the message.

Cipher Mode Command. Sent by the network (on SDCCH). It indicates that the network will encipher and decipher (encrypt and decrypt) all subsequent information on SDCCH and TACH, and requests the MS to do the same.

Cipher Mode Complete. Sent by MS (on SDCCH), to acknowledge the above command.

Immediate Assignment. This is sent by the network (on PAGCH), in response to a channel request message. It indicates that an SDCCH has been allocated, and requests the MS to tune to this channel. The message includes the *channel description* IE, which indicates the absolute RF carrier number (ARFCN), the time slot number (0–7) and, since a SDCCH is not full-rate channel, information regarding the frames in which the channel appears. Also included is the *request reference* IE, which is equal to reference number received in the channel request message. A MS accepts an assignment message when it recognizes the reference number of its channel request message.

Paging Request. Sent by the network (on PAGCH). It informs a MS about an incoming call, and includes the TMSI of the called MS.

Paging Response. This is sent by the MS (on SDCCH), to confirm the receipt of a paging request. It includes TMSI and the *MS class mark* IE. The latter IE specifies the mobile's RF transmitter power class, and indicates whether the mobile has ciphering capability.

System Information Messages. These messages are broadcast by the network on BCCH. There are several information messages. Examples of included IEs are:

- Location area identity (LAI). This identifies the location area of the cell that is broadcasting the message.
- Cell options. This indicates whether the MS is allowed to use discontinuous transmission (DTX, see Section 12.4.2).
- RACH control. This indicates the maximum number of allowed MS attempts to access RACH, and the number of burst periods (BP) between successive attempts.
- Neighbor cells description. This lists the ARFCNs of RF channels in neighboring cells that carry broadcast channels (BCCH).

12.9.2 Messages for Mobility Management (MM)

Authentication Request. Sent by the network (on SDCCH), to check the authenticity of the MS. It includes the random number IE (RAND).

Authentication Response. Sent by a MS (on SDCCH), in response to the authentication request. It includes the *signed response* (SRES).

CM Service Request. Sent by a MS (on SDCCH), to request a connection management (CM) service. It includes the *CM service type* IE, which specifies the requested service (for example, the set-up of a circuit-switched connection, the activation of a supplementary service, etc.). Also included are the MS Class mark, and TMSI.

IMSI Detach Indication. Sent by MS (on SDCCH), just before it powers down. It indicates that the MS cannot receive new incoming calls. It includes the MS class mark, and TMSI.

Location Updating Accept. Sent by the network (on SDCCH), to indicate acceptance of the location updating request. It includes the location area of the cell that sends the message.

Location Updating Request. Sent by MS (on SDCCH). It includes the MS

Class mark, and the IMSI, or TMSI and the location area (LAI) stored in the SIM of the mobile.

TMSI Reallocation Command. Sent by the network, to inform the MS that a new TMSI has been allocated. It includes TMSI, and the location area of the cell that sends the message.

TMSI Reallocation Complete. Sent by MS, acknowledging that it has received the reallocation command, and has entered the data into its SIM.

12.9.3 Messages for Circuit Switched Call Control (CC)

We only consider messages in the CC part of the connection management (CM) layer. GSM intends to provide both PSTN and ISDN services, and the CC messages are the counterparts of the Q.931 messages of Digital Subscriber Signaling System No.1 (DSS1—see 10.3.2). The information elements are also patterned after those of DSS1 (10.3.4), but have been adapted for mobile communications. All messages are sent on a control channel of the TACH that is currently dedicated to the MS.

Alerting. On calls originated by MS, the message is sent from the network, and indicates that the called party is being alerted. On calls to MS, the message is sent by MS, and indicates that it is alerting its user.

Call Confirmed. This message is sent by a called MS, indicating that it is processing the received setup message. The message has no counterpart in Q.931.

Call Proceeding. This message is sent by the network, on calls originated by MS. It acknowledges the setup message, and indicates that the connection is being set up.

Connect. On calls originated by a MS, the message is sent by the network. On calls terminating at a MS, it is sent by the MS. It indicates that the called party has answered.

Connect Acknowledge. This message acknowledges the receipt of a connect message.

Disconnect. Indicates that the MS user, or the other party in the call, has disconnected. It requests the release of the connection. The message is sent by the MS or the network, depending on which party has cleared. It includes the *cause* IE, which indicates the disconnect reason (for example, normal call clearing).

Release. This message is sent in two situations. It either acknowledges a received disconnect message, or indicates a network-initiated release. In the latter case, the message includes a *cause* IE, which indicates why the call set-up has failed. For example: unassigned called number received, called party busy, network failure, etc. The message indicates that the originating entity is about to release the connection at its end.

Release Complete. The message acknowledges a release message. It indicates that its originator will release the connection at its end.

Set-up. Sent by a MS or the network, on respectively calls originating or terminating at the MS. The IEs that may be included are outlined below.

- Bearer capability. Mandatory in direction MS → MSC. Optional in the other direction. This is a compound IE, which includes several information items, for example:
  - Radio channel requirement (full rate, half rate).
  - Information transfer capability (speech, unrestricted digital information, 3.1 kHz audio, group 3 facsimile).
- Called party address. Mandatory in direction MS → MSC. Optional in the other direction.
- Calling party address. Optional in both directions.
- Called and calling party subaddresses. Optional in both directions.
- High and low layer compatibility. Optional in both directions.
- User-user information. Optional in both directions.

12.10 ACRONYMS

ACT	Action field
AMPS	Advanced mobile phone service
AUC	Authentication center
AUTH	Authentication indicator
AUTHR	Result of authentication calculation
BCCH	Broadcast channel
BCD	Binary coded decimal
BI	Busy/idle bit
BIS	Busy/idle status indicator
BP	Burst period
BS	Base station
BSS	Base station subsystem
BSSAP	Base station subsystem application part
BSSMAP	Base station subsystem management application process
BST	Base station trunk
BSTG	Base station trunk group

CAVE	Cellular authentication and voice encryption
CC	Call control
CCCH	Common control channel
CCITT	International Telegraph and Telephone Consultative Committee
CDMA	Code-division multiple access
CEPT	European Conference of Postal and Telecommunications Administrations
CFW	Control filler word
CHAN	Channel number
CM	Connection management
CMAC	Mobile attenuation code on control channel
CMAX	Number of access channels
CMN	Cellular mobile network
CPA	Combined paging and access indicator
DCC	Digital color code
DL	Data link
DOT	Dotting (synchronization) sequence
DSS1	Digital subscriber signaling system No.1
DTAP	Direct transfer application part
DTX	Discontinuous transmission
DVCC	Digital verification color code
E	Extended address indicator
EIA	Electronic Industries Association
ETSI	European Telecommunications Standards Institute
FACCH	Fast associated control channel
FCCH	Frequency correction channel
FOCC	Forward control channel
FVC	Forward voice channel
GAW	Global action word
GSM	Global system for mobile communications
IMEI	International mobile equipment identity
IMSI	International mobile station identity
IS-54	U.S. standard for TDMA mobile system
IS-95	U.S. standard for CDMA mobile system
ISDN	Integrated services digital network
Kc	Cipher key
Ki	Individual subscriber authentication key
LAC	Location area code
LAI	Location area identity
LT	Last try indicator
MAC	Mobile attenuation code
MCC	Mobile country code
MIN	Mobile identification number
MIN1	Exchange code and line number of MIN
MIN2	Area code of MIN

MM	Mobility management
MNC	Mobile network code
MPCI	Mobile protocol capability indicator
MS	Mobile station
MSC	Mobile switching center
MSCTG	Trunk group between MSCs
MSG TYP	Message type
MSIN	Mobile subscriber identity number
MSN	Mobile serial number
MSG TYP	Message type field in mobile-control messages
N	Number of paging channels
NAWC	Number of additional words coming
OHD	Overhead field
OPM	Overhead parameter message
ORD	Order field in mobile-control messages
ORDQ	Order qualifier field in mobile-control messages
PACH	Paging and access grant channel
PIN	Personal identification number
PLMN	Public land mobile network
PSCC	Color code of present voice channel
PSTN	Public switched telecommunication network
PW1	First word of overhead parameter message
PW2	Second word of overhead parameter message
RACH	Random access channel
RAND	Random number
RANDC	Eight leading bits of RAND
RCF	Read control filler indicator
RECC	Reverse control channel
REGH	Registration indicator for home mobiles
REGID	Registration ID
REGINCR	Registration increment
REGR	Registration indicator for roaming mobiles
RF	Radio frequency
RR	Radio resource management
RSVD	Reserved message field
RVC	Reverse voice channel
SACCH	Slow associated control channel
S	MSN indicator
SACCH	Slow associated control channel
SAT	Supervision audio tone
SCH	Synchronization channel
SCM	Station class mark
SDCCH	Stand-alone dedicated signaling channel
SID	System identification (15 bits)
SID1	14 leading bits of SID

SSD_A	Shared secret data for authentication
SSD_B	Shared secret data for voice privacy
ST	Signaling tone
TACH	Traffic channel and its associated signaling channels
TCH	Traffic channel
TDMA	Time-division multiple access
TG	Trunk group
TIA	Telecommunications Industry Association
TMSI	Temporary mobile station identity
VC	Voice channel
VMAC	Mobile attenuation code on voice channels
WFOM	Indicator: wait for overhead message
WS	Word synchronization pattern

12.11 REFERENCES

1. Z.C. Fluhr, P.T. Porter, "Advanced Mobile Phone Service: Control Architecture", *Bell Syst. Tech. J.*, **58**, no.1, January 1979.
2. K.J.S. Chadha, C.F. Hunnicutt, S.R. Peck, J. Tebes Jr, "Advanced Mobile Phone Service: Mobile Telephone Switching Office," *Bell Syst. Tech. J.*, **58**, no.1, January 1979.
3. W.C.Y Lee, *Mobile Cellular Telecommunication Systems*, Second Edn., McGraw Hill, New York, 1995.
4. *Public Land Mobile Network: General Aspects*, Rec. Q.1001, CCITT Blue Book, **VI.12**, ITU, Geneva, 1989.
5. *Public Land Mobile Network: Network Functions*, Rec. Q.1002, CCITT Blue Book, **VI.12**, ITU, Geneva, 1989.
6. *Mobile Station—Land Station Compatibility Specification*, EIA/TIA-553, Electronic Industries Association, Washington, D.C., 1989.
7. *Cellular Features Description*, TIA/EIA/IS-53, Telecommunications Industry Association, Washington, D.C., 1991.
8. D.J. Goodman, "Trends in Cellular and Cordless Communications," *IEEE Comm. Mag.*, **29**, no. 6, June 1991.
9. J.E. Padgett*et al.*, "Overview of Wireless Communications," *IEEE Comm. Mag.*, **33**, No.1, January 1995.
10. M.Rahnema, "Overview of the GSM System and Protocol Architecture," IEEE Comm. Mag., **31**, no.4, April 1993.
11. D.D. Falconer, F. Adachi, B. Gudmundson, "Time Division Multiple Access Methods for Wireless Personal Telecommunications," *IEEE Comm. Mag.*, **33**, no.1, January 1995.
12. *Cellular System Dual—Mode Mobile Station—Base Station Compatibility Standard*, EIA/TIA/IS-54-B, Telecommunications Industry Association, Washington, D.C., 1992.
13. *Mobile Station—Base Station Compatibility Standard for Dual—Mode Wideband*

Spread Spectrum Cellular System, TIA/EIA/IS-95, Telecommunications Industry Association, Washington D.C, 1993.

14. M. Mouly, M.B. Pautet, *The GSM System for Mobile Communications*, published by the authors, France, 1992.
15. *European Digital Cellular Telecommunication System (Phase 2), MS—BSS Interface Data Link Specification*, **GSM 04.06**, European Telecommunications Standards Institute, Sophia Antipolis, France, 1994.
16. *European Digital Cellular Telecommunication System (Phase 2), BSS—MSC Interface Principles*, **GSM 08.02**, European Telecommunications Standards Institute, Sophia Antipolis, France, 1994.
17. *European Digital Cellular Telecommunication System (Phase 2), BSS—MSC Interface Layer 1 Specification*, **GSM 08.04**, European Telecommunications Standards Institute, Sophia Antipolis, France, 1994.
18. *European Digital Cellular Telecommunication System (Phase 2), Signaling Transport Mechanisms between BSS and MSC*, **GSM 08.06**, European Telecommunications Standards Institute, Sophia Antipolis, France, 1994.
19. *European Digital Cellular Telecommunication System (Phase 2), MSC to BSS Layer 3 Specification*, **GSM 08.08**, European Telecommunications Standards Institute, Sophia Antipolis, France, 1994.
20. *Identification Plan for Land Mobile Stations*, Rec. E.212, CCITT Blue Book, **II.2**, ITU, Geneva, 1989.
21. *European Digital Cellular Telecommunication System (Phase 2), Mobile Radio Interface Layer 3 Specification*, **GSM 04.08**, European Telecommunications Standards Institute, Sophia Antipolis, France, 1994.
22. *Public Land Mobile Network: Location Registration Procedures*, Recommendation Q.1003, CCITT Blue Book, **VI.12**, ITU, Geneva, 1989.

13

INTRODUCTION TO TRANSACTIONS

The TUP and ISUP protocols (Chapters 9 and 11) are SS7 protocols for trunk-related exchange actions (mainly call control). The SS7 signaling takes place between the exchanges at the ends of a TUP or ISUP trunk.

Starting with this chapter, the focus is on SS7 signaling for operations that are not trunk-related.

13.1 DEFINITIONS AND APPLICATIONS

13.1.1 Definitions

Remote Operation. This is an operation that is not trunk-related, and is executed by one signaling point (SP), at the request of another SP.

Transaction. This is a dialogue consisting of signaling messages between two SPs, for the execution of one or more remote operations [1–3].

Transactions can involve two exchanges, an exchange and a network database, an exchange and a maintenance center, etc.

13.1.2 Applications

Transactions are the successors of the *direct signaling* procedures of common-channel interoffice signaling (CCIS)—see Section 6.5. The applications of present-day transactions fall into several groups:

Intelligent Network Services such as "800" and "900" calls (6.5.1). For these calls, an exchange needs to query a network database to obtain a routing number that corresponds to the received 800 or 900 number (6.5.1). Intelligent network (IN) services are discussed in Chapter 16.

Services for Mobile Telecommunications [3]. As one example, transactions are used to keep track of the present location of each active mobile station (MS). This information is used when routing a call to a MS. Services for mobile telecommunications are described in Chapter 17.

Centralized Operation, Administration, and Maintenance (OAM) of a telecommunication network, where an OAM center uses transactions to verify and change the data stored in exchanges, to request a test of network equipment, etc.

Bulk Data Transfer, for example, the daily transfer of billing records from an exchange to a centralized revenue accounting center.

The transactions in the first two groups support the set-up of calls, and are known as "on-line" transactions. They have to be executed with minimal delays, since they add to the overall time to set up a call, and usually require two short messages.

The transactions in the third and fourth groups are "off-line." They are less time-critical, and usually require more—and longer—messages.

13.2 SS7 ARCHITECTURE FOR TRANSACTIONS

13.2.1 Application Service Elements

We say that a signaling point (SP) performs a set of functions in a telecommunication network. In the CCITT/ITU-T model of signaling system No.7, each function (application) that requires remote operations has an *application service element* (ASE), which handles the signaling aspects for that application [4–6]. In Fig. 13.2-1, SP-A is an exchange equipped for two transaction types, and has two ASEs. INASE handles the signaling messages for IN services, and OMASE takes care of the messages in operation/administration/maintenance (OM) transactions. The INASE and OMASE in SP-A communicate with their respective "peers" in an IN database (SP-B), and in an OAM center (SP-C).

The messages exchanged between ASEs are known as *components*.

13.2.2 Infrastructure for Transactions

Figure 13.2-2 shows the SS7 entities at a signaling point that are involved in the transfer of transaction messages.

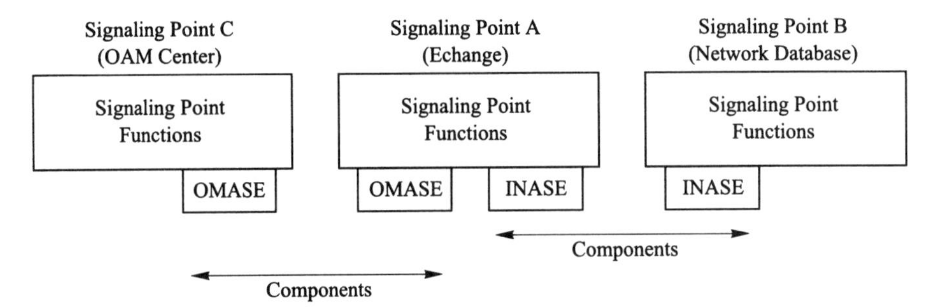

Figure 13.2-1 Application service elements.

The ASEs in a signaling point are the users of the *transaction capabilities application part* (TCAP) of signaling system No.7. In turn, TCAP is a user of the *signaling connection control part* (SCCP), and SCCP uses the services of the message transfer part (MTP) (Chapter 8).

In the CCITT/ITU-T model of SS7, an ASE passes its outgoing components to TCAP, in TC-request primitives. TCAP places one or more components in a TCAP message (or TCAP package), which is passed to SCCP in a N-unitdata request. SCCP then builds a SCCP message, and passes it to MTP in a MTP-transfer request. Finally, MTP forms a message signal unit (MSU), which is sent out on a signaling link (Section 8.8).

Incoming components arrive in MSUs on a signaling link. MTP extracts the SCCP message, and passes it to SCCP in a MTP-transfer indication. In turn,

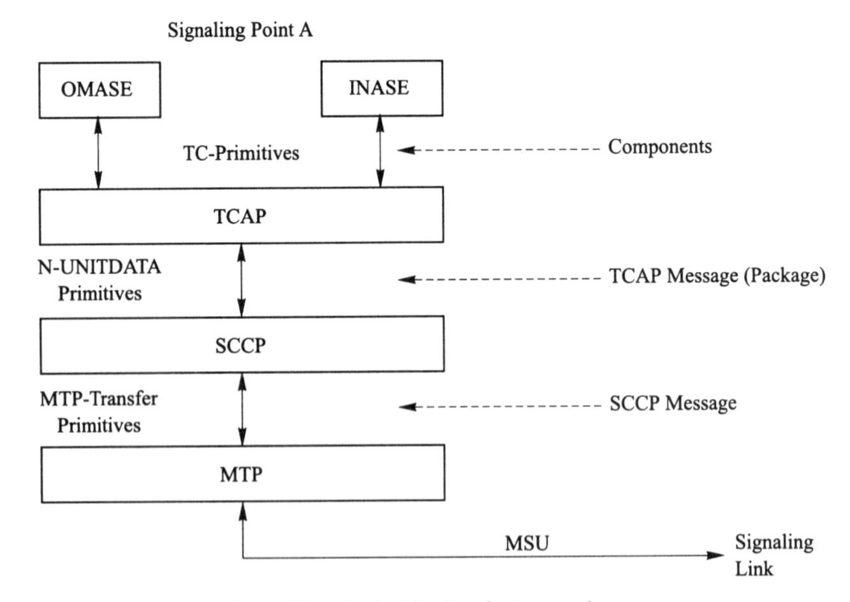

Figure 13.2-2 Architecture for transactions.

SCCP extracts the TCAP package, and passes it to TCAP in a N-unitdata indication. Finally, TCAP extracts the component(s) and passes them to the appropriate ASE, in TC-indications.

SCCP and TCAP are described in Chapters 14 and 15.

13.2.3 Application Independence of TCAP

In Fig. 13.2-1, the ASEs are dedicated to specific applications. TCAP serves all ASEs at a signaling point, and is specified as an application-independent protocol.

13.2.4 Identification of ASEs

Within a signaling point, an ASE is identified by a *subsystem number* (SSN). In a telecommunication network, an ASE is uniquely identified by the point code (PC) of its signaling point, and its SSN. The combination PC + SSN is known as the "SCCP address" of an ASE, and is used by SCCP to deliver transaction messages to their destination ASEs.

13.3 ACRONYMS

ANSI	American National Standards Institute
ASE	Application service element
CCIS	Common-channel interoffice signaling
CCITT	International Telegraph and Telephone Consultative Committee
IN	Intelligent network
INASE	ASE for intelligent network applications
ISDN	Integrated services digital network
ISUP	ISDN user part
ITU	International Telecommunication Union
ITU-T	Telecommunication standards department of ITU
MS	Mobile station
MSU	Message signal unit
MTP	Message transfer part
OAM	Operations, administration and maintenance center
OMASE	ASE for operational, administrative and maintenance applications
PC	Point code
SCCP	Signaling connection control part
SP	Signaling point
SSN	Subsystem number
SS6	Signaling system No.6
SS7	Signaling system No.7
TCAP	Transaction capabilities application part
TUP	Telephone user part

13.4 REFERENCES

1. A.R. Modaressi, R.A. Skoog, "Signaling System No.7: A Tutorial," *IEEE Comm. Mag.*, **28**, no.7, July 1990.
2. R. Manterfield, *Common Channel Signalling*, Peter Peregrinus Ltd, London, 1991.
3. T.W. Johnson, B. Law, P. Anius, "CCITT Signalling system No.7: Transaction Capabilities," *Br. Telecomm. Eng.*, **7**, April 1988.
4. *Specifications of Signalling System No.7*, Rec. Q.700, CCITT Blue Book, **VI.7**, ITU, Geneva, 1989.
5. *Specifications of Signalling System No.7*, Rec. Q.771, CCITT Blue Book, **VI.9**, ITU, Geneva, 1989.
6. *Ibid.*, Rec. Q.775.

14

SIGNALING CONNECTION CONTROL PART

The message transfer part (MTP) of SS7 (Chapter 8) was designed to transfer TUP (and later on, ISUP) messages between exchanges at the ends of a trunk. The *signaling connection control part* (SCCP), in combination with MTP, provides the transfer of messages that are not related to individual trunks, for example, transaction messages [1–3]. The combination of MTP and SCCP is known as the *network service part* (NSP) of SS7, and is the equivalent of layers 1, 2, and 3 of the OSI (open systems interconnection) protocols in data communication systems (7.1.1).

SCCP has been documented by CCITT [4], and later by ITU, in a series of Recommendations [5–8]. The U.S. version has been specified by ANSI [9].

14.1 INTRODUCTION

Figure 14.1-1 shows SCCP and its relations with the other parts of SS7. TCAP and ISUP are SCCP users. In turn, the ASEs at a signaling point are users of TCAP, and can be considered as "indirect" SCCP users. Each SCCP user at a signaling point has a *subsystem number* (SSN) that ranges 1 through 127. When SCCP receives an incoming message from MTP, it uses SSN to deliver the message to the appropriate SCCP user (in the case of ASEs, the message is passed to TCAP, which then delivers it). In this chapter, the term *subsystem* is used to denote an SCCP user.

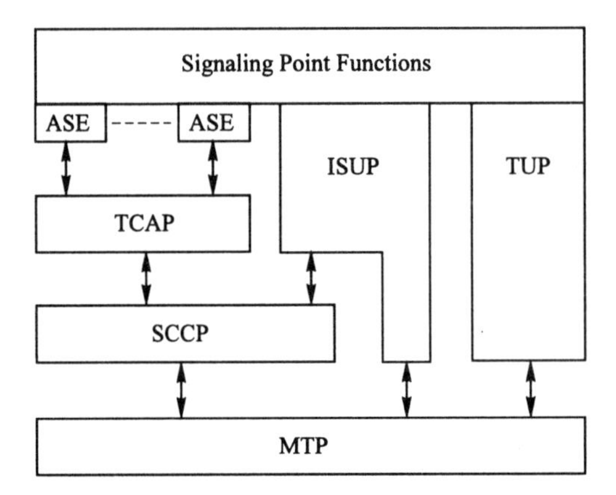

Figure 14.1-1 Position of SCCP in signaling system No.7. (From Rec. Q.700. Courtesy of ITU-T.)

14.1.1 Message Transfer Enhancements

SCCP enhances the message transfer capabilities of SS7 in two principal ways. In the first place, MTP uses the service indicator (SI) (8.8.3) in an incoming message to deliver it to the appropriate MTP user (SCCP is one of them). SI has a range 0 through 15, limiting the number of MTP users to 16. SCCP expands the addressing capability of SS7, allowing up to 127 subsystems at a signaling point.

In the second place, SCCP also allows a calling subsystem to address a called subsystem by a *global title* (GT). This is a functional address, in the form of a digit string, that cannot be used for message routing. The purpose of GTs is discussed in Section 14.3. SCCP includes provisions to translate a global title into an address that can be used to route a message to its destination subsystem.

14.1.2 SCCP Message Transfer Services

SCCP provides four service classes to its users:

- Class 0: Basic connectionless service.
- Class 1: Connectionless service with sequence control.
- Class 2: Basic connection-oriented service.
- Class 3: Connection-oriented service with flow control.

In connectionless services, the SCCPs at the signaling points of the two subsystems involved in a transaction are not aware of the transaction. In connection-oriented service, a (virtual) "signaling connection" between the SCCPs is established first, making both SCCPs aware of the transaction. At the end of the transaction, the signaling connection is released.

Messages for on-line transactions (Section 13.1) are usually transferred by connectionless SCCP service. Connection-oriented service is intended primarily for the transfer of messages in off-line transactions that involve a large amount of data. In addition, one group of on-line transactions in the Global System for Mobile Communications (GSM) uses connection-oriented SCCP—see Chapter 17.

14.1.3 Structure and Interfaces of SCCP

SCCP consists of the parts shown in Fig. 14.1-2. Connectionless control (SCLC) and connection-oriented control (SCOC) handle the message transfer in the corresponding services.

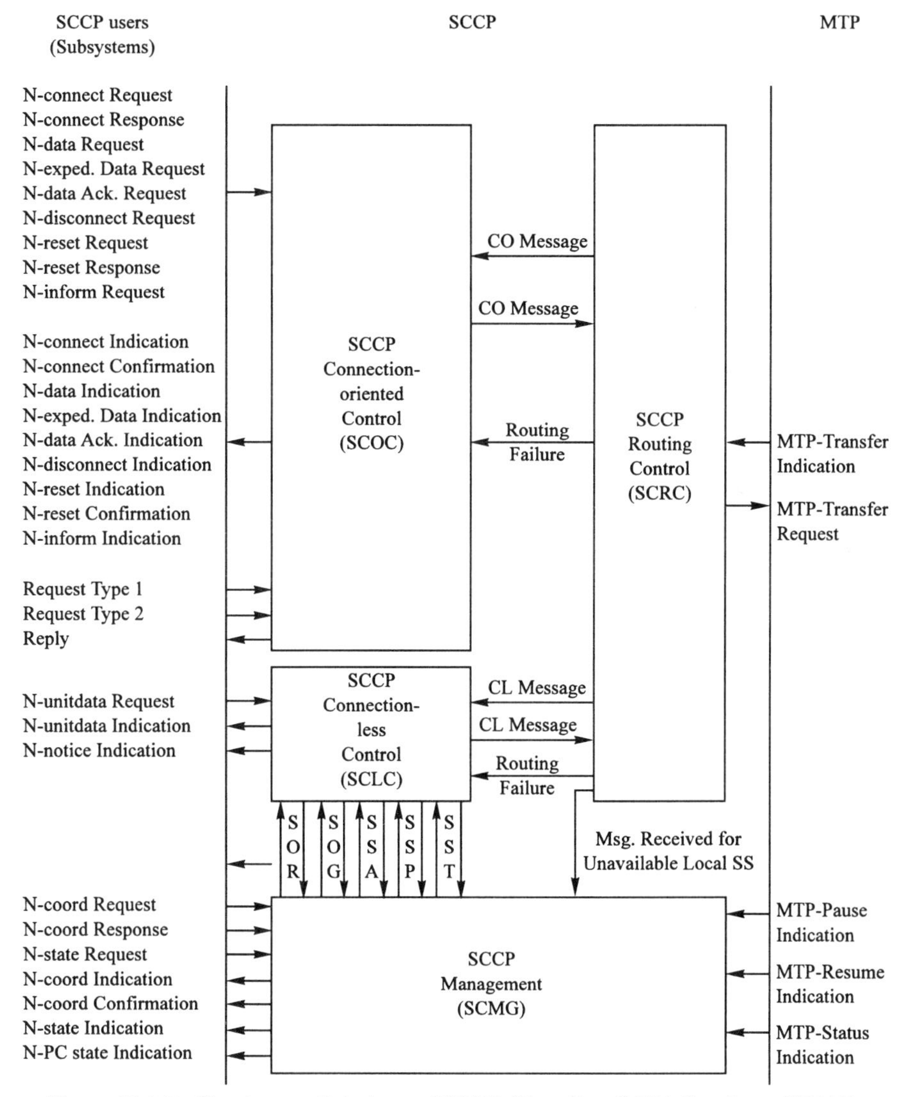

Figure 14.1-2 Structure and interfaces of SCCP. (From Rec. Q.714. Courtesy of ITU-T.)

Routing control (SCRC) determines the destinations of outgoing messages, and routes incoming messages to SCLC or SCOC, which then deliver them (directly, or via TCAP) to the destination subsystems.

SCCP management (SCMG) has functions comparable to those of MTP3 signaling network management (Section 8.8). It aims to maintain the message traffic of the SCCP users under congestion and failure conditions in the signaling network, signaling points, and subsystems.

The interfaces between SCCP and MTP are as described in Sections 8.8 and 8.9. MTP-transfer primitives are the interface between SCCP routing control and MTP for message transfer. MTP management passes MTP-pause, MTP-resume and MTP-status indications to SCCP management.

N-Primitives form the interface between SCCP and its local subsystems (either directly, or via TCAP). We distinguish three groups of primitives. The group to/from connection-oriented control, the group to/from connectionless control, and the group to/from SCCP management.

14.2 SCCP MESSAGES AND PARAMETERS

This section discusses a number of SCCP messages for the transfer of information between subsystems. SCCP messages are located in the user message (UM) fields of MTP3 messages (Fig. 8.8-1).

14.2.1 General Message Format

The format of SCCP messages is very similar to that of ISUP messages (see Fig. 11.2-2, without the CIC octets). A message can have mandatory (M) parameters with fixed or variable lengths, and optional (O) parameters. As in ISUP messages (Section 11.2), the mandatory fixed-length parameters appear with their *value* (*contents*) field only. The mandatory variable-length parameters have a pointer, and a length and a value field. Each optional parameter has a name, length, and value field. The name field is also known as *tag* field.

14.2.2 Message Types

The messages that will be discussed in this chapter are outlined below [4,6]. The first two apply to connectionless SCCP service, the others are used in connection-oriented service.

Unitdata Message (UDT). This is sent by an SCCP, to transfer subsystem data.

Unitdata Service Message (UDTS). This is sent to the SCCP that originated a UDT message, by an SCCP that cannot deliver a received UDT message to its destination.

Connection Request Message (CR). This is a request from a calling SCCP to a called SCCP, to set up a signaling connection.

Connection Confirm Message (CC). This is sent by called SCCP, indicating that it has set up the signaling connection.

Connection Refused Message (CREF). This is sent by the called SCCP, indicating that it is unable to set up the signaling connection.

Data Form 1 (DT1). This is a message sent by an SCCP at either end of a signaling connection, and containing subsystem data (used in class 2 operation).

Data Form 2 (DT2). This is a message sent by an SCCP at either end of a signaling connection. It contains subsystem data, and acknowledges the receipt of messages (class 3 operation).

Released Message (RLSD). This is sent by SCCP at one end of the signaling connection, and indicating that it wants to release the signaling connection.

Release Complete Message (RLC). This is sent in response to a RLSD message, and indicating that the sending SCCP has released the connection.

The message type codes of these messages are listed in Table 14.2-1.

14.2.3 Message Parameters

This section describes the most important parameters in the above messages. At this point is it suggested merely to peruse the descriptions, and to refer back to them when reading the later sections of this chapter.

To allow the reader to locate a parameter description quickly, each parameter has a reference number (for example, Par.1). In the sections that follow, a parameter is always identified by name and reference number.

Table 14.2-1 Message type codes of SCCP messages.

Acronym	Name	Message Type Code
CC	Connection confirm	0000 0010
CR	Connection request	0000 0001
CREF	Connection refused	0000 0011
DT1	Data form 1	0000 0110
DT2	Data form 2	0000 0111
RLC	Release complete	0000 0101
RLSD	Released	0000 0100
UDT	Unitdata	0000 1001
UDTS	Unitdata service	0000 1010

Source: Rec. Q.713. Courtesy of ITU-T.

Table 14.2-2 Parameters in SCCP messages.

Reference Number	Parameter Name	Message Acronyms								
		CC	CR	CREF	DT1	DT2	RLC	RLSD	UDT	UDTS
Par.1	Called party address (CDA)	O	M	O					M	M
Par.2	Calling party address (CGA)		O						M	M
Par.3	Destination local reference	M		M	M	M	M	M		
Par.4	Protocol class	M	M						M	
Par.5	Refusal cause		M							
Par.6	Release cause							M		
Par.7	Return cause									M
Par.8	Return option								M	
Par.9	Segmenting/reassembling				M					
Par.10	Sequencing/segmenting					M				
Par.11	Source local reference	M	M				M	M		
Par.12	SCCP user data (subsystem data)	O	O	O	M	M		O	M	M

Acronym	Message	Service Class			
		0 (CL)	1 (CL)	2 (CO)	3(CO)
CC	Connection confirm			X	X
CR	Connection request			X	X
CREF	Connection refused			X	X
DT1	Data form 1			X	
DT2	Data form 2				X
RLC	Release complete			X	X
RLSD	Released			X	X
UDT	Unitdata	X	X		
UDTS	Unitdata service	X	X		

CL: connectionless service. CO: connection-oriented service.
Source: Rec. Q.712. Courtesy of ITU-T.

Table 14.2-2 lists the parameters, and the messages in which they appear.

The focus is on parameter contents, and only the "value" fields are shown in the figures. The name (tag) fields of parameters that can appear as optional parameters are coded as follows:

Par.1	Called Party Address:	0000 0010
Par.2	Calling Party Address:	0000 0011
Par.12	User Data:	0000 1111

Par.1 Called Party Address (CDA). This variable-length parameter is a composite, consisting of several sub-parameters. It can contain any

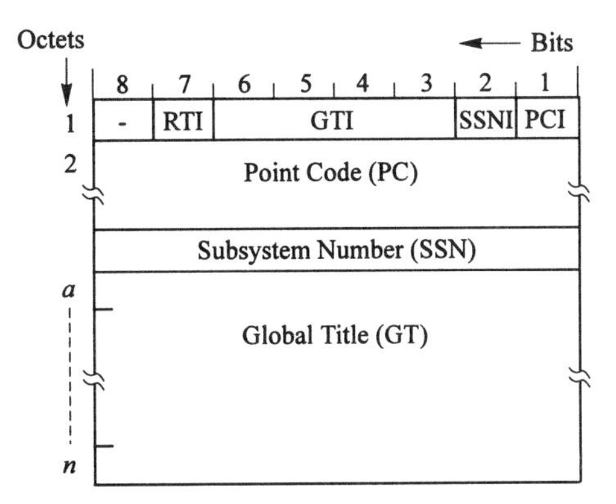

Figure 14.2-1 General format of: called party address (Par.1), calling party address (Par.2). PCI: point code indicator, SSNI: subsystem number indicator, GTI: global title indicator, RTI: routing indicator. (From Rec. Q.713. Courtesy of ITU-T.)

combination of point code (PC), subsystem number (SSN), and/or global title (GT). The general format is shown in Fig. 14.2-1.

Octet 1 indicates the presence or absence of PC, SSN, and GT in the address.

Point Code Indicator (PCI). This indicates the presence (1) or absence (0) of PC.

Subsystem Number Indicator (SSNI). This indicates the presence or absence of SSN.

Global Title Indicator (GTI). If GTI = 0000, no global title is present. If GTI is 0001, 0010, 0011, or 0100, a GT is present. The GT format depends on the value of GTI.

Routing Indicator (RTI). If the called address includes a global title, the sending SCCP specifies with RTI how the receiving SCCP should route the message:

RTI = 0 Global title translation should be performed, and routing should be based on the translation result.

RTI = 1 GT translation should not be performed, and routing should be based on the destination point code (DPC) in the MTP routing label of the message (Section 8.8).

Octets 2 through n contain PC, SSN, and GT, if included.

Point Code (PC). The point codes in CCITT No.7 signaling and ANSI No.7 signaling have respectively 14 and 24 bits (Section 7.2), and occupy three and two octets, respectively.

Subsystem Number (SSN). The one-octet SSN field identifies the SCCP user (subsystem). Examples of SSN codes standardized by CCITT/ITU:

Subsystem	8	7	6	5	4	3	2	1 ← Bits
SSN field not used	0	0	0	0	0	0	0	0
SCCP management	0	0	0	0	0	0	0	1
ISDN user part	0	0	0	0	0	0	1	1
Operation, maintenance, and administration part (OMAP)	0	0	0	0	0	1	0	0
Mobile application part (MAP)	0	0	0	0	0	1	0	1

Global Title. The GT contents consist of the *translation type* (TT) and the *global title address* (GTA)—see Fig. 14.2-2. We first consider GTA, which contains several parameters. The presence (P) or absence (A) of the octets with individual parameters is indicated by the value of GTI.

GT always includes GTA—consisting of a digit string (octets d through n) and may include the following parameters:

Encoding Scheme (octet b)	4	3	2	1 ← Bits
BCD (binary coded decimal), odd number of digits	0	0	0	1
BCD, even number of digits	0	0	1	0

Numbering Plan (octet b)	8	7	6	5 ← Bits
Telephony/ISDN numbering plan	0	0	0	1
Maritime mobile numbering plan	0	1	0	1
Land mobile numbering plan	0	1	1	0

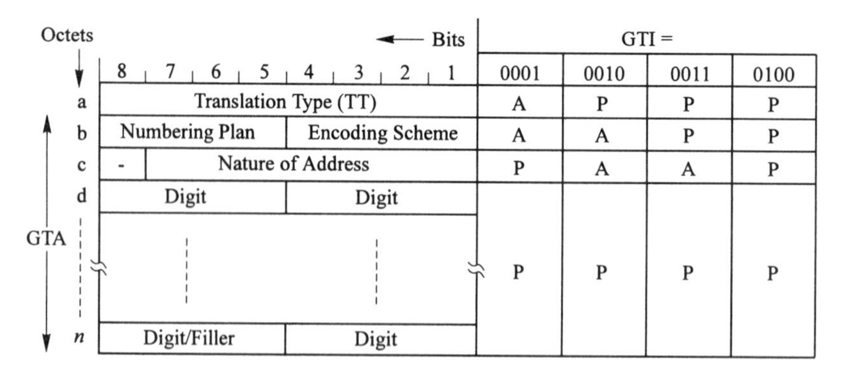

Figure 14.2-2 Global title format. P: octet is present. A: octet is absent. (From Rec. Q.713. Courtesy of ITU-T.)

Nature of Address (octet c)	← Bits 7	6	5	4	3	2	1
Subscriber number	0	0	0	0	0	0	1
National number	0	0	0	0	0	1	1
International number	0	0	0	0	1	0	0

If a GTA parameter is absent, the corresponding GT characteristic is implied. As an example, GTI is usually set to "0010" in U.S. networks. This implies that GTA is a "national number" (10 digits), that the digits are BCD coded, and that the numbering plan is the North American Telephony/ISDN numbering plan.

Translation Type (octet a)
Most GT translations yields the PC + SSN (point code and subsystem number) of a called subsystem. TT specifies the type of translation to be performed (see Section 14.3.3). The TT values in a particular network are established by the network operator.

Par.2 Calling Party Address (CGA). The format and coding are as in Par.1. PC and SSN are always included; GT is present only in transactions between signaling points in different networks.

Par.3 Destination Local Reference (DLR). This is a reference number (three octets) that identifies a signaling connection at the SCCP in the "called" (destination) signaling point of a connection-oriented message.

Par.4 Protocol Class (PRC). This represents the class of SCCP operation (Section 14.1.2), and occupies bits 4,...1 of one octet:

Protocol Class	← Bits 4	3	2	1
Class 0	0	0	0	0
Class 1	0	0	0	1
Class 2	0	0	1	0
Class 3	0	0	1	1

Par.5 Refusal Cause. This one-octet parameter appears in the connection refused (CREF) message only. It indicates the reason why a requested connection cannot be set up, for example:

Refusal Cause	← Bits 8	7	6	5	4	3	2	1
(a) Refused by called subsystem	0	0	0	0	0	0	1	1

(b) Unknown destination address	0	0	0	0	0	1	0	0
(b) Destination inaccessible	0	0	0	0	0	1	0	1
(b) Subsystem failure	0	0	0	0	1	0	1	0

Cause (a) indicates that the called subsystem has refused the connection. Causes (b) indicate that an SCCP along the message path was unable to transfer the connection request message.

Par.6 Release Cause. This is a one-octet parameter that appears in Released (RLSD) messages only. It indicates why a signaling connection is being released, for example:

							← Bits	
Release Cause	8	7	6	5	4	3	2	1
(a) Originated by subsystem	0	0	0	0	0	0	1	1
(b) Subsystem failure	0	0	0	0	1	0	0	0
(b) Network failure	0	0	0	0	1	0	1	0

Cause (a) indicates a normal release of a connection, by one of the subsystems. Causes (b) originate at an SCCP, and indicate that the connection has been released because of a failure.

Par.7 Return Cause. A one-octet parameter that appears in unitdata service (UDTS) messages only. It indicates why a unitdata message is being returned, for example:

							← Bits	
Return Cause	8	7	6	5	4	3	2	1
(a) No translation for an address of this nature	0	0	0	0	0	0	0	0
(b) No translation for this specific address	0	0	0	0	0	0	0	1
(c) Failure of called subsystem	0	0	0	0	0	0	1	1
(d) Subsystem not equipped	0	0	0	0	0	1	0	0
(e) Signaling network failure	0	0	0	0	0	1	0	1
(f) Other failures	0	0	0	0	0	1	1	1

Return causes (a) and (b) appear in UDTS messages originated at the SCCP in signal transfer point that has been unable to translate a global title. UDTS messages with return causes (c) and (d) are sent by the SCCP in the destination signaling point. UDTS messages, with return causes (e) and (f), can be sent by any SCCP in a signaling point along the message path.

Par.8 Return Option (RO). This appears in unitdata (UDT) messages only.

It indicates whether the UDT message should be returned to the originating subsystem if it cannot be delivered to its destination:

Return Option	← Bits 8	7	6	5
No return required	0	0	0	0
Return required	1	0	0	0

Par.9 Segmenting/Reassembling (one octet). This appears in data form 1 messages only (class 2 connection-oriented service). Only bit 1 is used:

Bit 1	
0	Last data form of transaction
1	More data forms follow

Par.10 Sequencing/Segmenting (Fig. 14.2-3). This appears in data form 2 messages (class 3 connection-oriented service) only. P(S) and P(R) are send and receive sequence numbers. Bit M indicates whether more data forms follow:

M = 0 Last data form in a sequence
M = 1 More data forms follow

Par.11 Source Local Reference (SLR). This is a reference number (three octets) that identifies a signaling connection at the SCCP in the "calling" (source) signaling point of a connection-oriented message.

Par.12 Subsystem Data. This is a variable-length parameter that passes information originated by—and destined to—a subsystem. The data are transferred transparently by SCCP.

14.3 CONNECTIONLESS SCCP

This section examines the role of connectionless SCCP in the transfer of messages between subsystems. The material in this section is illustrated by a transaction in which an exchange queries a network database to obtain the translation of an 800 number into a routing number (2.1.3).

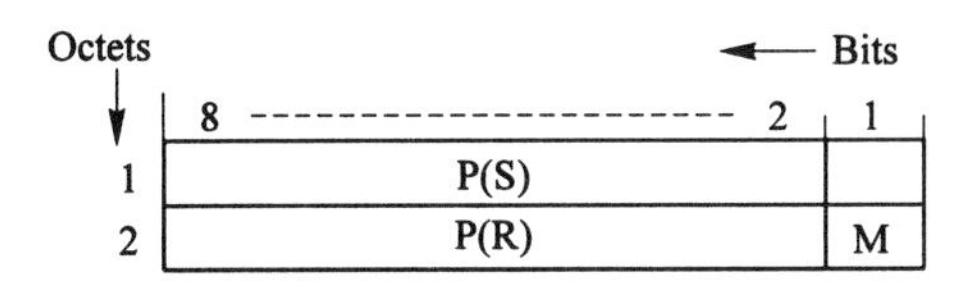

Figure 14.2-3 Sequencing/segmenting (Par.10).

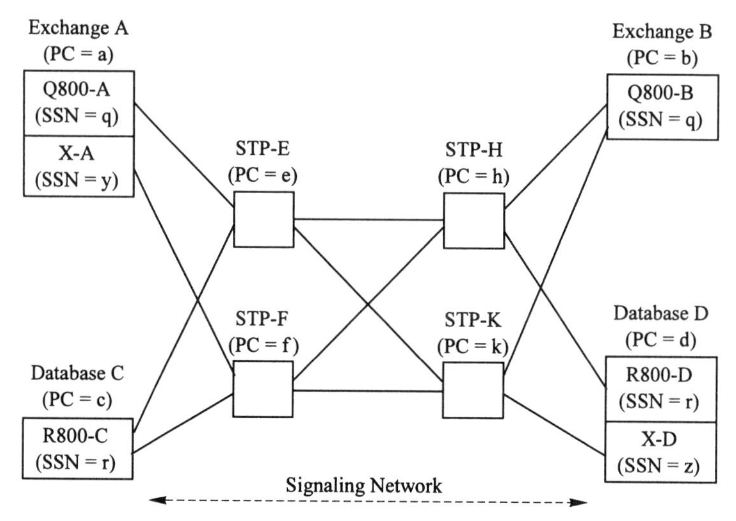

Figure 14.3-1 Transaction to obtain 800-number translation. STP: signal transfer point.

In Fig. 14.3-1, the subsystems (ASEs) denoted by Q800 and located at exchanges A and B, make the 800-number queries. The subsystems denoted by R800 are located at network databases C and D, and respond to these queries with messages that contain the national number of the called party, and routing and charging information. Databases C and D are "mates," and contain the same information. Normally, C and D are both in-service, and the network sends 50% of the query traffic to each of them. However, if C or D is out-of-service, all queries are sent to its mate.

When a R800 has received a query, it sends a response message with routing instructions for the 800 call.

Points E, F, H, and K are signal transfer points in the signaling network. The point codes (PC) in signaling points A through K are a through k. The subsystem numbers (SSN) for the Q800 and R800 ASEs are q and r.

When reading the material that follows, it is helpful to look up the parameter descriptions in Section 14.2.3.

14.3.1 Unitdata Messages

In connectionless service, transaction messages are passed between SCCP and MTP as SCCP *unitdata* messages—see Fig. 14.3-2 [4,6]. The message type code (MTYP) indicates: unitdata message. All message parameters are mandatory:

Par.1 Called Party Address (CDA). This is the address of the subsystem for which the message is intended.

Par.2 Calling Party Address (CGA). This is the address of the subsystem that originated the message.

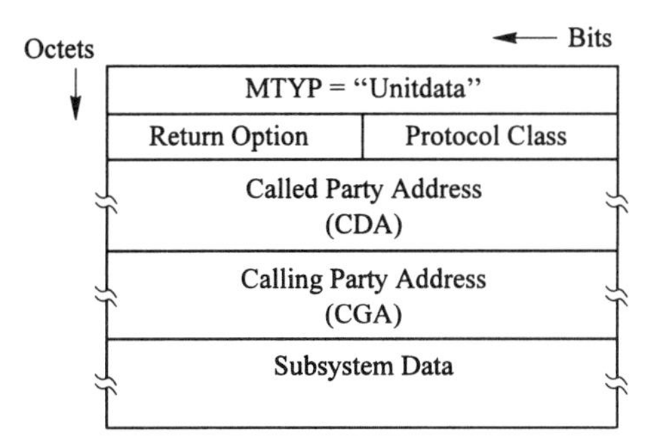

Figure 14.3-2 SCCP unitdata message. (From Rec. Q.713. Courtesy of ITU/T.)

Par.4 Protocol Class (PRC) indicates the requested service class: basic connectionless service (0), or connectionless service with sequence control (1).

Par.8 Return Option (RO). This indicates whether the calling subsystem wishes to be informed if its message cannot be delivered.

Par.12 Subsystem Data. This can be a TCAP message (package), or an ISUP message.

14.3.2 Primitives

In connectionless SCCP, subsystem data are passed in N-unitdata requests and indications. The parameters in the primitives are shown in Fig. 14.3-3 [4,5]. On receipt of a N-unitdata request, SCCP forms an unitdata message, using the information in the request. In connectionless service, SCCP does not maintain address information for the messages of individual transactions, and all N-unitdata requests therefore include a called and calling party address.

SCCP forms the unitdata message. SCCP routing control (14.1.3) determines the destination point code (DPC) and signaling link selector (SLS) for the message, and passes the outgoing unitdata message to MTP in a MTP-transfer request (see Fig. 14.1-2 and Section 8.8). Service indicator value (SI) = 0011 indicates an SCCP message.

Incoming messages with SI = 0011 are passed by MTP to SCCP, in MTP-transfer indications. SCCP routing control passes the subsystem data in N-unitdata indications to ISUP or TCAP, depending on the value of SSN in the called party address.

The N-unitdata indications include all information listed in Fig. 14.3-3, except the called party address, the protocol class and return option (they are not significant to the receiving TCAP or ISUP). TCAP then delivers the information in the package to the subsystem specified by SSN.

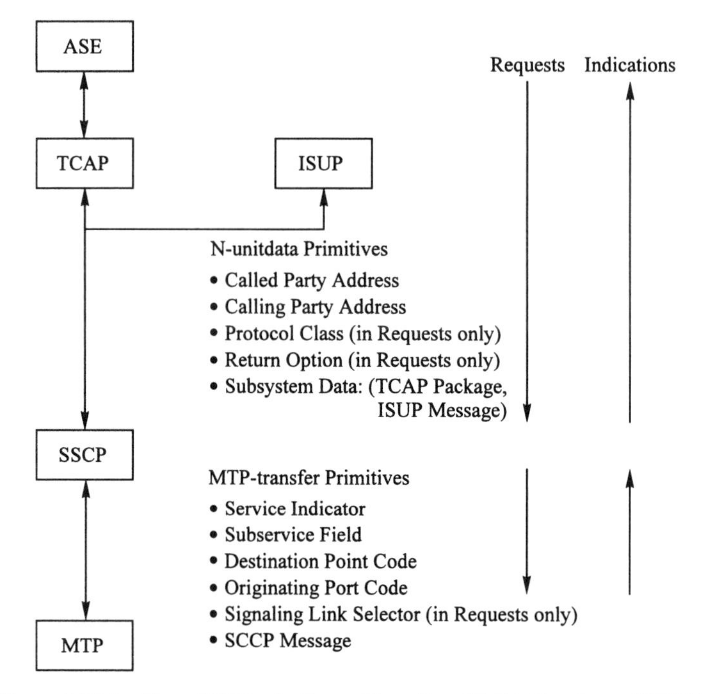

Figure 14.3-3 N- and MTP-primitives. (From Rec. Q.711. Courtesy of ITU-T.)

14.3.3 Global Title Translation

A SCCP called party address (Par.1) can contain various combinations of PC, SSN and/or GT. We now explore the reasons for global titles and GT translations.

A subsystem in a network is uniquely identified by the combination of the point code PC of its signaling point, and its SSN at that point. If an SCCP message has a PC + SSN called party address, the MTPs along the message path then use PC to route the message to its destination signaling point, and the SCCP at this point delivers it to the subsystem specified by SSN. For example, in Fig. 14.3-1, a SCCP message with called party address PC = c, SSN = r, is delivered to subsystem R800 in database C.

A global title is a "functional" address of a subsystem in an exchange or database. The reason for these functional addresses is illustrated with the example of Fig. 14.3-1.

Suppose that the set-up of a call with a called 800 number, say 800-123-4567, has reached exchange A. Subsystem Q800-A then has to send a query message to an R800 subsystem at a database that stores the translation for this number—in this example: databases C and D.

In principle, exchange A could translate the received 800 number into a PC + SSN called party address of an R800 in the appropriate database. However, this would require "800 number" translation tables in all exchanges, and would entail a large effort when an item in the tables has to be added or removed.

A better arrangement is to let the exchanges use the received 800 number as a global title (GT) address, which is the "functional" address of an R800 at a database with information on the 800 number, and install the GT translation capability in the SCCPs of the signal transfer points (STP) of the network. The SCCP at the originating exchange then routes the query message to a directly connected STP, whose SCCP translates GT into the PC + SSN address of the destination. From this point on, the message can be routed by MTP to its destination.

The number of exchanges in a network greatly exceeds the number of STPs, and placing the GT translation capability in the SCCPs at the STPs instead of in the exchanges greatly reduces the effort to update the translation data.

A GT consists of two parts, known as the GT address (GTA), which is a digit string received from a calling subscriber, and the translation type (TT), which indicates the desired translation (Fig. 14.2-2). For example, if GTA is the national number of a subscriber (S), one value of TT could indicate that GTA is to be translated into the PC + SSN address of an ASE in the maintenance center that covers S; another TT value could require a translation that yields the PC + SSN address of an ASE in the revenue accounting center that covers S, and so on.

14.3.4 Transfer of Unitdata Messages

We now examine the transfer of unitdata messages for an 800 number query—response transaction—see Fig. 14.3-1. We assume that all entities in the figure are part of the same telecommunication network.

Query Message. Suppose that the set-up of an 800 call has reached exchange A. The Q800-A then launches a query to a database, to obtain the routing number for the call. We assume that the call-routing information for this particular 800 number is stored in R800-C and R800-D (at databases C and D).

Figure 14.3-4 shows the MTP and SCCP address parameters in the primitives at signaling points A, E and D, and in the message signal unit (MSU) that carries the first unitdata message.

TCAP-A has received the calling and called address (CDA, CGA) from Q800-A, and includes them in the N-unitdata request to SCCP-A. The CDA is a GT in which the translation type = t, and the address = n (the called 800 number). The translation type indicates that n has to be translated into the SP + SSN address of an R800 at a database with information on that number.

SCCP-A enters GT in the CDA field of the unitdata message, PC = a, SSN = q in the CGA field, and sets the routing indicator (Fig. 14.2-1) to RTI = 0, indicating that a GT translation is needed. Since SCCP-A has received a GT called address, and knows that the SCCP-E can perform GT translations, it

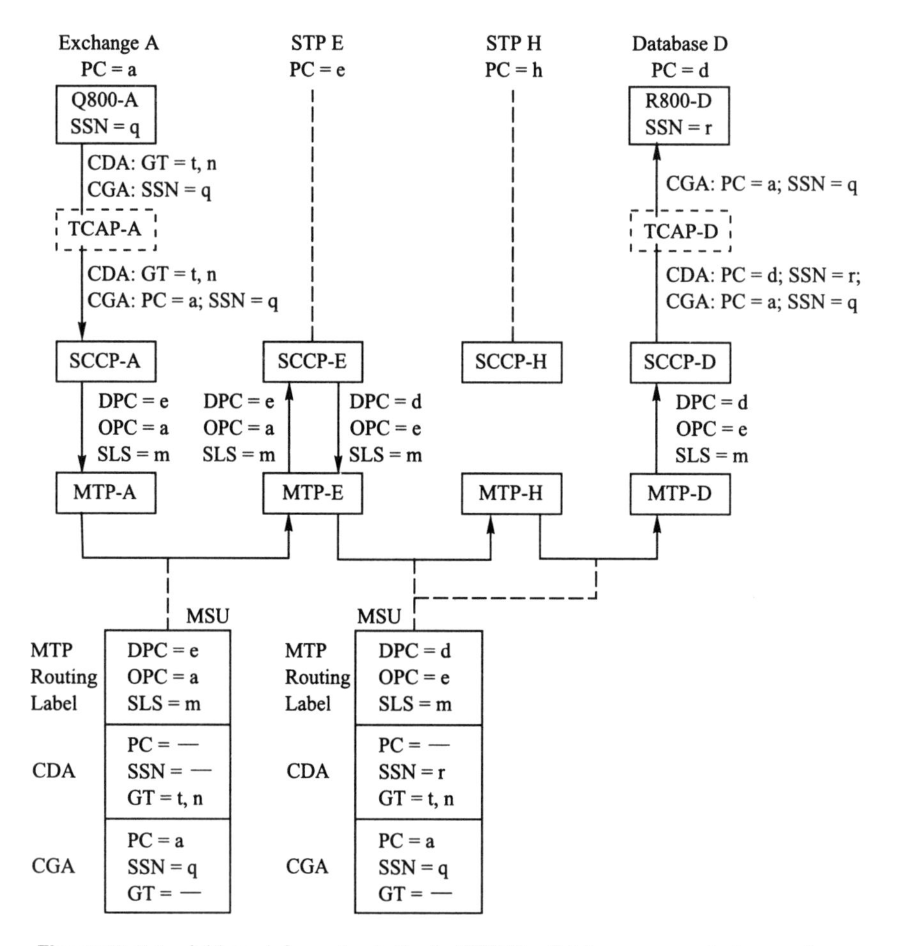

Figure 14.3-4 Address information in the first SCCP unitdata message of a transaction.

includes the MTP address of signaling point E (DPC = e) when passing the message. MTP-A then forms the MSU, and transfers it to STP-E.

SCCP-E translates the GT, and obtains the addresses PC= c, SSN = r, and PC = d, SSN = r, of the R800 units at databases C and D. Assuming that SCCP-E selects database D, it enters SSN = r in the CDA of its outgoing message, and passes it to its MTP-E, in a MTP-transfer request which includes destination point code DPC = d. It also sets the routing indicator to RTI = 1. The MTP-E routes the MSU to MTP-H, which routes it to database D.

SCCP-D receives the message from MTP-D, and passes it to TCAP-D in a N-unitdata indication, which includes the calling address PC = a, SSN = q. Finally, TCAP-D delivers the message to R800-D.

Response Message. R800-D uses the received CGA (PC = a, SSN = q) as the CDA for its response message. SCCP-D passes the message to MTP-D, in a

MTP-transfer request that includes DPC = a. The transfer of the message to SCCP-A is done by MTPs exclusively (no GT translation needed). At exchange A, subsystem number SSN = q in the CDA is used to deliver CGA and the routing number for the call (in the subsystem data field) to Q800-A. Exchange A then routes the call to its destination.

The transaction between Q800-A and R-800-D requires just two messages. This is the case for most transactions. However, after the second message, both R800-D and Q800-A know each other's PC + SSN address. If a transaction requires additional messages, the called addresses in these messages are always PC + SSN addresses, and can be transferred by MTPs along the signaling route (no GT translation required).

14.3.5 Final and Intermediate GT Translations

In the above example, SCCP-E makes a *final* GT translation, which yields the PC + SSN address of destination R800-D. This is possible because an SCCP in a signal transfer point has the necessary data to perform final translations for all destinations in its network, and we have assumed that all entities in Fig. 14.3-4 are in the same network.

When the originating and destination ASEs are in different networks, the SCCPs in the originating network have no data to do final GT translations. In this case, the message transfer involves one or more *intermediate* GT translations to route the message to an STP in the destination network, and a final GT translation by the SCCP at that STP.

Figure 14.3-5 shows an example. We consider long-distance networks in countries 1 and 2, and explore the transfer of an initial message by originating ASE-A to destination ASE-E.

Each national network has *international* STPs (STPI), whose SCCPs can do final GT translations for destinations in their respective countries, and intermediate GT translations to route a message to an STPI in the destination country.

In the figure, the point codes of signaling points in countries 1 and 2 are denoted by 1a, 1b, 1c, and 2d, 2e. The STPIs also have an international point code (ic, id).

We now examine the routing of the first unitdata message of an international calling-card verification transaction. An operator at exchange A has received a call from a caller who has a calling card issued in country 2, and wants the call to be charged to the card. We denote the calling card number by 2*n*. Before extending the call set-up, the operator needs to verify the validity of the calling card. Information about the validity of 2*n* is available only in country 2 (at ASE-E).

By entering the country code (cc2) of country 2 and 2*n*, the operator causes ASE-A to start the query. The ASE generates the global title address GTA = cc2-2*n* and translation type TT = 1t (indicating international calling card verification). SCCP-A sets RTI to 0, and routes the message to STP B.

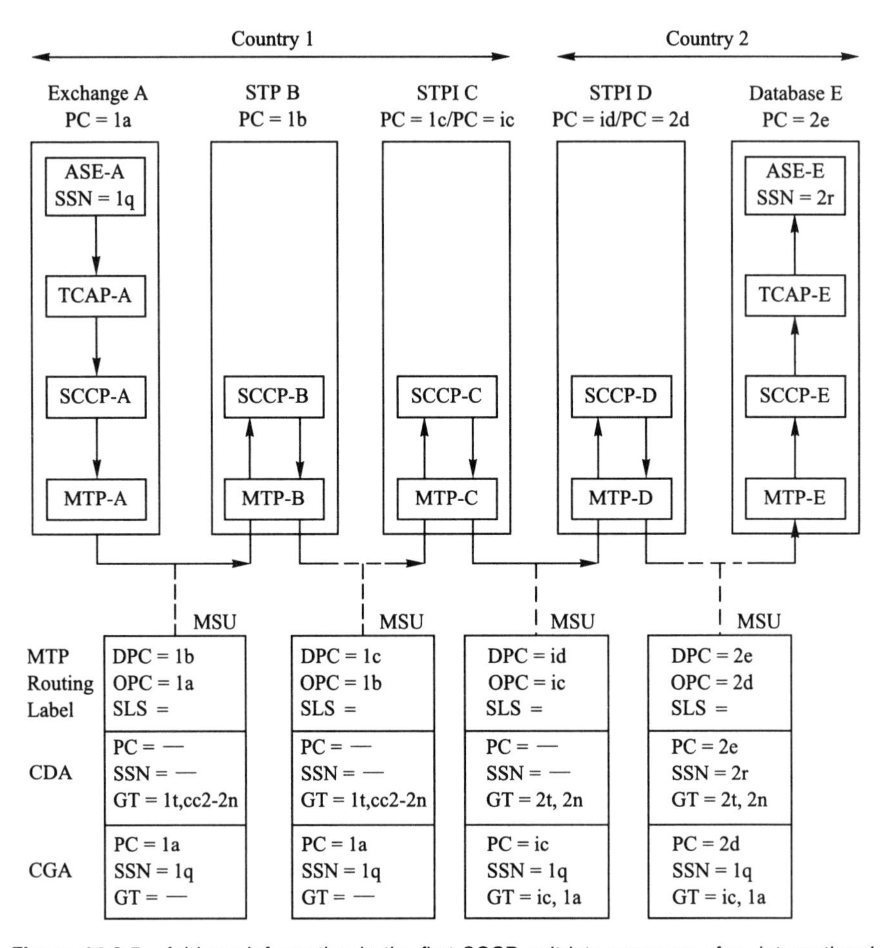

Figure 14.3-5 Address information in the first SCCP unitdata message of an international transaction.

SCCP-B performs an intermediate translation (based on 1t and cc2) which yields the point code (DPC = 1c) of STPI-C. The message reaches SCCP-C, which does another intermediate translation, based on 1t, cc2, and possibly the initial digits of 2*n*. This yields the international point code (id) of signal transfer point D, and translation type 2t (the value used in country 2 for calling card verification). SCCP-C also removes the country code from the GTA.

SCCP-D does the final GT translation—based on 2t and 2*n*—which yields the PC + SSN address of ASE-E.

Since SCCP-B and SCCP-C have done intermediate translations, their outgoing messages include RTI = 0. SCCP-D has done a final translation, and includes RTI = 1.

Note that SCCP-C enters ic (its international point code) and 1a (the point code of A) into the calling (CGA) global title field of its message. The N-unitdata indication passed from SCCP-E to ASE-E includes the CGA address parameters: 1a, 1c, and 1q (SSN of ASE-A), and 2d (the point code of D).

When responding, ASE-E passes the CDA global title parameters 1a, ic, and 1q, along with the appropriate translation type and the destination point code DPC = 2d, to SCCP-E.

The intermediate GT translations at D and C then simply consist of using ic and 1a as destination point codes for their outgoing unitdata messages.

14.3.6 Class 0 and Class 1 Service

A calling subsystem requests class 0 or class 1 service with Par.4 (protocol class). In class 0 (basic connectionless) service, the calling SCCP is free to assign any value of SLS (signaling link selector) to its outgoing messages. This means that two consecutive outgoing messages from a subsystem may traverse different routes in the signaling network, and could arrive out-of-sequence (8.8.5).

In class 1 (sequence-controlled connectionless) service, SCCP assigns the same SLS value to all outgoing messages of a particular subsystem. Consecutive outgoing messages of a subsystem then traverse the same links in the signaling network. The signaling links themselves maintain in-sequence delivery of MSUs, even when transmission errors occur (Section 8.3).

One way to associate the same SLS value to all outgoing messages of a subsystem is to use the low-order bits of the calling subsystem's SSN as SLS.

14.3.7 Unitdata Service Message

An SCCP in a signaling point along the path of a unitdata message may determine that it cannot transfer a received unitdata message to its destination. For example, an SCCP in a STP may not be able to translate the global title address of the called party, or the SCCP at the destination signaling point may find that there is no subsystem at the signaling point that corresponds to the SSN in the received message, or that the called subsystem is out of service.

If Par.8 (return option) in a unitdata message that cannot be delivered indicates that the message should be returned, the SCCP sends a *unitdata service message* (UDTS) to the calling SCCP. This message includes the subsystem data of the received message, and a Par.7 (return cause) that indicates why the message is being returned. On receipt of a UDTS message, SCCP alerts the calling subsystem with a N-notice indication (Fig. 14.1-2) that includes the called address, the subsystem data, and the return cause.

14.4 CONNECTION-ORIENTED SCCP

In this mode of operation, a (virtual) connection is set up before data transfer between two SCCP users takes place. We distinguish two connection types.

Temporary connections are established and released at the start and end of a transaction. *Permanent* connections are long-term connections that can be set up and released only by administrative or maintenance personnel.

Connection-oriented service is the preferred way for transactions that involve the transfer of large amounts of data, which puts a momentary heavy load on the involved subsystems and the signaling network. Transactions of this nature are not call-related and can be deferred, say for several minutes. Connection-oriented service gives the called subsystem an opportunity—at the time it receives a connection request—to determine whether it can handle the transaction at this time. If yes, it accepts the request. If not, it refuses. This avoids cluttering up the network with messages that cannot be processed anyway.

CCITT/ITU has defined two classes of connection-oriented service [4,5,8]: *basic connection-oriented service* (class 2) and *connection-oriented service with flow control* (class 3).

When reading this section, it is helpful to look up the message contents (Table 14.2-2), and the parameter descriptions (14.2.3).

14.4.1 Primitives

The primitives and connection-oriented service are shown in Fig. 14.1-2. The N-connect and N-disconnect primitives are used to establish and release connections. The N-data primitives are used for the transfer of user data during the connection.

In addition to the familiar *request* and *indication* primitives, SCCP also uses *response* and *confirmation* primitives. They play a role in the establishment of a connection.

14.4.2 Connection-oriented Class 2 Service

Figure 14.4-1 shows the SCCP messages and primitives during a temporary connection with class 2 service. The transaction involves subsystem P at signaling point A, and subsystem Q at signaling point B.

Establishing the Connection. Subsystem P initiates the transaction, and passes a N-connect request to SCCP-A. The SCCP sends a connection request message to SCCP-B, which passes an N-connect indication to subsystem Q. The subsystem decides to accept the connection request, and passes an N-connect response to SCCP-B, which now sends a connection confirm message. SCCP-A informs subsystem P that the connection request has been accepted, with an N-connect confirmation.

Data Transfer. The transfer of subsystem data begins. Subsystem P passes its data to SCCP-A in a N-data request. SCCP-A places the data in Par.12 (subsystem data) of a data form 1 message, and sends it to SCCP-B. The latter SCCP

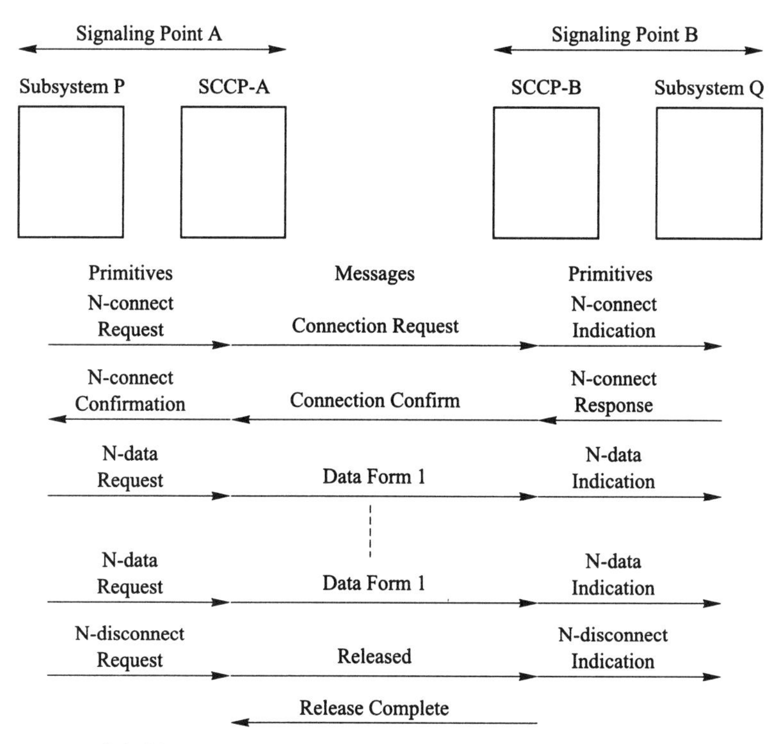

Figure 14.4-1 Primitives and messages in a class 2 connection. (From Rec. Q.711. Courtesy of ITU-T.)

extracts the subsystem data, and passes the data to subsystem Q in a N-data indication. Figure 14.4-1 shows data transfer in one direction only. However, data transfer in both directions is allowed.

Release of Connection. The connection can be released by either subsystem. In this example, P initiates the release, passing an N-disconnect request. SCCP-A releases the connection at its end, and sends a released message. SCCP-B then passes an N-disconnect indication to subsystem Q, releases the connection at its end, and sends a release complete message.

14.4.3 Records and References

During the exchange of the connection request and the connection confirm messages, the subsystems and SCCPs allocate reference numbers to identify the connection, and build records that store reference numbers, and address parameters that are associated with the connection. The records are consulted when sending—or receiving—a data form message, and are discarded at the end of the connection. The reference numbers are of the following types.

Connection Identifier CID. Subsystems and SCCPs can be involved with several simultaneous connections. A connection identifier uniquely identifies a connection at a signaling point. It is stored in the records of the subsystem and SCCP, and is included in the N-primitives between them.

Source Local Reference (SLR, Par.11). This identifies the connection, as known by the SCCP. It is included in SCCP messages to and from the other SCCP.

Destination Local Reference (DLR, Par.3). This identifies the connection, as known by the SCCP at the other end of the connection, and is included in SCCP messages.

Allocation of Connection Identifiers and Source Local References. The SCCP at each signaling point has a pool of available CIDs, and a pool of available SLRs. A subsystem, or an SCCP, can seize an available CID or SLR from a pool, and allocate it to the new connection by entering it into its connection record. At the end of the connection, CID and SLR are returned to their respective pools.

The records for the connection of Fig. 14.4-1 are shown in Fig. 14.4-2. A SCCP record is accessed with parameter CID or SLR. When a record has been accessed, its other parameters become available.

Building the Records. We explore the build-up of the records. Subsystem Q initiates the transaction. It establishes a connection record, and allocates the connection identification value (CID) = x. It then passes an N-connect request

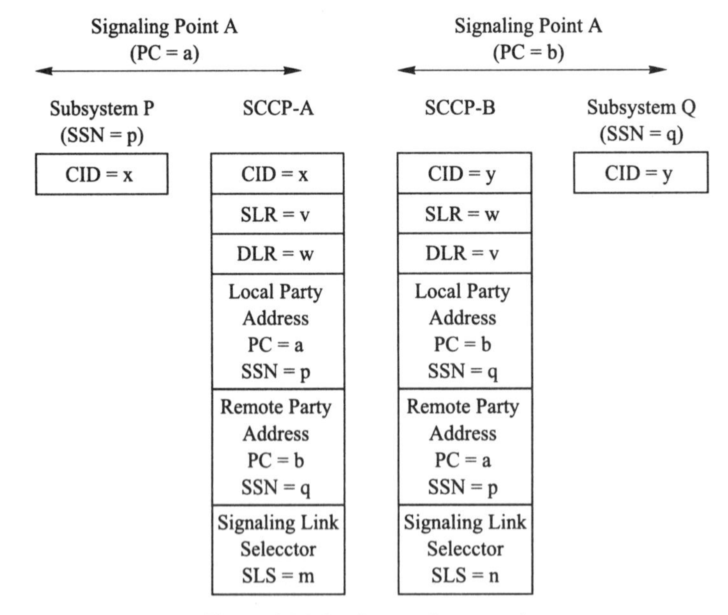

Figure 14.4-2 Connection records.

that includes the called and calling party addresses, and CID. The called party address may be a global title or a DPC + SSN address. The calling party address is always in the format PC + SSN (PC = a, SSN = p).

On receipt of the request, SCCP-A establishes a connection record. It enters the received connection identifier (CID = x) in the record. The calling party address (PC = a, SSN = p) is entered as the "local" party address. SCCP-A then allocates a source local reference value (SLR = v), and a signaling link selector value (SLS = m), and sends a connection request message that includes the called and calling party addresses (Par.1 and Par.3), and source local reference (Par.11).

On receipt of the message, SCCP-B establishes its record. It enters the received called party address (which may have undergone global title translation, and now is PC = b, SSN = q) as the "local" party address. The received calling party address (PC = a, SSN = p), and source local reference (SLR = v), are entered as the "remote" party address, and "destination" local reference (DLR). SCCP-B also allocates a connection identifier CID = y. It then passes an N-connect indication to subsystem Q (the "local" party) that includes the connection identifier.

Assuming that the subsystem accepts the connection request, it builds a connection record, and enters the received connection identifier (CID = y). It then passes an N-connect response to SCCP-B, including CID = y.

SCCP-B accesses the record in which CID = y, and allocates a source local reference (SLR = w), and a signaling link selector (SLS = n). It then sends a connection confirm message, using the "remote" party address (PC = a, SSN = p) as called party address, the "local" party address (PC = b, SSN = q) as calling party address (PC = b, SSN = q), the source local reference (SLR = w), and the destination local reference (DLR = v).

The SCCP message arrives at SCCP-A. It accesses the record whose source local reference matches the received DLR = v. It then enters the received calling party address as "remote" party address, and passes an N-connect confirmation that includes CID = x, to the "local" party identified by SSN = p. At this point all connection records are complete, and the transfer of subsystem data can start.

14.4.4 Transfer of Subsystem Data

Figure 14.4-3 shows the transfer of a data form 1 message on the established connection. Subsystem P passes an N-data request that includes CID = x, and the subsystem data. SCCP-A locates the record associated with the connection (which has CID = x), and determines the values of the destination local reference (DLR = w), and the "remote" point code (PC = b). It then creates a data form that includes the DLR and the subsystem data, and passes the form to MTP, in a MTP-transfer request that includes the destination point code DPC = b.

When SCCP-B receives the message, it locates the record whose source local

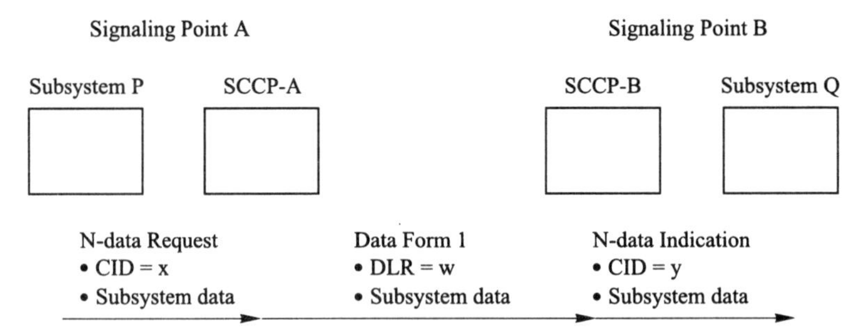

Figure 14.4-3 Transfer of a data form 1.

reference matches the received DLR = w. It then obtains the parameters SSN = q, CID = y, and passes an N-data indication to subsystem Q, which includes the CID and the subsystem data.

14.4.5 Connection-oriented SCCP Services

SCCP class 2 service provide the following services to its subsystems.

Connection Refusal. A subsystem may already be involved in one or more connections when it receives a new N-connect indication. If the already existing connections involve large data transfer activities, the subsystem can decide to refuse the new connection at this time. It does so by responding with an N-disconnect request. The local SCCP then sends a connection refused message to the SCCP of the subsystem that requested the connection, which is then informed by an N-disconnect indication.

In-sequence Delivery of DT1 Messages. Each SCCP allocates a SLS (signaling link selector) value when it establishes its connection record, and uses this value for all outgoing messages of the connection. In each direction all messages then follow a fixed signaling route, and are delivered in-sequence by MTP (8.8.5).

Segmentation and Reassembly of User Data. The maximum length of the signaling information field in a message signal unit is 272 octets (Fig. 8.8-1). After making allowances for the routing label, and for the overhead in a SCCP message, about 255 octets are available for subsystem data.

Class 2 service allows a calling subsystem to pass a much larger amount of data in one N-data request. The calling SCCP then segments the data in units of suitable size, and transmits them in successive DT1 messages. The called SCCP reassembles the data units, and passes them in one N-data indication to the called subsystem.

This service depends on the in-sequence delivery of the DT1 messages. Each message includes a Par.9 (segmenting/reassembling). The calling SCCP sets Par.9 = 1 in all DT1 (data form 1) messages, except the last one. The called SCCP reassembles the data units of the consecutive DT1s until it has received a message with Par.9 = 0. It then passes the reassembled subsystem data to the called subsystem, in one N-data indication.

Class 3 Services include the class 2 services, and *flow control*. This protects the called subsystem in a connection against being overloaded with incoming DT2 messages [3,4,8]. The calling SCCP assigns cyclically increasing (...0, 1,...126, 127, 0,...) *send sequence numbers* [P(S)] to consecutive outgoing DT2 (data form 2) messages of a connection. The called SCCP sends *data acknowledgment* messages which include a *receive sequence number* [P(R)] that represents the highest numbered (mod 128) accepted DT2. Parameters P(S) and P(R) are part of Par.10, which is included in DT2 and acknowledgment messages.

The calling SCCP may send additional DT2s as long as $P(\mathrm{S}) \leq P(\mathrm{R}) + W$, where $P(\mathrm{R})$ is the receive sequence number in the most recently received acknowledgment message, and window W represents a fixed number of DT2 messages. The value of W is negotiated during the establishment of a class 3 connection. In this way, the called SCCP controls the message flow.

14.5 SCCP MANAGEMENT

The purpose of SCCP Management (SCMG) is to maintain—if possible—the transfer of SCCP messages when failures occur in the signaling network and/or subsystems, and to inform SCCP users to stop sending messages that cannot be delivered. If a failed subsystem has a duplicate, SCMG requests its SCCP to reroute messages to the backup subsystem [4,7].

14.5.1 SCMG Interfaces

SCMG has interfaces with the subsystems, the MTP, and the SCCP connectionless control at its signaling point—see Fig. 14.1-2.

N-Primitives are the interface with the local subsystems and the local MTP. The subsystems pass information to SCMG in N-requests and N-responses, and SCMG passes information to the subsystems in N-indications and N-confirmations.

The interface with SCCP connectionless control allows SCMG to send and receive SCCP unitdata messages to/from SCMGs in other signaling points.

The interface with the local MTP consists of MTP-pause, MTP-resume, and MTP-status indications (8.9.1).

14.5.2 N-Primitives and their Parameters

Primitives. The N-primitives (Fig. 14.1-2) are outlined below:

N-State Primitives indicate a change in status of a subsystem. A subsystem indicates a status change to its SCMG in an N-state request, and SCMG informs its local subsystems about a status change of a subsystem at another signaling point with an N-state indication.

N-PC-State Indications are passed by SCMG to its local subsystems, to indicate the status of a signaling point.

N-Coord Primitives are used when a subsystem that has a backup subsystem (say, a network database) wants to go out of service.

Parameters. The parameters in N-primitives are listed in Table 14.5-1:

Subsystem Status. This indicates the new status of a subsystem ("in-service" or "out-of-service").

Affected Subsystem. This identifies the subsystem whose status has changed, by subsystem number SSN.

Subsystem Multiplicity. This indicates whether the affected subsystem has a duplicate copy in the network.

Affected Point Code. The point code of a signaling point whose status has changed, or where a subsystem has changed status.

Signaling Point Status. This indicates the new status of a signaling point ("inaccessible" or "accessible").

Table 14.5-1 SCCP-management primitives and parameters.

	Parameter Acronyms				
Primitives	AFSN	AFPC	US	SPS	SMI
N-State					
Request	X		X		
Indication	X	X	X		X
N-PC-State					
Indication		X		X	
N-Coord					
Request	X				
Indication	X	X			X
Response	X				
Confirmation	X	X			X

AFSN: affected subsystem number. AFPC: affected point code. US: user (subsystem) status. SPS: signaling point status. SMI: subsystem multiplicity indicator.
Source: Rec. Q.711.Courtesy of ITU-T.

14.5.3 SCMG Messages and Parameters

The SCMGs at different signaling points use the connectionless services of SCCP to send unitdata messages to each other. A SCMG is thus both a part, and a user, of its SCCP. A similar situation exists in the signaling network management part of MTP (Section 8.8). CCITT has specified SSN = 0000 0001 as the subsystem number of SCMG.

The contents of SCMG messages is shown in Fig. 14.5-1. Message type code (MTYP = 0000 1001) identifies the message as a unitdata message. Protocol class and return option are coded: PRC = 0000 (basic connectionless service) RO = 0000 (no return of messages that can not be delivered). Subsystem numbers (SSN) in the called and calling party address are coded 0000 0001. This identifies the message as an SCMG message. The *format indicator* identifies a particular SCMG message type:

Subsystem Allowed. The subsystem can be used.

Subsystem Prohibited. The subsystem cannot be used.

Subsystem Status Test. A request for subsystem status information.

Subsystem Out-of-service Request. A request by a subsystem to go out-of-service.

Subsystem Out-of-service Grant. A permission for a subsystem to go out-of-service.

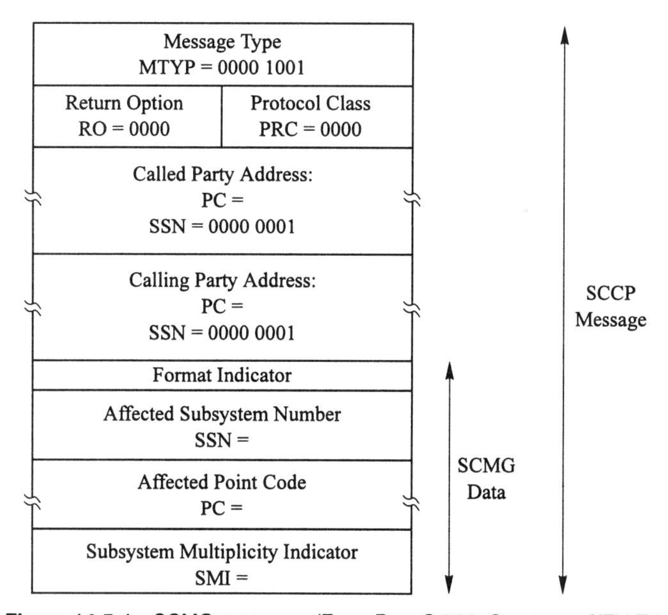

Figure 14.5-1 SCMG message. (From Rec. Q.712. Courtesy of ITU-T.)

Table 14.5-2 Format identifiers in SCMG unitdata messages.

Format Identifier Name	Code
Subsystem allowed	0000 0001
Subsystem prohibited	0000 0010
Subsystem status test	0000 0011
Subsystem out-of-service request	0000 0100
Subsystem out-of-service grant	0000 0101

Source: Rec. Q.713. Courtesy of ITU-T.

The format identifier codes are listed in Table 14.5-2. The message parameters have the same meanings as in the primitives.

14.5.4 SCMG Procedures

The N-state requests from its local subsystems, the MTP indications from the local MTP, and the received SCMG messages, keep an SCMG informed on status changes of subsystems and signaling points throughout the network. A status change can trigger the execution of a SCMG procedure.

There are two groups of SCMG procedures: subsystem management and point code management. The next sections give a number of examples, using the configuration of Fig. 14.3-1. The SCMGs and subsystems at the various signaling points are denoted as SCMG-A (SCMG at signaling point A), R800-D (R800 subsystem at signaling point D), and so on.

In the material that follows, a *concerned* signaling point or subsystem is a signaling point or subsystem that has to be informed immediately when the status of a subsystem or signaling point changes.

14.5.5 Examples of SCMG Procedures

Subsystem Status Test. This test is performed as a part of other procedures. In Fig. 14.3-1, when SCMG-A has received a subsystem-prohibited message for R800-C, it periodically audits the status of the subsystem, by sending subsystem-status-test messages. If SCMG-C determines that R800-C is in-service again, it responds with a subsystem-allowed message. Otherwise it does not respond.

Broadcast Procedure. This procedure is used when a status change occurs in a subsystem that has a number of concerned signaling points. Suppose that subsystem R800-D has gone out-of-service. It then passes an N-state out-of-service request to SCMG-D. The STPs in the network are concerned about R800-D, because they perform global title translations for 800-query messages that yield a primary and a backup R800 subsystem. Assume R800-D and R800-C have been designated as the primary and backup subsystems.

SCMG-D now broadcasts subsystem-prohibit messages for R800-D to the STPs. The messages indicate that R800-D has a duplicate. The SCMGs at the STPs then inform their local SCRC (SCCP routing control) functions to route messages for R800-D to the backup system R800-C. The SCMGs also start subsystem status tests of R800-D.

When R800-D goes in-service again, SCMG-D broadcasts subsystem-allowed messages to the SCMGs at the STPs. When a SCMG receives the subsystem-allowed message from SCMG-D (either as a broadcast message or in response to a subsystem-status-test message), it requests its SCRC to resume routing to R800-D.

Subsystem-prohibited Messages to Individual Subsystems. Suppose that subsystem X-D at signaling point D goes out-of-service, and that SCMG-D has determined that X-D has no concerned subsystems (Fig. 14.3-1). Therefore, subsystem-prohibited messages are not broadcast. However, when SCCP-D now receives an SCCP message for X-D from say X-A, it informs SCMG-D. The SCMG then sends a subsystem-prohibited message for X-D to SCMG-A. This SCMG informs X-A, by passing an N-state out-of-service indication for X-D. SCMG-A also starts a subsystem status test for subsystem X-D.

When SCMG-D determines that X-D is in-service again, it responds to the next subsystem-status-test message with a subsystem-allowed message, and SCMG-A then passes an N-state in-service indication to X-A.

Coordinated Status Changes. With this procedure, a subsystem which has a duplicate, and wishes to go out-of-service for a modification or scheduled maintenance, first checks whether its backup subsystem can take over the load. Consider the case that R800-D wishes to go out of service. It passes an N-coord request to SCMG-D, which then sends a subsystem-out-of-service request message to SCMG-C. This SCMG passes an N-coord indication to the backup subsystem R800-C.

If the subsystem agrees to take over the load, it passes an N-coord response to SCMG-C, which then sends a subsystem out-of-service-grant message to SCMG-D. This SCMG then passes an N-coord confirmation primitive to R800-D, to indicate that it can go out-of-service. It also broadcasts subsystem-prohibited messages to the concerned subsystems as described above.

If R800-C cannot take over the load of its mate, it does not respond to the N-coord indication, and no subsystem-out-of-service-grant is sent to SCMG-D. When R800-D does not receive the N-coord confirmation for its request within a certain time, it knows that its request has been denied.

14.5.6 Point Code Management Procedures

Point code management procedures are triggered by the receipt of MTP indications. Consider the receipt at SCMG-A of a MTP-pause indication for affected signaling point D in Fig. 14.3-1. SCMG-A then informs SCRC-A that destination D and its subsystems are prohibited, and SCRC-A stops routing messages to D. SCMG-A also locally broadcasts N-PC-state signaling-point-inaccessible indications for destination D, and N-state out-of-service indications for the subsystems at D, to its local subsystems.

When SCMG-A receives a MTP-resume indication for destination D, it informs SCRC-A that the destination is accessible again, and broadcasts N-PC-state signaling-point-accessible indications for destination D to its local subsystems. It then updates its status information on the subsystems at D, by sending subsystem-status-test messages. On receipt of a subsystem-allowed message for a subsystem at D, it passes N-state in-service indications for that subsystem to its local subsystems.

14.6 ACRONYMS

AFPC	Affected point code
AFSN	Affected subsystem number
ANSI	American National Standards Institute
ASE	Application service element
BCD	Binary coded decimal
CC	Connection confirm message
CCITT	International Telegraph and Telephone Consultative Committee
CDA	Called address
CGA	Calling address
CID	Connection identifier
CR	Connection request message
CREF	Connection refused message
DPC	Destination point code
DLR	Destination local reference
DT1	Data form 1 message
DT2	Data form 2 message
FI	Format indicator
GT	Global title
GTA	Address information in GT
GTI	Global title indicator
ISDN	Integrated services digital network
ISUP	ISDN user part of SS7
ITU	International Telecommunications Union
MSU	Message signal unit on signaling link
MTP	Message transfer part
MTYP	Message type
NSP	Network service part
OPC	Originating point code
OSI	Open systems interconnection

PC	Point code
PCI	Point code indicator
PRC	Protocol class
Q800	Subsystem (ASE) making 800-number queries
REF	Reference number
RLC	Release complete message
RLSD	Released message
RO	Return option
RTI	Routing indicator
SCCP	Signaling connection control part of SS7
SCLC	SCCP connectionless control
SCMG	SCCP management
SCOC	SCCP connection-oriented control
SCRC	SCCP routing control
SLR	Source local reference
SLS	Signaling link selector
SMI	Subsystem multiplicity indicator
SOG	Subsystem out-of-service grant message
SOR	Subsystem out-of-service request message
SREF	Source reference number
SSA	Subsystem allowed message
SSN	Subsystem number
SSNI	Subsystem number indicator
SSP	Service switching point
SSP	Subsystem prohibited message
SST	Subsystem test message
SS7	Signaling system No.7
STP	Signal transfer point
TCAP	Transaction capability application part
TT	Translation type
TUP	Telephone user part of SS7
UDT	Unitdata message
UDTS	Unitdata service message
UM	User message

14.7 REFERENCES

1. A.R. Modaressi, R.A. Skoog, "Signaling System No.7: A Tutorial," *IEEE Comm. Mag.*, **28**, no.7, July 1990.
2. R. Manterfield, *Common Channel Signalling*, Chapter 5, Peter Peregrinus Ltd, London, 1991.
3. P.G. Clarke, C.A. Wadsworth, "CCITT Signalling System No.7: Signalling Connection Control Part," *Br. Telecomm. Eng.*, **7**, April 1988.

4. *Specifications of Signalling System No.7*, Rec. Q.711–Q.714, CCITT Blue Book, **VI.7**, ITU, Geneva, 1989.
5. *Rec. Q.711, Signalling System No.7*, Functional Description of the Signalling Connection Control Part, ITU, Geneva, 1993.
6. *Rec. Q.712, Signalling System No.7*, Definition and Functions of SCCP Messages, ITU, Geneva, 1993.
7. *Rec. Q.713-Signalling System No.7*, SCCP Formats and Codes, ITU, Geneva, 1993.
8. *Rec. Q.714, Signalling System No.7*, SCCP Procedures, ITU, Geneva, 1993.
9. *American National Standard for Telecommunications—Signaling System No.7 (SS7)—Signaling Connection Control Part,* ANSI T.112, 1992, New York, 1992.

15

TRANSACTION CAPABILITIES APPLICATION PART

15.1 INTRODUCTION

The transaction capabilities application part (TCAP) of signaling system No.7, in conjunction with the signaling control connection part (SCCP) and the message transfer part (MTP) [1–3], enables application service elements (ASE) at two *nodes* (the TCAP term for *signaling points*) to conduct transactions.

TCAP is similar in many respects to the protocols defined by CCITT/ITU-T for data communication networks, which are documented in X-series recommendations.

The first CCITT Recommendations on TCAP were published in 1989 [4], and have been revised in 1993 by ITU-T [5–9].

The U.S. version of TCAP has been specified by ANSI [10]. The work on this version started before the publication of the initial CCITT recommendations. As a consequence, there are pronounced differences in terminology, and some differences in coding, between the two versions. Sections 15.1–15.4 of this chapter follow the CCITT version. Section 15.5 describes the main differences of the U.S. version.

15.1.1 Transactions and Remote Operations

During a transaction, one (usually) or more (rarely) remote operations are executed. The operation is requested by an ASE at one node, and executed by an ASE at another node. Transaction ASEs are application-specific, and involve "peer" ASEs (dedicated to the same application) at the two nodes.

Some remote operations provide information to the requesting ASE. For example, in the "800 number calling" application, an ASE at an exchange requests its peer to translate an 800 number into a routing number. Other remote operations provide instructions to the requesting ASE. In the chapters that follow, we shall encounter operations of both types.

15.1.2 Components and Messages

Fig. 15.1-1 shows the SS7 entities involved in a transaction between ASE-1 and ASE-2. The physical message path traverses the TCAPs, SCCPs, and MTPs at the nodes, and a path in the signaling network, which transfers the message signal units (MSU) between the nodes.

This section describes transactions at the *component* and the *message level.* A component is a communication between ASEs. It can contain a requested operation, or the results of the operation. A message (containing one or more components) is the unit of communication between two TCAPs—see Fig.15.1-2.

15.1.3 TCAP Interfaces

In the CCITT/ITU-T model of TCAP [4,8], TC-primitives are the interface between ASE and TCAP in a node, and N-unitdata primitives are the interface between TCAP and SCCP [4,9].

To send a message, an ASE passes a series of TC-requests to the TCAP at its node, and TCAP passes the message to its SCCP, in a N-unitdata request. When a TCAP receives a message from its SCCP, it passes the contents to the destination ASE in its node, with a series of TC-indications.

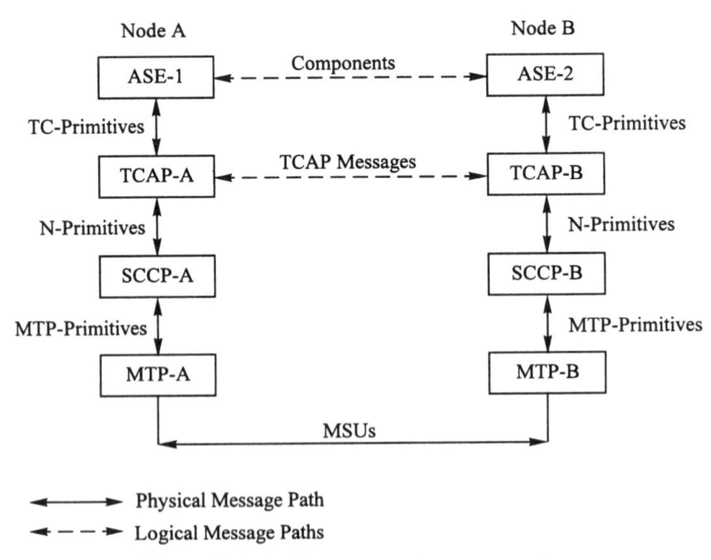

Figure 15.1-1 Messages and message paths.

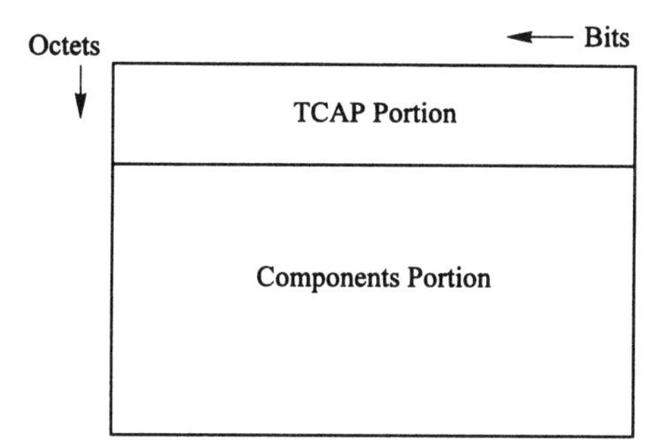

Figure 15.1-2 TCAP message.

15.1.4 TCAP Messages

As shown in Fig. 15.1-2, a TCAP message consists of a *TCAP portion*, which is processed by TCAP, and a *components portion* with one or more components, which is passed transparently. There are several TCAP message types. They represent the status of the transaction, as perceived by the sending ASE and TCAP.

Unidirectional. This is a message for one-way information transfer. The sender does not expect a response, and the message does not initiate a transaction.

Begin. This message initiates a transaction.

Continue. This message is sent in response of a begin or continue message, and continues the transaction.

End. This message is sent in response to a begin or continue message, and terminates the transaction.

Abort. This message is sent when a fatal problem has been encountered during the transaction, and terminates the transaction.

An abort message can originate at an ASE or a TCAP. The other messages always originate at an ASE.

15.1.5 Components

CCITT/ITU-T has specified the following components:

Invoke. This component specifies the requested operation. A Begin message always includes one (or more) invokes. Continue and End messages sometimes include invokes.

Return Result, Last (RR-L). This component is a final response to an invoke. It contains the information obtained in a successful operation.

Return Result, Not Last (RR-NL). This component is a non-final response to an invoke, and contains a part of the information generated by the operation. This component is necessary because the maximum length of a message signal unit (MSU) is 272 octets. Since the MSU contains the component, and information for TCAP, SCCP, and MTP, the maximum length of a component is about 240 octets. Most operations yield information that fits within one component. If the information is too voluminous, it is segmented, and carried in one or more RR-NL components, and a final RR-L component.

Return Error. This component is a final response to an invoke, and indicates that the requested operation was executed, and has failed.

Reject. This component is a final response to a received component. It indicates that the component cannot be processed, because its syntax is not correct.

15.1.6 Application-independence

TCAP serves all (application-specific) ASEs in a node. TCAP messages are specified in an application-independent manner [4–8 and 10]. TCAP specifies the structure (syntax) of components, but does not cover their information content.

We shall discuss the contents of components in the TCAP application examples of Chapters 16 and 17.

15.1.7 TCAP Message Sequences

Figure 15.1-3 shows examples of TCAP message sequences in transactions that involve ASE-1 and ASE-2. The transactions are initiated by ASE-1, which sends a Begin message with an invoke 1. The vast majority of transactions require just two messages (Begin and End)—see cases (a) and (b). In case (a), the invoked operation results in information, which is conveyed in an End message with a RR-L component. In case (b), the operation results in an instruction, which is conveyed in an End message with invoke 2.

In case (c), ASE-2 needs more information from ASE-1 before it can execute invoke 1. It therefore returns a Continue message, whose invoke 2 requests the information. After executing the invoke, ASE-1 sends a Continue message, in which RR-L 1 holds the requested information. ASE-1 now executes invoke 1, and returns a RR-L or an invoke in an End message.

Figure 15.1-4 shows a chain of two transactions, involving three ASEs. ASE-2 cannot execute invoke 1 without information from ASE-3. It therefore

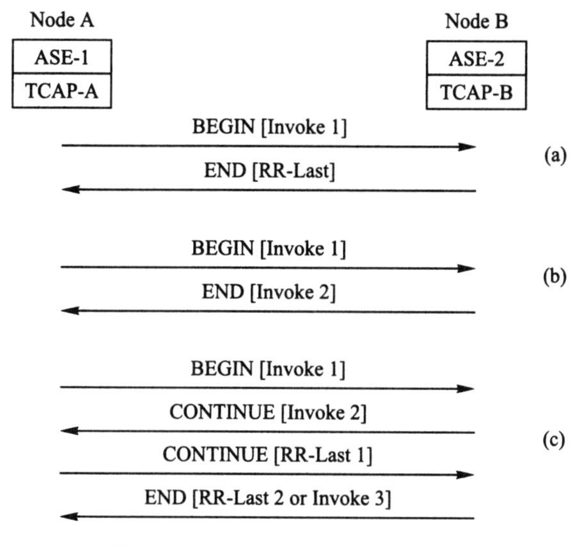

Figure 15.1-3 Transaction examples.

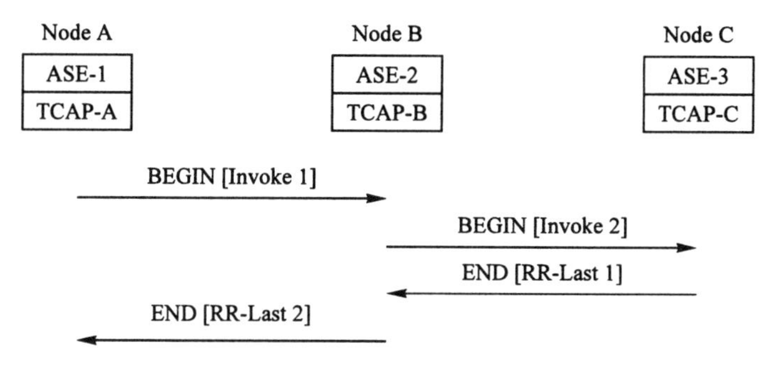

Figure 15.1-4 Chain transaction.

initiates a transaction with ASE-3, sending invoke 2. ASE-3 executes invoke 2, obtains a result, and includes it in the RR-last 1 of its End message. ASE-2 then executes invoke 1, and sends an End message with the result in RR-last 2.

15.2 TCAP INFORMATION ELEMENTS

TCAP messages consist of a number of mandatory (M) and/or optional (O) *information elements* (IE). We shall see that the term *parameter* is used also, but has a restricted meaning.

This section describes the TCAP information elements [7]. Each IE has a reference number (IE.1, etc). These numbers are used in the sections that follow.

Table 15.2-1 Information elements in TCAP portion of message.

			Message				
Information Element		Type	UNIDIR	BEGIN	CONT	END	ABORT
IE.1	Unidirectional	C	M				
IE.2	Begin	C		M			
IE.3	Continue	C			M		
IE.4	End	C				M	
IE.5	Abort	C					M
IE.6	Originating-TID	P		M	M		
IE.7	Destination-TID	P			M	M	M
IE.8	P-Abort-Cause	P					O
IE.9	U-Abort-Information	C					O
IE.10	Component-Portion	C	M	M	M	M	O

CONT: Continue. UNIDIR: Unidirectional. TID: Transaction-ID (identity). C: Constructor. P: Primitive.
Source: Rec. Q.773. Courtesy of ITU-T.

15.2.1 Information Elements in the Transaction Portion

Table 15.2-1 lists the IEs in the transaction portion of TCAP messages. IE.1 through IE.5 represent the individual message types.

The TCAP at a node has a pool of *transaction-identities* (TID), which are reference numbers that identify individual transactions at the node. When a transaction is initiated, a TID is allocated. When the transaction ends, the TID is returned to the pool.

IE.6 Originating-TID (OTID). In Begin and Continue messages, OTID specifies the TID of the transaction at the originating (sending) node.

IE7 Destination-TID (DTID). In Continue and End messages, DTID specifies the transaction at the destination (receiving) node.

IE.8 P-Abort-Cause. This IE appears in Abort messages originated by TCAP, and indicates why TCAP has aborted the transaction.

IE.9 U-Abort-Information. Appears in Abort messages originated by an ASE, and indicates why the ASE has aborted the transaction.

IE.10 Component-Portion. This IE separates the transaction and component portions of a TCAP message. All messages except Abort messages have a component portion.

Table 15.2-2 Information elements in components.

Information Element		Type	INVOKE	RETRSL	RETRSNL	RETERR	REJECT
IE.11	Invoke	C	M				
IE.12	Return-Result-Last	C		M			
IE.13	Return-Result-Not-Last	C			M		
IE.14	Return-Error	C				M	
IE.15	Reject	C					M
IE.16	Invoke-ID	P	M	M	M	M	M
IE.17	Linked-ID	P	M				
IE.18	Operation-Code	P	M				
IE.19	Error-Code	P				M	
IE.20	General-Problem	P					#
IE.21	Invoke-Problem	P					#
IE.22	Return-Result-Problem	P					#
IE.23	Return-Error-Problem	P					#
IE.24	Parameter-Sequence	C	@	@	@	@	@
IE.25	Parameter-Set	C	@	@	@	@	@
IE.26	Parameter	P	O	O	O	O	O

RETRSL: Return-Result-Last (RR-L). RETRSNL: Return-Result-Not-Last (RR-NL). RETERR: Return-Error. C: Constructor. P: Primitive. #: One of the IEs is mandatory in reject components. @: One of the IEs is mandatory if the component includes more than one parameter.
Source: Rec. Q.713. Courtesy of ITU-T.

15.2.2 Information Elements in Components

The mandatory (M) and optional (O) IEs in the various components are listed in Table 15.2-2. Elements IE.11 through IE.15 represent the component types.

IE.16 Invoke-ID. This IE is mandatory in all component types. In invoke components, it is a reference number that identifies an invoke in a transaction. A return-result, return-error, and reject component is a response to a received invoke, and has the IE.16 value of that invoke.

IE.17 Linked-ID. This IE is mandatory in an invoke component that is sent as the response to a received invoke. In this case, IE.16 identifies the sent invoke, and IE.17 identifies the received invoke.

IE.18 Operation-Code. This is a mandatory IE in invoke components, and specifies the requested operation.

IE.19 Error-Code. This IE is mandatory in return-error components, and indicates why a requested operation has failed.

IE.20 General-Problem. This IE indicates that a received component cannot be processed.

IE.21 Invoke-Problem. This IE indicates that a received invoke cannot be processed.

IE.22 Return-Result-Problem. This IE indicates that a received return-result cannot be processed.

IE.23 Return-Error-Problem. This IE indicates that a received return-error cannot be processed.

A reject component has to include IE.20, or IE.21, or IE.22, or IE.23.

IE.24 Parameter-Sequence. An ordered sequence of parameters.

IE.25 Parameter-Set. A set (not ordered) of parameters.

If a component contains more than one parameter, it has to include an IE.24 or IE.25.

IE.26 Parameter. A parameter contains additional information in a component.

In invoke components, parameters specify the operands for the requested operation.

In return-result components, parameters hold the results of the operation.

In return-error and reject components, the parameters are the information elements that respectively caused the failure of the operation, and the rejection of a received component.

15.3 TCAP FORMATS AND CODING

This section describes the formats and coding of the TCAP messages defined by CCITT/ITU-T [7].

15.3.1 Information Elements

The information elements (IE) in TCAP messages have three fields—see Fig. 15.3-1. The *tag* field indicates the IE type, and the *length* field indicates the number of octets in the *value* (or contents) field.

A TCAP IE can be a *primitive* IE or a *constructor* IE. The value field of a primitive IE contains the information of the IE. Primitive IEs should not be confused with the primitives that pass information between the SS7 protocols in a node (Section 7.3.2).

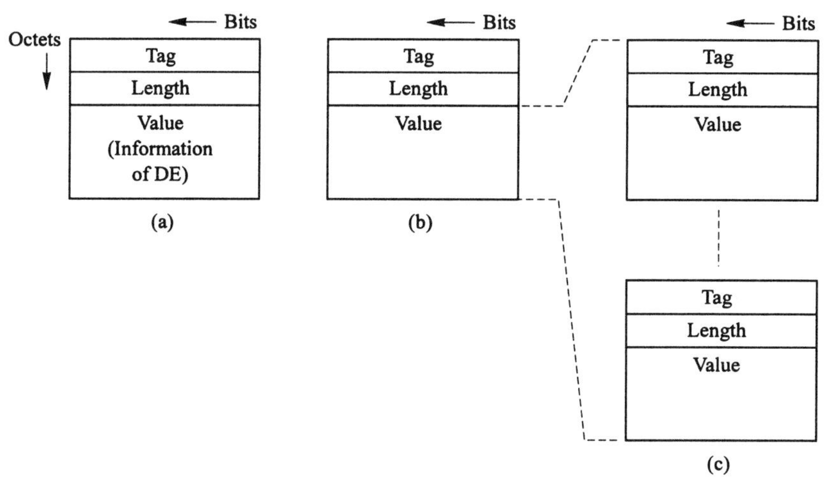

Figure 15.3-1 Structure of information elements. (a): primitive IE. (b) constructor IE. (From Rec. Q.773. Courtesy of ITU-T.)

The value field of a constructor IE holds one or more other IEs, which may be primitives or constructors.

The concepts "primitive" and "constructor" have been carried over from CCITT Recommendation X.209 (data communication networks).

In Tables 15.2-1 and 15.2-2, the primitive and constructor IEs are marked (P) and (C). The constructor IEs, and the information in their value fields, are listed below.

IE.1–IE.5. The tag fields of these IEs identify a message type, and the value fields hold all other IEs in the message.

IE.9 U-Abort-Information. The value field of this IE holds application-specific IEs with information on why the ASE (TCAP user) has aborted the transaction.

IE.10 Component-Portion. The value field of this IE holds the IEs of all components in the message.

IE.11–IE.15. The tag fields of these IEs identify a component type, and the value fields hold all other IEs of the component.

IE.24 Parameter-Sequence. The value field of this IE holds all parameter IEs in the component, in a predetermined order.

IE.25 Parameter-Set. The value field of this IE holds all parameter IEs in the component, in any order.

As an example, Fig. 15.3-2 shows the format of a begin message with one

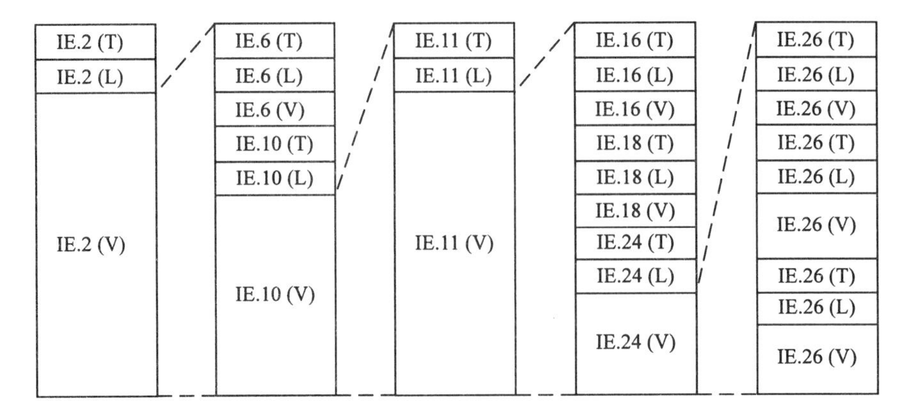

Figure 15.3-2 Information elements in Begin message. IE.2: Begin message. IE.6: Originating transaction ID. IE.10: Components-portion. IE.11: Invoke component. IE.16: Invoke ID. IE.18: Operation code. IE.24: Parameter sequence. IE.26: Parameters.

invoke. The TCAP portion consists of IE.2 (begin message), whose value field holds an IE.6 (originating-TID), and an IE.10 (component-portion).

The value field of IE.10 holds an IE.11 (invoke component), whose value field holds an IE.16 (invoke-ID), an IE.18 (operation-code), and an IE.24 (parameter-sequence). Finally, the value field of IE.24 holds three IE.26s (parameters).

15.3.2 Coding of Tag Fields

An IE tag field consists of one, two, or three octets, and has three fields—see Fig. 15.3-3.

Class Field. Bits H,G of octet 1 indicate the IE class:

Class	H,G
Universal	0,0
Aplication-wide	0,1
Context-specific	1,0
Private use	1,1

The class implies where the IE is specified. Universal IEs are specified in CCITT/ITU-T Recommendation X.409. IEs in this class are used primarily in data communication networks, and some are also used in TCAP. Application-wide IEs are specified by CCITT/ITU-T in Recommendation [7]. They are used in all TCAP applications defined by CCITT/ITU-T. Context-specific IEs are specified within the context of the next higher constructor IE. A context-specific tag value thus has different meanings, which depend on the class of the constructor. Private use IEs are defined by various national organizations.

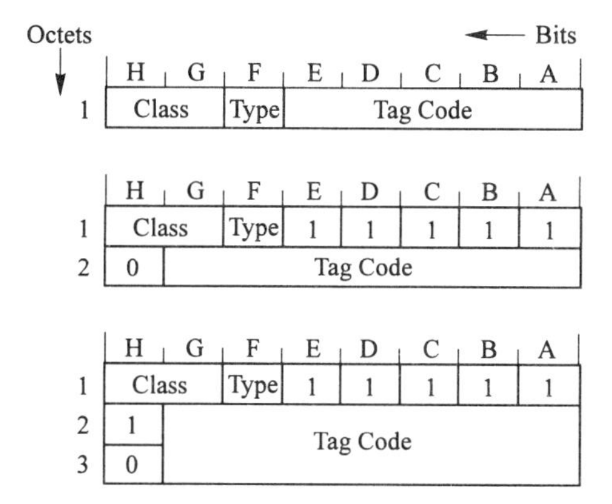

Figure 15.3-3 Tag fields. (From Rec. Q.773. Courtesy of ITU-T.)

The "class" concept enables the various standards organizations to define IEs independently of each other.

Type Field. Bit F of octet 1 indicates whether the IE is a constructor (F = 1) or a primitive (F = 0).

Tag Code Field. The tag code identifies the IE (Fig. 15.3-3). In one-octet tags, bits E...A hold a tag code that ranges from 00000 through 11110 (decimal 0 through 30). In two-octet tags, bits E...A in octet 1 are set to 11111, and bits G...A in octet 2 hold a tag code that ranges from 31 through 127. In three-octet tags, the tag code (bits G...A of octets 2 and 3) ranges from 128 through 16383.

Table 15.3-1 lists the tag codes of the IEs discussed in Section 15.2. The tags of IEs in the transaction portion are coded "application-wide." The tags of most IEs in the components portion are coded "context-specific," i.e. in the context of IE.10 (which is application-wide). A few tags are coded as "universal". Note that the tag codes of IE.16, IE.18, and IE.19 are identical. Their meanings are implied by the component types which hold these IEs.

Parameter IEs are application-dependent, and their tag codes can be found in the specifications of individual applications.

15.3.3 Coding of the Value Fields

This section gives examples of IE value field (V) coding for the primitive IEs. More details can be found in Ref. [7].

IE.6(V) and IE.7(V). The value fields of the originating and destination TIDs hold reference numbers (binary coded integers) that identify a TID at a node.

IE.8(V) P-Abort-Cause. The one-octet field indicates why a TCAP has aborted the transaction.

- Received message has an unrecognized message type   0000 0000
- Received message has an unrecognized transaction-ID   0000 0100

Table 15.3-1 Tag codes.

		HG	F	E	D	C	B	A
								← Bits
	Information Elements in TCAP Portion							
IE.1	Unidirectional	01	1	0	0	0	0	1
IE.2	Begin	01	1	0	0	0	1	0
IE.3	Continue	01	1	0	0	1	0	1
IE.4	End	01	1	0	0	1	0	0
IE.5	Abort	01	1	0	0	1	1	1
IE.6	Originating-TID	01	0	0	1	0	0	0
IE.7	Destination-TID	01	0	0	1	0	0	1
IE.8	P-Abort-Cause	01	0	0	1	0	1	0
IE.9	U-Abort-Information	01	1	0	1	0	1	1
IE.10	Component-Portion	01	1	0	1	1	0	0
	Information Elements in Components Portion							
IE.11	Invoke	10	1	0	0	0	0	1
IE.12	Return-Result-Last	10	1	0	0	0	1	0
IE.13	Return-Result-Not-Last	10	1	0	0	1	1	1
IE.14	Return-Error	10	1	0	0	0	1	1
IE.15	Reject	10	1	0	0	1	0	0
IE.16	Invoke-ID	00	0	0	0	0	1	0
IE.17	Linked-ID	10	0	0	0	0	0	0
IE.18	Operation-Code	00	0	0	0	0	1	0
IE.19	Error-Code	00	0	0	0	0	1	0
IE.20	General-Problem	10	0	0	0	0	0	0
IE.21	Invoke-Problem	10	0	0	0	0	0	1
IE.22	Return-Result-Problem	10	0	0	0	0	1	0
IE.23	Return-Error-Problem	10	0	0	0	0	1	1
IE.24	Parameter-Sequence	00	1	1	0	0	0	0
IE.25	Parameter-Set	00	1	1	0	0	0	1
IE.26	Parameter							

Source: Rec. Q.773. Courtesy of ITU-T.

IE.16(V) and IE.17(V). The value fields of invoke-IDs and linked-IDs hold references numbers (binary coded integers) that identify an invoke in a transaction.

IE.18(V), IE.19(V), and IE.26(V). The operation-code, parameter, and error-code IEs are application-specific. The coding of their value fields are defined in the specifications of the individual ASEs.

IE.20(V). Examples of value field codes in the general-problem IE:

- Unrecognized component type 0000 0000
- Mistyped component (the component is not properly structured) 0000 0001

IE.21(V). Examples of value field codes in the invoke-problem IE:

- Duplicate invoke-ID 0000 0000
- Unrecognized operation-code 0000 0001
- Mistyped parameter (tag code not recognized) 0000 0010

IE.22(V). Examples of value field codes in the return-result-problem IE:

- Unrecognized invoke-ID (the invoke-ID value in the received component is not recognized) 0000 0000
- Mistyped parameter 0000 0010

IE.23(V). Examples of value field codes in the return-error problem IE:

- Unrecognized invoke-ID 0000 0000
- Unrecognized error (the error-code value in the received component is not recognized) 0000 0010

15.4 TRANSACTION AND INVOKE IDENTITIES

At any point in time, a number of transactions can be active at a node. Moreover, an ASE can be involved in several concurrent transactions, and a transaction can consist of more than one remote operation.

The transfer of transaction messages has been described in Section 14.3.4. We now discuss the transaction and component IDs, which are used to keep track of individual transactions and operations.

15.4.1 Transaction Identifiers

A transaction identifier (TID) is a reference number that identifies a transaction to the involved ASE, and to the TCAP of its node. A TCAP has a pool of TIDs. When ASE-1 at node A initiates a transaction with ASE-2 at node B, TCAP-A seizes an available TID—say TID $= p$—from its pool, and allocates it to the transaction. TCAP-A also establishes a record that associates ASE-1 with local TID $= p$. It also passes TID $= p$ to ASE-1.

The begin message from TCAP-A includes an *originating TID* field (OTID) which has the value OTID $= p$—see Fig. 15.4-1. TCAP-B allocates TID $= q$ to the new transaction at its node, and establishes a record that associates ASE-2, the local TID $= q$, and the remote TID $= p$. It also passes TID $= q$ to ASE-2, along with the component(s) in the begin message.

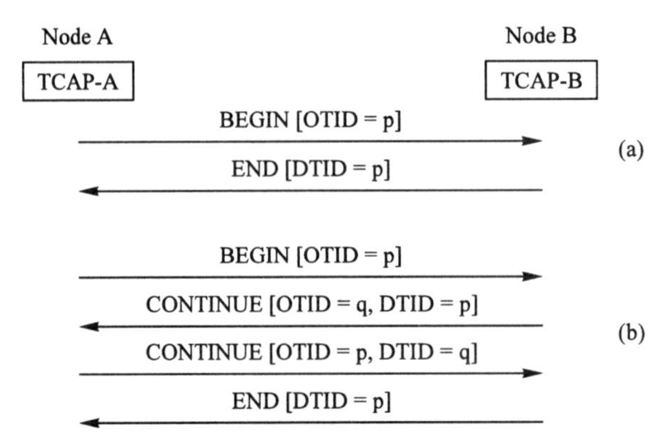

Figure 15.4-1 Transaction identifiers. See text and Section 15.6 for acronyms.

The transaction is now identified for ASE-1 and TCAP-A by TID $= p$, and for ASE-2 and TCAP-B by TID $= q$.

In case (a), ASE-2 requests TCAP-B to send an end message, which terminates the transaction. TCAP-B then returns TID $= q$ to its TID pool. Its end message includes DTID $= p$. This identifies the transaction for TCAP-A, which then passes the component(s) in the message to ASE-1 (the ASE that is currently associated with TID $= p$). Since the transaction has ended, TCAP-B then erases its record for the transaction, and returns TID $= p$ to its TID pool.

In case (b), ASE-2 continues the transaction. The continue message from TCAP-B includes OTID $= q$ and DTID $= p$. TCAP-A then adds remote TID $= q$ to the record that associates ASE-1 with local TID $= p$. In this example, ASE-1 also continues the transaction, and the continue message from TCAP-A includes OTID $= p$ and a destination transaction identifier DTID $= q$.

The procedures for the end message of TCAP-B are as described above.

15.4.2 Component Identifiers

A transaction can involve several invoke components. An invoke is identified by an *invoke-ID* (IID). Each ASE has a pool of IIDs, and allocates an IID when it originates an invoke. IID identifies a particular operation in the transaction to both ASEs.

In Fig. 15.4-2(a), the allocated IID $= x$ is included in the invoke component of the begin message from TCAP-A.

The same IID value is included in the return-result of the end message sent by TCAP-B. In general, a return-result error, or reject component includes an IID that "reflects" the IID of the received invoke.

In example (b), the begin message contains two invokes (IID $= x$, and IID $= y$). ASE-B executes the first operation, and sends a return-result component with IID $= x$. It then executes the second operation, and the return-result component includes IID = y.

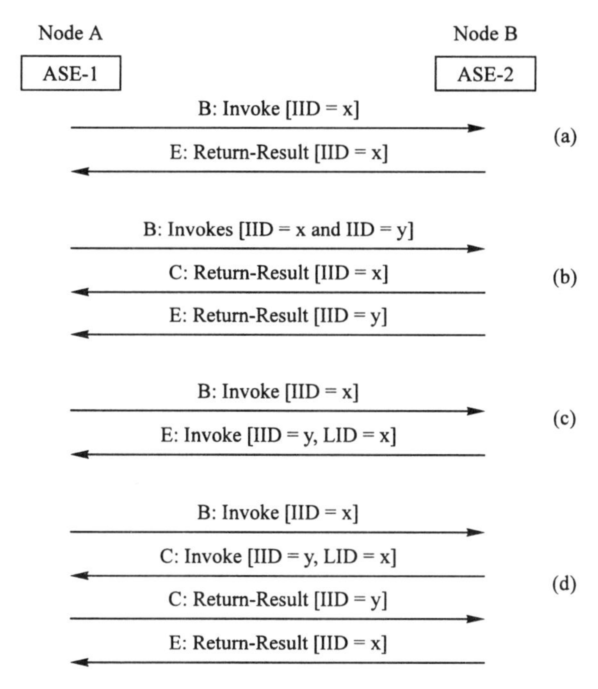

Figure 15.4-2 Invoke identifiers. See text and Section 15.6 for acronyms.

In example (c), the invoke of ASE-1 is a request for instructions, and includes IID $= x$. ASE-2 responds with an invoke of its own, which specifies the requested instructions. The invoke includes IID $= y$, assigned by ASE-2 to its invoke, and *linked* identifier LID $= x$, which refers to the invoke of ASE-1.

Finally, in example (d), ASE-2 responds to the invoke of ASE-1 with an invoke which requests more information. In the conversation message from TCAP-B, IID $= y$ and LID $= x$ refer to the invokes of ASE-2 and ASE-1. In the return-result component from ASE-1, IID $= y$ refers to the invoke of ASE-2, and the return-result from ASE-2 includes IID $= x$.

15.5 U.S. NATIONAL TCAP

The version of TCAP defined by the American National Standards Institute (ANSI) for use in fixed networks in the U.S. is very similar to the TCAP defined by CCITT/ITU-T. The differences, which are relatively small, are outlined below.

15.5.1 Terminology

Table 15.5-1 lists a number of CCITT/ITU-T TCAP terms and their ANSI equivalents, or near equivalents. In ANSI documents, *data element* (DE), *package*, and *identifier* replace *information element*, *message*, and *tag*. The term "identifier" can be awkward; for example, in *component-ID identifier.*

Table 15.5-1 TCAP terms.

CCITT/ITU-T		ANSI
Information Element (IE)		Data Element (DE)
Message		Package
Tag		Identifier
	Transaction Portion	
	Information Elements	Data Elements
IE.1	Unidirectional (C)	Unidirectional (C)
IE.2	Begin (C)	Query (C)
IE.3	Continue (C)	Conversation (C)
IE.4	End (C)	Response (C)
IE.5	Abort (C)	Abort (C)
IE.6	Originating-TID (P)	TID (P)
IE.7	Destination-TID (P)	TID (P)
IE.8	P-Abort-Cause (P)	P-Abort-Cause (P)
IE.9	U-Abort-Information (C)	U-Abort-Information (P)
IE.10	Component-Portion (C)	Component-Sequence (C)
	Component Portion	
	Information Elements	Data Elements
IE.11	Invoke (C)	Invoke (C)
IE.12	Return-Result-Last (C)	Return-Result-Last (C)
IE.13	Return-Result-Not-Last (C)	Return-Result-Not-Last (C)
IE.14	Return-Error (C)	Return-Error (C)
IE.15	Reject (C)	Reject (C)
IE.16	Invoke-ID (P)	Component-ID (P)
IE.17	Linked-ID (P)	Component-ID (P)
IE.18	Operation-Code (P)	Operation-Code (P)
IE.19	Error-Code (P)	Error-Code (P)
IE.20	General-Problem (P)	Problem-Code (P)
IE.21	Invoke-Problem (P)	Problem-Code (P)
IE.22	Return-Result-Problem (P)	Problem-Code (P)
IE.23	Return-Error-Problem (P)	Problem-Code (P)
IE.24	Parameter-Sequence (C)	Parameter-Sequence (C)
IE.25	Parameter-Set (C)	Parameter-Set (C)
IE.26	Parameter (P)	Parameter (P)

15.5.2 Transaction and Component IDs

ANSI does not have separate DEs for originating and destination transaction IDs. In query and response packages, the transaction-ID (TID) value field holds one 32-bit integer, which represents the originating or destination TID value. In conversation packages, the TID field holds two 32-bit integers. The first integer represents the originating TID, and the second one represents the destination TID.

ANSI uses the *component-ID* data element in lieu of the invoke-ID and linked-ID information elements. If an ASE sends an invoke in response to a received invoke, the sent invoke includes two component-IDs, representing respectively the invoke-ID and the linked-ID.

15.5.3 Identifier Coding

U.S. TCAP uses *context-specific* (H,G = 1,0) and *private-use* (H,G, = 1,1) identifier (tag) classes of CCITT/ITU-T (Section 15.3.2). However, the private use class is redefined as *national TCAP/private TCAP.*

National TCAP refers to TCAP applications in "fixed" telecommunication networks, which are specified by ANSI and Bellcore.

Private TCAP applications are defined by the Electronic Industries Association (EIA) and the Telecommunications Industry Association (TIA), for use in "mobile" networks.

For identifiers (tags) with H,G, = 1,1 in octet 1 (see Fig. 15.3-3), the values in the identifier code field are divided into separate ranges, for national and private use. All values in one-octet identifier fields are assigned to national TCAP. In two- and three-octet fields, the value of bit G in octet 2 indicates whether the code is assigned to national TCAP (G = 0), or private TCAP (G = 1).

In general, if a CCITT/ITU-T information element is coded as "application-wide", the corresponding ANSI data element is coded as "national TCAP".

Further details on identifier coding of national TCAP data elements can be found in Ref. [9].

15.5.4 National TCAP Error-code DE

ANSI has specified a one-octet value field for the error-code DE in return-error packages. The value indicates the reason for sending the return-error:

Unexpected component sequence	0000 0001
Unexpected data value	0000 0010
Unavailable network resource	0000 0011
Reply overdue	0000 0101
Data not available	0000 0110

15.5.4 National TCAP Problem DE

National TCAP has one problem-code DE instead of the four problem IEs of CCITT/ITU-T—see Table 15.5-1. The DE has a two-octet value field. The first octet (known as *problem code*) indicates the nature of the problem (general problem, invoke problem, etc.). The second octet (*problem specifier*) indicates a specific problem (unrecognized component, duplicate invoke-ID, incorrect parameter, etc.).

Further details on the coding of value fields for national TCAP data elements can be found in Ref. [9].

15.5.5 National TCAP Constructor and Primitive DEs

Table 15.5-1 indicates the type of the individual IEs and DEs, with one exception: If an IE is a constructor (C), the corresponding DE is also a constructor, and the same holds for primitive (P) IEs and DEs.

However, the U-abort-information IE is a constructor, and the corresponding DE is a primitive.

15.6 ACRONYMS

ANSI	American National Standards Institute
ASE	Application service element
C	Constructor information element
CCITT	International Telegraph and Telephone Consultative Committee
DE	Data element
DTID	Destination transaction identifier
ID	Identifier
IID	Invoke-identifier
ISUP	ISDN user part
ITU	International Telecomunication Union
ITU-T	Telecommunication standardization sector of ITU
LID	Linked identifier
MSU	Message signal unit
MTP	Message transfer part
OTID	Originating transaction identifier
P	Primitive information element
RETERR	Return-error
RETRES	Return-result
RR-L	Return-result-last
RR-NL	Return-result-not last
SCCP	Signaling connection control part
SS7	Signaling system No.7

SSN	Subsystem number
TC	Transaction capabilities
TCAP	Transaction capability application part
TC-primitives	Primitives between ASE and TCAP
TID	Transaction identifier

15.7 REFERENCES

1. A.R. Modaressi, R.A. Skoog, "Signaling System No.7: A Tutorial," *IEEE Comm. Mag.*, **28**, No.7, July 1990.
2. R. Manterfield, *Common Channel Signalling*, Peter Peregrinus Ltd., London, 1991.
3. T.W. Johnson, B. Law, P. Anius, "CCITT Signalling System No.7: Transaction Capabilities," *Br. Telecomm. Eng.*, **7**, April 1988.
4. *Specifications of Signalling System No.7*, Rec. Q.771–Q.775, CCITT Blue Book, **VI.7**, ITU, Geneva, 1989.
5. *Rec. Q.771, Signalling System No.7*, Functional Description of Transaction Capabilities, ITU, Geneva, 1993.
6. *Rec. Q.772, Signalling System No.7*, Transaction Capabilities Information Element Definition, ITU, Geneva, 1993.
7. *Rec. Q.773, Signalling System No.7*, Transaction Capabilities Formats and Encoding, ITU, Geneva, 1993.
8. *Rec. Q.774, Signalling System No.7*, Transaction Capabilities Procedures, ITU, Geneva, 1993.
9. *American National Standard for Telecommunications—Signaling System No.7 (SS7)—Transaction Capabilities*, ANSI T.114, 1992, New York, 1992.

16

TRANSACTIONS IN INTELLIGENT NETWORKS

This chapter describes an important application of transactions, and includes information on the operations and parameters for this application.

16.1 INTRODUCTION TO INTELLIGENT NETWORKS

16.1.1 History

The introduction of stored-program controlled exchanges in telecommunication networks has enabled the telecoms to offer call-forwarding, call-waiting, and other supplementary services to their customers [1]. These services are *exchange-based*: the (software) logic and data for the services reside in the exchanges.

Telecommunication networks are evolving into *intelligent networks* that offer, in addition to exchange-based services, a number of services whose logic and data reside in a small number of centralized information sources that can be queried (interrogated) by the exchanges.

The first steps toward intelligent networks (IN) were made by the Bell System in the early 1980s, for 800-call services [1]. 800-numbers contain no routing information, and have to be translated into "routing" numbers that have the North American numbering plan (NANP) format *NPA-NXX-XXXX* (Section 1.2.1). The translation data are stored in network databases. This approach was taken because the translations require a large amount of data that has to be updated frequently. It is much simpler to add or change an entry in a few centralized databases than to enter these changes into every local exchange.

The queries by the exchanges, and the responses from the databases, were transferred in CCIS "direct signaling" messages (Section 6.5.1).

After the divestiture of the Bell system, the new regional telecoms started to deploy *service control points* (SCP) with databases for 800-calling, and *alternate billing services* [2]. Messages between exchanges and SCPs are transferred by the SS7 signaling network (Section 14.3.4).

The evolution to the present intelligent networks, known in the U.S. as *advanced intelligent networks* (AIN), started in the 1980s [3,4]. The logic and data for AIN services is stored in AIN SCPs, and exchanges obtain their call-handling instructions by executing transactions with these SCPs.

The definition of AIN is the result of the combined efforts of Bellcore, the former Bell system regional telecoms, and equipment manufacturers. During the same time frame, ITU-T (International Telecommunication Union) started a phased standardization process for IN architecture and capabilities [5]. Each phase is known as an IN Capability Set CS. The definition of CS-1 has been completed [6,7,8].

AIN Releases. AIN Release 1 outlines the general long-term objectives for AIN in the U.S. [9,10]. The detailed specifications have been published as a sequence of AIN releases (0.0, 0.1, and 0.2). Release 0.0 [11,12,13] has been used mainly for laboratory and service trials. Compared with its predecessor, AIN 0.1 is more closely aligned, in capabilities and terminology, with the international (ITU-T) standards. It is the first true step towards AIN 1. The specifications for AIN 0.1 have been published by BellSouth [14,15] and by Bellcore [16,17]. In this chapter, the focus is on AIN 0.1. The initial version of AIN 0.2 has recently become available [18,19].

16.1.2 AIN 0.1 Architecture

The principal entities in an AIN 0.1 network are shown in Fig. 16.1-1.

Service Control Point. This is a data processor that stores the logic and data for AIN services. It responds to queries (requests for instructions) from SSP exchanges. The queries and responses are transferred by the SS7 signaling network.

Service Switching Point (SSP). This is an exchange that provides AIN services, by querying an SCP.

An SSP can be a local, tandem, or a combined local/tandem exchange. The present trend is to install SSP capabilities in all local exchanges, and this chapter limits the discussion to SSP *local* exchanges.

The SSP in Fig. 16.6-1 serves analog and digital (ISDN) subscribers. We shall refer to them as *customers* (C). SSP also serves private branch exchanges (PBX—see Section 1.1.2). The PBXs are attached to SSP by trunk groups that are known as *private-facility groups* (PFG). Other PFGs are the *tie-trunk*

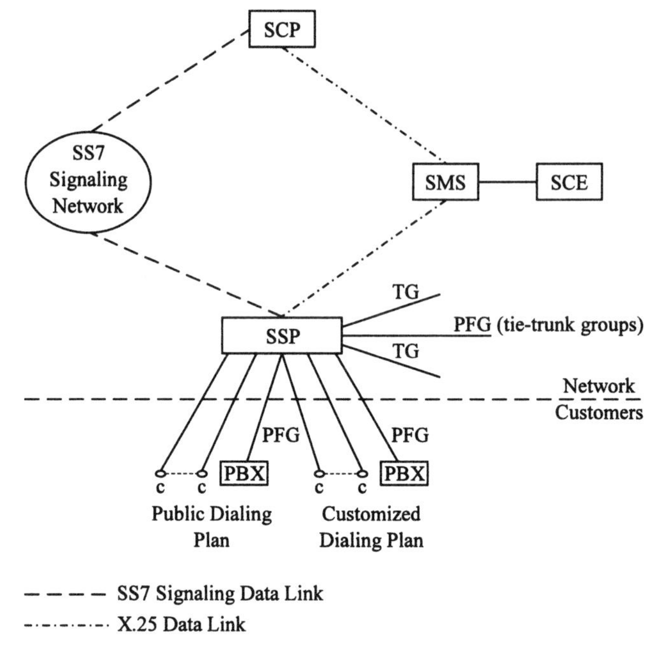

Figure 16.1-1 AIN 0.1 architecture.

groups, leased by the telecom to business customers who have several business locations. These trunk groups interconnect exchanges of the public network, but are dedicated to calls made by lines of the individual business customers.

Numbering Plans. Most customers and PBXs use the public North American numbering plan (NANP). However, at the request of a multi-location business customer, a telecom can provide a *customized* dialing plan that is "tailored" to the customer's requirements.

AIN provides services to customers using the NANP, and to customers who have a customized dialing plan.

Service Management System (SMS). This system enables a telecom to update and maintain the service logic and data in its SCPs. The SMS validates inputs from the telecom, and then updates the affected SCPs. SMS is also capable to test the AIN, and to collect AIN traffic data from SSPs. The SMS communicates with the SCPs and SSPs in its network by X.25 data links.

Service Creation Environment (SCE). This system interfaces with SMS, and contains software tools that enable the telecoms to develop and test new AIN services.

16.1.3 AIN Service Switching Point

The conversion of an exchange into an SSP requires software and hardware additions.

SSP Software. Three main software functions have to be installed. In the first place, an SSP needs an AIN *application service element* (ASE), to communicate with a "peer" ASE in an SCP (Section 15.1.1).

In the second place, an SSP has to determine whether a call requires assistance from an SCP. This is done in a software procedure known as *triggering*.

In the third place, AIN transactions end with a response message from SCP that contains call-handling instructions. Software for the execution of these instructions has to be installed in the SSPs.

The triggers, and the call-handling instructions, are defined in a "generic" manner (not specific to a particular AIN service). An SSP does not know anything about the individual services: it merely informs SCP when it has encountered a trigger, and executes the instructions received from SCP.

SSP Hardware. An SSP has to be equipped with a group of *intelligent network service circuits* (INSC)—see Fig. 16.1-2. These circuits have two functions. In the first place, they play customized recorded announcements, whose contents are specified by SCP. In the second place, they can receive *dial-tone multi-frequency* (DTMF) digits (Section 3.3.4). This enables an SCP to interact with the calling party. On instructions from SCP, the SSP processor connects an INSC to a calling line or incoming trunk, and orders it to play a specified announcement, which prompts the calling party to respond with one or more digits. The SSP processor collects the received digits, and then sends them to SCP.

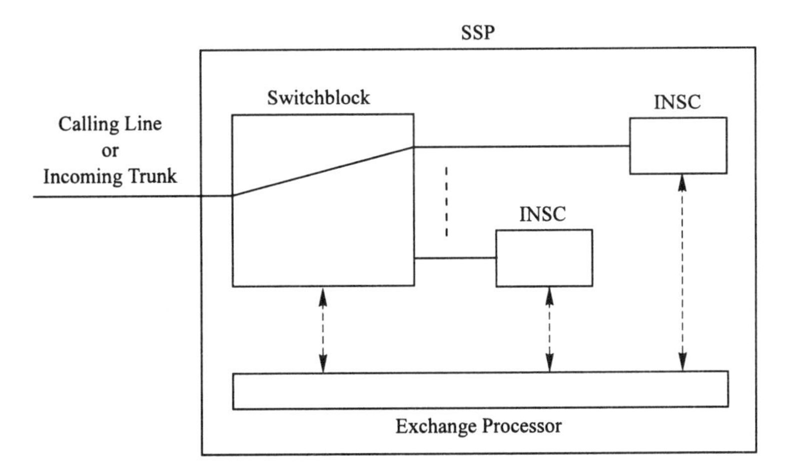

Figure 16.1-2 Intelligent network service circuits.

16.1.4 AIN Service Control Point

The AIN SCP is a data processor which has the logic and data needed for AIN services. It includes one or more ASEs for communications with the SSPs in its network.

The advantages of SCPs to a telecom are two-fold:

Telecom Programming. Before AIN, when new features needed to be added in a telecommunication network, the telecom had to purchase a new version of exchange software from the equipment manufacturer, and then install and test it in all exchanges.

AIN services can be programmed by telecom personnel, using the service creation environment. This gives the individual telecoms the freedom to design their own service features.

Also, the tests of a new AIN feature can be done with the SCP and just one of the SSPs in the network.

Flexibility. SCPs are designed to provide great flexibility in the definition of AIN services. A service can be designed to take into account the time-of-day, and day-of-week, of the call, and the locations of the calling and called parties.

Caller-interaction is another important contributor to service flexibility. SCP can prompt the caller to send DTMF digits, which indicate a particular service option for the call, or an authorization code that is checked at SCP, to verify that the caller is entitled to receive the requested service.

16.1.5 AIN 0.1 Services

The services defined for AIN 0.1 involve transactions during the set-up of connections only, and apply to calls between two parties. Services requiring transactions during the conversation phase of a call, and services for multi-party calls, are being planned for later AIN releases.

The emphasis is on capabilities to screen outgoing and incoming calls, and to route and charge calls.

16.2 CALL MODELS AND TRIGGERS

When an SSP exchange receives a call that involves an AIN service, it initiates a transaction with SCP, to obtain call-handling instructions. The mechanism by which the SSP determines that it needs to query SCP is known as *triggering*. We say that an SSP *launches a query* (initiates a transaction with an SCP) when it "encounters a trigger" during the processing of a call. Triggers can be encountered by SSP at various points in the processing of a call. In order to explore the various trigger types, we need a call-processing model.

16.2.1 Basic Call Model

The call-processing models used in documents on intelligent networks are known as *basic call models* (BCM), and consist of an originating and a terminating part [5,6,13].

Originating BCM (O-BCM). This chapter uses the simplified originating BCM shown in Fig. 16.2-1. The model consists of a number of call states, starting with the detection of an origination attempt by SSP (A). For simplicity, the model ends at points (E), which are the points where the call destination—either a line attached to the SSP or a route set to the destination exchange—has been determined.

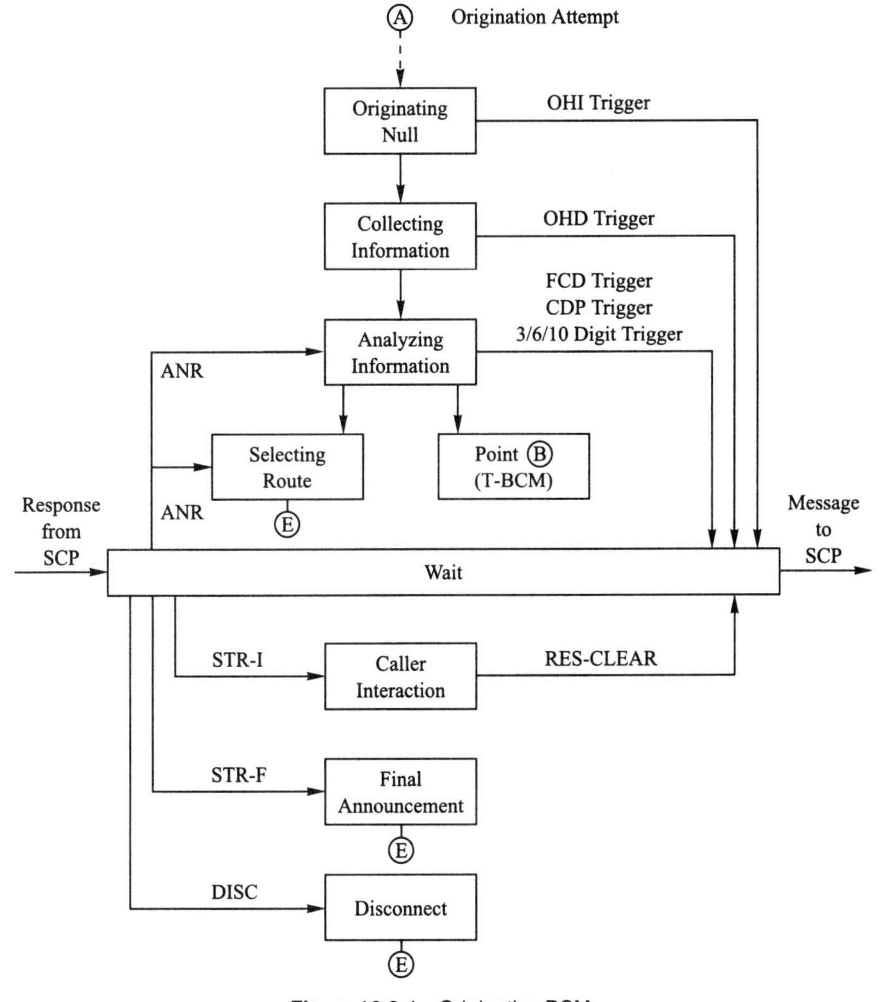

Figure 16.2-1 Originating BCM.

We first describe "normal" call processing (no trigger encountered). Idle lines and trunks are supervised by the SSP. When an origination attempt is detected, say an off-hook on an analog subscriber line (point A), the call process starts at the *originating null* state, and moves to the *collecting information* state. In this state, SSP collects the information received from the calling customer (the digits dialed by an analog subscriber, or the digits in the SETUP message from an ISDN user).

The call then moves to the *analyzing information* state, where the call destination is determined, from the analysis of the digits of the called number. If the destination is "local" (the called customer is served by SSP), the call moves to point B in the terminating BCM (see below). If the destination is at another exchange, the route set to the destination exchange is determined (see Section 1.3). The call then moves to *selecting route* state, in which SSP attempts to find an available trunk in one of the trunk groups belonging to the specified route set.

Triggers can be encountered in the originating null, the collecting information, and analyzing information states. If this happens, call processing is suspended, and the call moves into the *wait* state.

The trigger types, which are shown along the transitions to the wait state, are discussed later. When the call enters the wait state, SSP orders its AIN-ASE to initiate a transaction with SCP, to obtain call-handling instructions. The message from SCP that ends the transaction includes one of the following instructions:

Analyze Route (ANR). This message instructs SSP to resume the call processing at the analyzing information state, or the selecting route state, depending on a parameter in the message (to be discussed later).

Disconnect (DISC). This instruction is sent when SCP has determined that the call should not be set up, and that the calling party should receive a standard tone (say, reorder-tone), or announcement. SSP then resumes the call processing at the *disconnect* state. SSP connects the calling line or incoming trunk to a tone or announcement circuit, and the caller hears a standard tone (e.g. reorder-tone), or an announcement. When SSP receives a disconnect signal from the line or the trunk, it clears the connection.

Send to Resource, Final Announcement (STR-F). This instruction is sent when SCP has determined that the call should not be set up, and that the calling line should be connected to an INSC (resource) which should play the AIN announcement specified in the instruction. SSP then resumes the processing at the *final announcement* state. It connects the calling line or incoming trunk to an INSC, plays the specified announcement, and awaits the disconnect signal from the line or trunk.

Send to Resource, Caller Interaction (STR-I). This instruction indicates that

SCP needs to interact with the calling party. It requests SSP to resume processing at the *caller interaction* state. SSP connects the calling line or incoming trunk to an INSC, and plays the specified announcement. The announcement contains a question, and prompts the calling party to respond with one or more digits. The INSC collects the digits, and SSP sends them to SCP in a *resource clear* message. The call reenters the wait state, and stays there until a new message with instructions from SCP has been received.

Terminating BCM (T-BCM). This consists of the states of calls with called party numbers that identify SSP as the terminating exchange—see Fig. 16.2-2. The calls may have originated by a customer served by SSP, or have reached SSP on an incoming trunk.

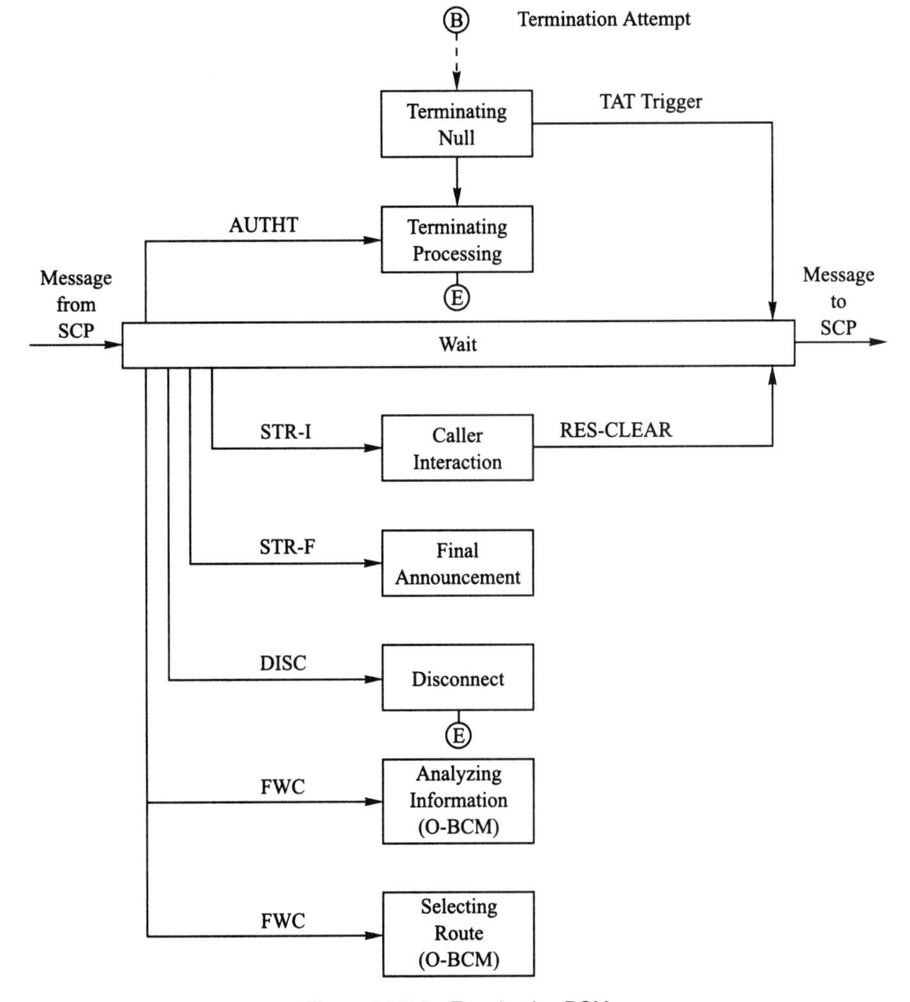

Figure 16.2-2 Terminating BCM.

The call model is entered at point B (*terminating null* state). If no trigger is encountered, the call moves to the *terminating processing* state. If a trigger has been encountered, the call moves to the wait state, and SSP initiates a transaction with SCP. If SCP determines that the call destination is at SSP, it returns an *authorize termination* (AUTHT) message, and SSP resumes processing, reentering the terminating processing state.

If SCP determines that the call destination is at another exchange, it returns a *forward call* (FWC) message. SSP then resumes call processing at the analyzing information state, or the selecting route state, of the originating BCM.

The SSP actions on receipt of a disconnect (DISC), a send to resource, final (STR-F), or a send to resource, interactive (STR-I) message are the same as in the originating BCM. However, depending on where the call originated, the standard tone or announcement source or INSC is now connected to the calling line, or to the incoming trunk.

16.2.2 Trigger Types and Call States

We begin our examination of the various triggers by listing the call states in which they can be encountered—see Figs. 16.2-1 and 16.2-2.

Off-hook Immediate Trigger (OHI). This trigger can be encountered when a call moves out of the originating null state (i.e. when a customer served by SSP goes off-hook).

Off-hook Delay Trigger (OHD). This trigger can be encountered in the collecting information state, when the digits from a calling customer, or from a private facility trunk, have been received.

Feature Code Dialing Trigger (FCD). This trigger can be encountered in the analyzing information state, if the caller has dialed a *feature code* of the public office dialing plan.

Customized Dialing Plan Trigger (CDP). This trigger can be encountered in the analyzing information state, if the caller using a customized dialing plan has dialed an *access code* belonging to the plan.

3/6/10 Digit Triggers (3/6/10D). This trigger can be encountered in the analyzing information state, if the caller has dialed a called number of the NANP.

Termination Attempt Trigger (TAT). This trigger can be encountered when a call leaves the terminating null state, and the digits received from the calling customer or incoming trunk represent the NANP number of a customer served by SSP.

16.2.3 Subscribed, Group, and Office-based Triggers

The OHI, OHD, FCD, and TAT triggers are *subscribed* triggers. They are assigned to the lines, private facility groups and/or directory numbers on customer request. The customers are charged a monthly, or a per-use, fee.

The CDP triggers are *group* triggers, and are assigned, on request, to the lines of a business group. They also require a fee.

3/6/10 Digit triggers are *office-based*. They do not involve charges to the customers whose calls encounter them.

16.2.4 Encountering a Trigger

An SSP encounters a trigger (and initiates a transaction with SCP) when the following conditions are met:

1. The trigger is either:
   - office based, or
   - group based, and the call is originated by a line or a private facility trunk associated with a business group, or
   - subscribed, and the call is originated by a line that has subscribed to the (OHI, OHD, FCD) trigger, or terminates at a called number that has subscribed to the TAT trigger.
2. The trigger is active. The OHI, OHD, FCD, CDP, and TAT triggers can be activated and deactivated on command from SCP.
3. The criteria for the trigger type have been met.

We now examine the assignments and criteria for the various triggers.

16.2.5 Off-hook Immediate Trigger

Off-hook immediate (OHI) triggers can be assigned to analog lines attached to SSP. When the line goes off-hook, the trigger is encountered, and SSP immediately initiates a transaction.

16.2.6 Off-hook Delay Trigger

Off-hook delay (OHD) triggers can be assigned to analog or digital (ISDN) subscriber lines, or private facility trunks, attached to SSP. The trigger is detected when an analog subscriber line goes off hook, a setup message is received from an ISDN line, or a seizure signal has been received on a private facility trunk.

The primary criterion for OHD is that the customer is making a call. This can be determined only after the customer's dialed digits have been received.

If the customer has dialed a valid called number, say NPA-*NXX-XXXX* or *NXX-XXXX*, the OHD trigger is usually encountered. However, a telecom can designate certain called numbers as "escape codes." A call to an escape code—

for example, 911 (emergency)—escapes the OHD trigger, and no transaction is initiated.

Next, suppose that the customer has dialed an access code to an exchange-based feature, followed by a called party number. This may—or may not—mean that a call is being originated. For example, *67-*NXX-XXXX* indicates that the customer is originating a call, and does not want his number to be presented to the called party (Section 3.7.1). In this case, the OHD trigger is encountered, unless the called number is an escape code. However, the digit string 74#-3-*NXX-XXXX* is a request by a customer—who has "speed calling" service—to change the third entry on his speed-calling list to *NXX-XXXX*. Since no call is originated, the OHD trigger is not encountered.

16.2.7 Feature Code Dialing Trigger

The feature code dialing (FCD) trigger can be assigned to analog and digital subscriber lines, and to private trunk groups. This enables them to invoke an AIN service, by dialing a FCD feature code.

In addition, one of the following criteria, which depend on the particular FCD code, has to be met.

- no digits follow the FCD code
- the digits following the FCD code represent a valid NANP number
- the digits following the FCD code represent a number of another type, say an authorization code, or a personal identification number (PIN).

The FCD codes are considered as belonging to the "public" dialing plan. On calls originated by a line or private facility trunk using a customized dialing plan, SSP considers the received digits as CDP numbers, and does not recognize FCD codes. However, a calling CDP customer can "escape" to the public numbering plan, by dialing an access code in his numbering plan—say *9.

The caller then receives a second dial-tone, and SSP interprets the subsequent received digits as belonging to the public dialing plan. This enables the CDP customer to invoke the AIN services in the manner described above, provided that an FCD trigger has been assigned the line or trunk.

16.2.8 Custom Dialing Plan Trigger

Customer dialing plan (CDP) triggers can be assigned to lines and private facility trunks of business customer groups using a customized dialing plan. In these plans, the caller dials a called number (in the particular CDP format), or an *access code* defined for the CDP, possibly followed by additional digits. Some access codes invoke exchange-based services. Other codes invoke AIN services that have been defined for the business group, and require the assignment of a CDP trigger, one of the following criteria, which depend on the particular access code, has to be met:

- no digits follow the access code
- the digits following the access code represent a valid number in the CDP of the business group
- the digits following the access code represent a number of another type, say an authorization code, or a personal identification number (PIN).

16.2.9 3/6/10 Digit Trigger

The telecom can assign these triggers to the first three, the first six, or all digits of NANP numbers:

NPA	Area code (three digits)
NPA-*NXX*	Area and exchange code (six digits)
NPA-*NXX-XXXX*	Complete national number
SAC	Service access code (700, 800, 900, 976, etc.).
SAC-*NXX*	First six digits of a 700, 800, 900, etc. number
SAC-*NXX-XXXX*	Complete (ten-digit) 700, 800, 900, etc. number

The criterion for this trigger is met when a call is made to any NANP number to which the trigger has been assigned. Calls made by any originator (line, private facility trunk) using the NANP can encounter the trigger.

16.2.10 Terminating Attempt Trigger (TAT)

Terminating attempt (TAT) triggers can be assigned to directory numbers covered by the SSP exchange. Any TAT trigger detected during processing in the terminating null state is considered to be encountered.

16.3 AIN MESSAGES AND TRANSACTIONS

This section explores the most important messages and transactions of AIN 0.1 [16,17,20].

16.3.1 Messages

AIN messages are grouped into a number of message families—see Table 16.3-1.

Request Instructions Family. The messages are sent by SSP when a trigger has been encountered, and request SCP to provide instructions for handling the call.

The names of the messages in this family represent the SSP event during which the trigger has been encountered—see Figs. 16.2-1 and 16.2-2.

Table 16.3-1 AIN 0.1 messages.

	Sender		Package			Component	
Message	SSP	SCP	QRY	CON	RES	INV	RR
Request Instructions Family							
Origination_Attempt	X		X			X	
Info_Collected	X		X			X	
Info_Analyzed	X		X			X	
Termination_Attempt	X		X			X	
Connection Control Family							
Analyze_Route		X			X	X	
Authorize_Termination		X			X	X	
Disconnect		X			X	X	
Send_To_Resource_Final		X			X	X	
Connectivity Control Family							
Forward_Call		X			X	X	
Caller Interaction Family							
Send_To_Resource_Interaction		X		X		X	
Resource_Clear	X			X		X	
Status Notification Family							
Monitor_For_Change		X	X			X	
Monitor_For_Success	X			X		X	
Status_Reported	X				X	X	
Information Revision Family							
Update_Request		X	X	X		X	
Update_Data	X				X		X

QRY: Query package. CON: Conversation package. RES: Response package. INV: Invoke component. RR: Return-result component.
Source: TR-NWT-001285. Reprinted with permission of Bellcore. Copyright © 1992.

Origination_Attempt. This message requests instructions for a call that has encountered a trigger (OHI) in the originating null state of O-BCM.

Info_Collected. This message requests instructions for a call that has encountered a trigger (OHD) in the collecting information state of O-BCM.

Info_Analyzed. This message requests instructions for a call that has encountered a trigger (FCD, CDP, or 3/6/10D) in the analyzing information state of O-BCM.

Termination_Attempt. This message requests instructions for a call that has encountered a trigger (TAT) in the null state of T-BCM.

Connection Control Family. These messages are sent by SCP, and contain call-control instructions.

Analyze_Route. This message instructs SSP to resume call processing, at the analyzing information, or the selecting route state of O-BCM (the choice depends on parameters in the message).

Authorize_Termination. This message instructs SSP to resume processing at the terminating call processing state of T-BCM.

Disconnect. This message instructs SSP to resume processing the call at the disconnect state of O-BCM or T-BCM. SSP then connects the calling line or incoming trunk to a (non AIN) tone or announcement source, to inform the calling party that the connection cannot be set up.

Send_To_Resource_Final. This message instructs SSP to resume the processing at the final announcement state of O-BCM or T-BCM. SSP then connects the calling line or incoming trunk to an INSC, and plays the announcement specified by SCP.

Connectivity Control Family. This family consists of just one message, which is invoked sent by SCP.

Forward_Call. This message instructs SSP to resume the processing at the analyzing information state, or the selecting route state, of O-BCM.

Caller Interaction Family. This family consists of two messages.

Send_To_Resource_Interaction. This message is sent by SCP, and requests SSP to resume the processing at the caller interaction state in O-BCM or T-BCM. SSP then connects the calling line or incoming trunk to an INSC (the resource), plays an announcement specified by SCP, and collects the digits sent by the calling party.

Resource_Clear. This message is sent by SSP, after it has collected the digits received by INSC, and has released the resource. The message includes the collected digits.

Status Notification Family. This family consists of three messages.

Monitor_For_Change. This message is sent by SCP, and requests SSP to determine the status (idle or busy) of a specified *facility* (a line, a group of lines, or a public or private trunk group).

Monitor_For_Success and Status_Reported. These messages are sent by SSP, in response to a monitor for change message (16.3.4).

Information Revision Family. These messages enable SCP to activate or deactivate an OHI, OHD, FCD, CDP, or TAT trigger at an SSP.

Update_Request. This message is sent by SCP, and requests SSP to change the status of a specified trigger.

Update_Data. This message is sent by SSP, in response to an update request.

In addition to listing the messages and the message senders, Table 16.3-1 shows that, with one exception, all messages are invokes. It also lists the TCAP packages (query, conversation, response) that can contain the various messages.

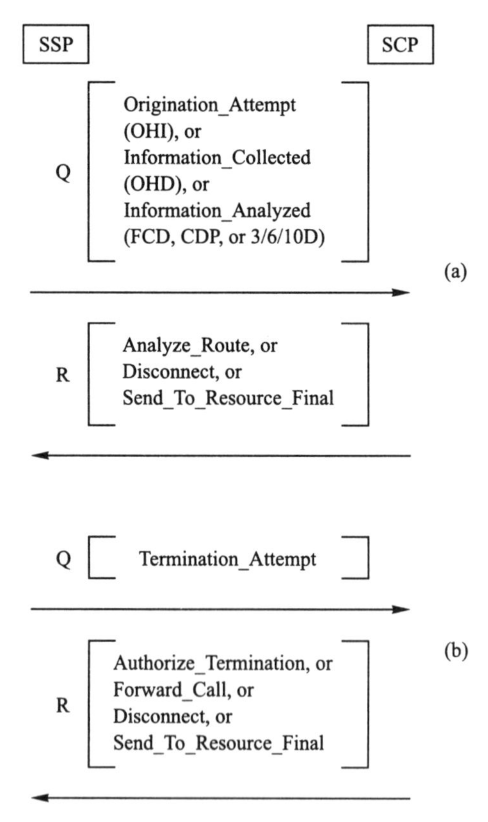

Figure 16.3-1 Basic call-related transactions. Q: Query package. R: Response package.

We next explore a number of AIN transactions. We distinguish call-related transactions (initiated by SSP when a trigger has been encountered), and transactions that are not call-related (and initiated by SCP).

16.3.2 Basic Call-related Transactions

A basic call-related transaction involves a TCAP *query* package sent by SSP, containing a message of the request instructions family, and a *response* package, sent by SCP, and containing a message of the connection control, or connectivity control, family.

Figure 16.3-1(a) shows basic transactions that are initiated because a trigger has been encountered in the originating BCM. The query and response packages are denoted by Q and R. Their messages are shown inside square brackets and, in query packages, the acronyms of the triggers are shown in parentheses. For example, a query package with an info_analyzed message can be the result of encountering an FCD, CDP, or 3/6/10D trigger.

Figure 16.3-2 shows basic transactions initiated as a result of encountering a TAT trigger (in the terminating BCM).

16.3.3 Interactive Call-related Transactions

Any of the transactions described above can include one or more pairs of conversation (C) packages. In the example of Fig. 16.3-2, SCP has received a query package with an info_collected message, and returns a conversation package with a send_to_resource_interaction (STR-I) message.

After connecting the calling line or incoming trunk to an INSC, playing the specified announcement, and collecting the caller's DTMF digits, SSP returns a conversation package with a resource_clear (RES-CLR) message, which includes the collected digits.

The SCP uses this information to formulate the call handling instructions, and ends the transaction with a response package, which in this example contains an analyze_route message.

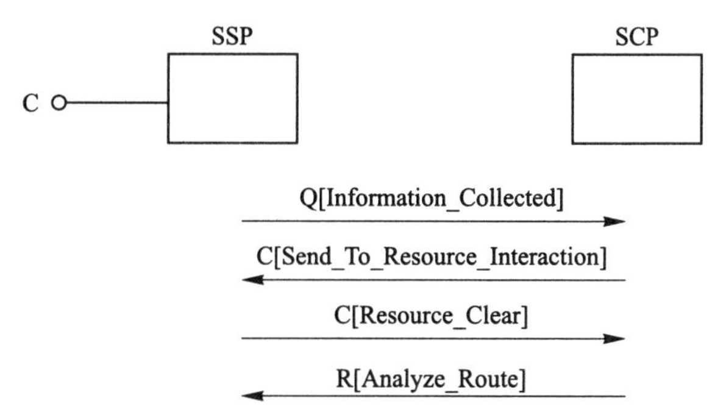

Figure 16.3-2 Call-related transaction with caller interaction. Q: Query package. C: Conversation package. R: Response package.

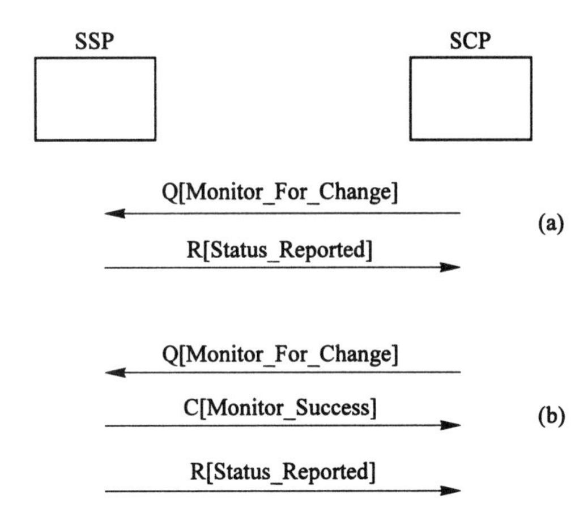

Figure 16.3-3 Status monitoring transactions.

16.3.4 Status Monitoring Transaction

The status monitoring transaction is not call-related, and enables an SCP to determine the status (busy or idle), of a "facility" (line or trunk), of a "facility group" (line group or trunk group) attached to a SSP.

In the example of Fig. 16.3-3, SCP needs to know the status of a line, and initiates the transaction with the SSP to which the line is attached. The query package holds a monitor_for_change message, which specifies the line to be monitored, and a line state. In case (a), SSP finds that the line is in the state specified by SCP. It returns a response package with a status_reported message. This ends the transaction.

In case (b), SSP finds that the line is not in the specified state. It then returns a conversation package with a monitor_success message, and keeps monitoring the line. When the line changes to the specified state, SSP ends the transaction, sending a response package with a status_reported message.

16.3.5 Trigger Activation and Deactivation

This transaction enables an SCP to activate or deactivate a trigger in an SSP. When SCP has determined that the status of a trigger has to be changed, it opens a transaction with the SSP where the trigger is located—see Fig. 16.3-4. The query package contains an update_request message, in which the trigger and its activity status are specified. After executing the operation, SSP returns a response package. This package is somewhat of an exception: the *update_data* message is not an invoke, but a return-result.

16.3.6 SCCP Addresses

Call-related transactions are initiated by the intelligent network (IN) ASE at an SSP. For a query package (which is the first package of the transaction),

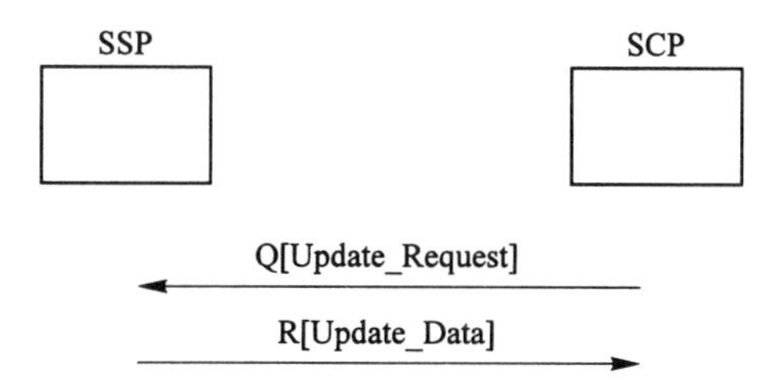

Figure 16.3-4 Information revision transaction.

ASE provides a global title (GT) called address (Section 14.3.3). The meaning of GTA (address) part of GT (ten BCD digits) depends on the type of the encountered trigger. For queries resulting from OHI, OHD, FCD and CDP triggers, GTA is the calling customer's charge number. For the TAT and 3/6/10D triggers, GTA is the called party number.

The translation type (TT) of GT specifies that the number in GTA is to be translated into the PC-SSN address of the IN-ASE in an SCP that stores information for the number.

Transactions that are not call-related are initiated by an SCP. The GTA of the global title is a ten-digit number that represents the line, multi-line group, or private facility group to be monitored, or the line or private facility trunk to which the trigger to be activated/deactivated is assigned. The TT value indicates that GTA is to be translated into the PC-SSN address of the IN-ASE in an SSP.

In the subsequent messages of a transaction, a GT translation is not necessary, because the ASEs then know each other's PC-SSN address.

16.4 AIN 0.1 PARAMETERS

This section outlines the most important parameters in AIN 0.1 messages [17,20]. At this point, it is suggested that these descriptions are merely perused. Each parameter has a reference number, e.g. Par.1. This enables the reader to quickly locate a parameter description when reading the material in the later sections of this chapter.

16.4.1 Introduction

Before describing the parameters, we need to make a few introductory remarks.

Parameter Names. The names of most parameters consist of several words. This section follows the AIN convention for denoting parameter names. Multi-word names are written without spacings, and with the first letter of each word in upper case, e.g. CallingPartyID.

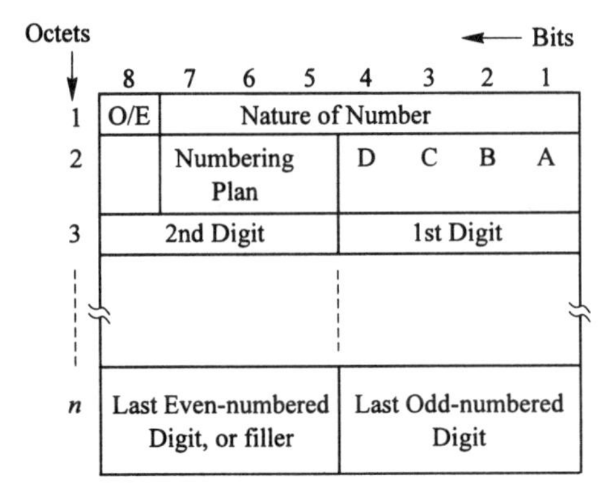

Figure 16.4-1 AINDigits format. (From Rec. Q.763. Courtesy of ITU-T.)

Primitives and Constructors. Most parameters are "primitive" data elements (Section 15.3.1). However, Par.1 (UserID) and Par.29 (STRParameterBlock) are "constructor" DEs.

Contents Fields. The contents fields of primitive parameters mostly consist of octets holding an integer and octets holding two BCD digits.

The format of the contents field shown in Fig. 16.4-1, which is known as the *AINdigits* format, has been carried over from ISUP signaling (see Fig. 11.2-6). It is used for several AIN parameters that represent "numbers" coded as BCD digits—for example, the CallingPartyID.

Bit O/E indicates whether the number of digits is even (0) or odd (1). The digits are in octets 2–*n* (if the number of digits is odd, bits 8–5 of octet *n* contain a filler code).

The *numbering plan* (NP) field indicates whether the number belongs to a *public* (001) or *private* (101) numbering plan. In this context, "public" means the PSTN/ISDN numbering plan of ITU-T Recommendation Q.164 [21] or, more specifically, the North American Numbering Plan NANP (Section 1.2.1). Private numbers are numbers belonging to a customized dialing plan (CDP).

16.4.2 Parameter Descriptions

Par.1 UserID. This parameter identifies the "user" of the encountered trigger. It is a "constructor"—see Fig. 16.4-2. Its value (contents) field V1 holds another parameter, for example:

Par.2 DirectoryNumber (DN)
Par.3 TrunkGroupID
Par.4 PrivateFacilityGID.

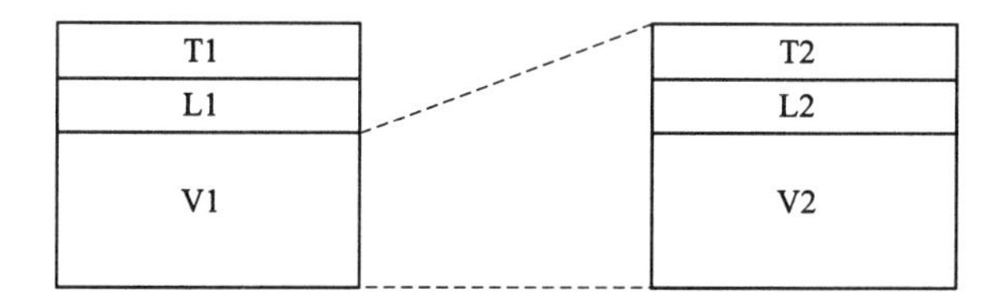

Figure 16.4-2 Format of UserID.

In queries caused by OHI, OHD, or FCD triggers, UserID identifies a calling entity that is using thc public numbering plan to which the encountered trigger is assigned.

In queries resulting from CDP triggers, UserID identifies the calling entity (using a private numbering plan) to which the encountered trigger is assigned.

In queries caused by a TAT trigger, UserID identifies the called customer to whom the trigger is assigned.

In queries with 3/6/10D triggers, UserID identifies the calling entity that has dialed a NANP number that caused an encounter with a 3, 6, or 10 digit trigger.

Par.2 DirectoryNumber (DN). The five-octet contents field of this parameter holds ten BCD digits that represent the NANP number of a customer.

Par.3 TrunkGroupID. This is an integer that specifies a public trunk group.

Par.4 PrivateFacilityGID. This is an integer that specifies a private facility trunk group.

Par. 5 BearerCapability. This parameter indicates the bearer capability requested by the calling party (Section 10.3.5). The contents field is coded as follows:

0000 0000: speech
0000 0001: 3.1 kHz audio
0000 0011: 56 kb/s data
0000 0100: 64 kb/s data

Par.6 TriggerType. This parameter indicates the type of a trigger:

0000 0001 Feature dialing code trigger (FCD)
0000 0010 Customized dialing plan trigger (CDP)
0000 0100 Three-digit trigger (area code)
0000 0101 Six-digit trigger (area and exchange code)
0000 1000 Ten-digit trigger (national number)
0000 1111 Termination attempt (TAT) trigger
0001 0000 Off-hook immediate (OHI) trigger
0001 0001 Off-hook delay (OHD) trigger

Par.7 CallingPartyID. This parameter represents the number of the calling party. Its contents field has the AINDigits format.

Numbering Plan (NP): public or private

Nature of Number (NN)

000 0001	Subscriber number
000 0011	National number
000 0100	International number

Bits BA indicate whether the calling customer allows the presentation of his number to the called party:

00	Presentation allowed
01	Presentation restricted

Bits DC indicate whether the calling number was provided by a calling ISDN user, or by the exchange:

01	Provided by user, not screened by exchange
01	Provided by user, passed screening by exchange
11	Provided by exchange

Par.8 ChargeNumber. This parameter identifies the charge number of the calling customer. If the customer has one line, the charge number is usually the caller's number. If a customer has a group of lines, a common number can be designated as the charge number for all lines of the group, and the customer receives one consolidated monthly bill. The contents field has the AINDigits format.

Nature of Number (NN)

000 0001	Subscriber number (calling party)
000 0011	National number (calling party)
000 0101	Subscriber number (called party)
000 0111	National number (called party)

Numbering Plan (NP): public or private

Bits C,D,B,A are not used.

Par.9 LATA. This parameter represents the LATA network of the calling party (Section 1.1.2), and consists of three BCD digits. The contents field has the AINDigits format. The NN and NP fields are set to "not applicable" (all zeros).

Par.10 PrimaryCarrier. This parameter specifies the first-choice long-distance (interexchange) carrier for the call. Octet 1 of the contents field indicates whether the carrier is the regular (presubscribed) long-distance carrier of the calling customer, or has been selected by the customer, by dialing a 10XXX prefix (Section 3.7.1):

0000 0000	No indication
0000 0001	Regular (subscribed) carrier
0000 0100	Dialed by calling party

Octets 2 and 3 contain the three or four BCD digit carrier identification.

Par.11 AlternateCarrier. This parameter specifies the second-choice long-distance carrier for the call. The contents field is coded as in Par.10.

Par.12 SecondAlternateCarrier. This parameter specifies the third-choice long-distance carrier for the call. The contents field is coded as in Par.10.

Par.13 CalledPartyID. This represents the number of the called party, and indicates whether the calling party has requested operator assistance for the call. The contents field has the AINDigits format.

Nature of Number (NN)

Bits 7–5	
000	No assistance requested
111	Assistance requested

Bits 4–1	
0001	Subscriber number
0011	National number
0111	International number

Numbering Plan (NP): public or private

Par.14 FeatureCode (VerticalServiceCode). This is a code, dialed by a caller who is using the public numbering plan, to request an AIN service.

The contents field is in the AINDigits format. The NN and NP fields are set to "not applicable" (all zeros).

The code is usually an * (asterisk), followed by up to four digits.

Par.15 AccessCode. This parameter is a code, dialed by a caller who is using a customized dialing plan, to request an AIN service. The contents field is as described for Par.14.

Par.16 CollectedAddressInfo. This parameter contains address information collected from the calling party. It has the same format and coding as Par.13 (CalledPartyID).

Par.17 CollectedDigits. This parameter holds the digits collected from the user, which may be a called party number, or a number with a different meaning (e.g. authorization code, personal identification code, etc.). The contents field has the AINDigits format.

The NN and NP fields are set to "unknown" (000 0000 and 000).

Par.18 CallingPartyBGID. This parameter is included in queries, if the calling line is a member of a business group. It identifies the group, and also indicates whether the line is restricted from making or receiving certain calls. The contents and coding have been carried over from U.S. ISUP signaling—see Section 11.9.2.

Par.19 PrimaryBillingIndicator. There are several "billing indicator" parameters. They contain information required by the telecom accounting centers, for the calculation of the call charge.

The contents field of these parameters have four octets, holding two BCD digits each.

Octets 1 and 2 specify the *automatic message accounting (AMA) call type.* Octets 3 and 4 contain a *service feature identifier.*

Par.19 is the billing indicator to be used if SSP routes the call to a trunk of the first-choice private trunk group.

Par.20 AlternateBillingIndicator. The meaning and contents of this parameter are as in Par.19.

Par.20 is the billing indicator to be used if SSP routes the call to a trunk of the second-choice private trunk group.

Par.21 SecondAlternateBillingIndicator. The meaning and contents of this parameter are as in Par.19.

Par.21 is the billing indicator to be used if SSP routes the call to a trunk of the third-choice private trunk group.

Par.22 OverflowBillingIndicator. The meaning and contents of this parameter are as in Par.19.

Par.22 is the billing indicator to be used if SSP has tried unsuccessfully to route the call to a private facility trunk, and is now routing the call to a trunk in the public network.

Par.23 AMAAlternateBillingNumber. This parameter is included in a SCP response, if the call is to be charged to a number other than the ChargeNumber (Par.8). The contents field is coded as in Par.8.

Par.24 PrimaryTrunkGroup. This parameter specifies the first-choice private-facility trunk group for the call, and the digits to be outpulsed (sent out). Octets 2–5 of the contents field contain eight BCD digits that specify the group. Bit H in octet 1 indicates the parameter that holds the digits to be outpulsed:

H = 0 Par.27 (OutpulseNumber)
H = 1 Par.13 (CalledPartyID)

Par.25 AlternateTrunkGroup. This parameter specifies the second-choice private-facility trunk group for the call. Its contents are coded as in Par.24.

Par.26 SecondAlternateTrunkGroup. This parameter specifies the third-choice private-facility trunk group for the call. Its contents are coded as in Par.24.

Par.27 OutpulseNumber. This parameter holds the digits to be outpulsed on a private facility trunk. The contents field has the AINDigits format. The NN and NP fields are set to "not applicable" (all zeros).

Par.28 ResourceType. This parameter differentiates the "final" and "interaction" send to resource operations (STR-F and STR-I). The contents field is coded as follows:

0000 0000 Final STR operation
0000 0001 Interaction STR operation

Par.29 StrParameterBlock. This parameter specifies the announcement to be played by an INSC.

The parameter is a constructor. Its contents field holds one or more *announcement elements*—see Fig. 16.4-3. *AnnouncementID* is an integer which specifies a phrase (e.g. "please enter your authorization code"). A phrase may be followed by one or more spoken *information digits*, which are specified by BCD numbers.

For interactive operations, the element also has a *MaximumDigits* field which is an integer that specifies the number of digits to be collected:

Integer	Number of Digits
0–31	As specified by the integer
253	"Normal number of digits"
254	"Any number of digits"

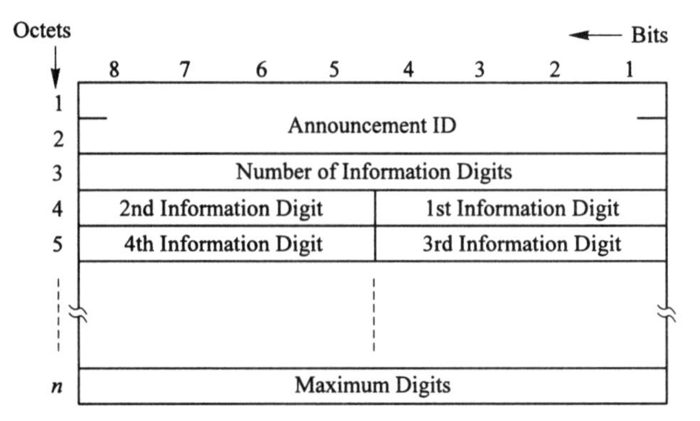

Figure 16.4-3 Announcement element. (From TR-NWT-001285. Reprinted with permission of Bellcore. Copyright © 1992.)

Par.30 DisconnectFlag. The presence of this parameter in a message indicates that the calling party should be disconnected after the resource (INSC) has played the announcement. The parameter has no contents field.

Par.31 ClearCause. This indicates why the SSP has cleared a resource (INSC):

0000 0000	Normal clearing
0000 0010	Timeout
0000 0110	Caller disconnect

Par.32 FacilityStatus. This parameter indicates the status of a line, a private facility, a public or private trunk group, etc.:

0000 0000	Busy
0000 0011	Idle

Par.33 FailureCause. This indicates why SSP has been unable to execute a requested operation:

0000 0001	Rate too high
0000 0010	Unavailable resources

Par.34 MonitorTime. This parameter specifies the length of time that SSP has to monitor the status of line (or line group, etc.).

The contents field consists of three octets, each of which holds two BCD digits. Octets 1, 2, and 3 specify the number of hours, minutes, and seconds, respectively.

Par.35 StatusCause. This indicates why an SSP has ended the monitoring of a line or facility group, and is sending a status_reported message:

0000 0000	The status of the monitored line or facility group matches the specified state.
0000 0001	The status does not match the specified state, and the monitoring time has elapsed

Par.36 TriggerTypeFlag. This parameter identifies a trigger type, and specifies whether the trigger is to be activated. The parameter has a two-octet contents field. Octet 1 specifies the trigger status:

0000 0000	Do not activate trigger
0000 0001	Activate trigger

Octet 2 identifies the trigger type, and is coded as in Par.6.

16.5 CODING OF DATA ELEMENTS

This section outlines the coding of AIN 0.1 data elements (DE) in the components part of TCAP packages [17,20].

16.5.1 Tag Codes

All data elements other than parameter DEs are "national TCAP" DEs. In their one-octet tags (specifiers), bits H,G are set to 1,1 (Section 15.3.3).

The parameter DE tags are coded "context specific" (H,G = 1,0), in the context of the parameter set DE (which itself is a national TCAP DE).

The tag fields of parameter DEs consist of one or two octets. The integers in the tag code fields (Fig. 15.3-3) are listed in Table 16.5-1.

16.5.2 Contents of the Operation Code DE

Operation code DEs have two-octet contents fields. Octet 1 identifies the operation family, and octet 2 represents an operation within that family—see Table 16.5-2.

16.5.3 Contents of Parameter DEs

The contents fields of parameter DEs have been outlined in Section 16.4.2. Additional details can be found in Refs. [17, 20].

Table 16.5-1 AIN 0.1 parameter tags.

Reference	Parameter	Tag (integer)
Par.1	UserID	53
Par.2	DN (DirectoryNumber)	- [1]
Par.3	TrunkGroupID	27 [5]
Par.4	PrivateFacilityGID	28 [6]
Par.5	BearerCapability	13
Par.6	TriggerType	52
Par.7	CallingPartyID	18
Par.8	ChargeNumber	19
Par.9	LATA	35
Par.10	PrimaryCarrier	41
Par.11	AlternateCarrier	4
Par.12	SecondAlternateCarrier	47
Par.13	CalledPartyID	15
Par.14	FeatureCode	54
Par.15	AccessCode	1
Par.16	CollectedAddressInfo	22
Par.17	CollectedDigits	23
Par.18	CallingPartyBGID	17
Par.19	PrimaryBillingIndicator	40
Par.20	AlternateBillingIndicator	3
Par.21	SecondAlternateBillingIndicator	46
Par.22	OverflowBillingIndicator	38
Par.23	AMAAlternateBillingNumber	6
Par.24	PrimaryTrunkGroup	42
Par.25	AlternateTrunkGroup	5
Par.26	SecondAlternateTrunkGroup	48
Par.27	OutpulseNumber	37
Par.28	ResourceType	45
Par.29	StrParameterBlock	50
Par.30	DisconnectFlag	25
Par.31	ClearCause	21
Par.32	FacilityStatus	61
Par.33	FailureCause	34
Par.34	MonitorTime	65
Par.35	StatusCause	66
Par.36	TriggerTypeFlag	68

Note: [1], [5], and [6] are the tag values of Par.2, Par.3, and Par.4 when the parameter is held in Par.1 (UserID).

Source: TR-NWT-001285. Reprinted with permission of Bellcore. Copyright © 1992.

16.6 MESSAGES AND PARAMETERS

This section revisits the messages described in Section 16.3, this time including their most important mandatory (M) and optional (O) parameters. It is

Table 16.5-2 Contents of operation code DEs in AIN 0.1 messages.

Message	Octet	← Bits HGFE DCBA
Request Instructions Family	1	0110 0100
Origination_Attempt	2	0001 1000
Info_Collected	2	0000 0010
Info_Analyzed	2	0000 0011
Termination_Attempt	2	0000 0101
Connection Control Family	1	0110 0101
Analyze_Route	2	0000 0001
Authorize_Termination	2	0000 0010
Disconnect	2	0000 0011
Send_To_Resource_Final	2	0000 0001
Connectivity Control Family	1	0110 1010
Forward_Call	2	0000 0001
Caller Interaction Family	1	0110 0110
Send_To_Resource_Interaction	2	0000 0001
Resource_Clear	2	0000 0010
Status Notification Family	1	0110 0111
Monitor_For_Change	2	0000 0001
Monitor_For_Success	2	0000 0011
Status_Reported	2	0000 0101
Information Revision Family	1	0110 1000
Update_Request	2	0000 0001
Update_Data	—	—

Notes:

(1) Send_To_Resource_Final and Send_To_Resource_Interaction have the same operation code. The messages are differentiated by ResourceType (Par.28).

(2) Update_Data is a Return-Result component.

Source: TR-NWT-001285. Reprinted with permission of Bellcore. Copyright © 1992.

helpful to look up the parameter descriptions of Section 16.4 when reading the material that follows.

With the exception of Update_Data, the messages invoke operations specified by an *operation code* DE, and the message parameters are the "operands" for the operation.

16.6.1 Messages in the Request Instructions Family

The messages in this family are sent by SSP, and invoke SCP operations. The messages include parameters with attributes of the calling line or trunk, which

Table 16.6-1 Parameters in messages of the request instructions family.

Parameter		Message			
		(1)	(2)	(3)	(4)
Par.1	UserID	M	M	M	M
Par.5	BearerCapability	M	M	M	M
Par.6	TriggerType	O	O	O	O
Par.7	CallingPartyID	O	O	O	O
Par.8	ChargeNumber	O	O	O	O
Par.9	LATA	O	O	O	O
Par.10	PrimaryCarrier	O	O	O	
Par.13	CalledPartyID			O	O
Par.14	FeatureCode		O	O	
Par.15	AccessCode		O	O	
Par.16	CollectedAddressInfo		O	O	
Par.17	CollectedDigits		O	O	
Par.18	CallingPartyBGID			O	

Messages:
(1) Origination_Attempt.
(2) Info_Collected.
(3) Info_Analyzed.
(4) Termination_Attempt.
Source: TR-NWT-001285. Reprinted with permission of Bellcore. Copyright © 1992.

are stored in semipermanent memory of SSP, and parameters with information "dialed" by the caller. The parameters are listed in Table 16.6-1.

For calls originated by analog or ISDN lines attached to SSP, UserID (Par.1), ChargeNumber (Par.8), LATA (Par.9), and CallingPartyBGID (Par.18) are stored attributes.

Some parameters are stored attributes of the calling analog subscriber line, or the incoming non-ISUP trunk, but are parameters in SETUP messages received from ISDN users (Section 10.3.2), or in IAM (initial address) messages for trunks with ISUP signaling (Section 11.2.4). For example, BearerCapability (Par.5) for analog lines always indicates "3.1 kHz audio," but the BearerCapability received in SETUP and IAM messages can indicate various capabilities. The same holds for CallingPartyID (Par.7).

PrimaryCarrier (Par.10) is an attribute of analog and ISDN lines that can be overwritten by the caller, by dialing a 10*XXX* code.

CalledPartyID (Par.13), FeatureCode (Par.14), AccessCode (Par.15), CollectedAddressInfo (Par.16), and CollectedDigits (Par.17) are parameters that always hold information received from the calling party. Table 16.6-2 shows the combinations of these parameters in messages caused by the various trigger types. For example, the info_collected message, sent when an OHD trigger has been encountered, may include CollectedAddressInfo only

Table 16.6-2 Combinations of parameters holding dialed digits.

Parameter		Trigger Type (1)	(2)	(3)	(4)
Par.13	CalledPartyID				H
Par.14	FeatureCode	BCD	BCD		
Par.15	AccessCode			EFG	
Par.16	CollectedAddressInfo	A D	D	G	
Par.17	CollectedDigits	C	C	F	

Trigger types:
(1) Off-hook delay (OHD).
(2) Feature code dialing (FCD).
(3) Customized dialing plan (CDP).
(4) 3/6/10 Digits and termination attempt (TAT).

(A), FeatureCode only (B), FeatureCode and CollectedDigits (C), or FeatureCode and Collected AddressInfo (D).

16.6.2 Messages in the Connection and Connectivity Control Families

These messages are sent by SCP, and contain call-handling instructions for SSP. The message parameters are listed in Table 16.6-3.

Analyze_Route and Forward_Call. If SCP determines that a call set-up can proceed, it ends the transaction with one of these messages, depending on the message received from SSP (see Fig. 16.3-1). The messages contain routing and charging information for the call.

If SCP does not provide other routing information, SSP resumes the call processing at the analyzing_info state (Fig. 16.2-1), and analyzes the called number. If the number represents a line served by SSP, the call moves to the termination_attempt state of T-BCM. Otherwise, SCP determines the route set for the call destination. On long-distance calls, it takes into account the IC carrier selected by the calling party.

SCP can include AlternateCarrier (Par.11), and possibly SecondAlternate Carrier (Par.12). In this case, SSP also resumes processing at the analyzing-information state. If the call cannot be routed to the caller's selected IC carrier, SSP routes the call to the alternate, or second alternate, IC carrier.

If the analyze_route message includes PrimaryTrunkGroup (Par.24), SSP resumes its call processing at the selecting-route state, bypassing the determination of the (public) route set for the call. This is because Par.24 specifies a private-facility trunk group, owned or leased by a business customer (Section 1.1-2), as the first-choice trunk group for the call. If Par.24 is present, the message may also include Par.25 and Par.26 (second-choice and third-

Table 16.6-3 Parameters in messages of the connection control and connectivity control families.

Parameter		Message				
		(1)	(2)	(3)	(4)	(5)
Par.7	CallingPartyID	O	O	O		
Par.8	ChargeNumber	O	O			
Par.10	PrimaryCarrier	O	O			
Par.11	AlternateCarrier	O	O			
Par.12	SecondAlternateCarrier	O	O			
Par.13	CalledPartyID	O	O			
Par.19	PrimaryBillingIndicator	O	O	O	O	O
Par.20	AlternateBillingIndicator	O	O			
Par.21	SecondAlternateBillingIndicator	O	O			
Par.22	OverflowBilling Indicator	O	O			
Par.23	AMAAlternateBillingNumber	O	O	O	O	O
Par.24	PrimaryTrunkGroup	O	O			
Par.25	AlternateTrunkGroup	O	O			
Par.26	SecondAlternateTrunkGroup	O	O			
Par.27	OutpulseNumber	O	O			
Par.28	ResourceType					O
Par.29	StrParameterBlock					O
Par.30	DisconnectFlag					O

Messages:
(1) Analyze_Route.
(2) Forward_Call.
(3) Disconnect.
(4) Authorize_Termination.
(5) Send_To_Resource, Final.
Source: TR-NWT-001285. Reprinted with permission of Bellcore. Copyright © 1992.

choice private-facility trunk group), and Par.10, Par.11, and Par.12 (first, second, and third choice IC carrier). SSP tests these trunk groups for available trunks, in the order listed above.

We now turn to the charging information in the analyze_route and forward_call messages. If the call is to be charged to a party other than the calling party, the charge number of this party is identified by AMAAlternate BillingNumber (Par.23). For example, in 800-calls, Par.23 holds the charge number of the business customer who "owns" the 800-number.

Par.19, Par.20, Par.21, and Par.22 (primary, alternate, second alternate, and overflow billing indicator) contain billing information to be included in the billing record for the call, if the call is routed over respectively the first, second, or third choice private-facility trunk group, or a public (IC carrier) trunk group.

Authorize_Termination Message. This message is sent when SCP has determined that the call destination is a line attached to SSP, and that the set-up can proceed. Routing instructions are not needed, but the message can include charging information (Par.19 and/or Par.23).

Disconnect and Send_To_Resource (Final) Messages. These messages are sent when SCP has determined that the call should not be set-up.

Both messages can include charging information (Par.19, Par.23). In send_to_resource messages, ResourceType (Par.28) is set to "play announcement," and the announcement is specified by StrParameterBlock (Par.29). DisconnectFlag (Par.30) instructs SSP to disconnect the caller, after the announcement has been played.

16.6.3 Messages in the Caller Interaction Family

This family includes two messages. The message parameters are listed in Table 16.6-4.

Send_To_Resource (Interaction) Message. This message is sent by SCP, when it needs to interact with the calling party.

ResourceType (Par.28) is set to "interaction STR operation." StrParameter Block (Par.29) contains one or more announcement elements, which specify a spoken phrase to be played, and the number of DTMF digits to be collected from the caller.

Resource_Clear Message. If the digit collection is successful, ClearCause (Par.31) indicates "normal clearing." If SCP requested address information, the collected digits are in CollectedAddressInfo (Par.16); otherwise, they are in CollectedDigits (Par.17).

Table 16.6-4 Paramaters in messages of the caller interaction family.

Parameter		Message	
		(1)	(2)
Par.16	CollectedAddressInfo		O
Par.17	CollectedDigits		O
Par.28	ResourceType	M	O
Par.29	StrParameterBlock	M	
Par.31	ClearCause		M
Par.33	FailureCause		O

Messages:
(1) Send_To_Resource_Interaction.
(2) Resource_Clear.
Source: TR-NWT-001285. Reprinted with permission of Bellcore. Copyright © 1992.

If the digit collection fails, ClearCause indicates "timeout" (the expected number of digits has not been received within say 5 seconds), or "caller disconnect."

If no intelligent-network service circuit (INSC) is available, SSP sends a resource_clear message, in which FailureCause (Par.33) indicates "unavailable resources."

16.6.4 Multiple Trigger Encounters

More than one trigger can be encountered during the processing of a call. For example, suppose that a call attempt is made from a line with an OHD trigger, and to a 900-number. The call first encounters the OHD trigger (in the collecting_info state) and—assuming that 900-calls from the line are allowed—next encounters a three-digit trigger, in the analyzing_info state (Fig. 16.2-1).

Now suppose that a call has encountered a trigger in the analyzing_info state. In most cases, SCP instructs SSP to resume call processing at same state. In certain situations, this can cause another encounter with the same trigger. SSP programs include defenses against entering a loop that keeps on encountering a trigger. For example, if SSP has encountered a predetermined maximum number of triggers in a call, subsequent triggers are ignored.

16.6.5 Messages in the Status Notification Family

Status monitoring is a transaction in which an SCP requests an SSP to monitor the status (busy, idle) of a "facility" (line, trunk)—see Fig. 16.3-3. The parameters in the messages for this transaction are listed in Table 16.6-5.

Monitor_For_Change Message. This message, sent by SCP, initiates the transaction. CalledPartyID (Par.13) specifies the line to be monitored. The state to be monitored is indicated by FacilityStatus (Par.32). MonitorTime (Par.34) sets a limit on the monitoring time.

Status_Reported Message. This message ends a monitoring transaction. If the line is initially in the specified state, SSP immediately returns this message, and StatusCause (Par.35) indicates that the line status matches the specified state.

The SSP also immediately returns a status_reported message if it is unable to perform the monitoring operation. The message then includes FailureCause (Par.33). If SSP has determined that the CallingPartyID is invalid, FailureCause indicates "resource unavailable." If SSP is unable to monitor because of processor overload, FailureCause indicates "rate too high."

Table 16.6-5 Parameters in operations of the status notification and information revision families.

Parameter		Message				
		(1)	(2)	(3)	(4)	(5)
Par.1	UserID				M	
Par.13	CalledPartyID	O				
Par.32	FacilityStatus	M	M	O		
Par.33	FailureCause			O		O
Par.34	MonitorTime	M				
Par.35	StatusCause			O		
Par.36	TriggerTypeFlag				O	

Messages:
(1) Monitor_For_Change.
(2) Monitor_Success.
(3) Status_Reported.
(4) Update_Request.
(5) Update_Data.
Source: TR-NWT-001285. Reprinted with permission of Bellcore. Copyright © 1992.

Monitor_For_Success Message. This message is an immediate reply to SCP, in the case that the line is initially not in the specified state.

The transaction continues, and SSP keeps monitoring the line, until either the line has changed to the specified state, or the specified monitoring time has elapsed, whichever happens first. SSP then sends a (delayed) status_report message in which StatusCause indicates either a "status match" or a "time-out."

Snapshot Monitoring. If SCP needs a snapshot of the line status, it sets MonitorTime to "zero." SSP then immediately sends a status_report, in which StatusCause indicates either a "status match" or a "timeout."

16.6.6 Messages in the Information Revision Family

These messages enable an SCP to activate or deactivate a trigger in an SSP. The message parameters are listed in Table 16.6-5.

Update_Request Message. This message is sent by SCP. TriggerTypeFlag (Par.36) specifies the trigger type, and the state of the trigger. In an SSP, a trigger is uniquely identified by the combination of its type and its UserID (Par.1).

Update_Data Message. This message is the SSP response to an update_request. If SSP has honored the request, the message does not contain a parameter. If SSP has been unable to set the trigger to the specified state, the message includes FailureCause (Par.33).

16.7 AIN SERVICES

Compared with conventional "switch-based" services, the SCP-based AIN services can be designed with much more flexibility.

In the first place, the service creation environment (SCE) systems usually allow the telecom to define services that take into account the calling line—which is identified by UserID (Par.1), CallingPartyID (Par.7), or ChargeNumber (Par.8) in all messages of the request instructions family—and the time-of-day and day-of-week that the call is made.

Caller-interaction also increases the flexibility of services. In the first place, services can be designed such that the calling party can select a particular option from a service menu. Moreover, a service can be made available to "authorized" callers only, by requiring the caller to enter a telecom-provided authorization code, which is checked by SCP.

This section outlines a number of services that can be implemented with the triggers, messages, and parameters that have been discussed in this chapter [20]. The examples have been chosen as illustrations for the uses of the various trigger types and messages, and regardless of their availability in U.S. networks. They are grouped into a number of categories (call management, number translation, etc.).

16.7.1 AIN Outgoing Call Management

Outgoing call management services enable residential and business customers to restrict the calls made on their lines. Some exchange-based features of this nature are already in existence. As an example, a customer can specify that calls from his line to 900/976 numbers are to be blocked. The service is inflexible in that all 900/976 calls are either blocked or allowed.

Call Screening. Call screening is a service that enables individual customers to specify a set of numbers that should be blocked when dialed from his line. The set may contain groups of numbers, say all 976 numbers, all 011 (international) numbers, etc., and individual numbers.

To make the service available, the telecom assigns an off-hook delay (OHD) trigger to the line of the customer, and enters a list with dialing restrictions for this line into SCP semipermanent memory.

When the line originates a call, the OHD trigger is encountered after the dialed digits have been collected (except when the customer has dialed an escape code (16.2.6).

The info_collected message sent by SCP identifies the calling party, and includes TriggerType (Par.6) set to "OHD," and CollectedAddressInfo (Par.16), containing the called number.

Based on this information, SCP either allows or blocks the call. If the call is allowed, SCP returns an analyze_route message. If not, the SCP response is a disconnect, or a send-to-resource-final message.

Scheduled Call Screening. In this service, screening is in effect on a predetermined schedule, specified by the customer. For example, calls are screened before 8 am and after 7 pm on weekdays, and all day on weekends.

Call Screening with Override. In this service, the call screening of a line can be overridden for calls made by an "authorized" person, who identifies himself (herself) by dialing a feature code defined for the service, say *921, followed by an authorization code (AC) that has been provided by the telecom.

When SSP receives a call without feature code, the OHD trigger is encountered, normal call screening is in effect. If an authorized caller dials *921, followed by AC, the OHD trigger is also encountered. SSP then sends an info_collected message, in which the calling line is identified, and TriggerType (Par.6) is set to "OHD." The message also includes FeatureCode (Par.14), set to *921, and CollectedDigits (Par.17), which holds the AC.

SCP checks whether the received AC is associated with the calling line.

If so, SCP returns a send_to_resource_interaction message, in which StrParameterBlock (Par.29) indicates that dial-tone is to be sent to the caller, and that address digits are to be collected. When hearing the dial-tone, the caller dials the digits of called number. When the digit collection is complete, SSP sends a resource_clear message, in which CollectedAddressInfo (Par.16) contains the called number, and SCP ends the transaction with an analyze_route message, in which CalledPartyID (Par.13) holds the called number received from SSP.

If SCP determines that an incorrect AC has been dialed, it returns a disconnect, or a send_to_resource_final, message.

Call Screening with Activation/Deactivation. In this service, call screening can be activated and deactivated on command from an authorized person. The service requires the assignment of an OHD and a FCD trigger to the line.

We assume that the telecom has defined feature codes *922 and *923 to activate and deactivate the screening.

The activation and deactivation of the screening is accomplished by activating and deactivating the OHD trigger.

Suppose that the OHD trigger—and therefore the service—has been activated. To deactivate the service, the authorized person dials *922, followed by AC. The OHD trigger is encountered during the collecting_info state, and results in an information_collected message, in which the calling line is identified. Also included are FeatureCode (Par.14), set to *922, and CollectedDigits (Par.17) which holds the AC.

If SCP determines that AC is associated with the calling line, it returns an update_request message, in which the trigger to be deactivated is specified by UserID (Par.1) and TriggerTypeFlag (Par.36). SSP deactivates the OHD trigger, and terminates the transaction with an update_data message, which indicates that the trigger has been deactivated.

If SCP determines that the AC is not valid, it ends the transaction with a disconnect, or a send-to-resource_final message.

Now suppose that the service is currently deactivated. To activate the service, the authorized customer dials *922, followed by AC. Since the OHD of the line is deactivated, it cannot be encountered. However, the FCD trigger is now encountered, in the analyzing_info state, and SSP sends an info_analyzed message, with the same parameters as the info_collected message described above, except that TriggerTypeFlag now indicates that the OHD trigger has to be activated. The rest of the procedure again involves an update_request message from SCP, and an update_data message from SSP.

Hotline Telephones. A *hotline* is a telephone without a dial or keypad. Examples of hotlines are the emergency telephones along highways. When a caller lifts the handset, the exchange sets up a connection to a predetermined DN, of a police station, or a nearby automobile service station. This feature is already available as an exchange-based service.

AIN hotline services can be more versatile. They require the assignment of an OHI trigger to the lines. SCP stores a list with directory numbers of service stations along the highway.

When receiving an origination_attempt message from one of the phones, SCP bases the selection of the directory number on the identity of the calling line. In this way, the call can be routed to the nearest service station. In addition, the directory number selection can be based on time-of-day and day-of-week.

16.7.2 AIN Called Number Translation

AIN number translation requires the assignment of an (office-based) 3/6/10 digit trigger on certain directory numbers. The triggers have to be assigned in all SSPs. When an SSP encounters the trigger, it sends an info_analyzed message. The TriggerType (Par.6) is set to "3, 6, or 10 digits." The calling party is identified, and the called number is in CalledPartyID (Par.13).

Translation of 800, 900, and 976 Numbers. Translation of 800-numbers into routing numbers (unpublished 10-digit NANP numbers) has been available since 1980, as an *intelligent network 1* (IN/1) service [1,2]. Today, most 800-number translations are still made by IN/1 databases. However, telecoms can move these translations to SCPs.

It is also possible to assign the translations of some 800-numbers to SCPs, and leave the translations of other 800-numbers to the older databases. As an example, the split can be based on the *NXX* code in the 800-*NXX-XXXX* number. This is done by assigning six-digit triggers to the 800-*NXX* codes to be translated by SCP.

Translations of 900 and 976-numbers is usually done by SCPs, and requires the assignment of triggers to these numbers at the SSPs.

When SSP encounters one of these triggers, it sends an info_analyzed message in which TriggerType (Par.6) is set to "three, six, or ten digits," the called number is in CalledPartyID (Par.13), and the calling line is identified.

Translations of 800, 900, etc. numbers can be tailored to the requirements of the individual business customers.

For example, a multilocation business that has an 800, or 900, etc. number may require that incoming calls are routed to its nearest business office. Usually, the business customer specifies individual routing numbers for calls originated by lines in particular numbering plan areas (identified by NPA), or in a particular exchange areas (identified by NPA-*NXX*).

In addition, the determination of routing numbers can be made to depend on the time-of-day and day-of-week of the call.

In the analyze_route message from SCP, the routing number is in CalledPartyID (Par.13). Since the calls involve special billing and charging, the message also includes AMAAlternateBillingNumber (Par.23), and AlternateBillingIndicator (Par.20).

Personal Numbers. A personal number is a number by which a customer can be reached, regardless of his (her) present location. We assume that the numbers have the format 500-*NXX-XXXX*. For this service, three-digit triggers are assigned to 500-numbers at all SSPs. For each 500-customer, SCP stores a "current address" by which the customer can be reached at this time.

On a call to a 500-number, SSP encounters the trigger, and sends an info_analyzed message in which TriggerType (Par.6) indicates "3-digit," and CalledPartyID (Par.13) holds the 500-number. The SCP translates the number into the current address, and its analyze_route message includes the address in CalledPartyID (Par.13).

Customers with 500-numbers can update their current addresses when moving to a new location; for instance, after entering their respective homes, cars, offices, or after arriving at their hotels.

One way to update a current address is described below. The telecom establishes an "updating" number, say 500-234-5678. To update the current address, the customer calls the updating number. SSP encounters the three-digit trigger, and sends an info_analyzed message in which TriggerType (Par.6) indicates "3-digit," and CalledPartyID (Par.13) holds the updating number. SCP recognizes the updating number, and executes a series of caller interactions. In the first and second interactions, SCP requests the caller's 500-number and authorization code (AC). It then checks whether the AC is associated with the caller's 500-number.

If this is the case, the third interaction prompts the caller to enter the new current address. The fourth interaction plays back the received address, and requests the caller to dial "1" if the address is correct, and "2" if it is not. If SCP receives a "1", it updates the customer's current address, and ends the transaction send_to_resource_final message that indicates that the new address is now in effect. If SCP receives a "2", it asks again for the new address.

16.7.3 AIN Incoming Call Management

This service enables residential or business customers to specify the callers who are allowed to call their lines. This is important for lines that are connected to the customer's dial-up computers. The customer can specify a list of calling lines that are allowed to call the line(s), and possibly a list of authorization codes. For this service, a termination attempt trigger (TAT) is assigned to the directory number (DN) of the customer's line, in the SSP to which the line is attached, and SCP stores the lists with calling lines and authorization codes.

If SSP encounters a TAT trigger, it sends a termination_attempt message, with TriggerType (Par.6) set to "TAT," CalledPartyID (Par.13) holding the called DN, and CallingPartyID (Par.7) identifying the calling line.

If the calling line is on the list of lines allowed to call the DN, SCP allows the termination. If the list shows that an authorization code (AC) is required, SCP initiates a caller interaction to obtain AC, and allows the call if AC is associated with the calling line.

16.7.4 AIN Terminating Services

The procedure for providing terminating services to a line is very similar to incoming call management. A TAT trigger is assigned to the DN in the SSP that covers the DN. If SSP encounters the trigger, it sends a termination_attempt message with Par.6 set to "TAT," and the called DN in CalledPartyID (Par.13). The terminating service to be provided is determined by DN. A few possible services are outlined below.

Foreign Exchange Line. A business customer B, residing in one city, desires a DN of a line in another city [20]. For example, B may live in a suburb of Chicago, and wants to advertise a Chicago DN, so that his Chicago clients can reach him by making a local call. In the past, this required the installation of a physical subscriber line from B's premises to the local Chicago exchange that covered the advertised DN.

In AIN, the service can be implemented without a physical line. B's line is attached to his local exchange, where its directory number is DN_1. The number advertised by B is DN_2, which is covered by a Chicago SSP, where a TAT trigger is assigned to that number.

When a client dials DN_2, the SSP encounters the trigger, and sends a termination_attempt message, with DN_2 in CalledPartyID (Par.13). SSP returns a forward_call message, with DN_1 in CalledPartyID, and SSP forwards the call to B's local exchange.

Call Distribution. Suppose that business customer B employs n "work at home" agents [20]. The directory numbers of the agents, whose lines may be attached to different SSPs, are DN_1, DN_2, ..., DN_n.

The advertised directory number of B's business is DN_A, a number covered by SSP_A, but which does not represent an actual line. In SSP_A, a TAT trigger is assigned to DN_A, and SCP has a list of DNs associated with DN_A.

A call to DN_A encounters the TAT trigger, and causes SSP_A to send a termination_attempt message, in which CalledPartyID holds DN_A.

The SCP then checks idle/busy status of the agents on the list, by executing status notification transactions (16.6.4) with the SSPs to which the agents are attached, and allocates the call to an idle agent. Suppose that the previous call has been allocated to DN_5. SCP then starts checking the status of DN_6, and continues until it has found an idle DN, or has come full circle without finding an idle DN. In the first case, it sends an analyze_route, or a forward_call, message to SSP_A, depending on whether the line identified by DN is attached to SSP_A. If all DNs are busy, SCP sends a disconnect, or send_to_resource_final message.

16.8 ACRONYMS

AC	Authorization code
AIN	Advanced intelligent network
AMA	Automatic message accounting
ANR	Analyze route message
ASE	Application service element
BCD	Binary coded decimal
BCM	Basic call model
CCIS	Common-channel interoffice signaling
CDP	Customized dialing plan trigger
DISC	Disconnect message
DN	Directory number
DTMF	Dual tone multi-frequency
FCD	Feature-code dialing trigger
FWC	Forward call message
GT	Global title
GTA	Global title address (part of GT)
IAM	Initial address message in ISUP signaling
IC	Interexchange carrier
ID	Identifier
IN	Intelligent network
IN-ASE	Intelligent-network application service element
INSC	Intelligent-network service circuit
ISDN	Integrated services digital network
ISUP	ISDN User Part of SS7
ITU	International Telecommunication Union
ITU-T	Telecommunication standardization sector of ITU (successor to CCITT)
LATA	Local access and transport area

NN	Nature of number
NP	Numbering plan
NANP	North American numbering plan
NPA	Numbering plan area code (three digits)
PC-SSN	Point code and subsystem number
O-BCM	Originating basic call model
OHD	Off-hook delay trigger
OHI	Off-hook immediate trigger
PBX	Private branch exchange
PFG	Private facility group
RES-CLR	Resource clear message
SCE	Service creation environment
SCCP	Signaling connection control part of SS7
SCP	Service control point
SMS	Service management system
SSN	Subsystem number
SSP	Service switching point
SS7	Signaling system No.7
STR-F	Send to resource, final message
STR-I	Send to resource, interaction message
TAT	Termination attempt trigger
T-BCM	Terminating basic call model
TCAP	Transaction capability application part
TT	Translation type
3/6/10D	Three, six, or ten digit trigger

16.9 REFERENCES

1. R.K. Berman, J.H. Brewster, "Perspective on the AIN Architecture," *IEEE Comm. Mag.*, **31**, no.2, February 1992.
2. S.M. Boyles, R.L. Corn, L.R. Moseley, "Common Channel Signaling; The Nexus of an Advanced Communications Network," *IEEE Comm. Mag.*, **28**, no.7, July 1990.
3. J.J. Garrahan, P.A. Russo, K. Kitami, R. Kung, "Intelligent Network Overview", *IEEE Comm. Mag.*, **31**, no.3, March 1993.
4. P.A. Russo, K. Bechard, E. Brooks, R.L. Corn, R. Gove, W.L. Honig, J. Young, "IN Rollout in the United States," *IEEE Comm. Mag.*, **31**, no.3, March 1993.
5. J.M. Duran, J. Visser, "International Standards for Intelligent Networks," *IEEE Comm. Mag.*, **31**, no.2, February 1992.
6. *General Aspects of the Intelligent Network Application Protocol*, Rec. Q.1208, ITU, Geneva, 1993.
7. *Introduction to Intelligent Network Capability Set 1*, Rec. Q.1211, ITU, Geneva, 1993.
8. *Interface Recommendation for Intelligent Network CS-1,* Rec. Q.1218, ITU, Geneva, 1993.

9. *Advanced Intelligent Network (AIN) Release 1 Switching Systems Generic Requirements*, TA-NWT-001123, Bellcore, Piscataway, NJ, 1991.
10. *Advanced Intelligent Network (AIN) Release 1 Switch—Service Control Point (SCP)/ Adjunct Application Protocol Interface Generic Requirements*, TA-NWT-001126, Bellcore, Piscataway, NJ, 1991.
11. *Ameritech Service Switching Point Functional Specification*, AM-TR-OAT-000042, Ameritech, Hoffman Estates, IL, 1989.
12. *Ameritech Service Control Point Functional Specification*, AM-TR-OAT-000043, Ameritech, Hoffman Estates, IL, 1989.
13. *Ameritech Service Control Point—Service Switching Point Interface Specificaton*, AM-TR-OAT-000044, Ameritech, Hoffman Estates, IL, 1989.
14. *BellSouth AIN Release 0.1 Service Switching Point Functional Specification*, TR-73562, BellSouth Telecomm. Inc, Birmingham, AL, 1991.
15. *BellSouth AIN Release 0.1 Service Switching Point—Service Control Point Application Protocol Interface Specification*, TR-73563, BellSouth Telecomm. Inc, Birmingham, AL, 1991.
16. *Advanced Intelligent Network (AIN) 0.1 Switching System Generic Requirements*, TR-NWT-001284, Bellcore, Piscataway, NJ, 1992.
17. *Advanced Intelligent Network (AIN) 0.1 Switch—Service Control Point (SCP) Interface Generic Requirements*, TR-NWT-001285, Bellcore, Piscataway, NJ, 1992.
18. *Advanced Intelligent Network (AIN) 0.2 Switching System Generic Requirements*, TR-NWT-001298, Bellcore, Piscataway, NJ, 1993.
19. *Advanced Intelligent Network (AIN) 0.2 Switch—Service Control Point (SCP)/ Adjunct Interface Generic Requirements*, TR-NWT-001299, Bellcore, Piscataway, NJ, 1993.
20. *Advanced Services Platform, Release 0.1B*. Document AT&T 235-190-126. AT&A Bell Laboratories Technical Publications Center, Naperville IL, 1995.
21. *International Telephone Operation*, Rec. E.164, CCITT Blue Book, **II.2**, ITU, Geneva, 1989.

17

MOBILE APPLICATION PART

Chapter 12 has described the signaling between a mobile station (MS) and a mobile switching center (MSC) in a cellular (public land) mobile network (CMN, PLMN). The present chapter describes the mobile application part (MAP) of signaling system No.7 (SS7). MAP defines a number of remote operations (transactions) that support mobile telecommunications.

Public switched telecommunication networks (PSTN) are *fixed* networks. A subscriber line is fixed (attached) to a local exchange, which has a record for each of its subscribers. The record contains semipermanent and temporary data. Semipermanent data in the record include the subscriber's directory number and a *service profile*. This is a list of features (for example, call waiting) which the subscriber has requested, and the telecom has agreed to supply, for a monthly or per-use charge.

The temporary data include information on the current subscriber state (busy, idle, denied service), and, during the times that the subscriber is involved in a call, details about the connection. The exchange consults these data when processing the subscriber's calls.

When a MSC is serving a MS, it needs similar data. However, a MS is "mobile." At different points in time, it can be in the service area of different MSCs. It is not practical to store the information on all mobiles at every MSC. Instead, this information is stored in centralized databases. One of the purposes of MAP transactions is to enable the MSCs to obtain information about a MS from these databases.

This chapter describes two versions of MAP. The U.S. version, denoted here as IS-MAP, is defined by the Electronic Industries Association (EIA) and the Telecommunications Industry Association (TIA). The international version, used in the Global System for Mobile Communications (GSM), and denoted here as GSM-MAP, is standardized by the International Telegraph and Telephone Consultative Committee (CCITT)/International Telecommunications Union (ITU-T) and the European Telecommunications Standards Institute (ETSI).

IS-MAP is discussed in Sections 17.1–17.5. The description of GSM-MAP starts in Section 17.6.

17.1 INTRODUCTION TO IS-MAP

This section introduces some general MAP concepts and terms.

17.1.1 Definitions [1]

First, a few definitions.

Home Network. This is the CMN selected for mobile communication services by the owner of MS.

Roaming. A MS can receive service in its home CMN and in other CMNs. When MS is being served by a CMN other than its home CMN, we say that the MS is roaming. A MS stores the system identifier (SID) of its home CMN—see Section 12.3. A MS knows that it is "at home" if the SID received in overhead parameter messages (12.4.2) matches its stored SID. If the SIDs do not match, the MS is roaming.

Visited Network. This is the CMN that is serving the MS when it is roaming.

Home MSC. This is the MSC in the home CMN of the MS to which the PSTN delivers calls to the MS.

Serving MSC. This is the MSC that is currently serving the MS. Depending on the MS location, this can be the home MSC, another MSC in the home CMN, or an MSC in another CMN.

17.1.2 Equipment Entities Involved in IS-MAP Transactions

The MAP-related equipment entities in a cellular mobile network (CMN) are shown in Fig. 17.1-1.

Mobile Switching Centers (MSC). In this example, CMN has three MSCs, each of which serves the MSs in its part of the CMN service area.

Home Location Register (HLR). This register stores semipermanent and temporary data on mobiles for which the CMN is the *home* network.

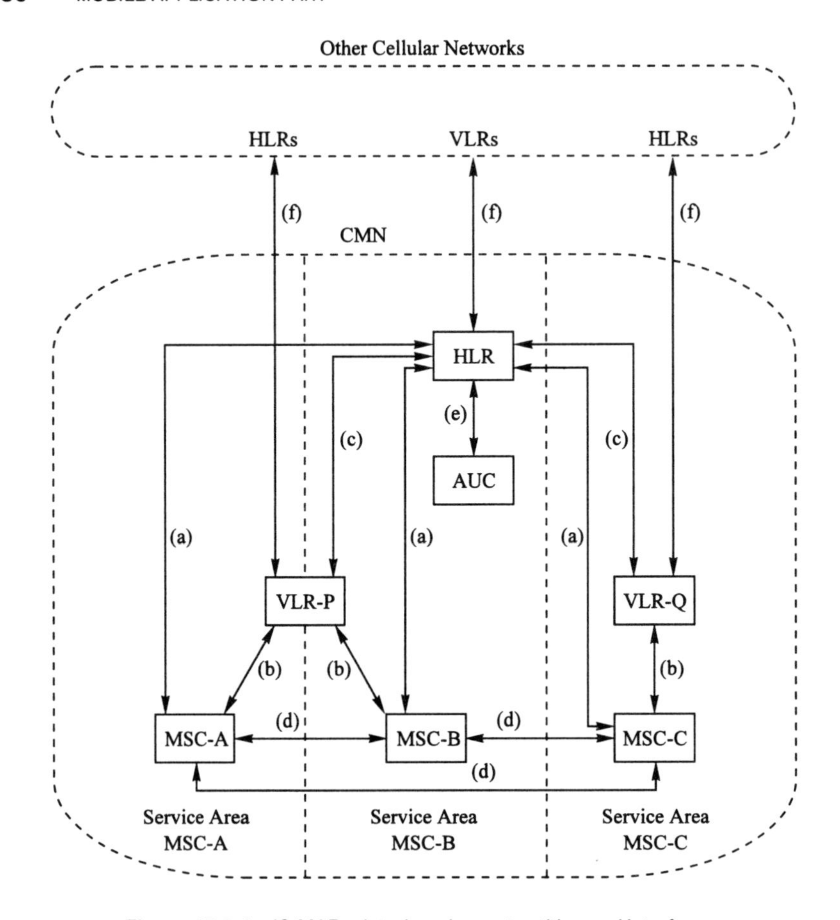

Figure 17.1-1 IS-MAP related equipment entities and interfaces.

Visitor Location Register (VLR). A VLR is associated with one or more MSCs. In the figure, VLR-P is associated with MSC-A and MSC-B, and VLR-Q is associated with MSC-C. A VLR stores information on roaming mobiles that are currently being served by an associated MSC.

Authentication Center (AUC). The AUC stores the authentication and voice-privacy information of mobiles for which CMN is the home network, and have authentication and voice privacy service (Sections 12.6-7 and 12.6-8).

In principle, all entities in the figure can be stand-alone units. They are SS7 signaling points, and are connected by signaling links to an SS7 signaling network, which transfers the transaction messages. This is the configuration assumed in the material that follows.

Actual equipment configurations can be slightly different. For instance, authentication center AUC, shown here as a stand-alone unit, can be implemented as an integral part of HLR. In this case, AUC is not a signaling point, and the transaction messages between HLR and AUC are internal messages which are passed in primitives. Also, a VLR is sometimes implemented as a part of its associated MSC (for example, VLR-Q and MSC-C).

17.1.3 Interfaces

The lines in Fig. 17.1-1 indicate the interfaces between equipment units for various transactions. Within a CMN, there are transactions involving a MSC and the HLR (a), a MSC and its associated VLR (b), a VLR and the HLR (c), two adjacent MSCs (d), and HLR and AUC (e). In addition, there are transactions involving a VLR in one CMN and the HLR of another CMN (f).

17.1.4 Identification of MS, CMN, MSC, and VLR

In the U.S., the numbering plan for mobiles is integrated with the PSTN numbering plan (Section 1.2.1). The mobile identification number (MIN) of a MS consists of a three-digit area code (AC), a three-digit exchange code (EC), and a four-digit station code. The MINs in this chapter are the counterparts of MIN1s and MIN2s (binary numbers) in messages on the RF channels between MS and base station (Section 12.4.9), but are formatted as ten BCD digits. When necessary, an MSC converts MIN to MIN1 + MIN2, or vice versa.

The PSTN regards a MSC as a local exchange which is identified by one or more AC-EC combinations. An MSC identified by say 708-371 and 708-456 is the home MSC of mobiles with MINs 708-371-*XXXX* and 708-456-*XXXX*. When a MS is called, the PSTN network routes the call to the home MSC of MS.

U.S. cellular networks are identified by a 15-bit *system identification* (SID—see Section 12.3). Within a CMN, a MSC or VLR is identified by an eight-bit *switch identification* (SWID). The combination SID-SWID is known as the *MSC identification* (MSCID), and uniquely identifies a MSC or VLR in the U.S.

17.1.5 Contents of HLR, VLR, and AUC

This section summarizes the most important parameters of the MS records in the various registers.

Home Location Register of CMN. This register has records for all mobiles for which CMN is the home network:

- Mobile identification number (MIN)
- Mobile serial number (MSN)—see Section 12.3

- MS status (qualified or not qualified for service)
- MS service profile
- If MS is currently served by a MSC in its home CMN, the MSCID (and optionally, the point code) of MSC.
- If MS is currently served by a MSC in a visited CMN, the MSCID (and optionally, the point code) of the VLR associated with the MSC.

All parameters are semipermanent except MSCID, which is temporary.

Visitor Location Register. This register has records for all roaming mobiles that are being served by an associated MSC:

- Mobile identification number (MIN)
- Mobile serial number (MSN)
- MS service profile
- MSCID, and possibly PC, of the associated MSC that currently serves MS

All parameters are temporary. They are entered when a MSC associated with VLR starts serving a roaming MS, and deleted when the service ends.

Authentication Center of CMN. This center has records for all mobiles for which CMN is the home network, and have authentication and voice privacy service:

- MIN
- SSD_A. Shared secret data for MS authentication (Section 12.6.7)
- SSD_B. Shared secret data for voice privacy (Section 12.6.8)

All parameters are semipermanent.

17.1.6 IS-MAP Operations

Like advanced intelligent network (AIN) operations (Chapter 16), most IS-MAP transactions consist of one operation, and involve a query and a response package. Unlike AIN operations, whose response packages contain invokes, IS-MAP response packages contain a return-result or a return-error.

At this time, over 30 IS-MAP operations have been defined. Sections 17.2–17.5 discuss operations for MS registration and authentication, for making calls to and from MS, and for MS handover.

17.1.7 Transfer of TCAP Packages

TCAP uses the connectionless service of the SS7 signaling connection control part (SCCP)—see Section 14.3.

To SCCP, each MSC, HLR, VLR, and AUC is a subsystem at a signaling point, and is identified by a point code (PC) and a subsystem number (SSN). In U.S. cellular networks, the SSN codes that identify the various equipment types have been standardized [1].

SSN	
5	Mobile application part (MAP)
6	Home location register (HLR)
7	Visitor location register (VLR)
8	Mobile switching center (MSC)
10	Authentication center (AUC)

SCCP Called Party Addresses. The ASE that initiates a transaction has to determine the SCCP address (CDA) of the called ASE. In transactions between ASEs in the same cellular network, the initiating ASE provides a PC-SSN (point code and subsystem number) called address.

In transactions involving called ASEs in different CMNs, the called address is a global title (GT) which has to be translated at a signal transfer point.

The examples of the IS-MAP transactions in Sections 17.2–17.4 include information on how the initiating ASE determines the CDA.

17.2 TRANSACTIONS FOR REGISTRATION AND AUTHENTICATION

This section discusses operations which are triggered when a MSC receives a registration or origination message from a MS.

The transactions involve the HLR in the home network of MS, the MSC which received the MS message, and, if MS is roaming, the VLR associated with MSC.

When a MSC of network CMN receives a message from a MS, it analyzes the AC-EC combination of the received MIN. If it recognizes AC-EC as one of the combinations for MINs of mobiles whose home is CMN, it concludes that CMN is the home network of MS; otherwise, MS is roaming.

17.2.1 Registration Notification

This operation is invoked by a MSC, and executed by the HLR in the home network of MS. It has several purposes:

- to determine whether MS is qualified to receive service. If yes,
- to transfer the MS service profile to MSC, possibly via its associated VLR, and
- to update HLR information about where the MS is receiving service.

We first consider the case that the mobile and/or the mobile network do not have authentication capability (Section 12.6.7). On receipt of a registration message from MS, the MSC initiates a "registration-notification" transaction [2,3].

In Fig. 17.2-1, the home network of the registering MS is CMN-1. MSC-A and HLR are in the home CMN of MS, and HLR is the source of information about MS. CMN-2 and CMN-3 are other cellular networks. VLR-P and VLR-Q are associated with respectively MSC-B and MSC-C.

The TCAP packages with an invoke and a return-result for this operation are denoted respectively by REGNOT and regnot.

Three cases are considered. In case (a), the mobile has registered at MSC-A, which then initiates a registration-notification transaction with HLR by sending a REGNOT. In case (b), the mobile has registered at MSC-B, which

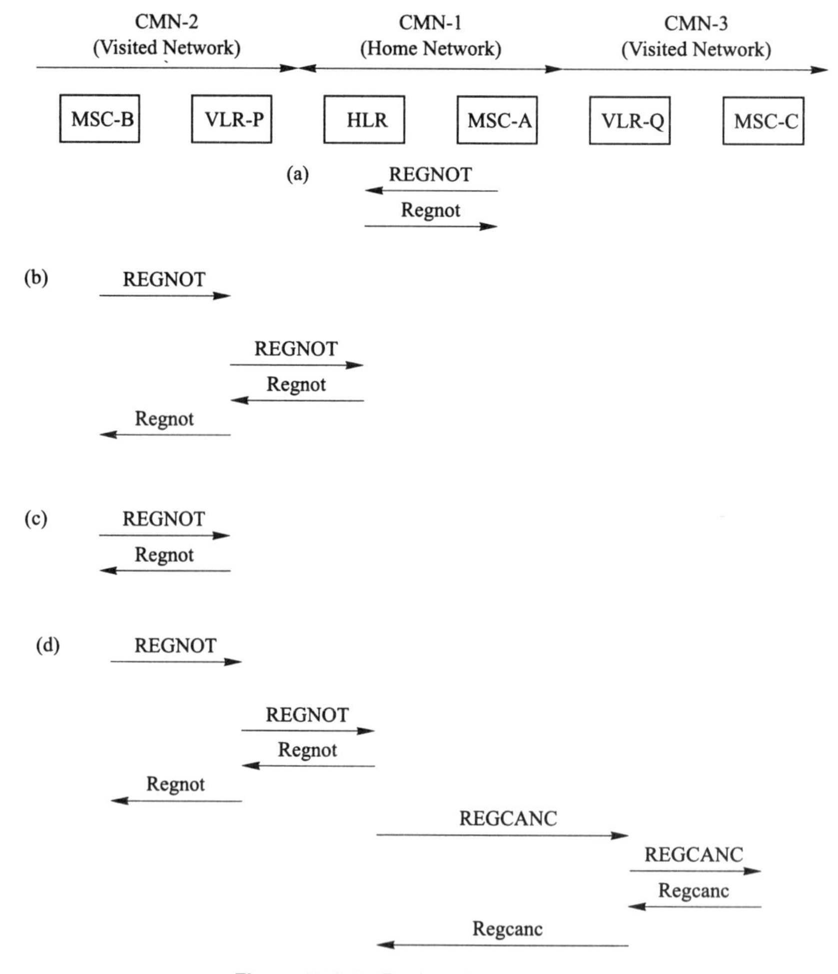

Figure 17.2-1 Registration procedures.

initiates a transaction with VLR-P. We assume that VLR-P has no information about MS, and therefore initiates a transaction with HLR. In case (c), VLR-P has information about MS, and does not need to contact HLR.

On receipt of REGNOT, HLR first checks whether it has a record for a mobile identified by MIN. If so, it checks whether the mobile is qualified to receive service and—if this is the case—authorizes the invoking MSC to serve the MS for a certain *authorization period*. The regnot return-result includes the authorization period and, if requested in REGNOT, the service profile of MS.

MSC-A or MSC-B and VLR receive the regnot and establish a temporary record with the authorization period and service profile of MS.

While processing REGNOT, HLR also updates its record, by entering the MSCID of MSC-A or VLR-P. The new serving MSC now has the information necessary to process calls for MS, and HLR has updated its information on the whereabouts of MS.

On receipt of regnot, MSC responds to MS, with a registration acknowledgment message.

If HLR determines that MS is not qualified to receive service, it indicates this in its regnot, and MSC then sends a release message to MS.

17.2.3 Expiration of Authorization Period

MSC-A, or MSC-B and VLR-P, periodically examine their records on MS, and check whether the authorization period has expired. If this is the case, they discard the record.

On expiration, HLR removes the MSCID, which identifies MSC-A or VLR-P, from its record.

17.2.4 Registration Cancellation

Now assume that MS has registered at MSC-B at a time that the authorization period for MS at the previous serving MSC-C has not yet expired—Fig. 17.2-1(d). This is known to HLR, because the MSCID of VLR-Q is in its MS record.

On receipt of the REGNOT, HLR executes the notification operation, and also initiates a registration cancellation transaction with VLR-Q, by sending a REGCANC invoke. VLR-Q initiates a cancellation transaction with MSC-B, which then discards its record for MS. On receipt of regcanc, VLR-Q returns a regcanc to HLR, and discards its record. When HLR receives the regcanc, it deletes the MSCID of VLR-Q in its MS record.

17.2.5 MS Authentication

Now consider the case that the MS and the MSC have authentication capability (Section 12.6.7), and possibly voice-privacy capability (Section 12.6.8). Figure 17.2-2 assumes that CMN-1 is the home network of MS, and that MS intends to register at MSC-B, in network CMN-2.

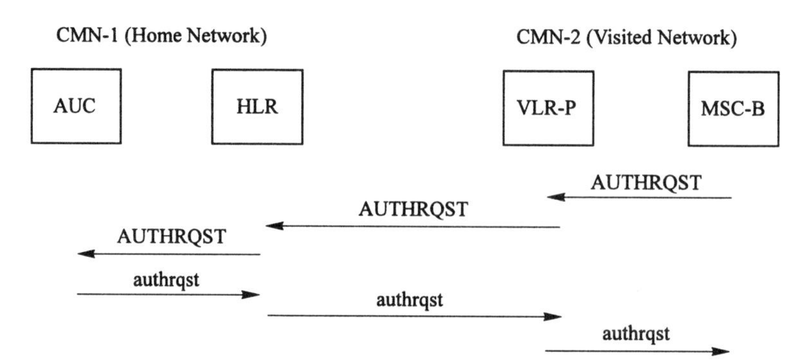

Figure 17.2-2 Authentication of a mobile. (From TSB-51. Reproduced with permission of TIA.)

Before registering, MS executes the CAVE algorithm, using a random number RAND received in an overhead parameter message, its identification and serial numbers (MIN and MSN), and the internally stored shared secret data parameter SSD_A (12.6.7). It includes the authentication result AUTHR in its registration message to MSC-B [4].

MSC-B then starts an authorization-request transaction with VLR-P, with an AUTHRQST invoke. In turn, VLR-P initiates a transaction with HLR, and HLR initiates a transaction with authentication center AUC.

The AUTHRQSTs include AUTHR and all parameters used by MS to calculate AUTHR, except SSD_A. AUC executes CAVE, using the received parameters, and the SSD_A value in its record for MS. If the result matches the received AUTHR, AUC concludes that the mobile stores the correct SSD_A, and is therefore authentic. In its authrqst response, AUC indicates whether the mobile is authentic, or should be denied access.

This information is passed to MSC-B. If MS is authentic, MSC-B returns a registration confirmation MS, and starts the registration operation—see Fig. 17.2-1(b). If the authrqst indicates that the mobile should not be served, MSC-B returns a release message to MS.

If the mobile is authentic and has voice-privacy service, the authrqst includes two voice-privacy masks (Section 12.6.8).

The same procedure, but without involving a VLR, applies when MS is served by an MSC in its home network.

17.2.6 Message Contents

This section lists the most important mandatory (M) and optional (O) parameters in the invokes and return-results discussed so far, and in the return-error message, which is returned to the invoker if the requested operation cannot be carried out [2].

In conformity with the EIA/TIA documents, the operation names and parameter tags are denoted as character strings in which the initial letter of each word is in upper case.

The references Par.1, etc. point to parameter format descriptions in Section 17.5.4. It is helpful to look up the descriptions while reading material that follows.

RegistrationNotification Invoke (REGNOT)—see Fig. 17.2-3. The MS is identified by Par.14: MobileIdentificationNumber (MIN) and Par.15: MobileSerialNumber (MSN). The QualificationInformationCode (Par.19) is used to request the validation of the MS, and/or the MS service profile from the HLR.

The invoker includes its identity MSCID (Par.16), and optionally its PC_SSN (point code + subsystem number) address (Par.18). The responder uses this when sending its return-result or return-error message.

RegistrationNotification Return-Result (regnot)—see Fig. 17.2-3. If Par.19 in the invoke has requested validation information, the return-result includes Par.4 (AuthorizationPeriod), or Par.3 (AuthorizationDenied). Authorization is denied when the MS record indicates a delinquent account, or a stolen unit, etc.

If the registration is accepted and Par.19 in the invoke requests the service profile of the MS, the return-result also includes Par.17 (OriginationIndicator), Par.26 (TerminationRestrictionCode), and Par.6 (CallingFeaturesIndicator).

Operation	**Timer**: 6 seconds	
RegistrationNotification (REGNOT)		
Invoke Parameters	**M/O**	**Reference**
MobileIdentificationNumber (MIN)	M	Par.14
MobileSerialNumber (MSN)	M	Par.15
MSCID (serving MSC or VLR)	M	Par.16
PC_SSN (serving MSC or VLR)	O	Par.18
QualificationInformationCode	M	Par.19
Return-Result Parameters	**M/O**	**Reference**
AuthorizationDenied	O	Par.3
AuthorizationPeriod	O	Par.4
CallingFeaturesIndicator	O	Par.6
OriginationIndicator	O	Par.17
TerminationRestrictionCode	O	Par.26
Return-Error Information	**M/O**	**Reference**
Error Code	M	Section 17.5.3
FaultyParameter	O	Par.12

Figure 17.2-3 Parameters in registration notification messages. (From IS-41.5-B. Reproduced with permission of TIA.)

RegistrationCancellation Invoke (REGCANC)—see Fig. 17.2-4. The message identifies the mobile by Par.14 and Par.15 (MIN and MSN).

RegistrationCancellation Return-Result (regcanc)—see Fig. 17.2-4. The message does not include parameters.

AuthenticationRequest Invoke (AUTHRQST)—see Fig. 17.2-5. The MS is identified by Par.14 and Par.15 (MIN and MSN), and the invoker includes its MSCID (Par.16) and PC_SSN address (Par.18). Also included are Par.20: RandomVariable (RAND), which is an input to CAVE, and Par.2: AuthenticationResponse (AUTHR), the result of the CAVE execution by the mobile. Authentication can take place when a MS registers, originates, or responds to a page message. On registrations or page responses, the CAVE inputs are MIN, MSN, RAND (received in the invoke), and SSD_A (stored at the AUC). On originations, the invoke also includes Par.11 (Digits), which holds the number dialed by the MS, and is used—in lieu of MIN—as an input to CAVE.

AuthenticationRequest Return-Result (authrqst)—see Fig. 17.2-5. If AUC determines that the MS is authentic, the message does not include parameters, except when the MS has voice privacy. In that case, the message includes Par.27: VoicePrivacyMask (VPMASK, see Section 12.6.8).

If AC determines that the MS is not authentic, the message includes Par.15 (DenyAccess).

Return-Error Messages. The error codes in return-error messages, which are common to all IS-MAP operations described in this chapter, indicate why a requested remote operation could not be executed. Some common reasons are unrecognized MIN (the responder has no record for the received MIN), and

Operation	**Timer**: 6 seconds	
RegistrationNotification (REGCANC)		
Invoke Parameters	**M/O**	**Reference**
MobileIdentificationNumber (MIN)	M	Par.14
MobileSerialNumber (MSN)	M	Par.15
Return-Result Parameters	**M/O**	**Reference**
None		
Return-Error Information	**M/O**	**Reference**
Error Code	M	Section 17.5.3
FaultyParameter	O	Par.12

Figure 17.2-4 Parameters in registration cancellation messages. (From IS-41.5-B. Reproduced with permission of TIA.)

Operation	**Timer**: 6 seconds	
AuthenticationRequest (AUTHRQST)		
Invoke Parameters	**M/O**	**Reference**
AuthenticationResponse (AUTHR)	M	Par.2
Digits (dialed by MS)	O	Par.11
MobileIdentificationNumber (MIN)	M	Par.14
MobileSerialNumber (MSN)	M	Par.15
MSCID (serving MSC or VLR)	M	Par.16
PC_SSN (serving MSC or VLR)	O	Par.18
RandomVariable (RAND)	M	Par.20
Return-Result Parameters	**M/O**	**Reference**
DenyAccess	O	Par.9
VoicePrivacyMask (VPMASK)	O	Par.27
Return-Error Information	**M/O**	**Reference**
Error Code	M	Section 17.5.3
FaultyParameter	O	Par.12

Figure 17.2-5 Parameters in authentication request messages. (From TSB-51. Reproduced with permission of TIA.)

unrecognized MSN (the responder's record shows a different MSN for the received MIN). Error codes are listed in Section 17.5.3.

If the error code indicates ParameterError, UnrecognizedParameterValue, or MissingParameter, the message also includes a Par.12 (FaultyParameter) that holds the name (tag) of the parameter that has caused the problem.

17.2.7 Timers

When beginning a transaction (sending an invoke), the invoker starts a timer. If the timer expires and no response has been received, the invoker concludes that a problem has occurred. It then tries to recover, for example, by repeating the invoke. Figures 17.2-2 through 17.2-5 show the timer values for the transactions described above.

17.2.8 Called Addresses of Invokes

We now outline how the initiators of the invokes determine the called party addresses [2].

We first consider the invokes of Fig. 17.2-1.

REGNOT Case (a). The REGNOT concerns a MS whose home CMN is also the CMN of MSC-A. MSC-A knows the PC of the HLR of its CMN.

***REGNOT* Cases (b, c, d).** The REGNOT sent by MSC-B concerns a roaming MS. MSC-B knows the PC address of its associated VLR-P.

***REGNOT* Cases (b, d).** VLR-P has to access the HLR of the roaming MS. It uses a global title (GT), in which the address GTA is the MIN of MS, and the translation type is set to TT = 3, to indicate a MIN to HLR translation.

***REGCANC* Case (d).** HLR has the MSCID, and possibly the PC, of VLR-Q, which is currently serving MS. It either uses PC, or derives PC from MSCID. VLR-Q has the MSCID, and, possibly the PC of MSC-C, which is currently serving MS.

In conclusion, we consider the invokes of Fig. 17.2-2.

AUTHRQST. MSC-B is serving a roaming MS, and sends its AUTHRQST to its associated VLR-P. This register addresses HLR with a global title, as in REGNOT cases (b, d). HLR knows the PC of the AUC in its network. Also, AUC is often implemented as a part of HLR. In this case, HLR passes its AUTHRQST to AUC in a primitive.

17.3 CALLS TO MOBILE STATIONS

Calls to a MS are routed by the PSTN to the "home" MSC of the mobile (17.1.3). This section discusses the operations needed to extend the call set-up from the home MSC to the MSC that currently serves the mobile [3].

17.3.1 Temporary Local Directory Numbers

We first introduce the concept of *temporary local directory numbers* (TLDN). Consider a MSC identified by a particular value of AC-EC. A MS for which MSC is the home MSC has a MIN in the format AC-EC-*XXXX*. When the MS is being called, its MIN is the called party address, and PSTN routes the call to the exchange identified by AC-EC, which is the home MSC of MS.

Each MSC identified by a particular value of AC-EC also has a pool of TLDNs which also have the AC-EC-*XXXX* format. It temporarily assigns a TLDN to a roaming MS which is being called while being served by MSC.

To distinguish TLDNs from MINs, *XXXX* is divided into two ranges, say 0000–0999 for TLDNs and 1000–9999 for MINs.

17.3.2 Set-up Example

The function of TLDN is illustrated by describing the set-up of a call to the MS with MIN = 708-357-8765. The MS homes on MSC-A, and is being served by MSC-B, which is identified by AC-EC = 615-443.

The PSTN routes calls to the MS to MSC-A, its "home" switching center. In Fig. 17.3-1, the call set-up has reached MSC-A on trunk T_1. MSC-A

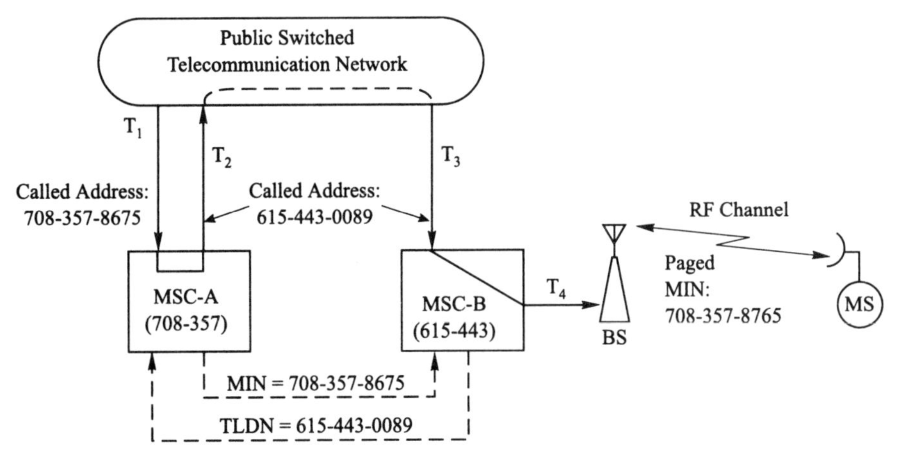

Figure 17.3-1 Connection for a call to mobile (MS).

determines that the received called address is the MIN of a mobile for which it is the home switching center. It then initiates a transaction with the HLR in its network to determine the MSC that is currently serving MS. One result of the operation is that serving MSC-B is informed about an incoming call to the mobile with MIN = 708-357-8765. MSC-B then allocates a TLDN, say TLDN = 615-443-0089, and records the association of the TLDN with the MIN of the called mobile. Another result of the operation is that MSC-A receives the allocated TLDN.

MSC-A has to route the call to MSC-B. If both MSCs belong to the same network, there may be a CMN trunk group between them, and MSC-A then seizes a trunk in this group.

Figure 17.3-1 assumes that there is no CMN trunk group between the MSCs. MSC-A therefore seizes trunk T_2 to a PSTN exchange, and a connection from MSC-A to MSC-B is set up in the PSTN network.

When the set-up reaches MSC-B, it determines that the called number 615-443-0089 is one of its TLDNs. It then finds the associated MIN = 708-357-8765, and sends paging messages for that MIN. On receipt of a page response by base station BS, MSC-B seizes trunk T_4, and connects it to T_3. The connection to MS is completed by T_4 and its associated RF voice channel.

MSC-B then discards the record that associates TLDN and MIN, and returns TLDN to its pool of temporary location direction numbers.

17.3.3 Operations

The operations involved in setting up a call to a mobile MS are shown in Fig. 17.3-2. Assuming again that MSC-A is the "home" of MS, the PSTN set-up reaches MSC-A, which starts a location-request operation with its HLR, sending a LOCREQ invoke. The subsequent operations depend on where the mobile has registered most recently, which is known by HLR.

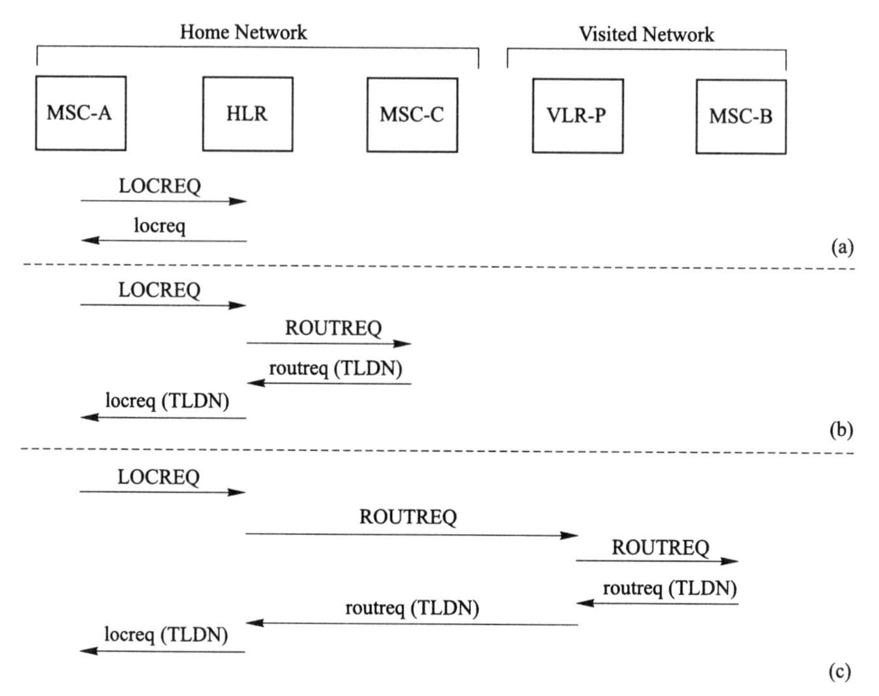

Figure 17.3-2 Operations for terminating calls. (From IS-41.3-B. Reproduced with permission of TIA.)

Case (a). HLR determines that the mobile is currently served by MSC-A, and reports this in its locrec return-result. MSC-A then pages the mobile.

Case (b). HLR determines that mobile is being served by MSC-C, a switching center in its home network. HLR then initiates a routing-request operation with MSC-C, by sending a ROUTREQ invoke that includes the MIN of the called mobile. MSC-C allocates a TLDN, associates it with the received MIN, and includes TLDN in its return-result (routreq).

The TLDN is also included in the return-result (locreq) sent by HLR. MSC-A then uses the TLDN to route the call to MSC-C. When the set-up arrives at MSC-C, it determines the MIN it has associated with the received TLDN, and pages the mobile.

Case (c). MS has registered at MSC-B, located in a visited network. In this case, HLR knows the identity of VLR-P, because it has received this information in the REGNOT transaction that was executed when MS registered.

HLR now opens a routing-request transaction with VLR-P, by sending a ROUTREQ invoke, and the VLR opens a routing-request transaction with MSC-B, with a ROUTREQ invoke. The TLDN allocated by MSC-B is included in the routreq return-result sent to VLR-P and HLR, and in the locreq return-result sent to MSC-A. This MSC then routes the call to MSC-B, and the MS is paged.

Case (d). HLR has received a LOCREQ from MSC-A, and determines that MS is not currently registered at any MSC. The call thus cannot be set up. HLR indicates this in its locrec return result, and MSC-A then connects the incoming trunk to a recorded announcement, to inform the calling party.

17.3.4 Message Contents

The main parameters in the invokes and return-results of the operations are listed in Figs. 17.3-3 and 17.3-4. When reading this section, it is helpful to look up the parameter descriptions (17.5.4).

LocationRequest Invoke (LOCREQ)—see Fig.17.3-3. The invoke is sent by the home MSC of the called mobile. Par.11 (Digits) holds the digits received by the MSC, which identify the mobile. The home MSC identifies itself by Par.16 (MSCID), and may include its PC_SSN address (Par.18). Also included is BillingID (Par.5). This is a reference number allocated to the call by the home MSC, for billing purposes. BillingID is also included in the ROUTEREQ invoke to the serving MSC, and appears on the call-billing records that are produced by the home and serving MSC at the end of the call. It is used by the accounting center to calculate the total charge for the call.

LocationRequest Return-Result (locreq)—see Fig. 17.3-3. The mobile is identified by Par.14 (MobileIdentificationNumber MIN), and Par.15

Operation	**Timer**: 12 seconds	
LocationRequest (LOCREQ)		
Invoke Parameters	**M/O**	**Reference**
BillingID	M	Par.5
Digits received (MIN)	M	Par.11
MSCID (home MSC)	M	Par.16
PC_SSN (home MSC)	O	Par.18
Return-Result Parameters	**M/O**	**Reference**
AccessDeniedReason	O	Par.1
Digits (TLDN)	O	Par.11
MobileIdentificationNumber (MIN)	M	Par.14
MobileSerialNumber (MSN)	M	Par.15
MSCID (serving MSC or VLR)	M	Par.16
PC_SSN (serving MSC or VLR)	O	Par.18
Return-Error Information	**M/O**	**Reference**
Error Code	M	Section 17.5.3
FaultyParameter	O	Par.12

Figure 17.3-3 Parameters in location request messages. (From IS-41.5-B. Reproduced with permission of TIA.)

Operation	**Timer**: 6 seconds	
RoutingRequest (ROUTREQ)		
Invoke Parameters	**M/O**	**Reference**
BillingID	M	Par.5
MobileIdentificationNumber (MIN)	M	Par.14
MobileSerialNumber (MSN)	M	Par.15
MSCID (home MSC)	M	Par.16
PC_SSN (home MSC)	O	Par.18
Return-Result Parameters	**M/O**	**Reference**
AccessDeniedReason	O	Par.1
Digits (TLDN)	O	Par.11
MSCID (serving MSC)	M	Par.16
PC_SSN (serving MSC)	O	Par.18
Return-Error Information	**M/O**	**Reference**
Error Code	M	Section 17.5.3
FaultyParameter	O	Par.12

Figure 17.3-4 Parameters in routing request messages. (From IS-41.5-B. Reproduced with permission of TIA.)

(MobileSerialNumber MSN). Par.16 and Par.18 hold the MSCID and PC_SSN address of the serving MSC. Par.11 (Digits) holds the allocated TLDN. If the mobile cannot be accessed, the message includes Par.1 (AccessDeniedReason) instead of Par.11.

RoutingRequest Invoke (ROUTREQ)—Fig. 17.3-4. Par.14 and Par.15 hold MIN and MSN of the called MS. The BillingID, assigned by the home MSC of the called MS, is in Par.5. The identity and PC-SSN address of the home MSC are in Par.16 and Par.18.

RoutingRequest Return-Result (routreq)—see Fig.17.3-4. Par.4 and Par.5 hold the MSCID and PC_SSN address of the serving MSC. If the MS can receive calls, the serving MSC has assigned a TLDN, which is in Par.11 (Digits). If the called MS cannot receive the incoming call, for example, when it is already involved in a call, the message includes Par.1 (AccessDenied Reason) in lieu of Par.1.

17.3.5 Called Addresses of Invokes

In conclusion, we outline how the initiators of the invokes of Fig. 17.3-2 determine the called party addresses [2].

LOCREQ Cases (a, b, c). The call for MS has been delivered to its home MSC-A, which knows the point code of the HLR in its network.

ROUTREQ Case (b). Since currently serving MSC-C of MS is in the same network of HLR, this register stores the MSCID, and possibly the point code, of MSC-C. It derives the PC for the called address from either MSCID, or uses the stored PC.

ROUTREQ Case (c). Since currently serving MSC-B of MS is not in the network of HLR, this register stores the MSCID, and possibly the point code, of VLR-P. It either derives the PC for the called address from MSCID, or uses the stored PC.

VLR-P stores MSCID, and possibly the point code, of currently serving MSC-B, and either derives the PC for the called address from either MSCID, or uses the stored PC.

17.4 OPERATIONS FOR INTER-SYSTEM HANDOFF

17.4.1 Introduction

Section 12.1.7 has described the handoff of a mobile that is involved in a call, and moves from a cell X to an adjacent cell Y, in the case that both cells are controlled by the same MSC. This section considers a MS handoff when cells X and Y are controlled by different MSCs. These handoffs are known as inter-system handoffs, even though the MSCs usually belong to the same cellular network [5].

In Fig. 17.4-1(a), adjacent cells X and Y are controlled by respectively MSC-A and MSC-B. It is assumed that the MS is in a conversation, using voice channel (VC-U) of base station (BS-X). The connection occupies trunk T_1 (to an exchange in the PSTN network), trunk T_2, and its associated voice channel (VC-U).

MSC-A periodically requests BS-X to monitor the strength of the signal received on VC-U. When the signal strength drops below a certain level, the MSC orders MS to increase its transmit power (12.1.6, 12.4.7). However, if MS is already transmitting at its maximum power, it has to be handed off to an adjacent cell. We consider the case that MSC-A has determined that none of its own cells receives a strong enough signal from MS, and therefore attempts to hand off the MS to a cell controlled by MSC-B.

Inter-system handoffs require remote operations between the MSCs. Moreover, handoffs from say MSC-A to MSC-B require a "handoff" trunk group between the switching centers. The control of the handoff trunk has to be coordinated with other handoff actions, and is therefore done as part of a remote operation (which is an exception to the general rule that transactions are not trunk-related). The handoff trunk groups and the inter-MSC trunk groups discussed in Section 17.3 are separate groups, because the latter groups use signaling systems of the types used in fixed networks (MF, ISUP, etc.).

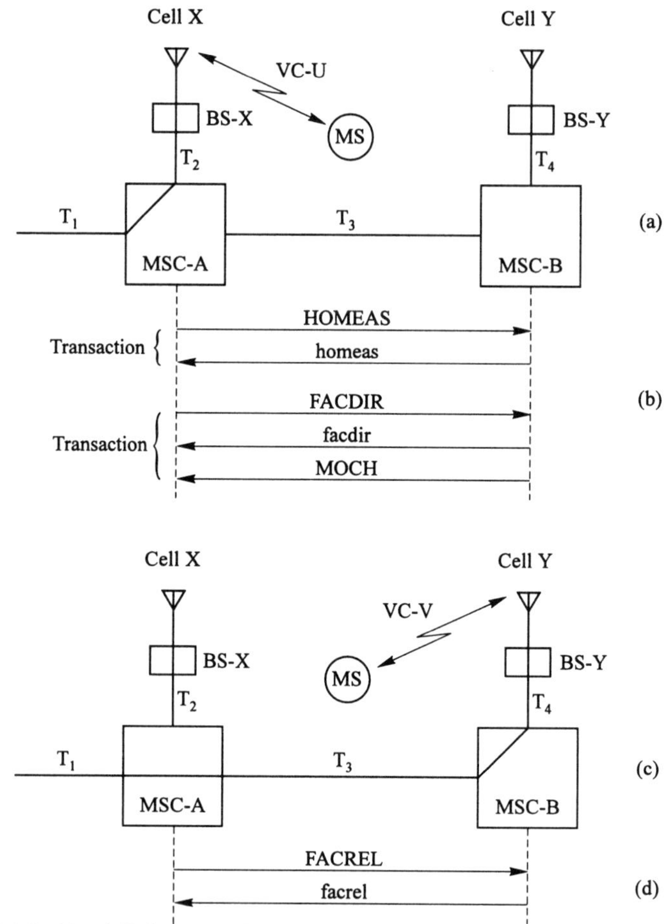

Figure 17.4-1 Handoff of mobile MS from MSC-A to MSC-B. (a): configuration prior to handoff. (b): handoff. (c): configuration after handoff. (d): release of trunk T_2. (From IS-41.2-B. Reproduced with permission of TIA.)

17.4.2 Handoff Example

Figure 17.4-1(b) shows the remote operations for the handoff of mobile station MS from MSC-A to MSC-B [4]. The handoff is done in a sequence of two transactions.

Handoff Measurement. The first transaction consists of one operation. It is started by MSC-A, with a handoff measurement invoke (HOMEAS). This requests MSC-B to measure the signal strengths of the MS (on the reverse voice channel of VC-U) received in the cells of MSC-B's service area that are adjacent to cell X. The return-result (homeas) contains a list of "target" cells (cells in which the signal from MS is received with adequate strength, and are therefore candidates for the handoff).

Facilities Directive. The second transaction consists of two operations. From the list of target cells, MSC-A selects cell Y. It then seizes a trunk T_3 in its handoff trunk group to MSC-B, and starts a facilities-directive operation with a FACDIR invoke that requests MSC-B to set up the connection between T_3 and a voice channel in cell Y.

MSC-B selects an available voice channel VC-V of base station BS-Y. It connects T_3 to trunk T_4 (associated with VC-V), and then sends a facdir return-result. The return-result is sent in a TCAP conversation message which ends the first operation, but continues the transaction (15.1.2).

On receipt of the facdir, MSC-A connects trunk T_1 to trunk T_3, signals the MS to tune to VC-V, and releases T_2 and its associated RF voice-channel (VC-U)—see Fig. 17.4-1(c).

The facdir return result is sent in a TCAP response package (15.2.1) and the transaction continues.

Mobile on Channel. When MSC-B detects the presence of a signal from MS on VC-V, it starts a mobile-on-channel operation with a MOCH invoke. This operation merely indicates that the handoff is successful, and does not require a response. The invoke is therefore sent in a TCAP response package that ends the transaction.

Facilities Release. At the end of the call, trunks T_1, T_3, and T_4, and the RF equipment of associated voice channel VC-V have to be released. In Fig. 17.4-1(d), MSC-A has received a disconnect signal on trunk T_1. It then releases trunks T_1 and T_3 at its end, and initiates a facilities-release operation, sending a FACREL invoke. This requests MSC-B to release trunk T_3, and the equipment in its part of the connection. MSC-B releases trunks T_3 and T_4, and the RF equipment of VC-V. It then confirms the release with a return-result (facrel), in a TCAP response package that ends the transaction.

If the MS user disconnects first, the MS starts sending signaling tone on voice channel (VC-Y) (Chapter 12). This is detected by MSC-B, which then initiates the release of the facilities.

17.4.3 Message Contents

This section introduces the main parameters in the invokes and return-results for the operations described above [5]. The reference numbers (Par.1, etc.) point to parameter descriptions in Section 17.5.4. It is helpful to look up the descriptions when reading this section.

HandoffMeasurementRequest Invoke (HOMEAS)—see Fig. 17.4-2. Par.22 (ServingCellID) identifies the cell that currently serves MS. Par.24 (StationClassMark) lists the transmission characteristics of the MS (12.4.9).

If the mobile is currently using an analog voice channel, the invoke includes Par.8 (ChannelData), which contains the channel number, and the attenuation and color codes with which the MS is transmitting.

Operation	**Timer**: 7 seconds	
HandoffMeasurementRequest		
Invoke Parameters	**M/O**	**Reference**
CallMode	O	Par.7
ChannelData (serving)	M	Par.8
DigitalChannelData (serving)	O	Par.10
ServingCellID	M	Par.22
StationClassMark	M	Par.24
Return-Result Parameters	**M/O**	**Reference**
List of target cells. For each cell:		
SignalQuality	M	Par.23
TargetCellID	M	Par.25
Return-Error Information	**M/O**	**Reference**
Error Code	M	Section 17.5.3
FaultyParameter	O	Par.12

Figure 17.4-2 Parameters in handoff measurement messages. (From IS-41.5-B. Reproduced with permission of TIA.)

If the mobile is currently using a digital traffic channel, the invoke also includes Par.10 (DigitalChannelData) and Par.7 (CallMode).

On receiving a HOMEAS invoke, an MSC determines which cells are neighbors to the current serving cell of the MS (from Par.22). These cells can be "target" cells for the handoff. The MSC orders the target cells to measure the signal strengths received from MS.

HandoffMeasurementRequest Return-Result (homeas)—Fig. 17.4-2. After receiving the measurement results, the MSC calculates the "quality" of the received signals. Cells in which the signal quality is above a predetermined level are target cells for the handoff. The return-result is a list of target cells. Each entry consists of a Par.23 (SignalQuality) and a Par.25 (TargetCellID),

FacilitiesDirective Invoke (FACDIR)—see Fig. 17.4-3. Par.14 and Par.15 (MIN and MSN) identify the mobile that will be handed off. Par.8, Par.22, and Par.24 (and possibly Par.7 and Par.10) of the HOMEAS invoke are included again. The target cell (selected from the list in the homeas return-result) and the seized inter-MSC trunk are in Par.25 and Par.13. The BillingID (Par.5) is included for billing purposes.

FacilitiesDirective Return-Result (facdir)—see Fig. 17.4-3. An MSC receiving a FACDIR invoke allocates an analog or digital channel in the selected target cell, and responds with a return-result that includes Par.8. If the allocated channel is a digital traffic channel, Par.10 is also included. These parameters contain information required by the mobile, and appear in the handoff message to the mobile that is sent by old serving MSC (12.4.7).

Operation	Timer: 12 seconds	
FacilitiesDirective		
Invoke Parameters	**M/O**	**Reference**
BillingID	M	Par.5
Callmode	O	Par.7
ChannelData (serving)	M	Par.8
DigitalChannelData (serving)	O	Par.10
InterMSCCircuitID	M	Par.13
MobileIdentificationNumber (MIN)	M	Par.14
MobileSerialNumber (MSN)	M	Par.15
ServingCellID	M	Par.22
StationClassMark	M	Par.24
TargetCellID	M	Par.25
Return-Result Parameters	**M/O**	**Reference**
ChannelData (target)	M	Par.8
DigitalChannelData (target)	O	Par.10
Return-Error Information	**M/O**	**Reference**
Error Code	M	Section 17.5.3
FaultyParameter	O	Par.12

Figure 17.4-3 Parameters in facilities directive messages. (From IS-41.5-B. Reproduced with permission of TIA.)

MobileOnChannel Invoke (MOCH)—see Fig. 17.4-4. This indicates that the new serving MSC has verified the presence of the signal from MS on the new channel. No parameters are included.

FacilitiesRelease Invoke (FACREL)—see Fig. 17.4-5. This indicates that the invoking MSC has released the inter-MSC trunk at its end. Par.14 (MIN) and Par.13 (InterMSCCircuitID) identify the mobile and the inter-MSC trunk. Par.21 (ReleaseReason) indicates why the connection is being released.

FacilitiesRelease Return-Result (facrel)—Fig. 17.4.5. This acknowledges the receipt of FACREL, and indicates that the responding MSC has cleared the inter-MSC trunk at its end. No parameters included.

17.4.4 Called Addresses of Invokes

In conclusion, we outline how the initiators of the invokes of Fig. 17.4-1 determine the called party addresses [2].

Operation	Timer: none	
MobileOnChannel		
Invoke Parameters	**M/O**	**Reference**
None		

Figure 17.4-4 Mobile on channel invoke. Note: no return-result or return-error is sent in this operation. (From IS-41.5-B. Reproduced with permission of TIA.)

Operation	**Timer**: 4-15 seconds	
FacilitiesRelease		
Invoke Parameters	**M/O**	**Reference**
InterMSCCircuitID	M	Par.13
MobileIdentificationNumber	M	Par.14
ReleaseReason	M	Par.21
Return-Result Parameters	**M/O**	**Reference**
None		
Return-Error Information	**M/O**	**Reference**
Error Code	M	Section 17.5.3
FaultyParameter	O	Par.12

Figure 17.4-5 Parameters in facilities release messages. (From IS-41.5-B. Reproduced with permission of TIA.)

HOMEAS and FACDIR. Serving MSC-A has stored information on the point codes of MSCs which have cells that are neighbors of cell X, which currently communicates with the mobile. The example assumes that MSC-B is one of these MSCs. MSC-A thus uses the PC of MSC-B as the called address for its HOMEAS and FACDIR invokes.

MOCH. MSC-B has received the PC of MSC-A in the HOMEAS and FACDIR invokes. It thus knows the PC of MSC-A, and uses it in its MOCH invoke.

FACREL. After the transactions of Fig. 17.4-1(b), the MSCs know each other's PCs. In the example, MSC-A initiates the release of the connection, and uses the point code of MSC-B as called address for its invoke.

17.5 IS-MAP FORMATS AND CODES

The coding principles of Section 15.3 apply, except that IS-MAP documents use the term *parameter* in lieu of *data element*. This section lists the formats and codes of the parameters in the messages discussed in Sections 17.2–17.4 [2].

17.5.1 Tags (Identifiers)

Operation code and error code tags are coded as "private TCAP:"

	← Bits							
	H	G	F	E	D	C	B	A
Operation code identifier	1	1	0	1	0	0	0	1
Error code identifier	1	1	0	1	0	1	0	0

Table 17.5-1 Operation-code specifiers.

Decimal Value	Operation
1	HandoffRequestMeasurement
2	FacilitiesDirective
3	MobileOnChannel
5	FacilitiesRelease
13	RegistrationNotification
14	RegistrationCancellation
15	LocationRequest
16	RoutingRequest
28	AuthenticationRequest

Source: IS-41.5-B. Reproduced with permission of TIA.

17.5.2 Operation-code Contents

The contents fields of operation codes consists of two octets. Octet 1 is coded as:

← Bits

H	G	F	E	D	C	B	A
0	0	0	0	1	0	0	1

This indicates that the operation belongs to the mobile applications part IS-MAP operations family.

Octet 2 specifies a particular operation—see Table 17.5-1.

17.5.3 Error-code Contents

The one-octet contents field indicates why a MSC, HLR, or VLR cannot execute a requested operation. Bits H,G,F,E are set to 1,0,0,0. Bits D–A contain an integer that indicates a specific error, for example:

Value	Meaning
1	*UnrecognizedMIN*. The received mobile identification number is not currently served by the HLR, VLR, or MSC.
2	*UnrecognizedMSN*. A HLR or VLR has received a combination of a MIN and a MSN and, according to its records, the received MSN is not associated with the received MSN.
3	*MIN/HLRMismatch*. The received mobile identification number is not known at the HLR.
4	*OperationSequenceProblem*. The requested operation is not allowed in the current state of the call.

8	*ParameterError*. Parameter is not expected, or is coded incorrectly.
10	*UnrecognizedParameterValue*. Parameter value is coded properly, but has an unrecognized value.
12	*MissingParameter*. Not all expected message parameters have been received.

When the error code is 8, 10, or 12, the return-error component also includes one or more Par.12 (FaultyParameter), which specify the parameter(s) that caused the problem.

17.5.4 Parameter Descriptions and Formats

The parameter tags (identifiers) are coded as "context specific" (bits H,G = 1,0; see Section 15.3.2), and are listed in Table 17.5-2. The contents fields of the parameters are described below.

Par.1 AccessDeniedReason (one octet). This is an integer indicating why the mobile should receive (terminate) a call:

Value	Meaning
1	Mobile identification number not recognized
2	Mobile inactive
3	Mobile busy
4	Terminations to mobile are denied

Par.2 AuthenticationResponse (AUTHR; three octets). This contains the 18-bit result of the CAVE algorithm executed by the MS (12.6.7).

Par.3 AuthorizationDenied (one octet). An integer indicating why the mobile should not receive service:

Value	Meaning
1	Delinquent account
2	Invalid mobile serial number
3	Stolen unit
5	Mobile identification number not recognized
6	Unspecified

Par.4 AuthorizationPeriod (2 octets). The parameter indicates that the MSC is authorized to serve the mobile station for a certain period. The length of the period is indicated by two integers. Octet 1 specifies a unit of time:

Table 17.5-2 Parameter type identifiers (tags).

Reference	Parameter Name	← Bits HGFE DCBA
Par.1	AccessDeniedReason	1001 0100
Par.2	AuthenticationResponse (AUTHR)	1001 1111 0010 0011
Par.3	AuthorizationDenied	1000 1101
Par.4	AuthorizationPeriod	1000 1110
Par.5	BillingID	1000 0001
Par.6	CallingFeaturesIndicator	1001 1001
Par.7	CallMode	1001 1101
Par.8	ChannelData	1000 0101
Par.9	DenyAccess	1001 1111 0011 0010
Par.10	DigitalChannelData	1001 1100
Par.11	Digits	1000 0100
Par.12	FaultyParameter	1001 1010
Par.13	InterMSCCircuitID	1000 0110
Par.14	MobileIdentificationNumber (MIN)	1000 1000
Par.15	MobileSerialNumber (MSN)	1000 1001
Par.16	MSCID	1001 0101
Par.17	OriginationIndicator	1001 0111
Par.18	PC_SSN	1001 1111 0000 0010
Par.19	QualificationInformationCode	1001 0001
Par.20	RandomVariable (RAND)	1001 1111 0010 1000
Par.21	ReleaseReason	1000 1010
Par.22	ServingCellID	1000 0010
Par.23	SignalQuality	1000 1011
Par.24	StationClassMark	1000 1100
Par.25	TargetCellID	1000 0011
Par.26	TerminationRestrictionCode	1001 1000
Par.27	VoicePrivacyMask (VPMASK)	1001 1111 0011 0000

Note: Tags of Par.2, Par.9, Par.18, Par.20, and Par.27 occupy two octets—see Section 15.3.2.
Source: IS-41.5-B. Reproduced with permission of TIA.

Value	Meaning
0	Per call authorization
1	Hours
3	Days
5	Weeks

Octet 2 indicates the number of time units (1–255) of the authorization period. When the value of octet 1 is 0, the value of octet 2 is also 0.

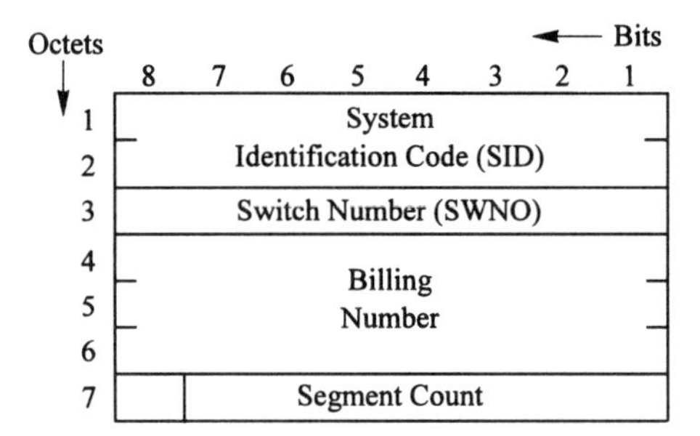

Figure 17.5-1 Format of Par.5 (BillingID). (From IS-41.5-B. Reproduced with permission of TIA.)

Par.5 BillingID (seven octets). This is a reference number for billing purposes. It is assigned to a call by the "anchor" MSC (the first MSC involved in the call). When a call involves several MSCs, because of roaming or intersystem handoffs, the BillingID is passed on to the next serving MSC. All MSCs involved in the call include the BillingId in their billing records. In this way, the billing records for the call can be correlated.

In Fig. 17.5-1, octets 1–3 identify the anchor MSC: octets 1 and 2 hold the 15-bit system identification (SID) that identifies a cellular system (Chapter 12), and octet 3 holds the switch number (SWNO) that identifies a MSC within the cellular system.

Octets 4–6 contain the billing number.

Par.6 CallingFeaturesIndicator (two octets). This contains a number of two-bit indicators, associated with calling features that can be offered to mobiles:

Indicator	Feature
a_1, a_2	Call forwarding—unconditional
b_1, b_2	Call forwarding if MS is busy
c_1, c_2	Call forwarding if MS is busy
d_1, d_2	Call waiting
e_1, e_2	Three-way calling
f_1, f_2	Call delivery
g_1, g_2	Voice privacy

The indicator values define the status of each feature for a particular mobile:

Value	Feature
0,1	Not authorized for MS
1,0	Authorized but deactivated by MS
1,1	Authorized and activated by MS

Par.7 CallMode (one octet). This indicates the mode of the current call: analog or digital.

Par.8 ChannelData (zero or three octets). If a mobile is using an analog voice channel, the contents field has three octets, and contains the color code (SCC) of the channel, the most recent mobile attenuation code (VMAC) sent to the mobile, and the channel number (CHAN) of the channel (see Section 12.4.9 for definitions of SCC, VMAC, and CHAN). If a mobile is using a digital traffic channel, the parameter appears in the message, but the length of its contents field is 0 octets.

Par.9 DenyAccess (one octet). This is an integer indicating the reason why an authentication center has determined that the mobile is not authentic, and should not be allowed to access the mobile network. The octet is usually coded "1" (reason not specified).

Par.10 DigitalChannelData (five octets). If a mobile is using an IS-54 TDMA traffic channel in a digital channel (Section 12.6.2), this parameter contains:

Channel Number (CHAN). This identifies a 48.6 kb/s digital channel (12.6.2).

Time Slot Indicator (TSR). A digital channel contains three TDMA traffic channels. TSR identifies the time-slot pair (1 and 4, 2 and 5, or 3 and 6) used by the traffic channel.

Digital Color Verification Code (DVCC). This is the color code of a digital channel—see Section 12.6.2.

Digital Mobile Attenuation Code (DMAC). This is the digital counterpart of VMAC (voice mobile attenuation code)—see Section 12.4.9.

Par.11 Digits (variable length). This holds a string of BCD coded digits—see Fig. 17.5-2. The integer in octet 1 indicates the type of the digit string:

Value	Meaning
1	Dialed number
2	Routing number
3	Destination number
4	Carrier identification code

Octet 2 is not used. The code in octet 3 indicates that the number has the format of the U.S. telephone numbering plan, and that the digits are BCD coded. The integer in octet 4 indicates the number of digits. The digits are located in octets 5 through n.

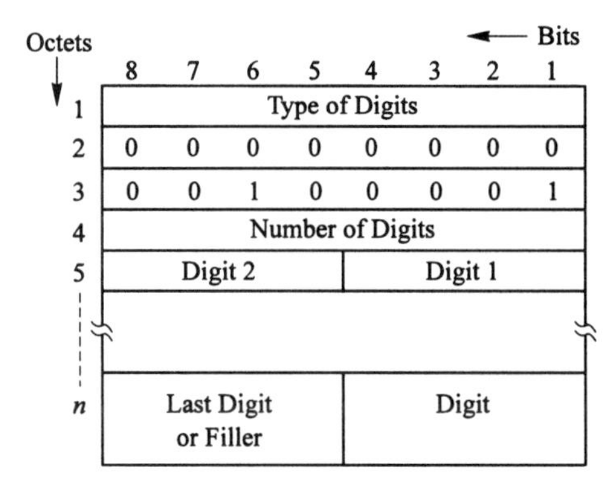

Figure 17.5-2 Format of Par.11 (Digits). (From IS-41.5-B. Reproduced with permission of TIA.)

Par.12 FaultyParameter (one or two octets). This contains the parameter tag (Table 17.5-2) of a missing, incorrect, or unexpected parameter.

Par.13 InterMSCCircuitID (two octets). This identifies an inter-MSC trunk. Octets 1 and 2 contain respectively the group and member number of the trunk.

Par.14 MobileIdentificationNumber (MIN). This consists of five octets. Each octet contains two BCD coded digits of a ten-digit mobile identification number (12.3).

Par.15 MobileSerialNumber (MSN). This consists of four octets, containing the 32-bit serial number of a mobile (12.3).

Par.16 MSCID (three octets). This identifies the MSC or VLR that has opened the transaction. Octets 1 and 2 hold the SID (system ID) that identifies a cellular mobile network (12.3). Octet 3 identifies an MSC or VLR in that network.

Par.17 OriginationIndicator (one octet). The integer that indicates the destinations allowed to be called by the mobile:

Value	Meaning
2	No originations allowed
3	Local calls only
4	Calls to selected area codes only
5	Calls to selected area codes and local calls
6	Calls to area codes of all destinations in world zone 1 (see Section 1.3.3) and local calls

Par.18 PC_SSN (five octets). This identifies a MAP subsystem in a signaling point. Octet 1 indicates the subsystem type:

Value	Meaning
1	Serving MSC
2	Home MSC
4	Home location register (HLR)
5	Visitor location register (VLR)
6	Authentication center (AUC)

Octets 2–4 hold the point code (PC); octet 5 holds the subsystem number (SSN).

Par.19 QualificationInformationCode (one octet). An integer indicating the requested data about a mobile:

Value	Meaning
1	No information requested
2	Validation only
3	Validation and service profile
4	Service profile only

Par.20 RandomVariable (RAND; four octets). This holds the 32-bit random number used by the MS as input to the CAVE algorithm (Section 12.6.7).

Par.21 ReleaseReason (one octet). This is an integer indicating why the inter-MSC trunk is being released:

Value	Meaning
0	Not specified
1	Call ended, clear-forward received
2	Call ended, clear-back received

Par.22 ServingCellID (two octets). This is an integer that identifies the cell that is currently serving the mobile.

Par.23 SignalQuality (one octet). This is an integer that indicates the quality (strength) of the received signal in a reverse voice or traffic channel:

Value	Meaning
1–8	Signal too weak
9–245	Acceptable signal
245–255	Signal too strong (may cause co-channel interference)

Par.24 StationClassMark (one octet). This parameter contains the transmission characteristics of the mobile (12.4.9).

Par.25 TargetCellID (two octets). This is an integer that identifies a cell to which a mobile can be handed off.

Par.26 TerminationRestrictionCode (one octet). This is an integer indicating the call types which the mobile station is allowed to receive:

Value	Meaning
1	No terminations allowed
2	No restrictions on terminations
3	Restricted to sent-paid calls (calls billed to the calling party)

Par.27 VoicePrivacyMask (VPMASK; 66 octets). This is a 528-bit field that holds two 260-bit masks (one for each direction of transmission). Used to scramble and unscramble the 260-bit user data blocks for IS-54 mobiles that have the voice privacy feature (Section 12.6.8).

17.6 INTRODUCTION TO GSM-MAP

The remainder of this chapter briefly describes GSM-MAP, the Mobile Application Part defined by the International Telegraph and Telephone Consultative Committee (CCITT)—now known as the telecommunications standards sector of the International Telecommunications Union (ITU-T)—and the European Telecommunications Standards Institute (ETSI).

GSM-MAP supports communications in mobile networks of the Global System for Mobile Telecommunications (GSM—see Section 12.7), which are installed in a large number of nations.

As in IS-MAP, the main functions of GSM-MAP support MS registration, MS roaming, MS authentication, and MS handover (the European term for handoff). International roaming is allowed. A customer who has a GSM service provider in one nation can roam in GSM networks of other nations.

17.6.1 GSM-MAP Terms

A number of GSM-MAP terms and acronyms are identical to those of IS-MAP, for example: mobile station (MS), mobile switching center (MSC), home location register (HLR), visitor location register (VLR), authentication center (AUC), roamer, and location area. Some differences in terminology are outlined below [6].

Public Land Mobile Network (PLMN). This is the equivalent of cellular mobile network (CMN).

Home PLMN. This is the PLMN selected by the MS owner to provide mobile communications. It is the equivalent of the home CMN.

Visited PLMN. This is the PLMN that currently serves a roaming MS. It is the equivalent of visited CMN.

Serving MSC. This is the MSC that is currently serving the MS.

Gateway MSC (GMSC). This is the MSC to which the fixed (PSTN/ISDN) network delivers a call to the MS. It is nearly equivalent to the home MSC. In U.S. mobile networks, calls to a particular MS are always delivered to its home MSC, which is determined by analyzing the MIN of the mobile.

GSM networks do not have home MSCs. Instead, the operator of a PLMN designates a number of MSCs as gateway MSCs (GMSCs), and calls to a particular MS are delivered by the fixed network to one of the GMSCs in the home PLMN of the mobile.

Mobile Station ISDN Number (MSISDN). This is the number dialed by a subscriber when calling the MS. It is used by the fixed network to route calls for MS to a nearby gateway MSC in the home PLMN of MS.

Mobile Station Roaming Number (MSRN). This is the equivalent of the temporary location directory number (TLDN).

International Mobile Equipment Identity (IMEI). This is the equivalent of the mobile serial number (MSN).

17.6.2 Equipment Entities

The equipment entities in a PLMN that are involved in GSM-MAP transactions are shown in Fig. 17.6-1 [6].

Mobile Switching Centers (MSC). In this example, the PLMN has two MSCs, each of which serves the mobiles in its service area.

Home Location Register (HLR). This register stores semipermanent and temporary data about the mobiles for which PLMN is the home network.

Visitor Location Register (VLR). The register has records on the mobiles that are currently being served by its associated MSCs. In this example, VLR-P and VLR-Q are associated with respectively MSC-A and MSC-B.

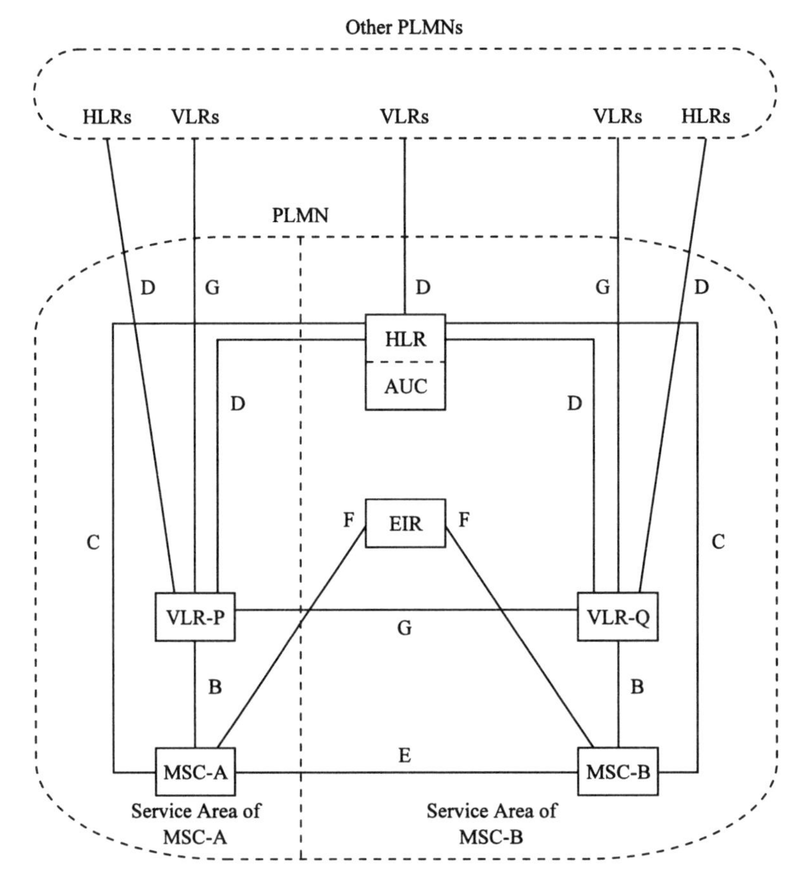

Figure 17.6-1 Equipment entities and interfaces. (From GSM.09.02-version 4.9.0. Courtesy of ETSI.)

Authentication Center (AUC). This register stores authentication and voice-privacy information about the mobiles for which PLMN is the home network.

Equipment Identity Register (EIR). This register stores the *international mobile equipment identity* (IMEI) of all mobile equipments (Section 12.7.4) for which PLMN is the home network.

17.6.3 Interfaces

We next explore the GSM-MAP interfaces between the equipment entities of Fig. 17.6-1 [6]. There are some differences from the IS-MAP interfaces.

The interface between a MSC and its associated VLR is known as the "B" interface. ETSI considers the B interface as "internal," and does not define it.

The message transfer does not involve the signaling network.

The C interface is between a gateway MSC and the HLR in its PLMN. The D interface is between a VLR and a HLR, in the same or different PLMNs.

The E interface is between adjacent MSCs in a PLMN, and is used in handover operations.

The F interface is an intra-PLMN interface, between a MSC and the EIR.

The G interface is between VLRs, in the same, or in different PLMNs. This interface does not exist in IS-MAP.

In GSM-MAP, the AUC is an integral part of HLR, and no interface is defined.

17.6.4 Numbering Plans

In the U.S., mobile networks use the numbering plan that was already in place for fixed networks. The MIN of a mobile has the format of national numbers (Section 1.2.1).

Nations with GSM mobile networks have a fixed network numbering plan, and a separate PLMN numbering plan.

In the fixed (PSTN/ISDN) network, a mobile is identified by a MSISDN, which has the format defined in CCITT Rec. E.164 [7]—see Section 1.2. MSISDN is an international number, consisting of country code CC and national number NN.

The mobile station roaming number (MSRN), which is used by the fixed network to extend the call set-up to the MSC currently serving the called MS, is a national or international E.164 number.

In the PLMN, a MS is idenfied by either an *international mobile station identity* (IMSI), or by the combination of a *location area identity* (LAI) and a *temporary mobile station identity* (TMSI) (Section 12.7.4).

The IMSI format has been defined in CCITT Rec. E.212 [8]:

$$\text{IMSI} = \text{MCC-MNC-MSIN}$$

where the combination of the mobile country code MCC and the mobile network code MNC identifies the home PLMN of MS, and MSIN is the mobile station identity which identifies a mobile in that PLMN—see Fig. 17.6-2. IMSI is an international number that uniquely identifies a mobile world-wide. In its

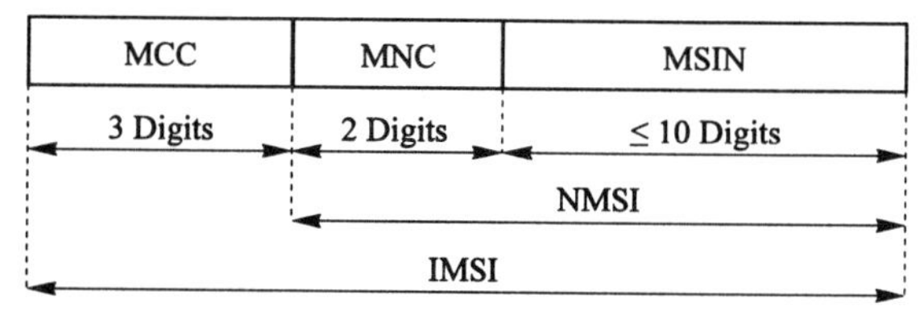

Figure 17.6-2 Format of IMSI. (From GSM.09.02-version 4.9.0. Courtesy of ETSI.)

home country, a MS can be uniquely identified by a national mobile station identity (NMSI):

NMSI = NMC-MSIN

MSISDN, IMSI, and NMSI permanently identify a MS. In addition, the combination of a *location area identity* (LAI) and *temporary mobile station identity* (TMSI) identifies a MS during the time it is served by the VLR that covers the location area. The format of LAI is:

LAI = MCC-MNC-LAC

where MCC-MNC identifies a particular PLMN, and the *local area code* (LAC) represents a location in the PLMN.

TMSI is a 32-bit binary number which identifies the MS while it is operating in a particular location area.

Because of the multiple MS identities and numbering plans, GSM-MAP requires a number of operations which are not encountered in IS-MAP.

17.6.5 Information in SIM, HLR, VLR, EIR, and AUC

The most important parameters stored in the subscriber identity module (SIM) of a MS (Section 12.7.4), and in the various GSM network entities, are listed below [9]. Semipermanent and temporary parameters are denoted by (S) and (T).

Information in SIM

- IMSI (S)
- TMSI (T)
- LAI (T)
- Ki (S), authentication key (Section 12.8.4)
- Kc (T), cipher key (Section 12.8.4)

A MS updates its temporary parameters on command from the serving MSC.

Information in HLR. The HLR of a PLMN has records for all MSs whose home network is PLMN.

- MSISDN (S) of MS
- IMSI (S) of MS
- Originating and terminating service profile of MS (S)
- Address of the VLR associated with the MSC that is currently serving MS (T)

Information in VLR. This register has a record for each MS currently served by one of its associated MSCs.

- MSISDN (T)
- Originating and terminating service profile of MS (T)
- IMSI (T)
- TMSI in the current location area (T)
- LAC of the current MS location area (T)
- MSRN that is currently assigned to the MS (T)

Information in AUC. The AUC in a PLMN has records with authentication and privacy information for all mobiles whose home network is PLMN.

- MSISDN (S)
- Ki (S)
- Kc (T)

17.6.6 SCCP Addresses

GSM-MAP messages are transferred by the message transfer part (MTP), and the signaling connection control part (SCCP) of the SS7 signaling networks that also transfer messages for fixed-network applications.

GSM-MAP uses the connectionless services of SCCP (Section 14.3.3).

We now explore the SCCP addresses of GSM-MAP entities. The application service entities (ASE) at a signaling point are addressed by subsystem numbers (SSN):

0000 0110	Home location register
0000 0111	Visitor location register
0000 1000	Mobile switching center
0000 1001	Equipment identity register

In SCCP called (CDA) and calling (CGA) addresses, SSN is always present. In addition, the addresses include a point code (PC) and/or a global title (GT). This depends on locations of the entities involved in the transaction:

- If both entities are in the same PLMN, the address is a point code.
- If both entities are in the same country, but in different PLMNs, the address is a global title with the format of an E.164 number, according to the numbering plan for the fixed network in that country.
- If the entities are in different countries, the address is a global title with a format defined in CCITT. Rec. E.214 [10].

The format is shown in Fig. 17.6-3(a). It consists of an E.164 part (country

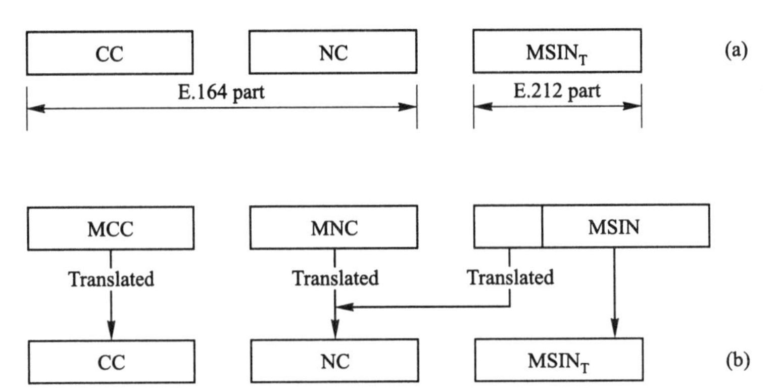

Figure 17.6-3 Format and derivation of E.214 global title. (From GSM.09.02-version 4.9.0. Courtesy of ETSI.)

code CC and network code NC), and an E.212 part ($MSIN_T$) which represents the leading digits of MSIN.

The derivation of this address from the IMSI of a mobile is shown in Fig. 17.6-3(b). Country code CC is derived from mobile country code MCC. Network code NC is derived from mobile network code MNC, and the leading digits of MSIN. $MSIN_T$ is truncated if necessary, in order to limit the length of the address to 15 digits.

17.6.7 Derivation of Called Address

The entity that initiates a transaction has to derive the address of the called entity from an input datum. We explore the address derivations in a number of situations that frequently occur in GSM-MAP.

A VLR initiates a transaction with the HLR of a mobile identified by IMSI. VLR derives the called address from IMSI.

A gateway MSC initiates a transaction with the HLR of a mobile identified by MSISDN. In this situation, the GMSC, mobile, and HLR belong to the same PLMN. Most PLMNs have one HLR, and the called address (a point code) is fixed. If a PLMN has several HLRs, GMSC derives the address from MSISDN.

VLR_1 initiates a transaction with VLR_2, which covers a particular location area (LAI). In this case, VLR_1 derives the address of VLR_2 from LAI. A VLR has no access to a VLR in another country.

A VLR initiates a transaction with its associated MSC, or vice versa. The interface between these two entities is internal (B interface), and addressing is not required.

An HLR initiates a transaction with the VLR serving a MS. This situation occurs only if the VLR initiated a previous transaction concerning MS with HLR. During the previous transaction, HLR has stored the VLR address in its record on MS, and uses this address when it initiates its transaction.

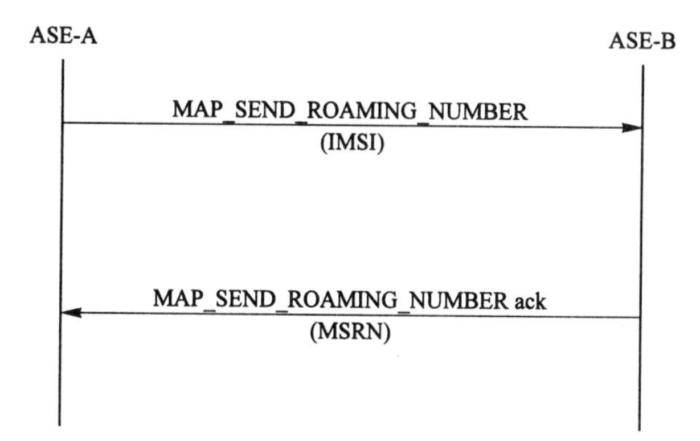

Figure 17.6-4 Messages in a GSM-MAP transaction.

17.6.8 GSM-MAP Transactions and Operations

As in IS-MAP, the GSM-MAP transactions consist of one operation, and require two messages. GSM-MAP uses the terms of CCITT-TCAP: a *begin* message (with one invoke) initiates the transaction, and an *end* message (with a return-result or return-error) ends the transaction.

In the sections that follow, the messages of a transaction are shown as in the example of Fig. 17.6-4, where ASE-A requests ASE-B to provide a roaming number for the mobile identified by IMSI. The begin message is denoted by MAP_SEND_ROAMING_NUMBER, which indicates the invoked operation. The end message is denoted by the same character string, followed by *ack* [9].

In what follows, the most important parameters in invokes and return-results are shown in parentheses, below the respective messages.

17.7 OPERATIONS RELATED TO LOCATION UPDATING

This section discusses location updating, which is counterpart of registration notification in IS-MAP (see Section 17.2). Its purposes are to verify that the MS qualifies for service (this may include MS authentication, etc.), and to update the HLR regarding the current MS location. Also, if MS registers in a location area covered by a "new" VLR, the VLR needs to establish a MS record, and therefore has to interrogate the HLR [9,11].

17.7.1 Description of Operations

The MAP operations for location updating, and the most important parameters in their invokes and return-results, are outlined below. The operations occur in the order they are executed during a typical location update.

In what follows, the "HLR of MS" is the home location register in the home

PLMN of MS that has a record for MS (17.6.5). MSC is the mobile switching center that has received a location update (LU) request from MS, and VLR is the visitor location register associated with that MSC. PVLR is the "previous" visitor location register, which has been serving the MS up to this point.

MAP_UPDATE_LOCATION_AREA. The operation is invoked by a MSC which has received a LU_request from a MS. It requests VLR to establish a record for MS if it does not already have one and, if desired by VLR, to authenticate the MS, and/or to allocate a new TMSI. The acknowledgment by VLR indicates that it has successfully performed these tasks.

MAP_UPDATE_LOCATION. The operation is invoked by a VLR whose MSC has received a LU_request from a mobile MS, and which has no record on MS. It requests the HLR of MS to enter the VLR address as the new current location of MS in its MS record.

MAP_CANCEL_LOCATION. The operation is invoked by the HLR of MS whose MS record includes a VLR address, and has received an update_location invoke about MS. The cancel_location invoke is made to PVLR, which is identified by the VLR address. The invoke requests PVLR to erase its MS record. After doing so, PVLR returns an acknowledgment.

MAP_INSERT_SUBSCRIBER_DATA. This operation is invoked by the HLR of mobile MS, after receiving an update_location invoke from a VLR. The invoke includes information items in the MS record at HLR (Section 17.6.5). VLR establishes a MS record that contains the information, and then acknowledges the invoke.

MAP_SEND_IDENTIFICATION. The operation is invoked by the VLR associated with a MSC which has received a LU_request from mobile MS in which MS has identified itself by the $TMSI_S$ and LAI_S that are stored in its SIM. The invoke, which includes these parameters, requests PVLR to provide the IMSI of MS. PVLR includes IMSI in its acknowledgment.

MAP_PROVIDE_IMSI. The operation is invoked by the VLR associated with the MSC which has received a LU_request from a MS in which the MS is identified by $TMSI_S$ and LAI_S. The invoke requests MSC to provide the IMSI of the mobile. MSC requests the IMSI from MS, and includes it in its return-result.

MAP_SEND_AUTHENTICATION_INFO. The GSM authentication procedures at the authentication center (AUC) and the mobile are shown in Fig. 17.7-1. The GSM authentication algorithm (A) has two inputs: random number RAND and authentication key Ki.

At AUC, a random number source generates a new RAND for each authentication. AUC stores a Ki_A for all mobiles that are "at home" in its

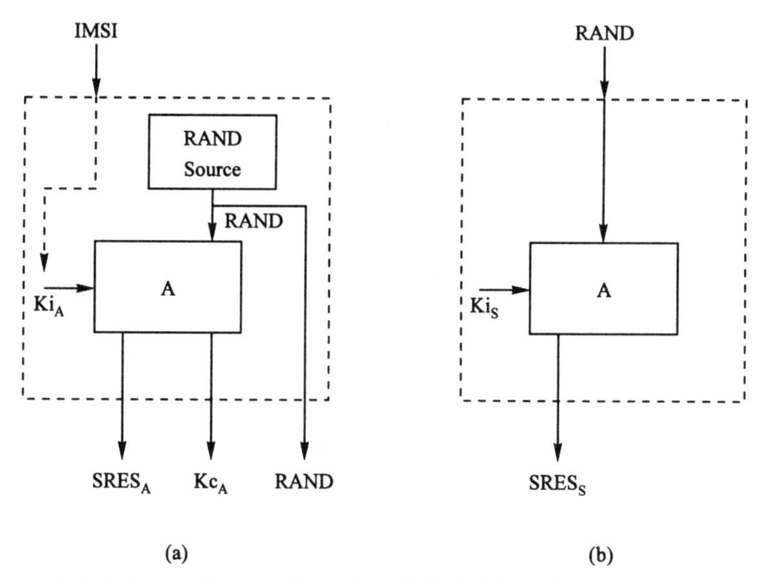

Figure 17.7-1 Authentication algorithm. (From GSM.09.02-version 4.9.0. Courtesy of ETSI.)

PLMN. The MS to be authenticated is identified by IMSI. AUC finds the Ki_A associated with IMSI, and then performs the algorithm. The results are the *signed result* ($SRES_A$) and the cipher key (Kc_A). Kc is used for data encryption on the radio interface.

At MS, the algorithm is performed using the same RAND value, and with the authentication key Ki_S (stored in its SIM). The result is $SRES_S$.

MS authentication is the responsibility of a VLR. It requests AUC/HLR and MS to perform the algorithm, and then compares the two results.

The invoke requests HLR/AUC to perform the algorithm for a mobile identified by its IMSI. The acknowledgment includes $SRES_A$, Kc, and the RAND value that was used.

MAP_AUTHENTICATE. The operation is invoked after VLR has obtained the authentication results of AUC. The invoke requests MSC to obtain authentication results from MS, and includes the RAND value used by AUC. MSC then requests MS to perform the authentication algorithm. After receiving the MS response, MSC acknowledges, and includes $SRES_S$.

VLR now compares $SRES_A$ and $SRES_S$. If the values match, the MS must have the proper Ki_S, and is therefore authentic. Otherwise, VLR informs MSC, which then sends a release message to MS.

MAP_SET_CIPHERING_MODE. This operation is invoked by VLR. It requests MSC to inform MS, and the BS that is serving MS, to initiate the ciphering (encryption) of information on the radio (RF) channel (Um interface). The invoke includes the cipher key (Kc), which is used by BS and MS to construct the encryption masks.

MAP_FORWARD_NEW_TMSI. The operation is invoked by a VLR which has decided to allocate a new TMSI to a MS. The invoke includes the new TMSI, and requests the associated MSC to forward the new TMSI to MS. MSC then informs MS and, after receiving confirmation, acknowledges the invoke.

17.7.2 Location Updating Examples [9]

We now explore a number of examples. In the figures of this section, MSC receives a location updating request from a MS, and VLR is associated with this MSC. HLR is the home location register of MS. PVLR is the "previous" visitor location register, associated with the MSC which had been serving MS up to this point.

In Fig. 17.7-2, MS does a location update in a location area covered by the same VLR that was involved in the previous location update of MS. VLR thus already has a record on MS. The update_location_area invoke includes the LAI_S and $TMSI_S$ stored in the SIM of MS, and LAI_B, the location area identity of the base station (BS) which received the LU_request. If LAI_S and LAI_B do not match, VLR updates its MS record, entering LAI_B as the new MS location area. It then acknowledges, and MSC confirms the MS request with a LU_confirm message which includes LAI_B. MS then updates its SIM, replacing LAI_S by LAI_B.

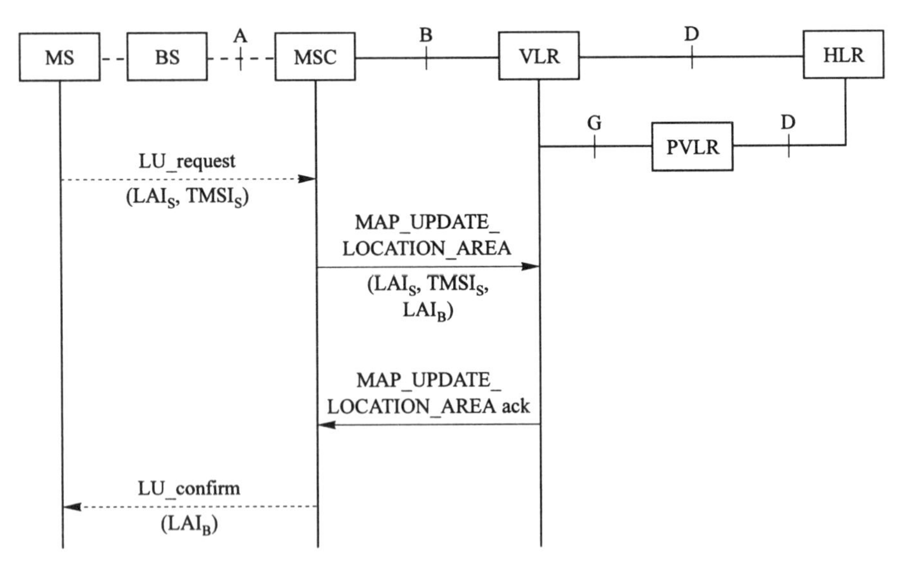

Figure 17.7-2 Operations for location area updating. Mobile is known at VLR. (From GSM.09.02-version 4.9.0. Courtesy of ETSI.)

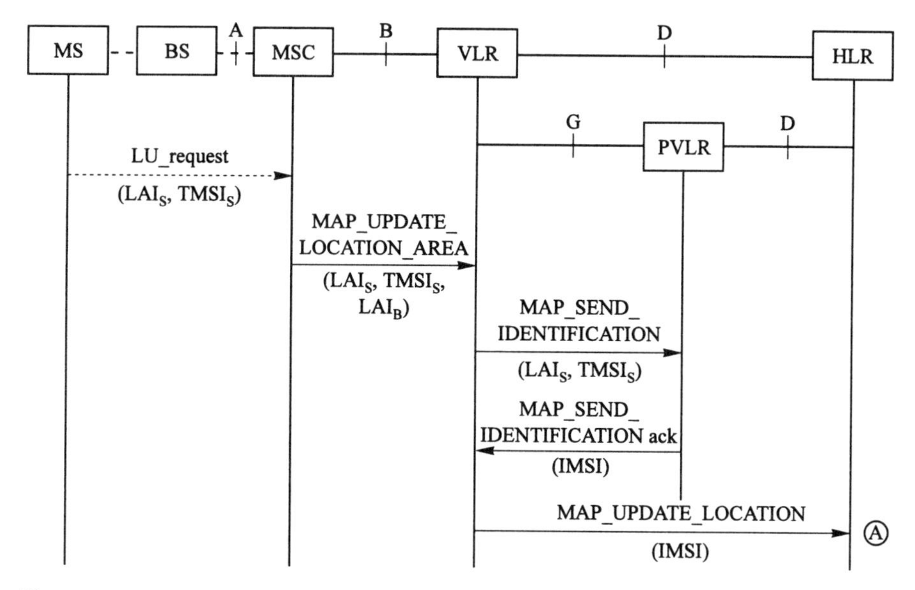

Figure 17.7-3 Initial operations for location area updating. Mobile is not known at VLR, and is identified by LAI_S and $TMSI_S$. (From GSM.09.02-version 4.9.0. Courtesy of ETSI.)

The next examples consider MS location updates in a location area covered by a new VLR. The VLR now has to establish a MS record. We first explore the operations that enable VLR to interrogate the HLR of MS.

In Fig. 17.7-3, MS sends a LU_request that again includes LAI_S and $TMSI_S$. VLR finds no record for MS, and has to interrogate HLR to obtain information for its new record. However, the HLR of MS cannot be determined from LAI_S and $TMSI_S$.

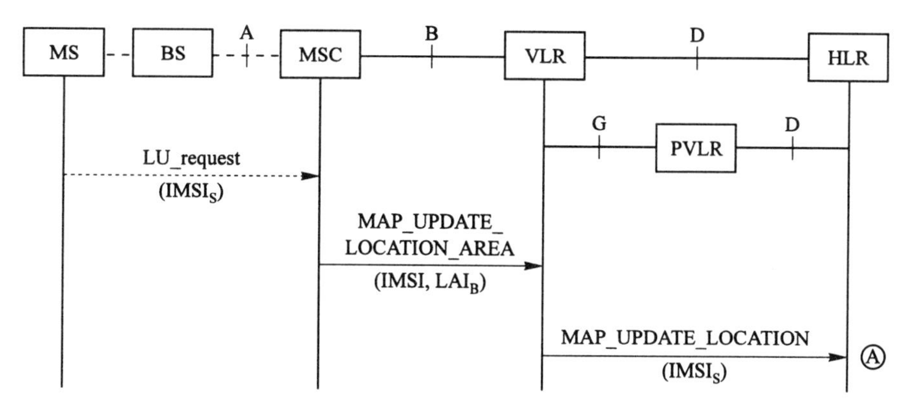

Figure 17.7-4 Initial operations for location area updating. Mobile is not known at VLR, and is identified by IMSI. (From GSM.09.02-version 4.9.0. Courtesy of ETSI.)

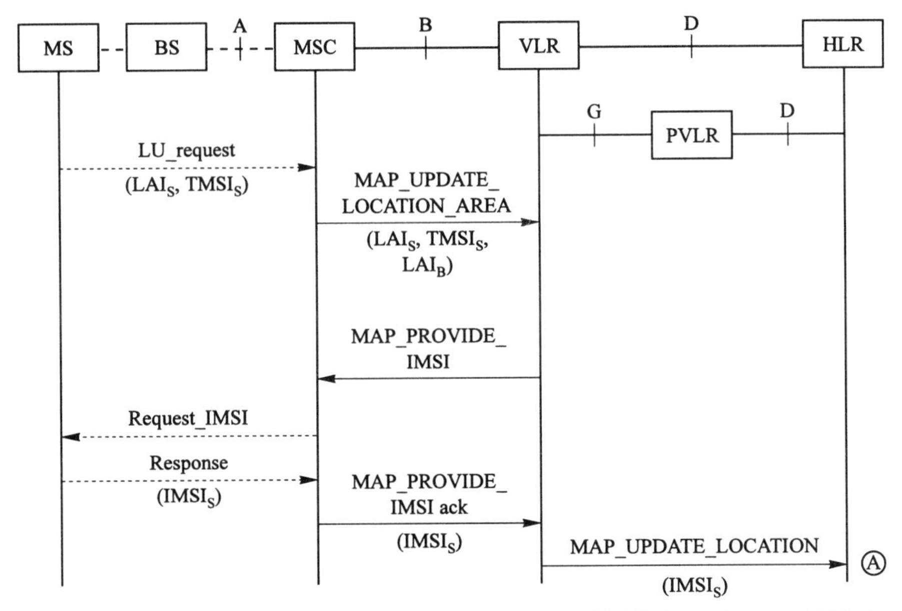

Figure 17.7-5 Initial operations for location area updating. Mobile is not known at VLR, is roaming in a foreign PLMN, and is identified by LAI_S and $TMSI_S$. (From GSM.09.02-version 4.9.0. Courtesy of ETSI.)

Therefore, VLR initiates a send_information operation with PVLR, requesting the IMSI of the MS identified by LAI_S and $TMSI_S$. VLR derives the address of PVLR from LAI_S. PVLR finds the associated MS record, and includes the IMSI in its acknowledgment.

VLR then derives the address of HLR from IMSI, and initiates an update_location operation (A) with the HLR.

In Fig. 17.7-4, MS has identified itself by the $IMSI_S$ stored in its SIM. This is done when the SIM is used for the first time, and when SIM has lost the LAI_S and/or $TMSI_S$. In this case, VLR can access HLR immediately (A).

In Fig. 17.7-5, MS has again identified itself by $TMSI_S$ and LAI_S. However, VLR now determines that LAI_S represents a location area in a foreign country. VLRs are not allowed to interrogate foreign VLRs. Therefore, VLR sends a provide_IMSI invoke to MSC, which then requests the IMSI from MS, and returns it to VLR. Visitor location registers are allowed to interrogate foreign HLRs, and VLR derives the HLR address from IMSI, and accesses HLR (A).

17.7.3 Subsequent Operations

We now explore the operations that follow the receipt of the update_location invoke by HLR (A). In Fig. 17.7-6, HLR initiates a cancel_location with PVLR, which now no longer serves MS. The address of PVLR is in the VLR field of the HLR record for the mobile. The invoke identifies the mobile by its IMSI. PVLR then deletes its MS record for the mobile.

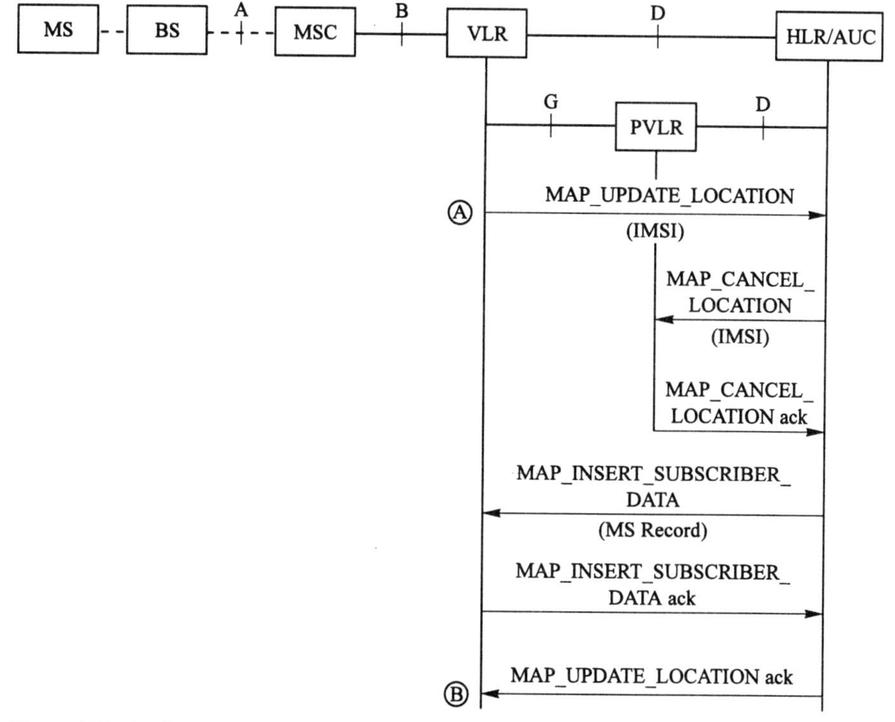

Figure 17.7-6 Subsequent operations for location area updating. (From GSM.09.02-version 4.9.0. Courtesy of ETSI.)

The HLR next updates its MS record, replacing the address of PVLR with the address of VLR. It then initiates and insert_subscriber_data operation with VLR. The invoke includes the information in the MS record at HLR. VLR establishes its MS record, and VLR and HLR then acknowledge each other's invokes (B).

The operations following (B) are shown in Fig. 17.7-7. VLR first requests the AUC and the mobile to execute the verification algorithm, and checks whether the results SRES are equal. We assume that this is the case. Otherwise, VLR aborts the location update.

If MS has ciphering (encryption) ability, VLR sends a MAP_SET_CIPHERING_MODE message, which includes Kc_A, to MSC. MSC sends a cipher_mode_command, including Kc_A, to the BS serving MS. BS then generates the encryption masks for its communications with MS. BS also repeats the command (without Kc_A) to MS, which generates its masks, using its stored Kc. MS then sends a cipher_mode_complete message. All subsequent information between MS and BS is encrypted.

If MS is new to VLR, the register allocates a new $TMSI_N$, and opens a forward_new_TMSI operation with MSC, which passes $TMSI_N$ to MS.

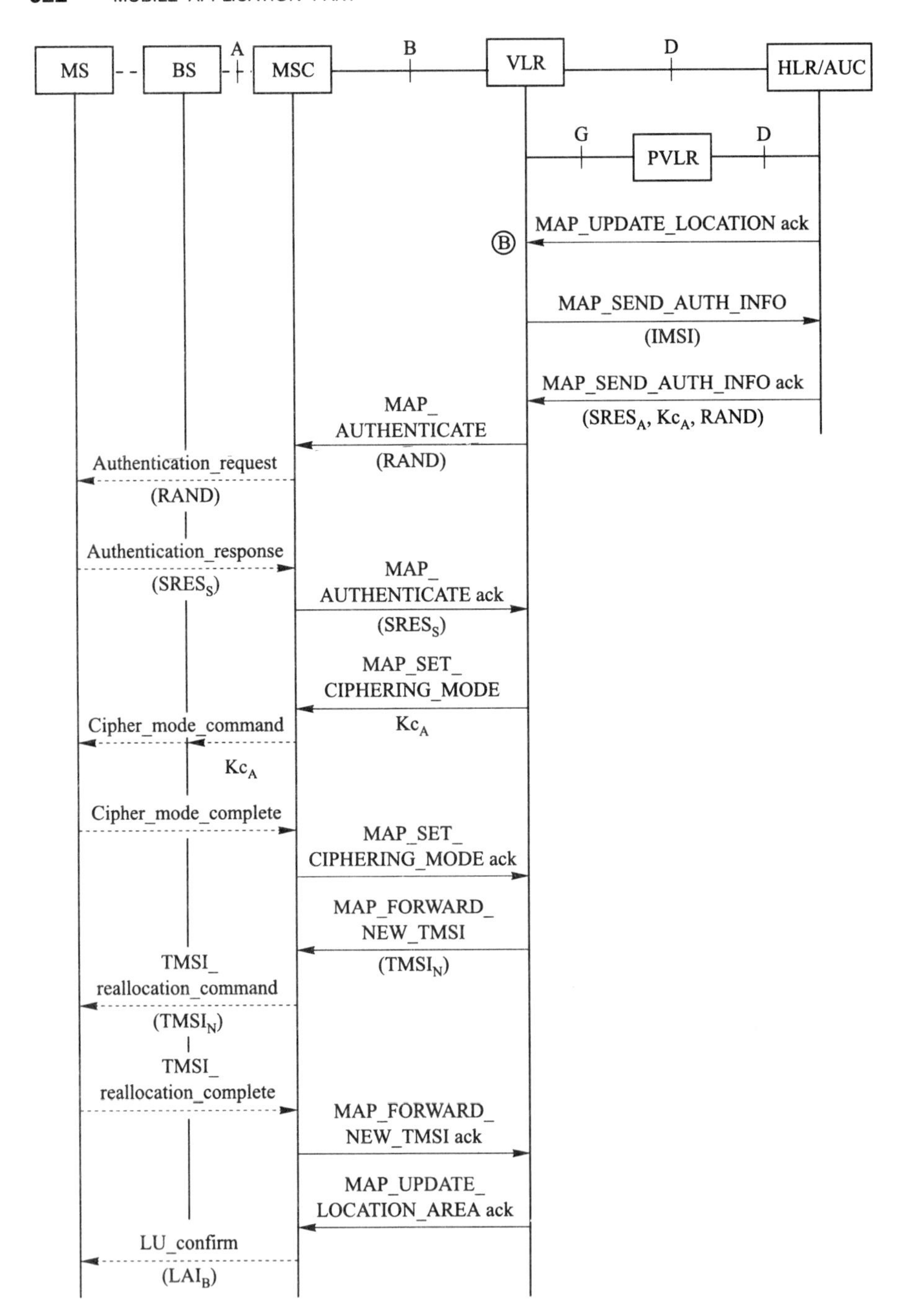

Figure 17.7-7 Final operations for location area updating. (From GSM.09.02-version 4.9.0. Courtesy of ETSI.)

After receiving a confirmation from MS, MSC acknowledges the forward_new_TMSI invoke.

Finally, VLR acknowledges the update_location invoke. This completes the updating procedure.

17.8 OPERATIONS FOR CALLS TERMINATING AT MS

17.8.1 Routing Calls to the Visited MSC

A call to a MS requires interworking between the fixed (PSTN/ISDN) and mobile (PLMN) networks. The fixed network extends the call set-up to a gateway mobile switching center (GMSC) in the home PLMN of MS, and the connection eventually reaches the MSC that is currently serving the MS.

There is a variety of possible configurations for this connection [12]. Figure 17.8-1 shows the simplest case. The calling party, the home PLMN of the called MS, and the PLMN that is currently serving MS, are all in country 1. The caller has dialed the MSISDN of MS, and the fixed network uses MSISDN to set up a connection to a gateway MSC (GMSC) in the home PLMN. GMSC interrogates HLR, and obtains the mobile station roaming number (MSRN), which is used to extend the connection to the MSC that is currently serving MS.

In the figure, the latter part of the connection is made in the fixed network. Some PLMNs have trunk groups between their MSCs. In that case, the connection from GMSC to MSC is made with PLMN trunks.

In Fig. 17.8-2, the calling party is in country 1, the home PLMN of MS is in

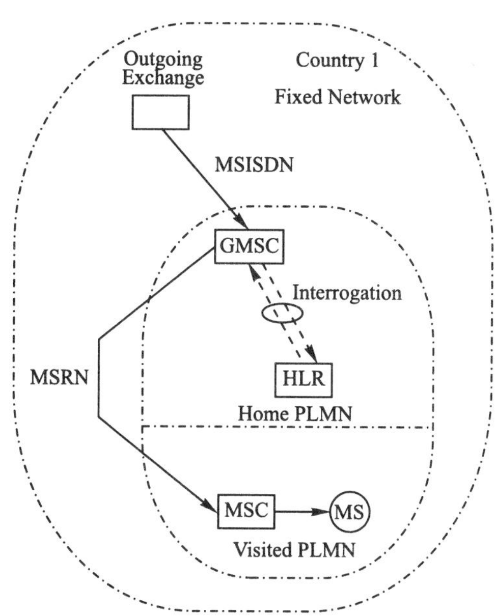

Figure 17.8-1 National connection to called mobile. (From GSM.09.02-version 4.9.0. Courtesy of ETSI.)

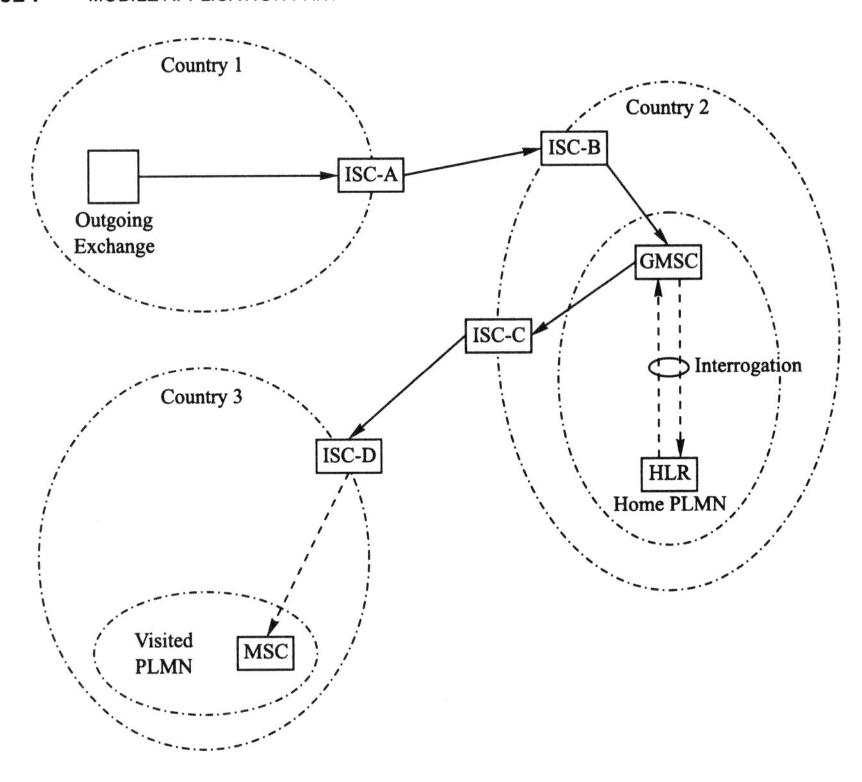

Figure 17.8-2 International connection to called mobile. (From GSM.09.02-version 4.9.0. Courtesy of ETSI.)

country 2, and the MS is currently visiting a PLMN in country 3. The calling party has dialed the country code of country 2, followed by MSISDN, and the call set-up arrives at GMSC.

GMSC interrogates HLR, and obtains the country code of country 3, and the MSRN. The fixed network uses these parameters to route the call to the MSC that is serving MS.

The resulting connection is not always an economic one. Let us consider the case that countries 1, 2, and 3 are respectively Belgium, Argentina, and France. A direct connection from Belgium to France would be more economical than a connection via Argentina.

Automatic rerouting (Section 1.3.4) can alleviate this problem in some instances. Suppose that in Fig. 17.8-3 the trunks between international switching centers ISC-A and ISC-B, and between ISC-B and GMSC, have SS7 signaling, that ISC-A has rerouting capability, and that GMSC can determine—from the calling party number received from ISC-A and the MSRN received from HLR—that the originating and terminating countries are Belgium and France. It then sends a backward unsuccessful message (BUM) to ISC-B. The message includes the country code of country 3 and MSRN as the

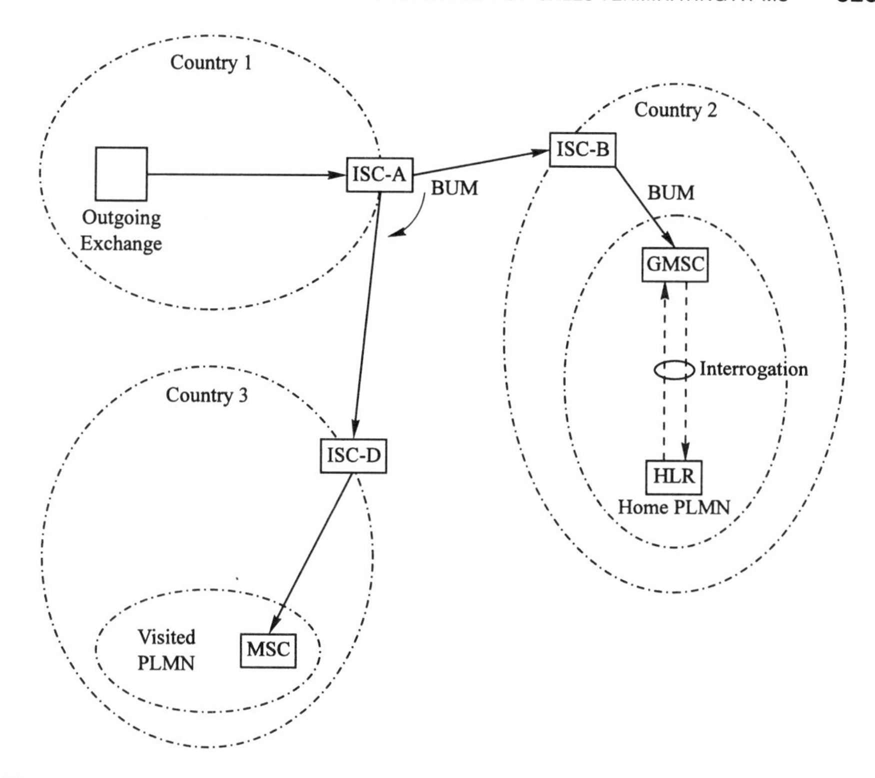

Figure 17.8-3 International connection with automatic rerouting. (From GSM.09.02-version 4.9.0. Courtesy of ETSI.)

redirecting address. ISC-B repeats the message to ISC-A. This exchange then drops the forward connection, and sets up a connection to ISC-D.

17.8.2 Operations for Terminating Calls

This section lists the operations involved in handling calls to a MS, and includes the most important parameters in the invokes and acknowledgments [9].

MAP_SEND_ROUTING_INFORMATION. This operation is initiated by a gateway MSC in the home PLMN of MS, which has received a call for MS from the fixed network. The send_routing_information invoke is a request to the HLR of MS to provide a MSRN for the call. The called MS is identified by its MSISDN. HLR includes MSRN—and possibly the country code of the destination country—in its acknowledgment.

MAP_PROVIDE_ROAMING_NUMBER. Roaming numbers are allocated by the VLR associated with the MSC that is currently serving MS. HLR has the address of this VLR in its MS record.

The HLR initiates a provide_roaming_number operation with VLR, identifying the MS by its IMSI. VLR includes MSRN—and possibly the country code of the destination country—in its acknowledgment.

MAP_SEND_INFO_FOR_INCOMING_CALL. The operation is initiated by a MSC that has received an incoming call to MS, which is identified by MSRN. It requests VLR to execute operations needed for handling the incoming call.

MAP_PAGE. The invoke is sent by VLR, and requests MSC to page the MS. It includes the TMSI and LAI (location area identity) of MS.

MAP_PROCESS_ACCESS_REQUEST. The operation is invoked by MSC, after it has received a page_response from MS. It requests VLR to execute those operations that are necessary to provide access by MS to the mobile network.

VLR may decide to execute one or more of the following operations: provide_IMSI, authentication, set_ciphering_mode, and forward_new_TMSI, and then acknowledges the invoke.

MAP_COMPLETE_CALL. The invoke for this operation is sent by VLR, and includes the terminating part of the MS service profile. MSC needs this information when processing the call to MS.

17.8.3 Terminating Call Procedures

In Fig. 17.8-4, the set-up of a connection for a call to a MS served by MSC has

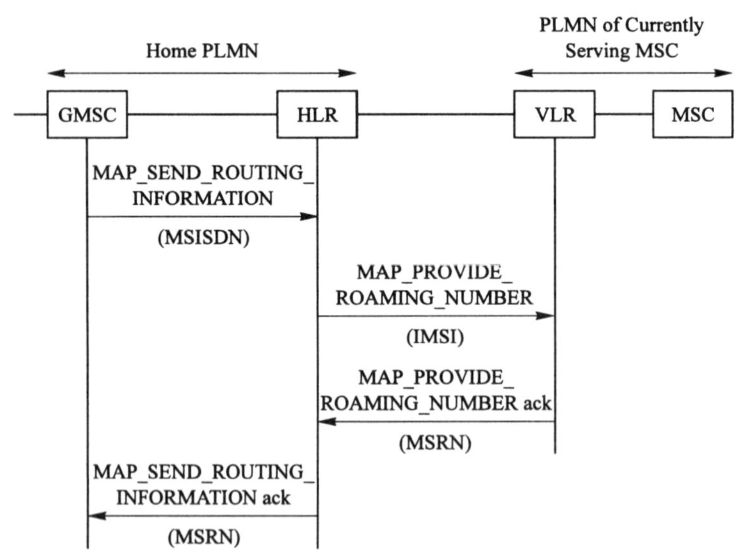

Figure 17.8-4 Obtaining a mobile station roaming number. (From GSM.09.02-version 4.9.0. Courtesy of ETSI.)

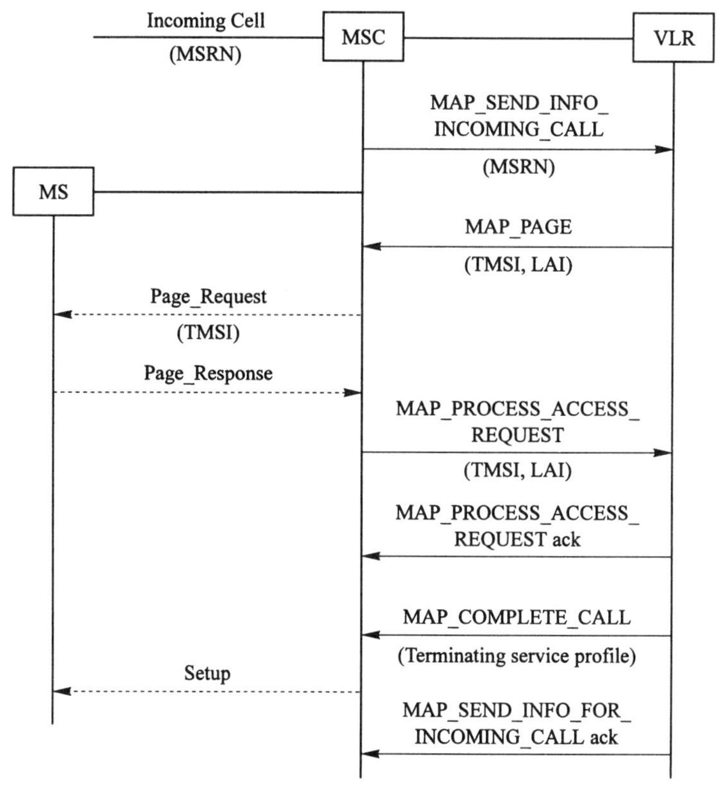

Figure 17.8-5 Operations for terminating calls. (From GSM.09.02-version 4.9.0. Courtesy of ETSI.)

reached a GMSC in the home PLMN of MS. GMSC has received the MSISDN of MS as the called party address, during interexchange signaling for the incoming trunk.

The send_routing_information invoke requests HLR to provide the MSRN for the MS. HLR derives the IMSI of MS from its MSISDN, and includes it in its provide_roaming_number invoke to VLR, which is currently serving MS. VLR allocates a MSRN, and enters it in its MS record. MSRN is included in the acknowledgments by VLR and HLR, and reaches GMSC. GMSC now has a new called number for the call, and seizes a trunk to (or towards) MSC.

In Fig. 17.8-5, the call set-up has arrived at MSC. MSC initiates a send_info_for_incoming_call transaction with its associated VLR. This operation triggers a series of operations. In the first place, VLR uses MSRN to find its record of MS, and retrieves TMSI and LAI. VLR then sends a page invoke to MSC that includes these parameters, and MSC broadcasts a page_request in the cells belonging to LAI.

On receipt of the page_response, MSC initiates a process_access_request operation with VLR. This register then may elect to execute one or more of the following operations: authenticate, set_ciphering_mode, and forward_new_TMSI which have been described in Section 17.7.3. The operations are not shown in the figure.

Having successfully accomplished this, VLR sends a complete_call invoke, which includes the terminating parameters of the MS service profile. MSC then starts processing the call, and VLR ends the send_info_for_incoming_call operation with an acknowledgment.

17.9 OPERATIONS AND PROCEDURES FOR ORIGINATING CALLS

17.9.1 Operations

We first explore the operations involved in handling calls originated by a MS, and include the most important parameters in the invokes and acknowledgments [9].

MAP_PROCESS_ACCESS_REQUEST. The operation is invoked by a MSC that has received a CM_service_request from a MS (Section 12.9.2). It requests VLR to execute those operations that are necessary to give MS access to the mobile network.

The VLR may then decide to execute one or more of the following

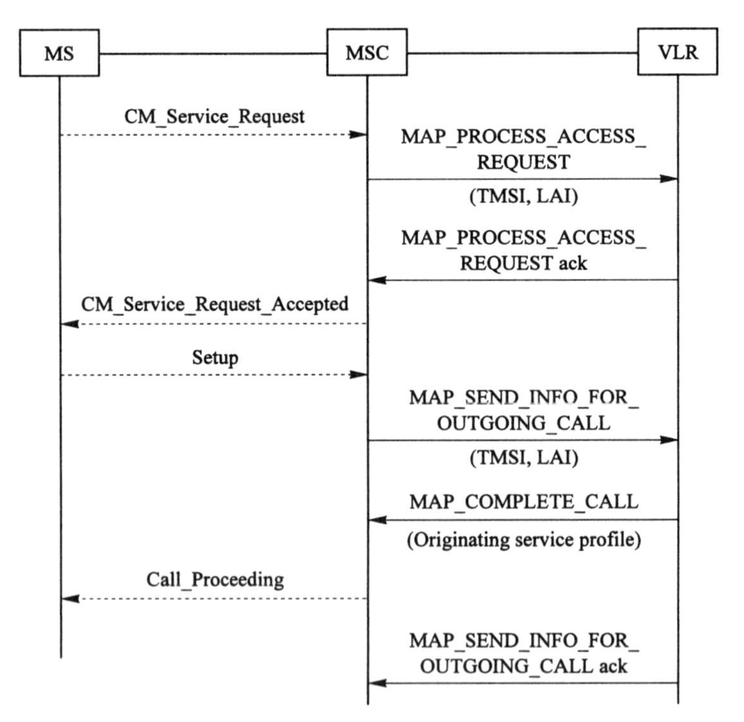

Figure 17.9-1 Operations for originating calls. (From GSM.09.02-version 4.9.0. Courtesy of ETSI.)

operations: provide_IMSI, MS authentication, set_ciphering_mode, and forward_new_TMSI. It then acknowledges the invoke.

MAP_SEND_INFO_FOR_OUTGOING_CALL. This operation is invoked by MSC, and requests VLR to send information needed to process the call.

MAP_COMPLETE_CALL. The invoke for this operation is sent by VLR, and includes the originating part of the MS service profile.

17.9.2 Originating Call Procedures

The procedure for handling MS-originated calls is shown in Fig. 17.9-1. On receipt of a CM_service_request, MSC initiates a process_access_request to VLR. The VLR may then elect to execute the MS authentication operations and, if MS has ciphering capability, the set_ciphering_mode operation (see Fig. 17.7-7). After completing these operations (which are not shown in Fig. 17.9-1), VLR acknowledges the process_access_request.

MSC then sends a CM_service_request_accepted message to MS. On receipt of a set-up message from MS, MSC initiates a send_information_for_outgoing_call operation with VLR. VLR returns a complete_call invoke which includes the origination-related parameters of the MS service profile, and then acknowledges the invoke of MSC.

MSC now has sufficient information to start the call processing, and sends a call_proceeding message to MS.

17.10 ACRONYMS

AC	Area code
AIN	Advanced intelligent network
AUC	Authentication center
AUTHR	Result of the execution of CAVE
AUTHRQST	Authentication request
BCD	Binary coded decimal
BS	Base station
CAVE	Cellular authentication and voice encryption algorithm
CCITT	International Telegraph and Telephone Consultative Committee
CDA	Called party address
CHAN	Channel
CMN	Cellular mobile network
DMAC	Digital mobile attenuation code
DTX	Discontinuous transmission
DVCC	Digital verification color code
EC	Exchange code
EIA	Electronic Industries Association

EIR	Equipment identity register
ETSI	European Telecommunications Standards Institute
FACDIR	Facilities directive
FACREL	Facilities release
GMSC	Gateway MSC
GSM	Global system for mobile telecommunications
GT	Global title
GTA	Global title address
HLR	Home location register
HOMEAS	Handoff measurement
IMEI	International mobile equipment identity
IMSI	International mobile station identity
IS-54	Interim standard #54
ISC	International switching center
ISDN	Integrated services digital network
Ki	Authentication key
Kc	Cipher key
LAC	Location area code
LAI	Location area identity
LOCREQ	Location request
LU	Location updating
MAP	Mobile application part
MCC	Mobile country code
MIN	Mobile identification number (ten digits)
MNC	Mobile network code
MOCH	Mobile on channel
MS	Mobile station
MSC	Mobile switching center
MSCID	Identity of MSC or VLR
MSIN	Mobile station identification number
MSISDN	Mobile station ISDN number
MSN	Mobile serial number
MSRN	Mobile station roaming number
PC_SSN	Point code and subsystem number
PLMN	Public land mobile network
PSTN	Public switched telecommunication network
PVLR	Previous visitor location register
RAND	Random number
REGCANC	Registration cancellation
REGNOT	Registration notification
RF	Radio frequency
ROUTREQ	Routing request
SCCP	Signaling connection control part
SID	System identification (15 bits)
SRES	Signed result

SSD_A	Shared secret data for authentication
SSD_B	Shared secret data for voice privacy
SS7	Signaling system No.7
SWNO	Switch number
TCAP	Transaction capability application part
TDMA	Time-division multiple access
TG	Trunk group
TIA	Telecommunictions Industry Association
TLDN	Temporary location directory number
TMSI	Temporary mobile station identity
TN	Translation type
VC	Voice channel
VLR	Visitor location register
VMAC	Mobile attenuation code on voice channel
VPMASK	Voice privacy mask

17.11 REFERENCES

1. *Cellular Radiotelecommunications Intersystem Operations: Functional Overview*, Interim Standard EIA/TIA/IS-41.1-B, Electronic Industries Association, Washington, D.C., 1991.
2. *Cellular Radiotelecommunications Intersystem Operations: Data Communications*, Interim Standard EIA/TIA/IS-41.5-B, Electronic Industries Association, Washington, D.C., 1991.
3. *Cellular Radiotelecommunications Intersystem Operations: Automatic Roaming*, Interim Standard EIA/TIA/IS-41.3-B, Electronic Industries Association, Washington, D.C., 1991.
4. *Cellular Radiotelecommunications Intersystem Operations: Authentication, Signaling Message Encryption, and Voice Privacy*, Telecommunications Systems Bulletin TSB51. Telecommunications Industries Association, Washington, D.C., 1993.
5. *Cellular Radiotelecommunications Intersystem Operations: Intersystem Handoff*, Interim Standard EIA/TIA/IS-41.2-B, Electronic Industries Association, Washington, D.C., 1991.
6. *General Aspects of Public Mobile Land Networks*, Rec. Q.1001, CCITT Blue Book **VI.12**, ITU, Geneva, 1989.
7. *Numbering Plan for the ISDN Era*, Rec. E.164, CCITT Blue Book **II.2**, ITU, Geneva, 1989.
8. *Identification Plan for Land Mobile Stations*, Rec. E.212, CCITT Blue Book **II.2**, ITU, Geneva, 1989.
9. *European Digital Cellular Telecommunication System (Phase 2), Mobile Application Part (MAP)*, **GSM 09.02**, European Telecommunications Standards Institute, Sophia Antipolis, France, 1994.
10. *Structure of the Land Mobile Global Title for the Signaling Connection Control Part (SCCP)*, Rec. E.214, CCITT Blue Book **II.2**, ITU, Geneva, 1989.
11. *Mobile Application Part*, Rec. Q.1051, CCITT Blue Book **VI.13**, ITU, Geneva, 1989.
12. *General Signaling Requirements on Interworking between the ISDN or PSTN and the PLMN*, Rec. Q.1051, CCITT Blue Book **VI.12**, ITU, Geneva, 1989.

INDEX